Microbiology of Fermented Foods

Microbiology of Fermented Foods

Second edition
Volume 2

Edited by

Brian J.B. Wood

Department of Bioscience and Biotechnology
University of Strathclyde
Glasgow, UK

BLACKIE ACADEMIC & PROFESSIONAL
An Imprint of Chapman & Hall
London · Weinheim · New York · Tokyo · Melbourne · Madras

Published by Blackie Academic and Professional, an imprint of Thomson Science, 2–6 Boundary Row, London SE1 8HN, UK

Thomson Science, 2–6 Boundary Row, London SE1 8HN, UK

Thomson Science, 115 Fifth Avenue, New York, NY 10003, USA

Thomson Science, Suite 750, 400 Market Street, Philadelphia, PA 19106, USA

Thomson Science, Pappelallee 3, 69469 Weinheim, Germany

First edition published by Elsevier Applied Science Publishers 1985

Second edition 1998

Thomson Science is a division of International Thomson Publishing I(T)P®

Typeset in 10/12 pt Times by Best-set Typesetter Ltd, Hong Kong
Printed in Great Britain by T.J. International Ltd, Padstow, Cornwall

ISBN 0 7514 0216 8

A catalogue record for this book is available from the British Library

Library of Congress Catalog Card Number: 97-73963

∞ Printed on acid-free text paper, manufactured in accordance with ANSI/NISO Z39.48-1992 (Permanence of Paper)

To the memory of my parents,
who encouraged and supported me
through a long education

Contents

Volume 2

Contributors

Martin R. Adams	School of Biological Sciences, University of Surrey, Guildford, Surrey GU2 5XH, UK
J.G. Anderson	Department of Bioscience and Biotechnology, University of Strathclyde, 204 George Street, Glasgow G1 1XW, UK
C.G. Beddows	Faculty of Health and Social Care, Leeds Metropolitan University, Calverley Street, Leeds LS1 3HE, UK
Michael A. Borowitzka	School of Biological and Environmental Sciences, Murdoch University, Perth, Western Australia 6150, Australia
Jean-Louis Cordier	Nestec S.A., Vevey, Switzerland
T.J. Elliott	Horticultural Research International, Wellesbourne, Warwick CV35 9EF, UK
Graham H. Fleet	Department of Food Science and Technology, University of New South Wales, Sydney, New South Wales 2052, Australia
Mark S. Fowler	Nestec S.A., Vevey, Switzerland. Also at Nestle R&D Centre, PO Box 204, York YO1 1XY, UK
Michael Gänzle	Universität Hohenheim, Institut für Lebensmitteltechnologie (ALT), Garbenstr. 28, D-70599, Stuttgart, Germany
A. Godfrey	E.T. Consulting, 'Glenview', Upper Common Road, Gilwern, Abergavenny, Gwent NP7 0DR, UK
W.P. Hammes	Universität Hohenheim, Institut für Lebensmitteltechnologie (ALT), Garbenstr. 28, D-70599, Stuttgart, Germany
Linda J. Harris	Department of Food Science and Technology, University of California, Davis, California 95616, USA

Linda M. Harvey	Department of Bioscience and Technology, University of Strathclyde, 204 George Street, Glasgow G1 1XW, UK
Martin Hoogland	Quest International, PO Box 2, 1400 CA Bursum, The Netherlands
Ian Jenson	Burns, Philp and Company Limited, PO Box 219, North Ryde, NSW 2113, Australia
Colja Laane	Agricultural University Wageningen, Dreijenlaan 3, 6703 HA Wageningen, The Netherlands
Patrick Leheup	Nestec S.A., Vevey, Switzerland
Zdzislawa Libudzisz	Institute of Fermentation Technology and Microbiology, Technical University, Lódz, Poland
Napha Lotong	Department of Microbiology, Faculty of Science, Kasetsart University, 50 Paholyothin Road, Jatujak, Bangkok 10900, Thailand
Friedrich-Karl Lücke	Fachbereich Haushalt und Ernährung, Fachhochschule Fulda, Marquardstrasse 35, D-36039 Fulda, Germany
B. McNeil	Department of Bioscience and Biotechnology, University of Strathclyde, 204 George Street, Glasgow G1 1XW, UK
S. Neelakantan	Dairy Microbiology Division, National Dairy Research Institute, Karnal, Haryana 132001, India
Helena Oberman	Institute of Fermentation Technology and Microbiology, Technical University, Lódz, Poland
S.A. Odunfa	Federal Institute of Industrial Research, Oshodi, Off Agege Motor Road, Oshodi, PMB 21023, Ikeja, Lagos, Nigeria
O.B. Oyewole	Federal Institute of Industrial Research, Oshodi, Off Agege Motor Road, Oshodi, PMB 21023, Ikeja, Lagos, Nigeria
Günter Pahlow	Institut für Grunland – und futterpflanzenforschung, Bundesalle 50, D-38116 Braunschweig, Germany

N.J. Rowan	Department of Bioscience and Biotechnology, University of Strathclyde, 204 George Street, Glasgow G1 1XW, UK
Rugmini Sankaran	89 Gokulam II Stage, Near J.K. Nursing Home, V.V. Mohalla, Mysore 570 002, India
Masaoki Sasaki	Research and Development Division, Kikkoman Co. Ltd, 339 Noda, Noda-Shi, Chiba-Ken, 278 Japan
Richard J. Scrase	38 Eastbourne Avenue, Bath BA1 6EN, UK
Kishan Singh	Dairy Microbiology Division, National Dairy Research Institute, Karnal, Haryana 132001, India
John E. Smith	Department of Bioscience and Biotechnology, University of Strathclyde, 204 George Street, Glasgow G1 1XW, UK
Hein Stam	Quest International, PO Box 2, 1400 CA Bursum, The Netherlands
Glyn Stanley	R.P. Texel Ltd, Poleacre Lane, Woodley, Stockport, Cheshire SK6 1PQ, UK
W.R. Stanton	73 Main Street, Stanbury, Keighley, West Yorkshire BD22 0HA, UK
K.H. Steinkraus	Institute of Food Science, Cornell University, 15 Cornell Street, Ithaca, New York 14850, USA
Brian J.B. Wood	Department of Bioscience and Biotechnology, University of Strathclyde, 204 George Street, Glasgow G1 1XW, UK
Michael Woolford	Alltech UK Ltd, 15–17 Abenbury Way, Wrexham Industrial Estate, Wrexham, Clwyd LL13 9UZ, UK
Tamottsu Yokutsuka	Research and Development Division, Kikkoman Co. Ltd, 339 Noda, Noda-Shi, Chiba-Ken, 278 Japan

Preface

When I undertook the production of the First Edition of this book it was my first foray into the world of book editing, and I had no idea of what I was undertaking! I was not entirely alone in this, as in asking me to produce such a book the commissioning Editor, Mr George Olley of Elsevier Applied Science Publishers, had pictured a text of perhaps 300 pages, but on seeing my list of chapter titles realized that we were talking about a 20-chapter, two-volume work. We eventually decided to go ahead with it, and the result was more successful than either of us had dared to hope could be the case. It was therefore with rather mixed emotions that I contemplated a second edition at the suggestion of Blackie Press, who had taken over the title from Elsevier. On the one hand, I was naturally flattered that the book was considered important enough to justify a second edition. On the other hand, I was very well aware that the task would be even greater this time. Indeed, so much had changed in the decade since the first edition, it appeared that this would be in many ways a totally new book. The subject matter has grown greatly in that period, and so has my knowledge of it. The appearance of glossaries of names associated with fermented foods (by Hesseltine & Wang, and by Campbell-Platt) had helped to place the subject on a much firmer foundation, but had compelled me to recognize the true extent of the subject. Moreover, it was clear that the regional dimension would warrant fuller consideration this time. The appearance of Dirar's 552-page text devoted to the indigenous fermented foods of one country, The Sudan, was a further reminder that the subject is immensely complex. Dirar subtitled his book *A Study in African Food and Nutrition*, and it is easy to justify this, but in fact he demonstrates that The Sudan's geography means that the country's food fermentations, and its food in general, is a rich mingling of African, Mediterranean, Islamic, and even British (from the time of the colonial mandate) influences. I therefore planned that the two specifically 'regional' chapters in the first edition (Odunfa on African food fermentations, and Stanton on the Tropics in general) should be accompanied by chapters on India, the Middle East and South America, but in practice I only succeeded in adding India to the list. However, so many other topics presented themselves that there is a nearly 50% increase in chapters in this edition, and I am already aware that there is adequate scope for reviewers to chide me for omissions from the subjects covered here.

Despite this, I hope that the finished book will be seen as rather more than just an updating of the first edition.

Some of the contributors to that first edition were able to contribute to the second edition, and in all cases the changes have been such that the new chapter is considerably expanded over the first version. In other cases, retirement from active involvement in scientific work or, sadly, the death of contributors has required me to find new authors. In all cases my authors have exceeded all that I could reasonably ask of them, and one or two of them did so at very short notice moreover. I am therefore greatly indebted to **all** of the authors who have contributed chapters to this book. It also gives me great pleasure to record my sincere gratitude to Dr Lesley Anderson of Blackies in Glasgow, who began this process, and to Ms Rose Gilliver who has seen it through to the end. Both of them have exhibited great patience and forbearance when faced with what must have seemed like endless prevarication on my part.

I conclude with some pleas to the Reader. First, I point out that I have deliberately used authors from many countries and cultures. As always, I am amazed at the fine English which they command, even when it is not their first language. I want to retain the vigour and diversity which their individual styles of writing bring to the book, so I have tried to exercise restraint in editing, limiting myself to changes which I considered to be demanded on grounds of clarity or the rules of English grammar. If the result is less harmonious than you might wish, then I accept full responsibility for that, and hope that you will understand the reasons for it; of course I also hope that you will share my pleasure in the authors' individuality. Secondly, if you find factual errors or you feel that there is some significant omission from the materials covered herein, please feel free to communicate with me. I am responsible for the Index to this edition, and have tried to model it on the Index which was prepared professionally for the first edition. However, indexing is an art, and I admit to lacking skill in that art, so bear with me please if I fall short of your standards. Finally, I hope that you will find working with this text rewarding, interesting, and even enjoyable.

Brian J.B. Wood,
Lenzie, Scotland
April 1997

14 Fermented sausages

F.-K. LÜCKE

14.1 Introduction

This review deals with meat products which at least partially owe their microbiological stability and organoleptic properties to a fermentation carried out by lactic acid bacteria. These products are usually made from comminuted meat and fat, mixed with salt, curing agents, sugar and spices, and filled into casings, and are referred to as 'fermented sausages'.

The other category of fermented meats includes raw unground products such as raw hams. These products, however, owe most of their characteristic sensory properties to the action of salt, curing agents, and proteolytic muscle enzymes only. Therefore, the microbiology of ham curing is not discussed in this article in any detail, even if microorganisms may contribute to the appearance and flavour of some raw hams by reducing nitrate and forming aroma precursors (Leistner, 1958), and lactic acid bacteria along with fermentable sugars may be added to the injection brine in order to lower the meat pH and to accelerate the salting proces (Jaud & Fischer, 1993).

Fermented sausages have their longest tradition in Southern and Central Europe. Production and per capita consumption figures are highest in Germany, Italy, Spain and France. From data compiled by Lücke *et al.* (1990) and Fisher & Palmer (1995), production figures of 1995 within the European Union may be estimated to be about 750000 tonnes. In the Federal Republic of Germany, for example, the annual per capita consumption of fermented sausages is of the order of 4.5 kg, which comprises about 7% of the total meat consumption. With few exceptions – for example nham, a Thai-style raw sausage – other parts of the world have adapted formulations and production methods from the European countries. Little, if any, microbial activity is involved in the preparation of Chinese-style raw sausages which are mainly preserved by drying. These sausages, nham and comparable products indigenous to countries outside Europe are normally eaten after cooking.

Leistner (1986) and Zeuthen (1995) gave accounts of the history of meat preservation through fermentation. Fermented sausages probably originated in the Mediterranean. The Romans knew that ground meat with added salt, sugar and spices turns into a palatable product with a long shelf life if prepared and ripened properly. Apparently, the normal winter climate in the Mediterranean countries, with its moderate temperatures and frequent rainfall, is favourable for sausage ripening. In contrast, salting and drying of unground meat was the traditional way of meat preservation in Germany and other European countries. In Germany, the manufacture of fermented sausages commenced only some 160 years ago, and most of the fermented sausages are smoked while in Italy, France and Spain, air-dried, spicy sausages predominate. Other types of fermented sausages emerged later as a consequence of advanced meat processing techniques and the availability of refrigeration. Such products include semi-dry as well as spreadable, undried sausages.

The official classification of fermented sausages varies from country to country (see Adams, 1986; Zeuthen, 1995): it may be based on moisture content, protein content, moisture/protein ratio, weight loss, etc. From a microbiologist's point of view, fermented sausages are best subdivided on the basis of water activity (a_w) and surface treatment (Table 14.1). In this article, sausages with a_w between 0.90 and 0.95 are termed 'semi-dry', and those with a_w below 0.90 as 'dry'. Additional criteria for classification include the casing diameter, the degree of comminution of the ingredients, the animal species used for obtaining the raw material, the fat content and type of tissue used, as well as the spices, seasonings and other non-meat ingredients used. The term 'sausage ripening' will be used to describe changes occurring between case filling and the time when the product is ready for sale. Sausage ripening is subdivided into 'fermentation' (lactic acid formation and concomitant processes) and 'ageing' (drying, aroma formation etc.).

Table 14.1 Classification of fermented sausages

Category	Ripening times	Final water activity (a_w)	Application of smoke	Examples
Dry, mould-ripened	>4 weeks	<0.90	No	Genuine Italian salami; French 'saucisson sec'
Dry, mould-ripened	>4 weeks	<0.90	Yes (during fermentation)	Genuine Hungarian salami
Dry, no mould growth	>4 weeks	<0.90	Yes or no	German 'Dauerwurst'
Semi-dry, mould-ripened	<4 weeks	0.90–0.95	No	Various French and Spanish raw sausages
Semi-dry, no mould growth	<4 weeks (usually 10–20 days)	0.90–0.95	Yes (with exceptions)	Most fermented sausages in Germany, The Netherlands, Scandinavia, USA, etc.
Undried, spreadable	<2 weeks	0.94–0.96	Yes or no	German 'Streichmettwurst'; Spanish 'sobrasada'

As with many other fermented foods, intensive research into the microbiology and chemistry of sausage ripening was triggered when traditional empirical methods of manufacture no longer met the requirements of large-scale, low-cost industrial production, i.e. short ripening times and highly standardized products. It is, therefore, not surprising that such research commenced in the United States in the 1930s, whereas in Europe the first systematic studies on the microbiology and chemistry of sausage ripening were published in the 1950s.

14.2 Ingredients and additives

14.2.1 Meat

Fresh raw sausage mixture contains 50–70% lean meat, that is mammalian (sometimes poultry) skeletal muscle tissue, the approximate composition of which is given in Table 14.2. After slaughter, important biochemical changes occur in meat (reviewed by Hamm, 1977; Lawrie, 1991). Regeneration of ATP by respiration stops as the oxygen supply is cut off, and mitochondria are damaged. Consequently, muscle glycogen is degraded to lactic acid and the ATP pool is maintained at the *in vivo* level (ca. 5 µmol/g fresh weight) until glycogen stores are exhausted or glycolytic enzymes lose their activity. Thereafter, the ATP content drops to 1 µmol/g, and irreversible links are formed between the actin and myosin filaments of the muscle, a state known as 'rigor mortis'. In beef muscle, post-mortem

Table 14.2 Approximate composition (g/100 g fresh weight) of mammalian skeletal muscle after rigor

Water	75
Protein	
Myofibrillar	11.5
Sarcoplasmatic	5.5
Connective tissue and organelle	2.0
Total	18
Lipids	3.0
Di- and trinucleotides	0.1
Other non-protein nitrogen-containing compounds	1.65
Carbohydrates	
Glycogen	0.1
Glucose	0.05–0.15
Glucose-6-phosphate	0.15
Total	0.3–0.4
Lactic acid	0.9
Inorganic constituents	0.65

From Lawrie (1991), modified.

glycogenolysis is normally complete within 24–36 hours after slaughter, whereas in pig muscle, it generally takes only a few hours. If muscle pH drops to about 5.8 while the muscle temperature is still about 40°C (i.e. within about 45 minutes after slaughter), changes in muscle proteins render pork pale, soft, exudative (PSE). This happens, in particular, if genetically predisposed pigs are exposed to stress before and during slaughter.

After post-mortem glycogenolysis, the pH of red muscles is usually 5.5–5.9. In muscles that naturally contain little glycogen, or which have been depleted of their glycogen reserves by faulty treatment of the animals before slaughter (e.g. allowing bulls to fight during lairage), the final pH of the meat can be higher than 6.0. Such meat is not suitable for raw sausage production because it binds water more tightly and spoils more rapidly than meat of normal pH (Newton & Gill, 1981). In contrast, PSE meat can be processed into dry sausage without impairing its organoleptic quality, provided that it comprises not more than 20% of the total weight of the mixture (Wirth, 1972).

In Germany, fermented sausages usually contain pork and beef, while Hungarian and Italian salamis are manufactured from pork only, and Turkish 'soudjouk' from beef only, often in conjunction with fatty tissue from selected sheep breeds. There is a small but expanding market for fermented sausages made from poultry meat, and local specialities may contain mutton or even game meat. Obviously, eating habits, religious traditions and meat prices determine the animal species. From a technologist's point of view, the suitability of a meat for fermented sausage production depends on its final pH, its water-holding capacity and on the desired intensity of the curing colour. Many butchers prefer meat from old animals (e.g. sows) for the manufacture of high-quality dry sausages.

14.2.2 Fatty tissue

For the production of fermented, especially dry sausages for long-term storage, the fat must have a high melting point and accordingly, a low content of unsaturated fatty acids. Because sausages containing beef tallow or other fatty tissues from ruminants are less acceptable organoleptically, firm pork back fat is most frequently used for high-quality dry sausages. Use of soft fatty tissue (e.g. back fat from pigs fed a diet rich in unsaturated fatty acids) not only causes colour and flavour defects, but also interferes with the drying process and reduces the shelf life of dry sausages which is usually limited by autoxidative fat deterioration. For dry sausages with a long shelf life, polyunsaturated fatty acids should not make up more the 12% of the total fatty acids in the fat (Stiebing *et al.*, 1993). The peroxide content of fatty tissue must be kept minimal, e.g. by rapid freezing after slaughter and by avoiding long-term frozen storage.

14.2.3 Carbohydrates

During post-mortem glycogenolysis, a small portion of the glycogen is converted to glucose and glucose-6-phosphate. The glucose content of fresh, post-rigor beef and pork is of the order of 4.5 and 8 µmol/g fresh weight, respectively (cf. Dainty, 1986; Lawrie, 1991). This amount, in conjunction with glucose-6-phosphate in meat (about 5 µmol/g) does not allow for a reduction of pH sufficient to inhibit undesired bacteria. Therefore, fermentable carbohydrates are added to raw sausage mixtures. The variety and amount of carbohydrate added are important because they determine the rate and extent of lactic acid formation and the composition of the sausage microflora. If large amounts of rapidly metabolizable carbohydrates (e.g. glucose) are added, the pH reduction can be so rapid that desired metabolic activities of the non-lactic microflora are suppressed. On the other hand, addition of too little carbohydrate and/or selection of a carbohydrate which is degraded only slowly (e.g. oligosaccharides) may result in growth of undesired organisms during fermentation, particularly at high ripening temperatures. For most European-style, semi-dry smoked sausages, the target pH after fermentation is 4.8–5.0, ensuring a satisfactory microbiological stability and a rapid increase in firmness (Klettner & Rödel, 1978). To achieve this, 0.4–0.8% of rapidly fermentable carbohydrates (e.g. glucose) are usually added. Lower amounts (0.2–0.3%) are recommended if nitrate rather than nitrite is used as the curing agent, and if the target pH after fermentation is about 5.3, e.g. in the manufacture of Italian salami. Up to 2% fermentable sugar is sometimes used, especially in the production of US-style semi-dry fermented sausages. The pH may then drop to 4.5, and many consumers – particularly in Europe – will not readily accept such a 'sour' sausage. The formulation of European-style sausage commonly contains a combination of 'rapidly' and 'slowly' fermentable sugars. In countries where regulations permit such additives, quite large amounts of lactose (or milk powder) or starch (e.g. potato flour) are sometimes added to raw sausage mixtures. However, there is no microbiological reason for this.

14.2.4 Curing agents

Usually 2.4–3% sodium chloride (NaCl) is added to the raw sausage mixtures so that the initial a_w is 0.965–0.955, depending on the fat content in the formulation. This water activity inhibits or delays the growth of many undesired or dangerous microorganisms and favours the development of lactobacilli, staphylococci and micrococci. Apart from its effect on the microflora, NaCl interacts with the myofibrillar structure and solubilizes proteins which form a sticky film around meat particles. Obviously, it also contributes to the taste of meat products.

Nitrite – which is added to the mixture as nitrite curing salt, or which is formed from added nitrate by microorganisms – fulfils three functions in raw sausage, namely the development of the curing colour, the inhibition of autoxidative processes leading to rancidity, and the contribution to the establishment of the desired Gram-positive 'ripening microflora' (lactobacilli and non-pathogenic catalase-positive cocci). The principles of these processes are outlined in Section 14.4.2. Nitrate is still used for the manufacture of dry sausages with long ripening times, whereas for other fermented sausages, nitrite is usually preferred. The amount of sodium nitrite added should not exceed 150 mg/kg, though 50 mg/kg usually proves sufficient for the desired colour and flavour (Puolanne, 1977; Wirth, 1991). Sodium ascorbate (usually 300–500 mg/kg) is frequently added along with nitrite to speed the development of curing colour and flavour.

Nitrate is still often used in sausage varieties with long ripening times. Dry sausages made with nitrate or low amounts of nitrite frequently taste better than those made with the usual amount of nitrite (Wirth, 1991). This indicates that high concentrations of nitrite present at the beginning of the fermentation may suppress microorganisms active in synthesizing flavour components or their precursors. The amount of potassium nitrate added to raw sausage mixtures is usually between 200 and 600 mg/kg (in Germany, it is limited to 300 mg/kg), even for those products that have been traditionally made with more nitrate, e.g. Hungarian salami (Incze, 1986) and Lebanon bologna (Zaika *et al.*, 1976). Experimental data (e.g. Hechelmann *et al.*, 1974; Wirth, 1991) and practical experience did not show any technological and microbiological necessity for using high levels of nitrate.

14.2.5 Spices and other additives

Ground pepper is usually present in all types of fermented sausages at the 0.2–0.3% level. Other spices commonly used include paprika, garlic, mace, pimento and cardamon. Generally, air-dried sausages are more heavily spiced than smoked sausages. Some spices (e.g. red pepper, mustard, mace) have been found to stimulate the rate of lactic acid formation (Vandendriessche *et al.*, 1980; Nes & Skjelkvåle, 1982). This is due to manganese present in the spices (Zaika & Kissinger, 1984). Manganese is required by lactic acid bacteria for various enzyme activities, including the key enzyme of glycolysis, fructose-1,6-diphosphate aldolase (Kandler, 1983). Some spices (garlic, rosemary, sage) contain powerful antioxidants and may thus extend the shelf life of dry sausages (Hammer, 1977).

Up to 0.5% glucono-δ-lactone (GdL) is used in the manufacture of some fermented sausages. It acts as an acidulant, since it is hydrolysed to gluconic acid within a few hours and lowers the pH, thus contributing to the control of undesirable acid-sensitive microorganisms during the first days of ripening. However, it is broken down to lactic and acetic acid by various lactic acid bacteria involved in sausage fermentations, and especially acetic acid strongly interferes with the activities of bacteria reducing nitrate or involved in aroma formation or protection from rancidity. Hence, GdL is used mainly in the manufacture of 'fresh', undried or semi-dry fermented sausages. Starter cultures for fermented sausages are discussed in Section 14.4.4.

14.3 The manufacture of fermented sausages

In order to keep the temperature as low as practical during grinding, meat is chilled or partially frozen before it is comminuted in a meat grinder or cutter. Curing salt, carbohydrates and seasonings are then mixed in. Fatty tissue is comminuted in the frozen state and then added to the mixture. The size of the particles in the sausage determines the product type: in Germany, for instance, raw dry sausages with large particles are named 'Plockwurst', while those with very small particles are named 'Cervelatwurst'.

It is important to remove as much oxygen from the mixture as possible before it is filled into casings because oxygen interferes with the formation of the desired colour and flavour. This is achieved by vacuum-filling devices or by forming the mixture into a firm, round ball which is then placed into the filler. During filling, the temperature of the mixture should not exceed 2°C. This avoids smears at the cut meat surfaces which may later cause drying and rancidity problems. Natural casings as well as casings made from modified collagen and/or cellulose are most frequently used. They must allow evaporation of water from the sausages and penetration of smoke, and they must follow the shrinkage of the sausage during drying. For undried products, synthetic casings are also used.

To avoid moisture condensation on the sausage surface, the sausages are first equilibrated to the ripening temperature at low relative humidity (RH) before entering the ripening chamber. For dry sausages with long shelf life and for mould-ripened sausages, fermentation temperatures are usually below 22°C, and Hungarian 'téliszalámi' ('winter salami'; Incze, 1986) or certain regional specialities in Germany are fermented at even lower temperatures (8–12°C). Undried and semi-dry sausages are usually fermented at 22–26°C, and for American-style semi-dry sausages, even higher fermentation temperatures along with shorter fermentation times are applied. The higher the temperature, the more rapid the fermentation, but the higher the risk of growth of undesired microorganisms if the fermentation is not strictly controlled (see Sections 14.4 and 14.5). During fermentation, the RH in the ripening chamber should be 5–10% lower than the RH (water activity ×100) within the sausages (Stiebing & Rödel, 1988), i.e. about 85–90%, and the air velocity about 0.4 m/s. Some smoke can be applied during the fermentation period if care is taken not to inhibit the fermentation process.

After fermentation, raw sausages are treated in various ways. If the development of moulds and yeasts on the sausage surface during drying or storage is to be prevented, the sausages are smoked more heavily at the end of the fermentation period. Smoke contains phenols, carbonyl compounds and organic acids of low molecular weight, and many of these constituents have antimicrobial activity. The preservation of fermented sausage is also assisted by antioxidative phenol derivatives present in the smoke. Smoke is generated by the controlled combustion of wood at about 350–600°C; at higher temperatures, increasing amounts of polycyclic aromatic hydrocarbons are formed (Tóth & Blaas, 1972).

Sausages are usually aged at 12–15°C. The RH and air velocity during ripening should ensure a slow but steady drying of the sausages and prevent excessive mould growth on the surface, while avoiding uneven drying of the product and the formation of a dry, hard outer layer on the sausages. During ageing, the RH in the chamber is gradually lowered, depending on the pH and a_w of the product. During fermentation, the RH is adjusted to about 10% below the equilibrium RH of the sausages. Once the pH is at or below 5.3, protein solubility is so low that a gel is formed around meat and fat particles. This in turn leads to a sharp increase in firmness (Klettner & Rödel, 1978; see Figure 14.1) and facilitates the removal of water, so that the RH gradient between the chamber and the product may be steeper (Stiebing & Rödel, 1990). A sufficient movement of air (ca. 0.1 m/s) during ageing ensures homogeneous drying and the absence of wet patches on the sausages.

Today, most fermented sausages are ripened in controlled environment chambers throughout the year. However, many small manufacturers do not have such – rather expensive – equipment available. They have to restrict production to suitable seasons or to undried products, or use various 'tricks' to adjust the climate of the ripening room.

Most sliceable semi-dry fermented sausages (on the German market, about 80% of the products) have a_w values around 0.93. Assuming an initial fat content of 34% and an initial a_w of 0.96, this means a weight loss of about 18% and a moisture content of about 40%. To obtain dry sausages, they are dried further to a_w values of 0.90 or below, bacterial growth is inhibited, and, if surface condensation of water and mould growth is avoided, they may be stored at ambient temperature. Their water content is 35% or less which corresponds to a weight loss of at least 25%. Obviously, more water must be removed if the formulation includes less fat and more lean meat.

In the United States, sausages are frequently heated at the end of the fermentation: regulations require a core temperature of 58.3°C (137°F) if the presence of *Trichinella* cannot be excluded.

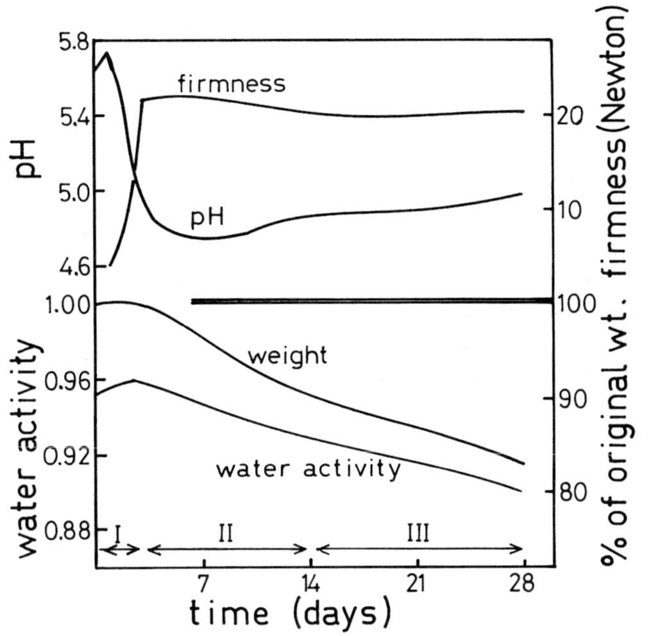

Figure 14.1 Ripening parameters during manufacture of German-type fermented sausage. Sausages (diameter 90 mm) were formulated with starter cultures (lactobacilli and staphylococci) and 0.7% carbohydrate. Values given apply to a zone approximately 20 mm below the surface. Ripening conditions: (I), 26°C, 90% relative humidity (RH); (II), 20°C, 88% RH; (III), 15°C, 82% RH. (Data from Klettner & Rödel, 1978.)

Measures to control pathogenic bacteria during sausage ripening are summarized in Section 14.5. A list of all control points for the production of good-quality fermented sausages is provided by Stiebing (1994).

14.4 Biochemical and microbiological changes during sausage ripening

14.4.1 Acid formation and lactic acid bacteria

After some time of refrigerated storage in air, the microflora of fresh meat largely consists of Gram-negative, oxidase-positive rods, particularly psychrotrophic pseudomonads (see Gill, 1982, for review). Psychrotrophic Enterobacteriaceae are also present, while Gram-positive organisms including lactic acid bacteria usually occur only in small numbers. If the meat is processed into raw sausage mixture, the water activity is reduced to 0.96–0.97, and the oxygen present within the mixture is rapidly consumed (Rödel *et al.*, 1993a). Thus, the pseudomonads, which require oxygen and are usually sensitive to salt and nitrite, are inhibited, as is the degradation of amino acids that would cause the typical off-odours of spoiled meat. Similarly, the competitiveness of the Enterobacteriaceae is reduced at low oxygen tension, low pH and in the presence of salt. This results in a shift in microflora composition towards the lactic acid bacteria and to staphylococci and micrococci ('Micrococcaceae', hereafter referred to as catalase-positive cocci) as illustrated in Figures 14.2 and 14.3. Figure 14.2 shows the development of the microflora in a smoked dry sausage as it was traditionally manufactured in Northern Europe and most parts of Germany. In this

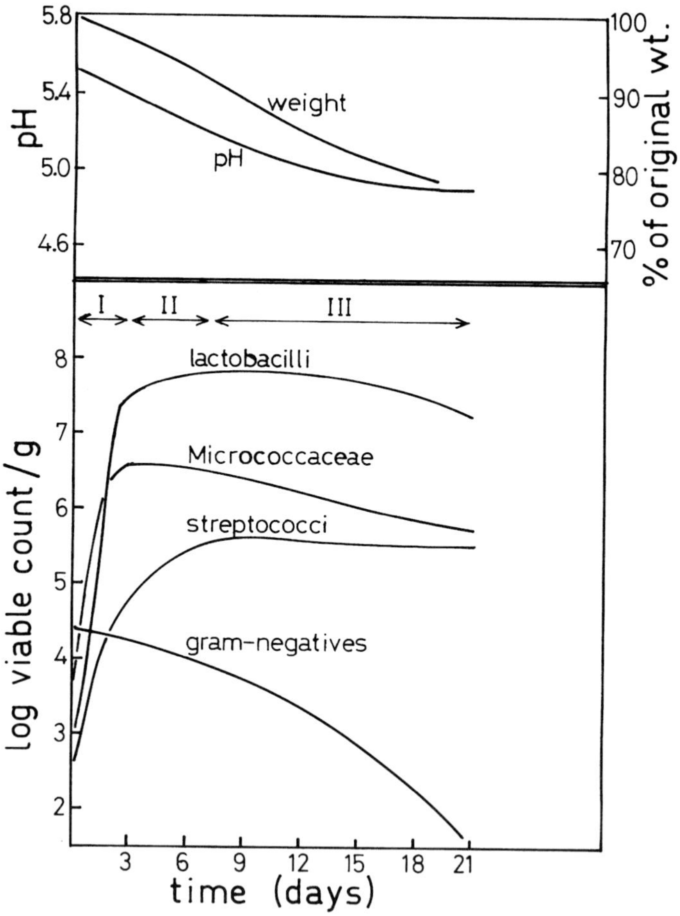

Figure 14.2 Development of the microflora (log colony-forming units/g fresh weight) during ripening of salami-type sausage by the 'natural' method (use of nitrate, low amounts of sugar, low fermentation temperature). Ripening conditions: (I), 18–21°C; (II); 20–22°C, smoke; (III), 15–17°C. (Data from Nurmi, 1966.)

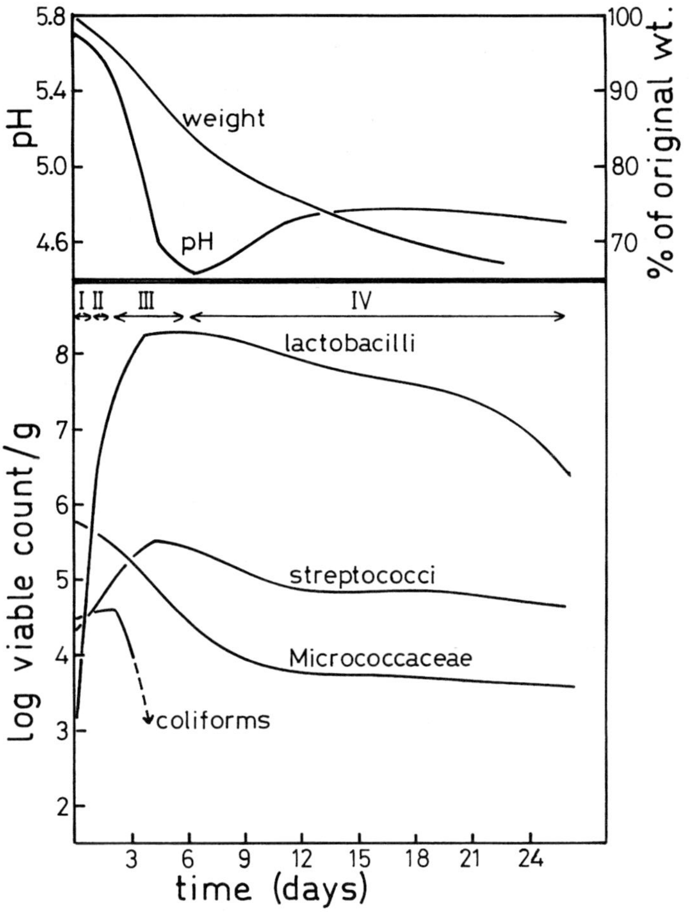

Figure 14.3 Development of the microflora (log colony-forming units/g fresh weight) during ripening of salami-type sausage by the 'rapid' method (use of nitrite, high amounts of sugar and elevated fermentation temperature). Ripening conditions: (I), 4°C; (II), 18°C; (III), 24°C, smoke; (IV), 18–20°C. (Data from Reuter *et al.*, 1968.)

'slow' method, low amounts of sugar and no lactic acid bacteria are added, nitrate rather than nitrite is used as the curing agent, and the sausages are ripened at low temperatures. Acid production is slow, and the activity of the acid-sensitive microorganisms (catalase-positive cocci, Gram-negative bacteria) is higher. In contrast, 'rapid' fermentation methods employing higher amounts of sugars, nitrite as curing agent and fermentation temperatures in the 22–25°C range, lead to lower counts of acid-sensitive bacteria (Figure 14.3). However, to make sure that undesired or hazardous acid-sensitive bacteria are suppressed even if very few lactic acid bacteria are present in the raw materials, such products are nowadays normally made with starters containing lactic acid bacteria.

The main product of carbohydrate fermentation in ripening sausages is lactic acid. The ratio of the two enantiomers varies somewhat, depending on the species of lactic acid bacteria present (Kagermeier, 1981), but normally, approximately equal amounts of D- and L-lactate are present in the final products (List & Klettner, 1978). Some acetic acid is also formed in fermenting sausages: in Belgian-style smoked semi-dry salami, for example, Deketelaere *et al.* (1974) and Demeyer (1982) found average molar ratios of lactic acid to acetic acid of 7 after fermentation to 12 after drying. This is similar to the ratio found in liquid cultures of meat lactobacilli (Reuter, 1971b; Nordal & Slinde, 1980). These lactobacilli are facultatively heterofermentative, which means that they ferment pentoses (such as ribose from nucleotides) to equimolar quantities of lactic and acetic acids, and many strains are capable of oxidizing lactic and/or pyruvic acid to acetic acid if oxygen is present (Kandler, 1983).

Minor fermentation products formed from carbohydrates by lactic acid bacteria and catalase-positive cocci include acetoin, diacetyl and/or 2,3-butanediol. The amounts found in fermented sausages are small but appear to contribute to the 'cheesy' component of the sausage aroma (see Dainty & Blom, 1995; also Section 14.4.3).

If oxygen was excluded from the sausages by vacuum-filling, Deketelaere *et al.* (1974) recovered 88–103% of carbohydrate fermented as lactic and acetic acid. Incomplete recovery of carbon indicated loss of carbon dioxide and other volatile compounds from the sausages, which is particularly significant in products of small diameter (Demeyer, 1992).

The variety and amount of added carbohydrate influences the course of the lactic acid formation during sausage fermentation. With glucose, sucrose or maltose, the rate and extent of lactic acid production is higher than with lactose, starch and dextrin (Pyrcz & Pezacki, 1974; Klettner & List, 1980). This is not surprising because lactic acid bacteria on meats do not ferment starch or dextrin, and few strains ferment lactose (Reuter, 1970; Kagermeier, 1981). A slow fermentation of carbohydrates usually stops at higher pH values because the water activity decreases during prolonged fermentation and becomes inhibitory to lactic acid formation at values of about 0.91 (List & Klettner, 1978).

For the manufacture of high-quality dry sausages with a long shelf life, the rate and extent of acid formation should be limited in order to allow acid-sensitive bacteria involved in developing and maintaining the desired sensory characteristics to 'keep up' with the lactobacilli to some extent (see Section 14.4.3). It must be emphasized that oversouring, colour defects and pore (CO_2) formation by lactic acid bacteria are, at present, the main microbial causes of spoilage of fermented sausages.

Most semi-dry sausages in Belgium and Germany have pH values of 4.8–5.0 after fermentation, which corresponds to a concentration of approximately 25 g lactic acid per kg dry weight (Deketelaere *et al.*, 1974; List & Klettner, 1978). In the manufacture of undried sausages, the extent of pH decrease is limited in order to keep them spreadable. During the fermentation of Italian and Hungarian salamis which are made with little sugar, the pH usually drops by not more than 0.5 units (Baldini *et al.*, 1983; Incze, 1986) whereas in many semi-dry fermented sausages produced in the United States, the pH is below 5.0 even at the time of sale (Acton & Dick, 1976).

Unless the mixture is inoculated with large amounts of other lactic acid bacteria, lactic acid formation during sausage fermentation is due to the activity of facultatively heterofermentative *Lactobacillus* spp., formerly designated as 'streptobacteria'. These bacteria ferment hexoses to lactate, and pentoses (many strains also gluconate) to lactate and acetate, and grow at 15°C. Early studies showed that the dominating strains could not be attributed to validly described species and were designed as 'atypical *Lactobacillus plantarum* strains' (Coretti, 1958, 1959; Urbaniak & Pezacki, 1975) or as 'unclassified' or 'atypical' (Deibel *et al.*, 1961a; Sharpe, 1962; Reuter, 1967; Laban *et al.*, 1978; Nordal & Slinde, 1980). Compared with 'typical' streptobacteria, these 'atypical' lactobacilli: (i) were capable of growth at lower temperature (most are able to grow at 4°C); and (ii) had a somewhat lower pH tolerance (no growth at pH 3.9); and (iii) had a higher tendency towards pleomorphism (growth as coccoid cells on low-sugar media) (Reuter, 1967, 1975). A taxonomic study by Kagermeier (1981) assigned the 'atypical' lactobacilli to *Lactobacillus sake* and *Lactobacillus curvatus*. This was later confirmed by Klein *et al.* (1996) using more advanced methods. Kagermeier (1981) also found that only in some sausages from countries with warm climate, *Lb. plantarum* was present in large numbers. *Streptococcus*, *Enterococcus* and *Pediococcus* species cannot compete well with the streptobacteria at fermentation temperatures commonly employed in Europe and, therefore, comprise only a minor part of the population (Reuter, 1967; Kagermeier, 1981). However, some pediococci proved to be good starters for various types of sausages (see Section 14.4.4). Generally, *Leuconostoc*, *Weissella* and heterofermentative lactobacilli made up less than 10% of the population. They are undesirable in fermented sausages because they form gas from sugars, and some strains produce slime from sucrose.

Since 1984, various studies on the biochemical characteristics of the lactic flora of fermented sausages from different countries have been published, e.g. from Spain (Sanz *et al.*, 1988; Hugas *et al.*, 1993) and Greece (Samelis *et al.*, 1994a). As in Kagermeier's study, the psychrotrophic species *Lb. sake* and *Lb. curvatus* predominated, and there was little effect of the geographical origin of the sample and type of product. Only in Turkish soudjouk, were the majority of strains mesophilic (*Lactobacillus alimentarius*; Gürakan *et al.*, 1995). It is safe to state that the reservoirs of psychrotrophic lactobacilli are cold rooms and/or cutting rooms that are usually kept at 12°C or below (in accordance with meat hygiene regulations). Another argument is that the species of lactic acid bacteria involved in sausage fermentation are also involved in the spoilage of vacuum-packed meat and pasteurized meat products.

Some properties of the *Lactobacillus* species predominant in fermented sausages and/or available as starters for these products are given in Table 14.3. All species listed form DL- lactate, except for some *Lb. sake* strains formerly named *Lactobacillus bavaricus*. *Lb. plantarum* possesses both D- and L-lactate dehydrogenases while *Lb. sake* and *Lb. curvatus* form L-lactate dehydrogenase and racemase.

Many lactic acid bacteria are capable of forming hydrogen peroxide. Peroxides attack haem-compounds and polyunsaturated fatty acids and thus lead to colour and flavour defects. Individual species and strains differ widely in their ability to accumulate and to destroy hydrogen peroxide (Lücke *et al.*, 1985; Condon, 1987). Strong producers are often found among strains of *Lb. sake* and *Lb. curvatus* because these species, unlike *Lb. plantarum*, possess flavin-dependent oxidases (Kandler, 1983). However, many strains of lactic acid bacteria from meats have been found to contain catalase activity when grown on myoglobin or haem-containing media (Nordal & Slinde, 1980), or haem-independent 'pseudocatalase' (Engesser & Hammes, 1994; see also Table 14.3). Thus, net hydrogen peroxide production in meat may be low even in the absence of catalase-positive, acid-sensitive microorganisms. However, practical experience indicates that oxygen should be excluded from raw sausage mixtures as far as possible, and enough peroxide-destroying activity should be present during fermentation. This is achieved, for instance, by slowing down the rate of acid formation and/or by the addition of appropriate starters (lactic acid bacteria that do not accumulate peroxides, along with suitable catalase containing microorganisms such as *Micrococcus* and *Staphylococcus* strains).

14.4.2 Curing

To 'cure' meats means to treat them with salt and nitrite and/or nitrate. Cured products have the characteristic pink curing colour and taste different to comparable products made without nitrite or nitrate. Fermented sausages made without nitrite or nitrate have a different colour and flavour, and tend to spoil rapidly due to oxidative rancidity, but with certain precautions (coarse particle size, small diameters, low ripening temperature) they may be produced in satisfactory quality (Wirth, 1991).

Table 14.3 Some important properties of lactic acid bacteria[a] involved in sausage fermentations[b]

Property	*Lb. curvatus*	*Lb. sake*	*Lb. plantarum, Lb. pentosus*[c]	*P. pentosaceus*	*P. acidilactici*
Cell shape	Rods	Rods	Rods	Cocci in tetrads	Cocci in tetrads
Growth at a_w 0.93	–(+)	+(–)	+	V	+
Growth at 4°C	+	+	–	–	–
Growth at 50°C	–	–	–	–	+
Peptidoglycan type	Lys-D-Asp	Lys-D-Asp	meso-DAP	Lys-D-Asp	Lys-D-Asp
Lactic acid enantiomer	DL	DL (L)[d]	DL	DL	DL
Nitrate reductase	–	–	V	–	–
Nitrite reductase					
+haem	–	–	V	V	–
–haem	–	V	V	–	–
Catalase					
+haem	–	+	V	–	+
–haem	–	V	V	+	–
Ammonia from arginine	–(+)	+	–	+	+
Fermentation of					
glucose	+	+	+	+	+
maltose	V	V	+(–)	+	–
sucrose	+(–)	+	+(–)	V	V
lactose	V	V	+(–)	V	V
melibiose	–(+)	+	+(–)	V	V
mannitol	–	–	+	–	–
ribose	+	+	+	+	+
gluconate	–(+)	+	+(–)	V	–
Minimal pH in MRS broth	ca. 4.0	ca. 4.0	ca. 3.7	ca. 3.8	ca. 3.8

[a] *Lb., Lactobacillus*; *P., Pediococcus.*
[b] +(–), positive with exceptions; –(+), negative with exceptions; V, variable (strain-dependent).
[c] *Lb. pentosus* is similar to *Lb. plantarum* but capable of fermenting D-xylose.
[d] Strains forming L-lactate only have been formerly classified as *Lb. bavaricus*.
Hammes *et al.* (1985); Weiss (1992); Hammes and Knauf (1994); Data from Klein *et al.* (1996).

To be effective, nitrate added as such (or formed from nitrite during curing; see below) must be reduced to nitrite by microorganisms. Some strains of *Lb. plantarum* and *Lb. pentosus* reduce nitrate *in vitro* when cultivated in media with high pH and little sugar (Coretti, 1958; Kagermeier, 1981). However, rates were insufficient for proper colour and flavour development (Hammes *et al.*, 1990), and acid-sensitive nitrate reducers must also be present and active. If nitrate is used as the sole curing agent, it is very important to restrict the rate of lactic acid formation: the pH should be above 5.4 until sufficient nitrate is reduced (Niinivaara, 1955). In European-type dry sausages, this was achieved by the use of low amounts of fermentable carbohydrate (0.2–0.3%; some genuine salamis are still made without any carbohydrate addition). To prevent growth of unwanted bacteria, fermentation temperatures must be lower, bearing in mind that nitrate may even enhance growth of these bacteria early in fermentation. In some traditional procedures, attempts were made to favour a nitrate-reducing flora by preincubating the sausage mix at low temperatures (Zaika *et al.*, 1976). Niinivaara (1955) partially overcame the problem by inoculating the mixture with a nitrate-reducing *Micrococcus* strain (see Section 14.4.4).

Whether added as such or formed by microbial nitrate reduction, nitrite is reduced further by many microorganisms present in the sausage, including the nitrate-reducing microflora as well as several strains of lactic acid bacteria (Coretti, 1958; Collins-Thompson & Rodriguez Lopez, 1981; Wolf *et al.*, 1990).

The formation of curing colour in fermented sausages requires several steps. First, oxygenated (red) myoglobin reacts with nitrous acid to give brown metmyoglobin and nitrate. Metmyoglobin then reacts with nitrous acid and/or nitric oxide (from acid-catalysed disproportionation of nitrous acid) and indigenous and/or exogenous reductants (e.g. ascorbate) to give red nitric oxide myoglobin (NOMb), the NO group occupying the sixth ligand position of the haem. The rate of NOMb formation increases with falling pH and is therefore accelerated by the activity of the lactic acid bacteria in fermented sausage. It is unlikely that microbial reduction of nitrite plays a significant role in curing colour formation: nitric oxide is not an intermediate of nitrite reduction to ammonia by haem-containing nitrite reductases as present in many bacteria including *Staphylococcus carnosus* (Hartmann *et al.*, 1995) which is frequently used as starter culture for fermented sausages. Moreover, sausages fermented with pediococci alone (e.g. US-type summer sausages) develop good curing colour even though the pediococci do not have nitrite reductases forming nitric oxide (Hammes *et al.*, 1990). In contrast, Garcia *et al.* (1992) did not observe curing colour formation in aseptically prepared sausage with 100 mg $NaNO_2$ added per kg, unless a *Micrococcus* strain had been added. However, it is possible that under their processing conditions, most nitrite was transformed into nitrate.

During ripening, the protein moiety of NOMb is partially denatured and nitric oxide myochromogen formed. This process improves colour stability because the nitric oxide dissociates less readily from the haem. However, at low pH values, the iron moiety in nitric oxide myochromogen can also be attacked by oxidants such as peroxide groups present in fatty tissue and peroxides formed by lactic acid bacteria in the presence of oxygen. This results in greyish or greenish discolorations and stresses again how important it is to use fresh firm fat and to introduce as little oxygen into the sausage mixture as possible.

Nitrite also retards lipid oxidation in fermented sausages and thus contributes to their flavour (Gray & Pearson, 1984) and chemical stability.

Experiments with cooked meats (MacDonald *et al.*, 1980) showed that nitrite acts by binding non-haem iron, the main pro-oxidant in meats. Attempts to replace nitrite by other antioxidants met with little success (Wirth, 1964; MacDonald *et al.*, 1980).

As will be explained further in Sections 14.4.3 and 14.4.4, catalase-positive cocci play an important role in forming and stabilizing the desired sensory properties of fermented sausages. Hence, strains from many fermented sausages have been isolated, characterized and identified. Most of them were found to be *Staphylococcus* spp.: Delarras & Laban (1981) assigned all of their strains to this genus, Rheinbaben & Hadlok (1979) did so with 156 of 166 strains, and Fischer & Schleifer (1980), Selgas *et al.* (1988), Comi *et al.* (1992) & Samelis *et al.* (1994b) found 50 of 82, 110 of 172, 118 of 132 and 122 of 140 strains studied, respectively, to be *Staphylococcus* spp. Differentiation into species revealed that *S. xylosus* and *S. saprophyticus* were widespread both in German and French dry sausage as well as in sausages on the British market (Fischer & Schleifer, 1980; Delarras & Laban, 1981; Seager *et al.*, 1986), while *S. xylosus* dominated in most products manufactured without starter, such as Italian salami (Comi *et al.*, 1992). In samples from Germany, *S. carnosus* (originally identified as *S. simulans*) and *Micrococcus varians* were also found frequently (Fischer & Schleifer, 1980). These strains may have originated from the starter cultures in use in Germany in the late 1970s.

The behaviour of catalase-positive cocci ('*Micrococcaceae*') in ripening sausages is illustrated in Figures 14.2 and 14.3. Because micrococci are obligate aerobes, and the facultatively anaerobic staphylococci grow much better in the presence of oxygen, there is little if any growth in the core of the sausage, but high numbers may be reached close to the surface, especially if its pH value rises because of mould growth.

Nitrite has little if any effect on the overall composition of the microflora of fermented sausages (Skjelkvåle *et al.*, 1974; Puolanne, 1977; Gerigk & Gossling, 1981; Lücke *et al.*, 1983), and pH change and temperature are much more important than nitrite in the control of pathogens (see Collins-Thompson *et al.*, 1984, and Section 14.5).

14.4.3 Development of aroma and flavour

A large variety of compounds are likely to contribute to the desired (and undesired) aroma and flavour of fermented sausages. Some are added to the sausage as such (salt, constituents of spices and smoke); others are formed during ripening. Abiotic reactions (in particular, lipid peroxidation), tissue enzymes (in particular, proteases and lipases) and microbial enzymes all contribute to the development of aroma and flavour, as well as to the spoilage of the products. There is much commercial interest in accelerating ripening processes and extending shelf life (Lücke *et al.*,

1990). A recent review on the flavour chemistry of fermented sausages has been published by Dainty & Blom (1995).

The main products of carbohydrate fermentation by lactic acid bacteria (lactic acid with small amounts of acetic acid) give the sausages an 'acid' flavour. In undried products, acid formation is restricted in order to keep them spreadable, and they retain more of the aroma and flavour of fresh meat. However, an acid flavour predominates in semi-dry products which are sold after less than about 2 weeks' ripening. The intensity of this flavour obviously depends on the pH value, but at a given pH, a high proportion of acetic acid gives the product a less 'pure' and more 'sour' flavour. High levels of acetic acid accumulate if glucono-δ-lactone (GdL) is added as an acidulant because this compound is fermented by many lactobacilli to lactic and acetic acid (Table 14.3).

The longer the ripening time, and the higher the activity of microorganisms other than lactic acid bacteria, the higher levels of volatile compounds with low sensory thresholds are found (Langner, 1969, 1972). Lipids are precursors of many unbranched aldehydes, 2-alkanones and short-chain fatty acids, whereas most branched-chain volatiles probably originate from amino acids (Halvarson, 1973). The respective roles of abiotic reactions, tissue enzymes and microorganisms in lipid degradation are not yet completely clear. Some strains of lactic acid bacteria from meats attack tributyrin in culture media (Reuter, 1971a; Papon & Talon, 1988) or even lard (Andersen *et al.*, 1995). However, activity under the conditions of sausage ripening is probably low, and there appears to be little formation of volatiles from lipids in rapidly fermented semi-dry products (Langner, 1972). In contrast, many catalase-positive cocci from fermented sausaqes actively attack pork fat (Nurmi & Niinivaara, 1964; Comi *et al.*, 1995). Also, many yeasts and moulds found on fermented sausages are lipolytic (Leistner & Ayres, 1967; Leistner & Bem, 1970; Comi & Cantoni, 1980) and contribute to the flavour of air-dried sausages where particularly high amounts of carbonyls (especially long-chain aldehydes) have been found (Langner, 1972). Cantoni *et al.* (1967a) suggested that catalase-positive cocci are the main lipolytic agents during sausage fermentation. The specificity of lipolysis observed by Demeyer *et al.* (1974) supported this. Stahnke (1994) found elevated amounts of long-chain fatty acids in sausages fermented with a lipolytic *Staphylococcus xylosus*, but this had no impact on flavour. Since lipolysis begins very early in fermentation (Dobbertin *et al.*, 1975), and appears to occur even in the absence of catalase-positive cocci or other bacteria (Garcia *et al.*, 1992), tissue lipases rather than microbial lipases are probably the main agents of lipolysis.

Free long-chain unsaturated fatty acids have little effect on aroma and detrimental effects on flavour. However, they are further degraded to carbonyls and short-chain fatty acids with peroxides as intermediates (Cerise *et al.*, 1973). This appears to be mainly an abiotic autoxidation process (Dainty

& Blom, 1995) but catalase-positive cocci may participate in these reactions (Smith & Alford, 1968), and the sausage microflora may further modify the carbonyl compounds generated by lipid autoxidation and form compounds which add to the desired aroma of the sausages. For example, Stahnke (1995) reported a positive correlation between the levels of 2-alkanones, salami odour and levels of *Staphylococcus xylosus.*

Nitrite addition to the sausage mixture results in less carbonyls being formed and in a shift of the spectrum of these compounds, and thereby in a different aroma ('curing aroma'; Gray & Pearson, 1984). Nitrite also inhibits oxygen consumption by tissue enzymes (Rödel *et al.*, 1993a) and the activity of Gram-negative bacteria early in fermentation. This may result in more aroma precursors being formed, particularly at low fermentation temperatures, and may explain why the use of nitrate rather than nitrite or of low instead of the usual levels of nitrite has been frequently observed to give a more intensive salami-like aroma (Wirth, 1991).

During sausage ripening, a significant portion of the protein is hydrolysed (Niinivaara *et al.*, 1964). Non-protein nitrogen increases by about 20% in smoked sausages (Dierick *et al.*, 1974). Experiments by Verplaetse *et al.* (1989; see also Demeyer, 1992) indicate that tissue proteases (in particular, cathepsin D) are the main agents of proteolysis early in fermentation. Cathepsin D is activated at pH values around 5.0 and produces peptides which are then further metabolized by the ripening flora. Montel *et al.* (1995) described a peptidase from *Lb. sake*, and Næs *et al.* (1995) were able to accelerate sausage ripening by addition of an peptidase isolated from *Lactobacillus paracasei.* Catalase-positive cocci may also be involved in peptide degradation, and during air-drying of sausages, moulds growing on the surface also hydrolyse proteins and catabolize amino acids, which is easily recognized though the ammonia smell of ripening rooms and the resulting products. The intensity of flavour of such sausages was found to be proportional to the degree of proteolysis (Geisen *et al.*, 1992).

Amino acids not assimilated by the microflora are further metabolized, both in smoked (Dierick *et al.*, 1974) and air-dried sausages. The ammonia and the amines thus formed – in conjunction with lactate oxidation at the surface and an increase in osmolarity – account for the rise in pH which is observed during drying of sausages. This rise is particularly marked in unsmoked products where moulds and yeasts colonize the surface and degrade not only proteins but also lactic acid. These products obviously taste quite different to the smoked 'sour' sausages. A mould layer on the sausage surface is also beneficial in air-dried products by keeping oxygen away from the sausage and providing a 'microclimate' which favours the correct drying of the product.

Amino acids and their nitrogen-containing degradation products may contribute to the taste of the sausage but less to the aroma since few if any nitrogen-containing volatiles have been found in fermented sausages

(Berger *et al.*, 1990; Dainty & Blom, 1995). Added flavour intensifiers such as glutamate and inosine monophosphate may be degraded during sausage fermentation (Cantoni *et al.*, 1967b; Dierick *et al.*, 1974) and thus be of little value as additives to these products.

Various esters (mainly ethyl esters) are found among the volatiles of fermented sausages (Dainty & Blom, 1995), and to contribute to salami odour. The data of Stahnke (1995) suggest staphylococci (e.g. *S. xylosus*) as main agents for their production.

While a certain degree of lipid and protein degradation undoubtedly improves the quality of fermented sausages, these processes must be carefully controlled, or, the products may become unacceptable. Fermented sausages have even been reported to be spoiled by lactobacilli capable of amino acid degradation (Lang, 1960).

14.4.4 Starter cultures

Excellent fermented sausages can be manufactured without addition of starter cultures or re-inoculation with finished sausages. Craftsmen have designed formulations and ripening conditions which favour the desired microorganisms so strongly that the products are safe and palatable, even if only few lactobacilli and micrococci are present in the fresh mixture.

The roles of, and benefits from, different groups of microorganisms available as starter cultures for sausage fermentation are compiled in Table 14.4. The interest in starter cultures arose parallel to the trend towards industrial production, short ripening times and standardization of product properties. The presence of 10^6–10^7 rather than 10^2 or so active lactic acid bacteria per gram of fresh sausage mixture leads to a more predictable and more rapid pH decrease and to earlier development of firmness, and may improve product safety (see Section 14.5). Adding catalase-positive cocci is advantageous in particular for products cured with nitrate, products with a long shelf life, or sausages very prone to lipid deterioration. Their mechanism of action is much less clear (see Section 14.4.3); they definitely are essential for nitrate reduction and aroma formation, and they play a role in protecting the product from detrimental effects of oxygen. The strongest evidence for this comes from the experience with GdL as acidulant. This compound is fermented to lactic and acetic acids, and catalase-positive cocci are severely inhibited. The sausages are very prone to rancidity and colour defects, and have only a short shelf life.

Yeast and mould strains are available as starter cultures for air-dried (not, or only lightly, smoked) sausages where these organisms readily colonize the surface. *Debaryomyces hansenii* or its imperfect form, *Candida famata*, appears to be the only yeast species in meat starter cultures because they grow at low water acitivity and have also been observed to affect

Table 14.4 Starter cultures for sausage fermentation

Microbial group	Species available as starters[a]	Desired metabolic activities	Benefits in sausage ripening
Lactic acid bacteria	*Lb. plantarum* *Lb. pentosus* *Lb. sake* *Lb. curvatus* *P. pentosaceus* *P. acidilactici*	Formation of lactic acid	Inhibition of pathogenic and spoilage bacteria; acceleration of colour formation and drying
Catalase-positive cocci	*St. carnosus* *St. xylosus* *M. varians*	Nitrate reduction	Colour formation and stabilization
		Nitrite reduction	Removal of excess nitrite
		Oxygen consumption, peroxide destruction	Delay of rancidity; colour stabilization
		Formation of carbonyls and esters?	Aroma and flavour development
Yeasts	*Deb. hansenii*, *Candida famata*	Oxygen consumption	Delay of rancidity; colour stabilization
		Not known in detail	Aroma and flavour development
Moulds	*Pen. nalgiovense* biotypes 2,3,6; *Pen. chrysogenum*	Surface colonization	Suppression of undesired moulds; facilitation of drying
		Oxygen consumption	Delay of rancidity; colour stabilization
		Lactate oxidation, degradation of proteins and amino acids	Flavour development

[a] *Lb., Lactobacillus; P., Pediococcus; St., Staphylococcus; M., Micrococcus; Deb., Debaryomyces; Pen., Penicillium.*

colour and flavour when added to the sausage mixture (Rossmanith *et al.*, 1972; Coretti, 1977; Miteva *et al.*, 1986; Gehlen, 1989).

As stated earlier, moulds contribute to the characteristic aroma, flavour and appearance of air-dried sausages. However, colonization of the sausage surface by the 'wrong' mould leads to unsatisfactory product quality and increases the risk of mycotoxin formation. The sausages should therefore be dipped into a suspension of conidia of an appropriate starter mould before ripening. Technological criteria for selection of mould starters include competitiveness on the sausage surface (i.e. in the presence of salt, curing agents and lactic acid, and at about 15°C), adhesiveness to casings, white or slightly yellowish colour (i.e., no formation of conidia during sausage drying), and formation of appropriate aroma and flavour. Strains fulfilling these requirements belong to the species *Penicillium nalgiovense* and *Penicillium chrysogenum*. However, most of these strains form the mycotoxin cyclopiazonic acid (El-Banna *et al.*, 1987) and many strains form the antibiotic penicillin (Andersen & Frisvad, 1994). Hence, strains must be carefully tested by chemical and biological assays for the absence of these compounds. Mintzlaff & Leistner (1972) selected a strain of *P. nalgiovense* as a starter culture, and later investigations (e.g. Fink-Gremmels *et al.*, 1988; Hwang *et al.*, 1993) showed that other strains of *P. nalgiovense* and *P. chrysogenum* are likewise beneficial to the appearance and flavour of air-dried sausages.

The (rather short) history of using starters for sausage fermentation is outlined by Jessen (1995). In the United States, the key role of lactobacilli in sausage fermentation was recognized early (Jensen and Paddock, US patent no. 2,225,783, Dec. 1940); however, early attempts to introduce *Lactobacillus* strains as starters failed because suitable strains did not readily survive lyophilization. *Pediococcus cerevisiae*, now classified as *Pediococcus acidilactici*, proved suitable for the manufacture of products like summer sausage or thüringer cervelat (Deibel *et al.*, 1961a,b) which are fermented at temperatures above 30°C. It can also be easily stored in a lyophilized or frozen form and is still widely used as a starter in the United States.

In Europe, where low fermentation temperatures are common and nitrate was used much more frequently than nitrite, sausage makers were faced with the problem that a rapid lactic acid fermentation suppressed the organisms containing catalase and nitrate reductase to such an extent that defects in colour and flavour were common. It is, therefore, not surprising that early European researchers regarded lactobacilli as detrimental, rather than beneficial, and a *Micrococcus* strain was first introduced as a starter culture by Niinivaara (1955). This strain (M 53) was selected because it rapidly reduced nitrate, improved colour and flavour, and inhibited undesired microorganisms (Pohja & Niinivaara, 1957). It was later replaced by a 'fermentative *Micrococcus*' isolated by Pohja (Licentiate Thesis, Helsinki 1960, cited by Nurmi, 1966) and by other catalase-positive cocci including *Staphylococcus carnosus* (renamed by Schleifer & Fischer, 1982). Nurmi (1966) isolated a strain of *Lactobacillus plantarum* from fermented sausage and showed that this strain was superior to '*Pediococcus cerevisiae*' in sausage fermentation at 20–22°C. He also showed that a combination of this strain with Pohja's 'fermentative *Micrococcus*' gave best results, i.e. a rapid reduction in pH value without inhibiting colour and flavour development. Early attempts to improve sausage manufacture and quality by using 'atypical lactobacilli' (*Lb. sake*, *Lb. curvatus*) as starters were of limited success (Reuter, 1972), even though these strains dominate in virtually all traditional European-style sausages. However, strains suitable for sausage ripening were selected later and became commercially available.

Nowadays, the majority of fermented sausages produced in Europe are made with starter cultures (see Lücke *et al.*, 1990). Starter cultures have, to a large extent, also replaced so-called 'back-slopping' methods where a fermented sausage from the previous batch is used to inoculate the following batch. The disadvantage of this method is that desired catalase-positive cocci will tend to become 'diluted out', and strains with undesirable metabolic properties may be selected.

Selection criteria for starter cultures for meats have been summarized by Jessen (1995) and Knauf (1995). Under the conditions (pH, a_w, temperature, concentrations of curing agents) prevailing during sausage fermentation, they must perform their desired activities (Table 14.4) at appropriate rates and after as short a lag as possible. Starter cultures should obviously be of no health concern, and appropriate tests for the absence of pathogenicity and toxin formation are necessary. In addition, they should not form amines from amino acids. Last but not least, they must be suitable for large-scale cultivation and preservation by freezing and lyophilization.

During *in vitro* screening of lactic acid bacteria, strains forming CO_2 from hexoses, amines from amino acids, volatile sulphur compounds from sulphur-containing amino acids, slime or larger amounts of hydrogen peroxide should be sorted out. However, *in vivo* tests (laboratory and pilot-plant sausage fermentations) are required to select strains of lactic acid bacteria or any other microbial group that form pleasant aroma and flavour notes.

The necessity and choice of starters depends on the type of fermented sausage to be produced. In the manufacture of traditional dry sausages, where the fermentation temperature is low and the ripening time long, there is only limited benefit from adding lactic acid bacteria. However, if nitrate is used as curing agent, addition of nitrate-reducing catalase-positive cocci is advisable. On the other hand, use of lactic acid bacteria in the manufacture of rapidly fermented semi-dry products is essential for microbial safety (see Section 14.5). In the temperature range between 20 and 25°C, acid formation by *P. acidilactici* is too slow to be of much benefit, and cultures containing *Lb. sake* strains tend to have the fastest rates and to be most competitive. Acidification rates of *P. pentosaceus* (Hechelmann *et al.*, 1988; Junker *et al.*, 1993) and *Lb. curvatus* (Landvogt & Fischer, 1991) depend much on the culture used, and may be equal to or higher than those of *Lb. plantarum* and *Lb. pentosus*. However, excessive lag phases after inoculation of the mix may abolish the advantages of using a culture with high acidification rate. With this in mind, a 'starter sausage' containing a strain of *Lb. curvatus* is produced centrally and distributed frozen, mainly in The Netherlands (see Krol, 1984). However, care must be taken to ascertain that there is sufficient activity of desired acid-sensitive bacteria under these conditions. *P. pentosaceus* is often stated to form a 'milder' aroma (Jessen, 1995). At any rate, it is essential to follow the instructions of the starter manufacturer, and not arbitrarily combine different starters and sugars.

Starter cultures are distributed frozen or freeze-dried on suitable carriers. Frozen cultures generally have a shorter lag phase, although this is somewhat less critical than in dairy fermentations to be finished after hours rather than days. An even distribution of the culture in the mix is important to ensure homogeneous fermentation (Katsaras & Leistner, 1988). Some sausage manufacturers resuscitate lyophilized cultures for several hours before use, but none of them propagates starters *in vitro* any further.

Combinations of lactic acid bacteria with catalase-positive cocci proved most useful for fermented sausages formulated with nitrite rather than nitrate, and are now most widely used throughout Europe (Lücke *et al.*, 1990).

Petäjä (1977) was able to improve the aroma of fermented sausages by combining an *Aeromonas* strain with *Lactobacillus plantarum*. However, the preservation, storage and distribution of such Gram-negative bacteria is difficult, as is the control of their activity during sausage fermentation. Furthermore, concern about public health may also prevent their use as starter cultures. *Streptomyces griseus* strains have also been reported to be beneficial to the quality of fermented sausages (Eilberg & Liepe, 1977; Rohde & Weckend, 1983).

14.5 Public health aspects

Most fermented sausages are not heated during manufacture or before consumption, and on first glance, it may be surprising that, at least in the past decade, they have a record of safety equal or even superior to other meats. However, outbreaks of salmonellosis, *Staphylococcus aureus* enterotoxicosis, and, recently, infections caused by enterohaemorrhagic strains of *Escherichia coli* have been traced back to fermented sausages. This section discusses the measures to minimize these as well as other possible health hazards from the ingestion of fermented sausages. How salmonellae, *Staphylococcus aureus* and *Listeria monocytogenes* are controlled during sausage ripening is summarized in Table 14.5. It is important to note that 'lowering' of one 'hurdle' may be compensated for by 'fortifying' another hurdle (Leistner, 1986). For example, increasing the initial water activity within certain limits may be compensated for by a more rapid rate of acid formation, or a lower fermentation temperature. This must be taken into account when setting up Hazard Analysis Critical Control Point (HACCP) systems for the manufacture of fermented sausages.

14.5.1 Salmonella and other food-borne pathogenic Enterobacteriaceae

During proper sausage fermentation, the Enterobacteriaceae count usually remains constant or increases by, at most, one log-cycle (compare Figure

Table 14.5 Preventive measures to control growth of *Salmonella enteritica*, *Staphylococcus aureus* and *Listeria monocytogenes* during the manufacture of fermented sausages

Process step	Additives	Preventive measures for: Semi-dry sausages	Dry and/or mould-ripened sausages	Traditional dry sausages
Carcass delivery		Internal temperature <7°C; pH ≤5.8 (6.0[a]); certified source		
Butchering		Meat temperature <7°C; no cross-contamination; proper cleaning and disinfection		
Chilling/freezing		Meat temperature <2°C		
Comminution and mixing	<----- Salt	Target water activity 0.955–0.965		
	<-- Curing agents	100–125 mg $NaNO_2$/kg[b]	50–70 mg $NaNO_2$/kg[b]	<300 mg KNO_3/kg or none
	<--- Sugars[c]	0.5–0.8%	0.3–0.5%	0.3%
	<-- Starters[d]	LAB +*Mic./Staph.*	LAB +*Mic./Staph.*	none required
	<--- Spices			
Filling	<--- Casings	Proper cleaning and preparation of natural casings		
Surface inoculum	<--- Starter	–	Use of tested/certified mould culture	
Fermentation		20–25°C, 2–3 days, target pH <5.3	18–22°C, target pH ≤5.3 (ca. 3 days)	15–18°C, 2 days
Smoking				
Ageing/drying		<15°C, 70–80% RH[f]; target a_w ca. 0.93	10–15°C[e], 70–80% RH[f]; target a_w ≤0.90	≤10°C, 70–80% RH[f]; target a_w ≤0.90
Storage		<15°C	<25°C	<25°C

[a] In exceptional cases, adjustment of sugar addition and fermentation conditions may be necessary.
[b] Along with max. 500 mg Na-ascorbate/kg.
[c] Rapidly fermentable by the starter used; 0.8% should be used if meat pH is significantly above 5.8.
[d] LAB, lactic acid bacteria; *Mic./Staph.*, *Micrococcus/Staphylococcus*. Starters must be active at the fermentation temperature selected.
[e] For mould-ripened products, the temperature should be lowered to about 10°C after 2 days of fermentation (Rödel *et al.*, 1993b).
[f] RH, relative humidity. Adjustment depending on pH and a_w of sausage (see text).

14.2). Factors which favour the growth of Enterobacteriaceae during sausage fermentation and thereby increase the risk of salmonellosis include a high initial water activity, a high initial pH value, a low concentration of fermentable carbohydrates, only low numbers of lactobacilli in the fresh sausage mixture, the use of nitrate or of very low levels of nitrite as the curing agent (Hechelmann *et al.*, 1974; Sirviö *et al.*, 1977), and high ripening temperatures. Table 14.5 gives examples of processes that safely prevent growth of salmonellae (and of *Yersinia enterocolitica*; see Asplund *et al.*, 1993; Kleemann & Bergann, 1996) during sausage fermentation.

In Germany, some outbreaks of salmonellosis could be traced to the consumption of fresh, spreadable sausages. To retain their 'fresh' consistency and 'meaty' taste, such sausages have high water activity, are often manufactured with little, or no, sugar or, acidulant (GdL), and are subjected only to 1–2 days fermentation at 20–25°C, which is sufficient for curing colour formation but does not always suppress the development of Enterobacteriaceae. In such sausages, increasing the amount of added salt to at least 2.5% and addition of appropriate lactic acid bacteria along with some fermentable carbohydrate or small amounts of GdL was most effective in the control of Enterobacteriaceae including salmonellae (Schmidt, 1988).

During drying, Enterobacteriaceae, including salmonellae and yersiniae (Kleemann & Bergann, 1996), are slowly inactivated. Reported rates for *Salmonella* destruction during sausage ripening differ between less than one log cycle (Scharner, 1968; Stecchini *et al.*, 1982) to about 4 log cycles (Smith *et al.*, 1975). For European-style smoked products, a reduction by one or two log cycles (Holley *et al.*, 1988a) is in agreement with the observation that salmonellae occur much less frequently in dry sausages (Boos, 1979; Schmidt, 1988) than in ground meat or fresh, only slightly fermented raw sausages. Inactivation is apparently more rapid in low-pH products stored at ambient temperatures.

An outbreak of salmonellosis from salami in Australia showed that salmonellae may grow to such high numbers during fermentation that consumers of the final, dry product still ingested more than the minimal infective dose. According to Leistner *et al.* (1982), the processor had combined very low input of nitrite (as common in Italian-type sausage formulations) with a fermentation temperature above 25°C (as common for US-style summer sausages).

Two severe outbreaks of haemorrhagic colitis from fermented sausages occurred in western states of the USA and in Australia (cf. *Morbidity and Mortality Weekly Report*, **44** (1995), 157–60; 550–1, 557, 558). In contrast to salmonellae, the infective dose of the causative agent, enterohaemorrhagic strains of *Escherichia coli* (EHEC), is generally low, and no multiplication during sausage fermentation is required to cause disease. Moreover, inactivation during sausage ripening is slow (about one log cycle in the experiments of Glass *et al.*, 1992), especially with acid-adapted cells (Leyer *et al.*, 1995). Fortunately, the prevalence of EHECs appears to be low and restricted to beef and lamb (Bülte *et al.*, 1996); however, as long as no effective control measures on the farm and slaughterhouse level exist, it is advisable not to include beef- or lamb-containing raw sausages with short ripening times into the diet of small children (the main risk group for haemorrhagic colitis and its life-threatening sequela disease, acute renal failure).

14.5.2 Staphylococcus aureus

Staphylococcus aureus is frequently found in raw meat and fermented sausages, but generally at low levels. Only if allowed to grow to cell densities above 10^7/g, it may form levels of enterotoxins sufficient to cause a – usually mild – form of intoxication. Growth of this organism has been studied during ripening of various types of raw sausages. The bacterium is little affected by salt and nitrite, but it is a poor competitor under anaerobic conditions, at low pH values and low temperatures. It may grow at high fermentation temperatures during the lag phase of the development of lactic acid bacteria and may reach high numbers in the outer layers of the sausages (Barber & Deibel, 1972). Niskanen & Nurmi (1976) and Metaxopoulos *et al.* (1981a,b) showed that the initial pH value and the initial activity of the lactic acid bacteria are critical for the control of *S. aureus*. If sausages are fermented at no more than 25°C for 2–3 days, and the initial count of *S. aureus* is below 10^4/g, the risk of enterotoxin formation is low (Holley *et al.*, 1988a), even if the sausages are not smoked (Metaxopoulos *et al.*, 1981a,b; Holley *et al.*, 1988b). However, fermentation temperatures up to 43°C are common in the manufacture of smoked, semi-dry sausages in the United States. Under such conditions, only a rapid pH drop in the sausages early in fermentation ensures inhibition of *S. aureus* (Raccach, 1981). Consequently, the American Meat Institute, in 1982, specified the maximum time allowed in 'good manufacturing practice for fermented dry or semi-dry sausage' to reach pH 5.3 (e.g. 80 hours at 24°C; see Bacus, 1984). Apparently, following such rules by introduction of appropriate process controls and starter cultures strongly reduced the incidence in the USA of outbreaks of *S. aureus* food poisoning due to the consumption of fermented sausages.

The effect of various formulations and ripening conditions on *S. aureus* in unsmoked sausages fermented at temperatures common in Europe has been studied by Hechelmann *et al.* (1988). If fermented at 23°C, a reduction in pH to 5.3 within 2–3 days also turned out to be critical in these products. This could be achieved by adding lactobacilli or *Pediococcus pentosaceus* as starters, along with 0.3% of rapidly fermentable sugars. It also proved to be important to adjust the initial water activity to above (!) 0.955, and to use ageing temperatures below 15°C until the water activity had decreased to below 0.90. At fermentation temperatures of 15°C, no growth of *S. aureus* was observed, irrespective of acid formation.

14.5.3 Listeria monocytogenes

Because of its wide distribution on meat and its ability to grow at low temperatures and relatively low water activity (>0.93), *Listeria monocytogenes* may be expected to represent a hazard to the consumer of fermented sausage. However, there is no epidemiological evidence for the involvement of fermented sausages in outbreaks of listeriosis, and little if any growth was observed during sausage fermentation under conditions resembling commercial practice (Farber *et al.*, 1993), even in undried sausages (Schillinger *et al.*, 1991; Karches & Teufel, 1988) and in mould-ripened salamis (Trüssel & Jemmi, 1989; Campanini *et al.*, 1993; Rödel *et al.*, 1993b; Hugas *et al.*, 1995). Conditions controlling salmonellae and *Staphylococcus aureus* were found also to prevent growth of listeriae. However, to make sure that listeriae do not grow on the surface of mould-ripened sausages, Rödel *et al.* (1993b) recommended lowering the aging temperature for Italian-type salami to 8–10°C.

During sausage ripening, listeriae die only slowly. Using lactobacilli that form a listericidal bacteriocin as starters reduced the *Listeria* count early in fermentation by about one log (Berry *et al.*, 1990; Foegeding *et al.*, 1992; Schillinger *et al.*, 1991; Hugas *et al.*, 1995).

14.5.4 Spore-forming bacteria

Raw sausage mixtures may contain considerable numbers of *Bacillus* spores of which spices are a major source. Although many *Bacillus* spp. produce spoilage symptoms like proteolysis in the appropriate nutrient media, they fail to do so when inoculated into raw sausage mixtures, even if faulty ripening conditions were chosen (Neumayr *et al.*, 1983). The authors concluded that *Bacillus* spp. are controlled in fermented sausages by a combination of low water activity, low pH and absence of oxygen.

There is no practical or epidemiological evidence that clostridia may grow and form toxins during ripening and storage of fermented sausages. Botulinum toxin was only detectable if a sausage with no glucose added was kept at high temperature (32.2°C) for extended time in which the pH did not fall below 5.5 (Christiansen *et al.*, 1975). This is far from any commercial practice. Growth and toxin formation by *Clostridium botulinum* is even less likely in 'European-type' fermented sausages (Nordal & Gudding, 1975; Lücke *et al.*, 1983). *C. botulinum* and other clostridia are controlled by low pH and low water activity, irrespective of nitrite addition.

14.5.5 Moulds and mycotoxins

Moulds colonize the surface of unsmoked or lightly smoked, air-dried fermented sausages. They can also grow on smoked sausages if these are stored under warm, moist conditions. This raises the question of whether mycotoxins can be formed on fermented sausages, and what can be done to control growth of undesirable moulds. Reviews on moulds and mycotoxins in meats were published by Leistner & Eckardt (1981), Leistner (1984), and Pestka (1995). On mouldy fermented sausages, *Penicillium* spp. predominate with *P. verrucosum*, *P. chrysogenum*, *P. frequentans*, *P. nalgiovense* and *P. variabile* being most frequently isolated. *Aspergillus* spp. can compete with the penicillia only at high storage temperature and low water activity, e.g. during the aging of dry Mediterranean-style hams. *Aspergillus glaucus* and *A. versicolor* were isolated. Fortunately, *A. flavus* is very seldom found on mouldy meats. *Scopulariopsis* spp. are present particularly on Hungarian salamis which receive some smoke during the fermentation period (Jirkovsky & Galgóczy, 1966).

Leistner and Eckardt (1981) and Leistner (1984) listed those mycotoxins that have been found on experimentally inoculated fermented sausages. Aflatoxins and other *Aspergillus* toxins were only found if such sausages were aged at elevated temperatures and relative humidities that are very unlikely to prevail in commercial practice (Alvarez-Barrea *et al.*, 1982). If, as in normal practice, sausages are dried at temperatures of 15°C or below, toxinogenic aspergilli are outgrown by penicillia and *Scopulariopsis* spp. (Incze *et al.*, 1976). Among the *Penicillium* toxins, small amounts of cyclopiazonic acid and verrucosidin and, possibly, citreoviridin and rugulosin may be formed in the outer layer of the sausage under these conditions (Leistner & Eckardt, 1981; Scheuer, 1995). Even though the levels found in experimentally inoculated sausages were small and restricted to no more than a few millimetres below the surface, growth of all undesirable moulds on fermented sausages should be prevented, not only to eliminate the risk of mycotoxin and antibiotic formation, but also to avoid defects in appearance and flavour.

In mould-ripened products, undesired moulds may be suppressed by inoculating the sausage surface with a suitable non-toxic mould strain. In other fermented sausages, mould growth may be delayed by various methods. Antifungal smoke constituents as well as acetic and propionic acids are volatile and cannot safely prevent mould growth during extended storage. Moulds may be inhibited by dipping the products into solutions of potassium sorbate or pimaricin at the end of the fermentation period. However, use of pimaricin for preservation of fermented sausage is prohibited in many countries because it is also used in chemotherapy of humans, and it is generally a sound principle to keep antibiotics out of foods.

If packaged products (like certain 'snack sausages') are to be stored at elevated temperatures for long periods, they should be dried to 22% moisture (corresponding to a water activity of 0.82 or less) and stored in the absence of oxygen. This results in the complete suppression of mould growth and a delay of rancidity (Tändler & Rödel, 1983).

14.5.6 Viruses

The potential viral hazards in meats have been reviewed by Mayr (1979) and Cliver (1988). Viruses already present in livestock at the time of slaughter normally do not affect humans. Foot-and-mouth-disease virus is acid sensitive and likely to be inactivated during sausage fermentation (Eissner *et al.*, 1966; McKercher *et al.*, 1982; Panina *et al.*, 1989). The causative agents of African swine fever, swine vesicular disease and hog cholera are inactivated only slowly during sausage ripening and are sometimes detectable even after three months (McKercher *et al.*, 1982). Hence, fermented meats should, like unprocessed meats, not be exported from areas where these diseases prevail. Hands contaminated by human faeces may be a vector of transmission to foods of viral agents affecting humans. Enterovirus (Herrmann & Cliver, 1973), and echovirus and poliovirus (Kantor & Potter, 1975) have been shown to retain their infectivity during fermentation and drying of sausage. Nevertheless, there is no epidemiological evidence that fermented sausages are a major vector of viruses.

14.5.7 Parasites

Toxoplasma is regularly found in pork, and there is a correlation between the ingestion of raw pork and the occurrence of toxoplasmosis (Grossklaus, 1979). Fortunately, *Toxoplasma* cysts are rapidly inactivated during manufacture of fermented sausage (Sommer *et al.*, 1965). Cysts of sarcosporidia occur in both pork and beef, but little is known about their persistence during sausage fermentation. Since they are sensitive to freezing (Grossklaus, 1979), it is reasonable to assume that they are inactivated in a similar way to the toxoplasmas.

The nematode *Trichinella spiralis* is a parasite of both humans and pigs and, therefore, a serious threat to the consumer of raw pork products if no effective precautions are taken. Different attitudes towards eating raw pork are related to the trust in the effectiveness of the measures taken by public health authorities and meat processors to ensure the absence of viable trichinae from pork. In Germany, for instance, routine inspection of the diaphragm of pig carcasses proved to be very effective in protecting the consumer who, accordingly, does not hesitate to eat raw meat products or even raw minced pork. If the absence of *Trichinella spiralis* from pork cannot be guaranteed, pork must be processed in such a way that

Trichinella larvae are inactivated. They are destroyed at low water activity, which can be attained by freezing or salting and drying processes. To inactivate 100–800 larvae per gram fresh mixture, German-type sausages (salami, cervelat, Teewurst) had to be ripened for 6 to 21 days and to attain a water activity (a_w) of 0.93–0.95 (Lötzsch & Rödel, 1974; Lötzsch & Leistner, 1979). Limiting a_w values between 0.91 and 0.94 were reported by Modic & Djordjevic (1980) for Yugoslavian-type dry sausage, and by Childers *et al.* (1982) for Genoa salami. It appears that the low pH of fermented sausages also contributes to the inactivation of the organisms, as in raw hams lower a_w values were required to kill the larvae than in fermented sausages (Lötzsch & Leistner, 1979). It may be concluded that a fully dry fermented sausage (a_w below 0.90) is very unlikely to contain viable *Trichinella* larvae, even if prepared from pork contaminated by trichinae. If the presence of the larvae cannot be excluded by meat inspection, semi-dry and undried fermented sausages must be made from pork which has been stored frozen, or be heated to 58.3°C (137°F).

There seems to be no systematic study on the survival of tapeworm cysticerci in raw sausages, but their sensitivity to freezing indicates that they, like *Trichinella*, are killed at low water activity.

14.5.8 Biogenic amines and nitrosamines

Fermented foods may contain considerable amounts of amines which may not only be precursors of nitrosamines but also elicit migraine or allergy-like symptoms in humans. Mean and median values of about 100 mg and maximum values of about 300 mg histamine/kg dry weight were reported (Vandekerckhove, 1977; Wortberg & Woller, 1982; Tschabrun *et al.*, 1990; Scheuer & Rödel, 1995), with a trend to higher levels in aged, mould-ripened products. Such levels, as well as the reported concentrations of other biogenic amines (e.g. up to 1000 mg/kg dry weight of tyramine) are unlikely to elicit scombroid-type poisoning or migraine-like symptoms in healthy individuals having normal activity of monoamino oxidase (Lüthy & Schlatter, 1983), but the published data also show that it is possible to produce high-quality fermented sausages with very low levels of biogenic amines. Hence, a better understanding of the factors affecting formation of biogenic amines during sausage fermentations is necessary.

The rate of amine formation is limited by the rate of protein degradation by tissue enzymes and, in particular in mould-ripened products, microbial enzymes (Eitenmiller *et al.*, 1978). Hence, more precursors (peptides, amino acids) must be expected in meat stored for long periods (Tschabrun *et al.*, 1990) or in mould-ripened products. Conditions enhancing the activity of acid-sensitive bacteria appear to increase the risk of biogenic amine formation (Maijala *et al.*, 1995; Scheuer & Rödel, 1995). By using good-quality, fresh raw material, appropriate starters, formulations and ripening conditions, formation of biogenic amines during sausage ripening may be minimized. Remarkably, certain strains of psychrotrophic lactic acid bacteria found on meat (e.g. *Carnobacterium* spp.; strains of *Lb. curvatus*) are also capable of decarboxylating certain amino acids (Maijala, 1994; Straub *et al.*, 1994). Hence, it is necessary to screen actual and potential starter cultures for absence of the capability to decarboxylate amino acids.

Carcinogenic nitrosamines may be formed from secondary amines and nitrite, particularly in highly acid environments or during severe heating (e.g. frying of bacon). In fermented sausages, only *N*-nitrosodimethylamine (NDMA) and, occasionally, *N*-nitrosopiperidin (probably from piperidine in added pepper) have been detected. Kühne (1995) found NDMA in 18 of 33 sausages tested, with only two samples above 2 μg/kg (maximum value, <5 μg/kg). The Nitrite Safety Council (1977) concluded from their data that fermented sausages are not a significant source of nitrosamines. However, nitrosamines may be formed in the stomach if nitrite plus certain amines are ingested. The nitrite content in fermented sausages is low because residual nitrite is reduced to ammonia by the sausage microflora. Usually, less than 10 mg $NaNO_2$/kg are found (Wirth, 1991). When sausages are formulated with nitrate rather than nitrite, considerable levels of nitrate may remain in the product. However, a better understanding of the fermentation and curing process along with legislative pressure and the availability of nitrate-reducing starter cultures have led to a marked reduction of the nitrate input and the levels of residual nitrate in the products: today, the nitrate content in dry sausages rarely exceeds 100 mg KNO_3/kg.

14.6 Future developments

In 1988/1989, the present author collected the opinions of various experts from industry and research on desirable improvements to sausage fermentation (Lücke *et al.*, 1990). Priority was given to shortening ageing times (i.e. obtaining better flavour in less time) and prolonging shelf life by inhibiting rancidity, while accelerating acid production ranked low. In fact, excessive acid formation, associated with colour defects, appeared to be the most frequent cause of defects in sausage ripening. However, in the past few years, suppression of pathogens (in particular, of enterohaemorrhagic *E. coli*) and of amine formation has become a major issue. In addition, there is continuous interest in making microbiological and chemical changes during sausage ripening more predictable: a prerequisite for standardization and automatization of the manufacture of these products.

14.6.1 Novel starter cultures

Possible developments in the field of starter cultures have been discussed by Lücke *et al.* (1990), Hammes *et al.* (1990), Hammes & Knauf (1994), and

Jessen (1995), and requirements for good starters are given in Section 14.4.4 and Table 14.4. Particularly within the 'era of molecular biology', it must be emphasized that starter cultures are just one factor affecting quality and safety of fermented sausage and cannot make up for incorrect formulations (e.g. use of rancid fat, addition of too much sugar) or incorrect ripening conditions. One difficulty is that, unlike milk and malt, meat cannot be pasteurized before inoculation without drastically changing its properties. This means that any starter culture for sausage ripening must be able to compete with the indigenous microflora of the sausage mixture. Moreover, selection and improvement of starters for 'better aroma in less time' and long shelf life is difficult because the chemical and biochemical reactions affecting sensory properties are very complex and insufficiently understood. This limits the possibility to screen a large number of strains in model systems, as well as the prospects to 'tailor' strains by means of genetic engineering. However, novel starters may increase product safety, and much work has been devoted to select strains that are antagonistic to pathogens. Bacteriocinogenic starter cultures may contribute to the inactivation of *Listeria monocytogenes* (see Section 14.5.5), and strains have been engineered to lyse *Staphylococcus aureus* (Geisen *et al.*, 1990). However, prospects of using such strains are limited. For example, bacteriocins in meat systems do not work reliably enough and are ineffective against Gram-negative bacteria (Lücke, 1992), and *Staphylococcus aureus* is well under control with existing technology, making it difficult to explain to the public and to regulatory bodies the necessity to use genetically engineered organisms.

14.6.2 Predictive modelling

A mass of data on the behaviour of pathogens is now available in electronic form (Ross & McMeekin, 1994), and predictions may also be made on the performance of lactic acid bacteria in sausages as affected by water activity, temperature and other parameters (Landvogt & Fischer, 1991). Mathematical models for sausage fermentation have been published by Demeyer's group (see Demeyer, 1992) and by Verlinden & Spliet (1994). These tools may be helpful in optimizing existing and developing new processes.

14.6.3 On-line process control

On-line sensors for various fermentation parameters (temperature, pH, relative humidity, air velocity, equilibrium humidity of the product) are nowadays available (see Rödel & Stiebing, 1988), and data may be used to control the fermentation (Stiebing & Rödel, 1992). This could lead to shorter ripening time and less labour caused by manually adjusting the ripening climate or transporting the products from one chamber to another. For example, a product pH below 5.3 means that the fermentation is nearly finished, and the temperature and relative humidity in the ripening chamber should be lowered.

14.6.4 Novel fermented meats

While there is a large variety of fermented milk products, fermentation has not been employed to any extent in the manufacture of meat products other than 'classical' fermented sausages. A better knowledge of the role of microorganisms in aroma and flavour formation may be used in developing raw ham and other non-fermented meat products with improved sensory properties. It may also be possible to improve the safety of non-fermented meats by addition of selected lactic acid bacteria, as shown for bacon (Tanaka *et al.*, 1980). However, to date, such attempts have met with limited success because of sensory deviation (sour taste is not tolerated in most meat products) and shortening of shelf life. Lastly, it is possible to extend the shelf life of pasteurized meat products by lowering the pH before heat treatment (Mol & Timmers, 1970; Roca & Incze, 1989), although the decrease in water binding capacity at low pH values is undesirable in pasteurized meats, and the use of acidulant (GdL) appears to be preferred to the use of sugar and lactic acid bacteria.

References

Acton, J.C. & Dick, R.L. (1976) Composition of some commercial dry sausages. *Journal of Food Science*, **41**, 971–2.

Adams, M.R. (1986) Fermented flesh foods, in *Micro-organisms in the Production of Food* (ed. M.R. Adams), Elsevier, Amsterdam, pp. 159–98.

Alvarez-Barrea, V., Pearson, A.M., Price, J.F., Gray, J.I. & Aust, S.D. (1982) Some factors influencing aflatoxin production in fermented sausages. *Journal of Food Science*, **47**, 1773–5.

Andersen, H.J., Østdal, H. & Blom, H. (1995) Partial purification and characterization of a lipase from *Lactobacillus plantarum* MF32. *Food Chemistry*, **53**, 369–73.

Andersen, S.J. & Frisvad, J.C. (1994) Penicillin production by *Penicillium nalgiovense*. *Letters in Applied Microbiology*, **19**, 486–8.

Asplund, K., Nurmi, E., Hirn, J., Hirvi, T. & Hill, P. (1993) Survival of *Yersinia enterocolitica* in fermented sausages manufactured with different levels of nitrite and different starter cultures. *Journal of Food Protection*, **56**, 710–12.

Bacus, J. (1984) Update: Meat fermentation, 1984. *Food Technology*, **38**(6), 59–63.

Baldini, P., Palmia, F. & Racynski, R.G. (1983) Influence of chemical and chemical-physical properties of the formulation on pH-lowering and weight loss in the ripening of Italian typical salami, in *Proceedings of the 29th European Congress of Meat Research Workers*, Salsomaggiore, pp. 189–97 (in Italian).

Barber, L.E. & Deibel, R.H. (1972) Effect of pH and oxygen tension on staphylococcal growth and enterotoxin formation in fermented sausages. *Applied Microbiology*, **24**, 891–8.

Berger, R.G., Macku, C., German, J.B. & Shibamoto, T. (1990) Isolation and identification of dry salami volatiles. *Journal of Food Science*, **55**, 1239–42.

Berry, E.D., Liewen, M.B., Mandigo, R.M. & Hutkins, R.W. (1990) Inhibition of *Listeria monocytogenes* by bacteriocin-producing *Pediococcus* during manufacture of fermented dry sausage. *Journal of Food Protection*, **53**, 194–7.

Boos, G. (1979) Vorkommen von Salmonellen in schnittfesten und streichfähigen Rohwürsten. *Fleischwirtschaft*, **59**, 1882–6.

Bülte, M., Heckötter, S. & Schwenk, P. (1996) Enterohämorrhagische *E. coli* (EHEC) – aktuelle Infektionserreger auch in der Bundesrepublik Deutschland? 2. Nachweis von VTEC-Stämmen in Lebensmitteln tierischen Ursprungs. *Fleischwirtschaft*, **76**, 88–91.

Campanini, M., Pedrazzoni, I., Barbuti, S. & Baldini, P. (1993) Behaviour of *Listeria monocytogenes* during the maturation of naturally and artificially contaminated salami: effect of lactic acid bacteria starter cultures. *International Journal of Food Microbiology*, **20**, 169–75.

Cantoni, C., Molnar, M.R., Renon, P. & Giolitti, G. (1967a) Lipolytic micrococci in pork fat. *Journal of Applied Bacteriology*, **30**, 190–6.

Cantoni, C., Molnar, M.R., Renon, P. & Giolitti, G. (1967b) Behaviour of nucleotides, nucleosides and bases in dry sausages during ripening. *Proceedings of the 13th European Meetinq of Meat Research Workers*, Rotterdam, manuscript no. D1.

Cerise, L., Bracco, U., Horman, I., Sozzi, T. & Wuhrmann, J.J. (1973) Veränderungen des Lipidanteils während des Reifungsprozesses von Salami aus reinem Schweinefleisch. *Fleischwirtschaft*, **53**, 223–5.

Childers, A.B., Terrell, R.N., Craig, T.M., Kayfus, T.J. & Smith, G.C. (1982) Effect of sodium chloride concentration, water activity, fermentation method and drying time on the viability of *Trichinella* spiralis in Genoa salami. *Journal of Food Protection*, **45**, 816–19, 823.

Christiansen, L.N., Tompkin, R.B., Shaparis, A.B., Johnston, R.W. & Kautter, D.A. (1975) Effect of sodium nitrite and nitrate on *Clostridium botulinum* growth and toxin production in a summer-style sausage. *Journal of Food Science*, **40**, 488–90.

Cliver, D.O. (1988) Virus transmission via foods. *Food Technology*, **42**, 241–8.

Collins-Thompson, D.L. & Rodriguez Lopez, G. (1981) Depletion of sodium nitrite by lactic acid bacteria isolated from vacuum-packed bologna. *Journal of Food Protection*, **44**, 593–5.

Collins-Thompson, D.L., Krusky, B. & Usborne, W.R. (1984) The effect of nitrite on the growth of pathogens during manufacture of dry and semi-dry sausage. *Canadian Insitute of Food Science and Technology Journal*, **17**, 102–6.

Comi, G. & Cantoni, C. (1980) Yeasts in matured raw sausages. *Industrie Alimentari*, **19**, 857–60 (in Italian).

Comi, G., Citterio, B., Mazano, M., Cantoni, C. & de Bertoldi, M. (1992) Bewertung und Charakterisierung der *Micrococcaceae*-Stämme in italienischen Rohwürsten. *Fleischwirtschaft*, **72**, 1693–7.

Comi, G., Boschelle, O., Conte, L. & Manzano, M. (1995) Lipolytische Aktivität aus Rohwürsten isolierter *Micrococcaceae*-Stämme. *Fleischwirtschaft*, **75**, 1153–60.

Condon, S. (1987) Responses of lactic acid bacteria to oxygen. *FEMS Microbiology Reviews*, **46**, 269–80.

Coretti, K. (1958) Rohwurstfehlfabrikate durch Laktobazillen. *Fleischwirtschaft*, **10**, 218–25.

Coretti, K. (1959) Fermentative Eigenschaften der aus Rohwurst isolierten Laktobazillen. *Archiv für Lebensmittelhygiene*, **10**, 27–9.

Coretti, K. (1977) Starterkulturen in der Fleischwirtschaft. *Fleischwirtschaft*, **57**, 386–94.

Dainty, R.H. (1986) Chemical changes associated with mirobial growth on meat stored at chill temperatures, in *Proceedings, 32nd European Meeting of Meat Research Workers*, Ghent, pp. 179–86.

Dainty, R.H. & Blom, H. (1995) Flavour chemistry of fermented sausages, in *Fermented Meats* (eds G. Campbell-Platt & P.E. Cook), Blackie, Glasgow, pp. 176–93.

Deibel, R.H., Niven, C.F. Jr & Wilson, G.D. (1961a) Microbiology of meat curing. III. Some microbiological and related technological aspects in the manufacture of fermented sausages. *Applied Microbiology*, **9**, 156–61.

Deibel, R.H., Wilson, G.D. & Niven, C.F. Jr (1961b) Microbiology of meat curing. IV. A lyophilized *Pediococcus cerevisiae* starter culture for fermented sausages. *Applied Microbiology*, **9**, 239–43.

Deketelaere, A., Demeyer, D., Vandekerckhove, P. & Vervaeke, I. (1974) Stoichiometry of carbohydrate fermentation during dry sausage ripening. *Journal of Food Science*, **39**, 297–300.

Delarras, C. & Laban, P. (1981) Distribution of the different *Staphylococcus* species according to their meat or dairy origin. *Zentralblatt für Bakteriologie und Hygiene, I. Abteilung Originale B*, **173**, 471–7.

Demeyer, D.I. (1982) Stoichiometry of dry sausage fermentation. *Antonie van Leeuwenhoek*, **48**, 414–16.

Demeyer, D.I. (1992) Meat fermentation as an integrated process, in *New Technologies for Meat and Meat Products* (eds F.J.M. Smulders, F. Toldrá, J. Flores & M. Prieto), Audet Tijdschriften, Nijmegen, pp. 21–36.

Demeyer, D., Hoozee, J. & Mesdom, H. (1974) Specificity of lipolysis during dry sausage ripening. *Journal of Food Science*, **39**, 293–6.

Dierick, N., Vandekerckhove, P. & Demeyer, D. (1974) Changes in nonprotein nitrogen compounds during dry sausage ripening. *Journal of Food Science*, **39**, 301–4.

Dobbertin, S., Siems, H. & Sinell, H.-J. (1975) Beiträge zur Bakteriologie der frischen Mettwurst. II. Mitteilung: Die Abhängigkeit lipolytischer Aktivität von der Keimdynamik in frischen Mettwürsten. *Fleischwirtschaft*, **55**, 237–42.

Eilberg, B.L. & Liepe, H.-U. (1977) Mögliche Verbesserungen der Rohwursttechnologie durch den Einsatz von Streptomyceten als Starterkultur. *Fleischwirtschaft*, **57**, 1678–80, 1683.

Eissner, G., Bauer, K., Ahl, R. & Reuter, H. (1966) Virus nachweis in Pökelwaren und Rohwurst aus dem Fleisch experimentell mit Maul- und Klauenseuche-Virus infizierten Tieren. *Fleischwirtschaft*, **46**, 41–4.

Eitenmiller, R.R., Koehler, P.E. & Reagan, J.O. (1978) Tyramine in fermented sausages: factors affecting formation of tyramine and tyrosine decarboxylase. *Journal of Food Science*, **43**, 689–93.

El-Banna, A.A., Pitt, J.I. & Leistner, L. (1987) Production of mycotoxins by *Penicillium* species. *Systematic and Applied Microbiology*, **10**, 42–6.

Engesser, D.M. & Hammes, W.P. (1994) Non-heme catalase activity of lactic acid bacteria. *Systematic and Applied Microbiology*, **17**, 11–19.

Farber, J.M., Daley, E., Holley, R. & Usborne, W.R. (1993) Survival of *Listeria monocytogenes* during the production of uncooked German, American and Italian-style fermented sausages. *Food Microbiology*, **10**, 123–32.

Fink-Gremmels, J., El-Banna, A.A. & Leistner, L. (1988) Entwicklung von Schimmelpilz-Starterkulturen für Fleischerzeugnisse. *Fleischwirtschaft*, **68**, 24–6.

Fischer, U. & Schleifer, K.-H. (1980) Vorkommen von Staphylokokken und Mikrokokken in Rohwurst. *Fleischwirtschaft*, **60**, 1046–8, 1051.

Fisher, S. & Palmer, M. (1995) Fermented meat production and consumption in the European Union, in *Fermented Meats* (eds G. Campbell-Platt & P.E. Cook), Blackie, Glasgow, pp. 217–33.

Foegeding, P.M., Thomas, A.B., Pilkington, D.H. & Klaenhammer, T.R. (1992) Enhanced control of *Listeria monocytogenes* by in situ-produced pediocin during dry fermented sausage production. *Applied and Environmental Microbiology*, **58**, 884–90.

Garcia, M.L., Selgas, M.D., Fernandez, M. & Ordóñez, J.A. (1992) Microorganisms and lipolysis in the ripening of dry fermented sausages. *International Journal of Food Science and Technology*, **27**, 675–82.

Gehlen, K.H. (1989) Einfluss der Technologie auf die Rohwurstreifung mit *Lactobacillus curvatus, Micrococcus varians* und weiteren Starterorganismen unter besonderer Berücksichtigung der Nitratreduktion. Dissertation, Universität Hohenheim, 297pp.

Geisen, R., Ständner, L. & Leistner, L. (1990) New mould starter cultures by genetic modification. *Food Biotechnology*, **4**, 497–504.

Geisen, R., Lücke, F.-K. & Kröckel, L. (1992) Starter and protective cultures for meat and meat products. *Fleischwirtschaft*, **72**, 894–8.

Gerigk, K. & Gossling, U. (1981). Bakteriologische und sensorische Untersuchungen von Rohwürsten mit verringertem Nitritzusatz. *Fleischwirtschaft*, **61**, 1124–8.

Gill, C.O. (1982) Microbial interactions with meats, in *Meat Microbiology* (ed. M.H. Brown), Applied Science Publishers, London, pp. 225–64.

Glass, K.A., Loeffelholz, J.M., Ford, J.P. & Doyle, M.P. (1992) Fate of *Escherichia coli*, O157:H7 as affected by pH or sodium chloride and in fermented, dry sausage. *Applied and Environmental Microbiology*, **58**, 2513–16.

Gray, J.I. & Pearson, A.M. (1984) Cured meat flavor. *Advances in Food Research*, **29**, 1–86.

Grossklaus, D. (1979) Aktuelle Zoonosen aus lebensmittelhygienischer Sicht. *Zentralblatt fur Bakteriologie und Hygiene, I. Abteilung Originale B*, **168**, 83–96.

Gürakan, G.C., Bozoglu, T.F. & Weiss, N. (1995) Identification of *Lactobacillus* strains from Turkish-style dry fermented sausages. *Lebensmittel-Wissenschaft und – technologie*, **28**, 139–44.

Halvarson, H. (1973) Formation of lactic acid, volatile fatty acids and neutral, volatile monocarbonyl compounds in Swedish fermented sausage. *Journal of Food Science*, **38**, 310–12.

Hamm, R. (1977) Postmortem breakdown of ATP and glycogen in ground muscle: a review. *Meat Science*, **1**, 15–39.

Hammer, G.F. (1977) The stabilizing effect of various natural spices in dry sausage during storage. *Fleischwirtschaft*, **57**, 1957–9, 1947.

Hammes, W.P. & Knauf, H.J. (1994) Starters in the processing of meat products. *Meat Science*, **36**, 155–68.

Hammes, W.P., Rölz, I. & Bantleon, A. (1985) Mikrobiologische Untersuchung der auf dem deutschen Markt vorhandenen Starterkulturpräparate für die Rohwurstbereitung. *Fleischwirtschaft*, **65**, 629–32, 635–6, 729–34.

Hammes, W.P., Bantleon, A. & Min, S. (1990) Lactic acid bacteria in meat fermentation. *FEMS Microbiology Reviews*, **87**, 165–74.

Hartmann, S., Wolf, G. & Hammes, W.P. (1995) Reduction of nitrite by *Staphylococcus carnosus* and *Staphylococcus piscifermentans*. *Systematic and Applied Bacteriology*, **18**, 323–8.

Hechelmann, H., Bem, Z. & Leistner, L. (1974) Mikrobiologie der Nitrat/Nitritminderung bei Rohwurst. *Mitteiungsblatt der Bundesanstalt für Fleischforschung, Kulmbach*, pp. 4059–64.

Hechelmann, H., Lücke, F.-K. & Schillinger, U. (1988) Ursachen und Vermeidung von *Staphylococcus aureus*-Intoxikationen nach Verzehr von Rohwurst und Rohschinken. *Mitteilungsblatt der Bundesanstalt für Fleischforschung Kulmbach*, part **100**, 7956–64.

Herrmann, J.E. & Cliver, D.O. (1973) Enterovirus persistence in sausage and ground beef. *Journal of Milk and Food Technology*, **36**, 426–8.

Holley, R.A., Jui, P.A., Wittmann, M. & Kwan, P. (1988a) Survival of *S. aureus* and *S. typhimurium* in raw ripened dry sausages formulated with mechanically separated chicken meat. *Fleischwirtschaft*, **68**, 194–201.

Holley, R.A., Lammerding, A.M. & Tittiger, F. (1988b) Microbiological safety of traditional and starter-mediated processes for the manufacture of Italian dry sausage. *International Journal of Food Microbiology*, **7**, 49–62.

Hugas, M., Garriga, M., Aymerich, T. & Montfort, J.M. (1993) Biochemical characterization of lactobacilli from dry fermented sausages. *International Journal of Food Microbiology*, **18**, 107–13.

Hugas, M., Garriga, M., Aymerich, T. & Montfort, J.M. (1995) Inhibition of *Listeria* in dry fermented sausages by the bacteriocinogenic *Lactobacillus sake* CTC494. *Journal of Applied Bacteriology*, **79**, 322–30.

Hwang, Han-Joon, Vogel, R.F. & Hammes, W.P. (1993) Entwicklung von Schimmelpilzkulturen für die Rohwurstherstellung. Technologische Eignung der Stämme und sensorische Bewertung der Produkte. *Fleischwirtschaft*, **73**, 327–8, 331–2.

Incze, K. (1986) Technologie und Mikrobiologie der ungarischen Salami. Tradition und Gegenwart. *Fleischwirtschaft*, **66**, 1305, 1308, 1311.

Incze, K., Mihályi, V. & Frank, H.K. (1976) Besteht eine Mykotoxingefahr bei der ungarischen Salami? III. Teil: Chemisch-analytische und biologische Untersuchungen an schnittfesten Salamiproben. *Fleischwirtschaft*, **56**, 1616–18.

Jaud, D. & Fischer, A. (1993) Modellversuche zur Verbesserung der Pökelbereitschaft von DFD-Fleisch durch pH-Wert-Senkung. *Fleischerei*, **44**(7), 545–8.

Jessen, B. (1995) Starter cultures for meat fermentation, in *Fermented Meats* (eds G. Campbell-Platt & P.E. Cook), Blackie, Glasgow, pp. 130–59.

Jirkovsky, M. & Galgóczy, J. (1966) Die Untersuchung der Schimmelpilzflora der ungarischen Wintersalami. *Fleischwirtschaft*, **46**, 128, 131.

Junker, M., Porobic, R., Sieber, W., Linhard, O. & Knauf, H.J. (1993) Rohwurstherstellung. Beschreibung einer neuen Starterkultur mit *Pediococcus pentosaceus*. *Fleischwirtschaft*, **73**, 325–6.

Kagermeier, A. (1981) Taxonomie und Vorkommen von Milchsäurebakterien in Fleischprodukten. Dissertation, Fakultät für Biologie, Ludwig-Maximilian-Universität München, 104 pp.

Kandler, O. (1983) Carbohydrate metabolism in lactic acid bacteria. *Antonie van Leeuwenhoek*, **49**, 209–24.

Kantor, M.A. & Potter, N.N. (1975) Persistence of echovirus and poliovirus in fermented sausage. Effects of sodium nitrite and processing variables. *Journal of Food Science*, **40**, 968–72.

Karches, H. & Teufel, P. (1988) *Listeria monocytogenes*: Vorkommen in Hackfleisch und Verhalten in frischer Zwiebelmettwurst. *Fleischwirtschaft*, **68**, 1388–92.

Katsaras, K. & Leistner, L. (1988) Topographie der Bakterien in der Rohwurst. *Fleischwirtschaft*, **68**, 1295–8.

Kleemann, J. & Bergann, T. (1996) Model experiments to establish behaviour of *Yersinia enterocolitica* σ:9 strains in various types of fresh dry sausage. *Journal of Applied Bacteriology*, **80**, 10–12.

Klein, G., Dicks, L.M.T., Pack, A., Hack, B., Zimmermann, K., Dellaglio, F. & Reuter, G. (1996) Emended descriptions of *Lactobacillus sake* (Katagiri, Kitahara, and Fukami) and *Lactobacillus curvatus* (Abo-Elnaga and Kandler): numerical classification revealed by protein fingerprinting, and identification based on biochemical patterns and DNA-DNA hybridizations. *International Journal of Systematic Bacteriology*, **46**, 367–76.

Klettner, P.-G. & List, D. (1980) Beitrag zum Einfluss der Kohlenhydratart auf den Verlauf der Rohwurstreifung. *Fleischwirtschaft*, **60**, 1589–93.

Klettner, P.-G. & Rödel, W. (1978) Überprüfung und Steuerung wichtiger Parameter bei der Rohwurstreifung. *Fleischwirtschaft*, **58**, 57–60, 63–4, 66.

Knauf, H. (1995) Starterkulturen für die Herstellung von Rohwurst und Rohpökelwaren: Potential, Auswahlkriterien und Beeinflussungsmöglichkeiten. *Fleischerei*, **46**(6), 4, 6, 8–14.

Krol, B. (1984) Progress and problems of fermented dry sausage, in *Trends in Modern Meat Technology – 1* (eds B. Krol, P.S. van Roon & J.H. Houben), Pudoc, Wageningen, pp. 55–7.

Kühne, D. (1995) Nitrosamine in Fleischerzeugnissen – derzeitiger Stand. *Mitteilungsblatt der Bundesanstalt für Fleischforschung, Kulmbach*, **34**, 220–5.

Laban, P., Favre, C., Ramet, F. & Larpent, J.P. (1978) Lactobacilli isolated from French saucisson (Taxonomic study). *Zentralblatt für Bakteriologie und Hygiene, I. Abteilung Originale B*, **166**, 105–11.

Landvogt, A. & Fischer, A. (1991) Dry sausage ripening. Targeted control of the acidification achieved by starter cultures. *Fleischwirtschaft*, **71**, 902–5, 1055–6.

Lang, K. (1960) Laktobazillen bei der Fäulnis von Rohwurst. *Fleischwirtschaft*, **12**, 461–3.

Langner, H.J. (1969) Zur Bildung von freien Aminosäuren, flüchtigen Fettsäuren und flüchtigen Carbonylen in reifender Rohwurst. *Fleischwirtschaft*, **49**, 1475–9.

Langner, H.J. (1972) Aromastoffe in der Rohwurst. *Fleischwirtschaft*, **52**, 1299–306.

Lawrie, R.A. (1991) *Meat Science*, 5th edn, Pergamon Press, Oxford.

Leistner, L. (1958) Bakterielle Vorgänge bei der Pökelung von Fleisch. II. Günstige Beeinflussung von Farbe, Aroma und Konservierung durch Mikroorganismen. *Fleischwirtschaft*, **10,** 226–8, 231–4.

Leistner, L. (1984) Toxinogenic penicilla occurring in feeds and foods: a review. *Food Technology in Australia*, **36**, 404–6.

Leistner, L. (1986) Allgemeines über Rohwurst. *Fleischwirtschaft*, **66**, 290–300.

Leistner, L. & Ayres, J.C. (1967) Schimmelpilze und Fleischwaren. *Fleischwirtschaft*, **47**, 1320–6.

Leistner, L. & Bem, Z. (1970) Vorkommen und Bedeutung von Hefen bei Pökelfleischwaren. *Fleischwirtschaft*, **50**, 350–1.

Leistner, L. & Eckardt, C. (1981) Schimmelpilze und Mykotoxine in Fleisch und Fleischerzeugnissen, in *Mykotoxine in Lebensmitteln* (ed. J. Reiss), Gustav Fischer Verlag, Stuttgart, pp. 297–341.

Leistner, L., Hechelmann, H. & Lücke, F.-K. (1982) Auswirkungen der neuen Nitrit-VO in der Mikrobiologie. *Mitteilungsblatt der Bundesanstalt für Fleischforschung, Kulmbach*, part **76**, 5001–5.

Leyer, G.J., Wang, L.-L. & Johnson, E.A. (1995) Acid adaptation of *Escherichia coli* O157: H7 increases survival in acidic foods. *Applied and Environmental Microbiology*, **61**, 3752–5.

List, D. & Klettner, P.-G. (1978) Die Milchsäurebildung im Verlauf der Rohwurstreifung bei Starterkulturzusatz. *Fleischwirtschaft*, **58**, 136–9.

Lötzsch, R. & Leistner, L. (1979) Überleben von *Trichinella spiralis* in Rohwurst und Rohschinken in Abhängigkeit von der Wasseraktivität (a_w-Wert). *Fleischwirtschaft*, **59**, 231–3.

Lötzsch, R. & Rödel, W. (1974) Untersuchungen über die Lebensfähigkeit von *Trichinella spiralis* in Rohwürsten in Abhängigkeit von der Wasseraktivität. *Fleischwirtschaft*, **54**, 1203–8.

Lücke, F.-K. (1992) Prospects for the use of bacteriocins against meat-borne pathogens, in *New Technologies for Meat and Meat Products* (eds F.J.M. Smulders, F. Toldrá, J. Flores & M. Prieto), Audet Tijdschriften, Nijmegen, pp. 37–52.

Lücke, F.-K., Hechelmann, H. & Leistner, L. (1983) Fate of *Clostridium botulinum* in fermented sausages processed with or without nitrite. *Proceedings of the 29th European Congress of Meat Research Workers,* Salsomaggiore, pp. 403–9.

Lücke, F.-K., Popp, J. & Kreutzer, R. (1985) Bildung von Wasserstoffperoxid durch Laktobazillen aus Rohwurst und Brühwurstaufschnitt. *Chemie Mikrobiologie Technologie der Lebensmittel*, **10**, 78–81.

Lücke, F.-K., Brümmer, J.-M., Buckenhüskes, H., Garrido Fernandez, A., Rodrigo, M. & Smith, J.E. (1990) Starter culture development, in *Processing and Quality of Foods, Vol. 2: Food Biotechnology: Avenues to Healthy and Nutritious Products* (eds P. Zeuthen, J.-C. Cheftel, C. Eriksson, T.R. Gormley, P. Linko & K. Paulus), Elsevier Applied Science, London, pp. 2.11–2.36.

Lüthy, J. & Schlatter, C. (1983) Biogene Amine in Lebensmitteln: Zur Wirkung von Histamin, Tyramin und Phenylethylamin auf den Menschen. *Zeitschrift für Lebensmittel-Untersuchung und -Forschung*, **177**, 439–43.

MacDonald, B., Gray, J.I. & Gibbins, L.N. (1980) Role of nitrite in cured meat flavor: antioxidant role of nitrite. *Journal of Food Science*, **45**, 893–7.

Maijala, R. (1994) Histamine and tyramine production by a *Lactobacillus* strain subjected to external pH decrease. *Journal of Food Protection*, **57**, 259–62.

Maijala, R., Nurmi, E. & Fischer, A. (1995) Influence of processing temperature on the formation of biogenic amines in dry sausage. *Meat Science*, **39**, 9–22.

Mayr, A. (1979) Tatsachen und Spekulationen über Viren in Lebensmitteln. *Zentralblatt für Bakteriologie und Hygiene I. Abteilung Originale B*, **168**, 109–33.

McKercher, P.D., Blackwell, J.H. & Callis, J.J. (1982) Residual viruses in processed meat products – a review. *Proceedings of the 28th European Congress of Meat Research Workers*, Madrid, pp. 291–3.

Metaxopoulos, J., Genigeorgis, C., Fanelli, M.J., Franti, E. & Cosma, E. (1981a) Production of Italian dry salami. I. Initiation of staphylococcal growth in salami under commercial manufacturing conditions. *Journal of Food Protection*, **44**, 347–52.

Metaxopoulos, J., Genigeorgis, C., Fanelli, M.J., Franti, E. & Cosma, E. (1981b) Production of Italian dry salami: II. Effect of starter culture and chemical acidulation on staphylococcal growth in salami under commercial manufacturing conditions. *Applied and Environmental Microbiology*, **42**, 863–71.

Mintzlaff, H.-J. & Leistner, L. (1972) Untersuchungen zur Selektion eines technologisch geeigneten und toxikologisch unbedenklichen Schimmelpilz-Stammes für die Rohwurst-Herstellung. *Zentralblatt für Veterinärmedizin B*, **19**, 291–300.

Miteva, E., Kirova, E., Gadjeva, D. & Radeva, M. (1986) Sensory, aroma and taste profiles of raw dried sausages manufactured with a lipolytically active yeast culture. *Die Nahrung*, **30**, 829–32.

Modic, P. & Djordjevic, M. (1980) Influence of certain physico-chemical factors on the possibilities of *Trichinella spiralis* larvae survival in meat products. *Proceedings of the 26th European Meeting of Meat Research Workers*, Colorado Springs, Vol. 2, pp. 309–12.

Mol, J.H.H. & Timmers, C.A. (1970) Assessment of the stability of pasteurized comminuted meat products. *Journal of Applied Bacteriology*, **33**, 233–47.

Montel, M.C., Seronie, M.P., Talon, R. & Hebraud, M. (1995) Purification and characterization of a dipeptidase from *Lactobacillus sake. Applied and Environmental Microbiology*, **61**, 837–9.

Næs, H., Holck, A., Axelsson, L., Andersen, H.J. & Blom, H. (1995) Accelerated ripening of dry fermented sausage by addition of a *Lactobacillus* proteinase. *International Journal of Food Science and Technology*, **20**, 651–9.

Nes, I.F. & Skjelkvåle, R. (1982) Effect of natural spices and oleoresins on *Lactobacillus plantarum* in the fermentation of dry sausage. *Journal of Food Science*, **47**, 1618–21, 1625.

Neumayr, L., Lücke, F.-K. & Leistner, L. (1983) Fate of *Bacillus spp.* from spices in fermented sausage. *Proceedings of the 29th European Congress of Meat Research Workers*, Salsomaggiore, pp. 418–44.

Newton, K.G. & Gill, C.O. (1981) The microbiology of DFD fresh meats: a review. *Meat Science*, **5**, 223–32.

Niinivaara, F.P. (1955) Über den Einfluss von Bakterienreinkulturen auf die Reifung und Umrötung der Rohwurst. Thesis, University of Helsinki. *Acta Agralia Fennica*, **85**, 1–128.

Niinivaara, F.P., Pohja, M. & Komulainen, S.E. (1964) Some aspects about using bacterial pure cultures in the manufacture of fermented sausages. *Food Technology*, **18**, 147–53.

Niskanen, A. & Nurmi, E. (1976) Effects of starter culture on staphylococcal enterotoxin and thermonuclease production in dry sausage. *Applied and Environmental Microbiology*, **31**, 11–20.

Nitrite Safety Council (1977) Responses for dry cured products and dry and semi-dry sausages. Supplement A: results of tests of dry and semi-dry sausages for nitrosamines. *Federal Register*, October 8, 1977.

Nordal, J. & Gudding, R. (1975) The inhibition of *Clostridium botulinum* type B and E in salami sausage. *Acta Veterinaria Scandinavica*, **16**, 537–48.

Nordal, J. & Slinde, E. (1980) Characteristics of some lactic acid bacteria used as starter cultures in dry sausage production. *Applied and Environmental Microbiology*, **40**, 472–5.

Nurmi, E. (1966) Effect of bacterial inoculation on characteristics and microbial flora of dry sausage. Thesis, University of Helsinki. *Acta Agralia Fennica*, **108**, 1–77.

Nurmi, E. & Niinivaara, F.P. (1964) Lipolytic changes of fat in dry sausage. *Proceedings of the 10th European Meeting of Meat Research Workers*, Roskilde, manuscript G8.

Panina, G.F., Civardi, A., Massirio, I., Scatozza, F., Baldini, P. & Palmia, F. (1989) Survival of foot-and-mouth-disease viruses in sausage meat products (Italian salami). *International Journal of Food Microbiology*, **8**, 141–8.

Papon, M. & Talon, R. (1988) Factors affecting growth and lipase production by meat lactobacilli strains and *Brochothrix thermosphacta. Journal of Applied Bacteriology*, **64**, 107–15.

Pestka, J. (1995) Fungal toxins in raw and fermented meats, in *Fermented Meats* (eds G. Campbell-Platt & P.E. Cook), Blackie, Glasgow, pp. 194–206.

Petäjä, E. (1977) The effect of some Gram-negative bacteria on the ripening and quality of dry sausage. *Journal of the Scientific Agricultural Society of Finland*, **49**, 107–66.

Pohja, M.S. & Niinivaara, F.P. (1957) Über die Reifung der Rohwurst. III. Mitteilung: Über die antagonistische Wirkung eines Mikrokokkenstammes gegen die in Rohwurst vorkommenden Bakterienstämme. *Zeitschrift für Lebensmittel-Untersuchung und -forschung*, **106**, 298–301.

Puolanne, E. (1977) Der Einfluss von verringerten Nitrit- und Nitratzusätzen auf die Eigenschaften der Rohwurst. Thesis, University of Helsinki. *Journal of the Scientific Agricultural Society of Finland*, **49**, 1–106.

Pyrcz, J. & Pezacki, W. (1974) Die technologische Steuerung der Rohwurstreifung. III. Einfluss verschiedener Arten von Kohlenhydraten auf die Dynamik der Prozesse einer Ansammlung von sauren Fermentationserzeugnissen. *Fleischwirtschaft*, **54**, 203–5, 208–11.

Raccach, M. (1981) Control of *Staphylococcus aureus* in dry sausage by newly developed meat starter culture and phenolic-type antioxidants. *Journal of Food Protection*, **44**, 665–9.

Reuter, G. (1967) Atypische Streptobakterien als dominierende Flora in reifer und gelagerter Rohwurst. *Fleischwirtschaft*, **47**, 397–402.

Reuter, G. (1970) Laktobazillen und eng verwandte Mikro organismen in Fleisch und Fleischwaren. 2. Mitteilung: Die Charakterisierung der isolierten Laktobazillenstämme. *Fleischwirtschaft*, **50**, 951–4.

Reuter, G. (1971a) Laktobazillen und eng verwandte Mikro organismen in Fleisch und Fleischwaren. 5. Mitteilung: Lipolytische Aktivitäten (Tributyrinspaltung). *Fleischwirtschaft*, **51**, 67–70.

Reuter, G. (1971b) Laktobazillen und eng verwandte Mikro organismen in Fleisch und Fleischwaren. 7. Mitteilung: Kohlenhydratstoffwechsel und antagonistische Aktivitäten gegenüber saprophytären und enterotoxischen Mikroorganismen. *Fleischwirtschaft*, **51**, 1237–45.

Reuter, G. (1972) Versuche zur Rohwurstreifung mit Lakt obazillen- und Mikrokokken-Starterkulturen. *Fleischwirtschaft*, **52**, 465–8, 471–3.

Reuter, G. (1975) Classification problems, ecology and some biochemical activities of lactobacilli of meat products, in *Lactic Acid Bacteria in Beverages and Foods* (eds J.G. Carr, C.V. Cutting & G.C. Whiting), Academic Press, London, pp. 221–9.

Reuter, G., Langner, H.J. & Sinell, H.-J. (1968) Entwicklung der Mikroflora in schnellreifender deutscher Rohwurst und analoge quantitative Aminosäureanalyse bei einer Salami. *Fleischwirtschaft*, **48**, 170–6.

Rheinbaben, K. von & Hadlok, R. (1979). Gattungsdifferen zierung von Mikroorganismen der Familie *Micrococcaceae* aus Rohwürsten. *Fleischwirtschaft*, **59**, 1321–4.

Roca, M. & Incze, K. (1989) Starterkulturen für die Herstellung von Kochdauerwurst. *Fleischwirtschaft*, **69**, 175–81.

Rödel, W. & Stiebing, A. (1988) Continuous measurement of the ripening pattern in dry sausage. *Fleischwirtschaft*, **68**, 1423–6.

Rödel, W., Scheuer, R., Stiebing, A. & Klettner, P.G. (1993a) Measuring the oxygen content in meat products. *Fleischwirtschaft*, **73**, 666–8.

Rödel, W., Stiebing, A. & Kröckel, L. (1993b) Ripening parameters for traditional dry sausages with a mould covering. *Fleischwirtschaft*, **73**, 848–53.

Rohde, R. & Weckend, B. (1983) Erste Ergebnisse beim Einsatz von Streptomyzeten und Hefen als Starterkulturen für die Rohwurstreifung. *Monatshefte für Veterinärmedizin*, **38**, 151–6.

Ross, T. & McMeekin, T.A. (1994) Predictive microbiology. *International Journal of Food Microbiology*, **23**, 241–64.

Rossmanith, E., Mintzlaff, H.-J., Streng, B., Christ, W. & Leistner, L. (1972) Hefen als Starterkulturen für Rohwürste. *Jahresbericht der Bundesanstalt für Fleischforschung 1972*, pp. I47–I48.

Samelis, J., Maurogenakis, F. & Metaxopoulos, J. (1994a) Characterisation of lactic acid bacteria from naturally fermented Greek salami. *International Journal of Food Microbiology*, **23**, 179–96.

Samelis, J., Stavropoulos, S., Kakouri, A. & Metaxopoulos, J. (1994b) Quantification and characterization of microbial populations associated with naturally fermented Greek dry salami. *Food Microbiology*, **11**, 447–60.

Sanz, B., Selgas, D., Parejo, I. & Ordóñez, J.A. (1988) Characteristics of lactobacilli isolated from dry fermented sausages. *International Journal of Food Microbiology*, **6**, 199–205.

Scharner, E. (1968) Rückgang des Infektionsrisikos mit Salmonellen bei Trockenlagerung von Rohwürsten. *Die Nahrung*, **12**, 201–2.

Scheuer, R. (1995) Untersuchungen zur Bildung von Verrucosidin auf Rohwurst. *Mitteilungsblatt der Bundesanstalt für Fleischforschung, Kulmbach*, **34**(127), 66–70.

Scheuer, R. & Rödel, W. (1995) Bestimmung von biogenen Aminen in fermentierten Fleischerzeugnissen. *Fleischwirtschaft*, **75**, 73–5.

Schillinger, U., Kaya, M. & Lücke, F.-K. (1991) Behaviour of *Listeria monocytogenes* in meat and its control by a bacteriocin-producing strain of *Lactobacillus sake*. *Journal of Applied Bacteriology*, **70**, 473–8.

Schleifer, K.H. & Fischer, U. (1982) Description of a new species of the genus *Staphylococcus: Staphylococcus carnosus*. *International Journal of Systematic Bacteriology*, **32**, 153–6.

Schmidt, U. (1988) Verminderung des Salmonellen-Risikos durch technologische Maßnahmen bei der Rohwurstherstellung. *Mitteilungsblatt der Bundesanstalt für Fleischforschung, Kulmbach*, **99**, 7791–3.

Seager, M.S., Banks, J.G., Blackburn, C. deW. & Board, R.G. (1986) A taxonomic study of *Staphylococcus* spp. isolated from fermented sausages. *Journal of Food Science*, **51**, 295–7.

Selgas, M.D., Salz, B. & Ordóñez, J.A. (1988) Selected characteristics of micrococci isolated from Spanish dry fermented sausage. *Food Micobiology*, **5**, 185–93.

Sharpe, M.E. (1962) Lactobacilli in meat products. *Food Manufacture*, **37**, 582–4, 587–9.

Sirviö, P., Nurmi, E., Puolanne, E. & Niinivaara, F.P. (1977) Der Einfluß von Starterkulturen und verschiedenen Zusatzstoffen auf das Wachstum von *Salmonella senftenberg* in Rohwurst. *Fleischwirtschaft*, **57**, 1007–8.

Skjelkvåle, R., Tjåberg, T.B. & Valland, M. (1974) Comparison of salami sausages produced with and without addition of sodium nitrite and sodium nitrate. *Journal of Food Science*, **39**, 520–4.

Smith, J.L. & Alford, J.A. (1968) Action of microorganisms on the peroxides and carbonyls of rancid fat. *Journal of Food Science*, **33**, 93–7.

Smith, J.L., Huhtanen, C.N., Kissinger, J.C. & Palumbo, S.A. (1975) Survival of salmonellae during pepperoni manufacture. *Applied Microbiology*, **30**, 759–63.

Sommer, R., Rommel, M. & Lewetzow, R. (1965) Die Überlebensdauer von Toxoplasmencysten in Fleisch und Fleischerzeugnissen. *Fleischwirtschaft*, **45**, 454–7.

Stahnke, L.H. (1994) Aroma components from dried sausages fermented with *Staphylococcus xylosus*. *Meat Science*, **38**, 39–53.

Stahnke, L.H. (1995) Dried sausages fermented with *Staphylococcus xylosus* at different temperatures and with different ingredient levels. III. Sensory evaluation. *Meat Science*, **41**, 211–23.

Stecchini, M., Caserio, G., Cossa, T. & Corsico, G. (1982) Further studies of *Salmonella* survival in dry salami during ageing. *Industrie Alimentari*, **21**, 521–3.

Stiebing, A. (1994) Kritische Kontrollpunkte bei der Herstellung von Rohwurst. *Fleisch*, **48**, 297–301.

Stiebing, A. & Rödel, W. (1988) Influence of relative humidity on the ripening of dry sausage. *Fleischwirtschaft*, **68**, 1287–91.

Stiebing, A. & Rödel, W. (1990) Influence of the pH on the drying pattern in dry sausage. *Fleischwirtschaft*, **70**, 1039–43.

Stiebing, A. & Rödel, W. (1992) Continuous measurement of the surface water activity in dry sausages. *Fleischwirtschaft*, **72**, 1547–9.

Stiebing, A., Kühne, D. & Rödel, W. (1993) Fettqualität. Einfluss auf die Lagerstabilität von schnittfester Rohwurst. *Fleischwirtschaft*, **73**, 1169–72.

Straub, B.W., Tichaczek, P.S., Kicherer, M. & Hammes, W.P. (1994) Formation of tyramine by *Lactobacillus curvatus* LTH 972. *Zeitschrift für Lebensmitteluntersuchung und – forschung*, **199**, 9–12.

Tanaka, N., Traisman, E., Lee, M.H., Cassens, R.G. & Foster, E.M. (1980) Inhibition of botulinum toxin formation in bacon by acid development. *Journal of Food Protection*, **43**, 450–7.

Tändler, K. & Rödel, W. (1983) Herstellung und Haltbarkeit von dünnkalibrigen Dauerwürsten. II. Haltbarkeit. *Fleischwirtschaft*, **63**, 150, 152–8, 160–2.

Tóth, L. & Blaas, W. (1972) Einfluss der Räuchertechnologie auf den Gehalt von geräucherten Fleischwaren an cancerogenen Kohlenwasserstoffen. II. Mitteilung: Einfluss der Glimmtemperatur des Holzes sowie der Kühlung, Wäsche und Filtration des Räucherrauchs. *Fleischwirtschaft*, **52**, 1419–22.

Trüssel, M. & Jemmi, T. (1989) Das Verhalten von *Listeria monocytogenes* während der Reifung und Lagerung von künstlich kontaminierter Salami und Mettwurst. *Fleischwirtschaft*, **69**, 1586–92.

Tschabrun, R., Sick, K., Bauer, F. & Kranner, P. (1990) Bildung von Histamin in schnittfesten Rohwürsten. *Fleischwirtschaft*, **70**, 448–52.

Urbaniak, L. & Pezacki, W. (1975) Die Milchsäure bildende Rohwurst-Mikroflora und ihre technologisch bedingte Veränderung. *Fleischwirtschaft*, **55**, 229–32, 235.

Vandekerckhove, P. (1977) Amines in dry fermented sausage. *Journal of Food Science*, **42**, 283–5.

Vandendriessche, F., Vandekerckhove, P. & Demeyer, D. (1980) The influence of some spices on the fermentation of a Belgian dry sausage, in *Proceedings of the 26th European Meeting of Meat Research Workers*, Colorado Springs, Vol. 2, pp. 128–33.

Verlinden, B.E. & Spliet, J.A. (1994) A limited mathematical model for the fermentation of dry sausage. *Journal of Food Engineering*, **21**, 1–17.
Verplaetse, A., Bosschere, M. de & Demeyer, D. (1989) Proteolysis during dry sausage ripening, in *Proceedings, 35th International Congress of Meat Science and Technology*, Copenhagen, Denmark, pp. 815–18.
Vogel, R.F., Pohle, B.S., Tichaczek, P.S. & Hammes, W.P. (1993) The competitive advance of *Lactobacillus curvatus* LTH 1174 in sausage fermentations is caused by formation of curvacin A. *Systematic and Applied Microbiology*, **16**, 457–62.
Weiss, N. (1992) The genera *Pediococcus* and *Aerococcus*, in *The Prokaryotes*, 2nd edn., vol. II, Chapter 68 (eds A. Balows, H.G. Trüper, M. Dworkin, W. Harder & K.H. Schleifer), Springer, Berlin, pp. 1502–7.
Wirth, F. (1964) Zur Lagerfähigkeit von Rohwurst (Dauerwurst). *Fleischwirtschaft*, **44**, 104–7.
Wirth, F. (1972) Qualitätsabweichungen bei Schweinefleisch – Konsequenzen für die Verarbeitung von wässrigem, blassem Schweinefleisch. *Fleischwirtschaft*, **52**, 212–16.
Wirth, F. (1991) Restricting and dispensing with curing agents in meat products. *Fleischwirtschaft*, **71**, 1051–4.
Wolf, G., Arendt, E.K., Pfähler, U. & Hammes, W.P. (1990) Heme-dependent and heme-independent nitrite reduction by lactic acid bacteria results in different N-containing products. *International Journal of Food Microbiology*, **10**, 323–30.
Wortberg, B. & Woller, R. (1982) Qualität und Frische von Fleisch und Fleischwaren im Hinblick auf ihren Gehalt an biogenen Aminen. *Fleischwirtschaft*, **62**, 1457–60, 1463.
Zaika, L.L. & Kissinger, J.C. (1984) Fermentation enhanced by spices: identification of active component. *Journal of Food Science*, **49**, 5–9.
Zaika, L.L., Zell, T.E., Smith, J.L., Palumbo, S.A. & Kissinger, J.C. (1976). The role of nitrite and nitrate in Lebanon bologna, a fermented sausage. *Journal of Food Science*, **41**, 1457–60.
Zeuthen, P. (1995) Historical aspects of meat fermentation, in *Fermented Meats* (eds G. Campbell-Platt & P.E. Cook), Blackie, Glasgow, pp. 39–52.

15 Protein-rich foods based on fermented vegetables

B.J.B. WOOD

15.1 Introduction

In industrialized societies there is, even among non-vegetarians, an increasing appreciation that meat-free dishes can be appetizing, enjoyable, and a generally welcome extension to the diet, quite apart from any question of a desire to cut down on meat intake. In the poorer areas of society, the emphasis is somewhat sharpened by the fact that meat is in general more expensive than vegetable foods. In developing countries this can become a matter of absolute necessity, with meat being completely beyond the reach of some sections of the population. Consequently, any foods which can combine being appetizing and vegetable-based might be expected to command a ready market, but in practice the identification of meat-eating with affluence leads to a downgrading or even rejection of traditional vegetable foods within an emerging middle class. On the other hand, within the community of industrialized nations there is increasing interest in such foods, as noted at the start of this chapter. Nowhere is this dichotomy more evident than in the case of the Indonesian food called 'Tempe(h)'. This chapter reviews the production and use of tempeh, and also looks briefly at another food based on a fermenting a protein-rich oil-seed with a fungus, specifically the groundnut.

15.2 The soybean (*Glycine max*)

Applications of this versatile seed are also dealt with in other chapters, notably that on soy sauce and miso (Chapter 12). Here it will be convenient to review the properties of the seed which make it such an important agricultural product. Although the soybean was originally farmed in China, Japan and South East Asia, its growth requirements made it an ideal alternate crop for those areas of the United States of America where maize (*Zea mais*) is a principal crop. The USA is now the leading world producer of this oil-seed, with Canada and Brazil among other major producers. Within Europe the most remarkable story with it must be the success of the French in developing strains suitable for every part of that country. It continues to be grown in its traditional homelands as well as in these new regions. The primary reason for growing it varies with region. The enormous American production is normally regarded as being driven in the first instance by the requirement for its oil, which comprises about 20% of the mature seed, and which is a major oil for cooking, and also finds diverse industrial applications including printing. The seed also contains about 40% by weight of protein, and much of the residue from oil extraction is used as a nitrogen source for animal feed. Until fairly recently it was believed that the protein's amino acid balance made it unsuitable for use as a human foodstuff without balancing input of sulphur-rich proteins from cereals. In practice, it is surely very unlikely that people would be so reliant on a single source of protein that serious malnutrition would result from consumption of even substantial amounts of soybeans. Nevertheless, this concern seems to have played a part in inhibiting the wider use of soybeans as a human food. In the past decade or so it has been appreciated that these fears were founded on the basis of animal feeding trials, and that in making deductions about human nutrition from these trials, the fact that animals have much more hair than do human beings had been ignored. This is not to suggest that diets in which soy proteins play an excessive part are suitable for all humans; baby and infant foods in particular are special cases where care about the amount of soy protein being used must be exercised. For older people in the West the soy bean also suffers from memories of its application in wartime foods, and this was not helped by well-intentioned attempts to sell textured vegetable proteins to the public in the 1960s and 1970s before the available technology was really capable of creating products which could meet the expectations engendered in consumers by the claims of advertisers.

A further source of difficulties with any legume seeds, including the soybean, is the presence of the so-called antinutritional factors. These are diverse, and may include proteins to which some people are allergic, phytohaemagglutenins, phytic acid and other naturally present components. Fortunately, most of these factors are destroyed by thorough cooking. More of a problem are some of the carbohydrates present in legumes. The soybean is rather unusual in that it contains no starch, and even more unusual in the amount of sucrose which it contains, and which contributes around 20% to its dry weight, but neither of these peculiarities is a particular problem to the consumer. On the other hand, the presence of oligosaccharides which contain glucose, fructose and galactose presents a real problem, as these compounds are not digested in, nor absorbed from the upper parts of the gastrointestinal tract. Bacteria in the lower part of the tract readily consume them however, and in so doing they generate a flatulence problem. Fermentations are an effective way to reduce or eliminate these oligosaccarides, and over the long time for which the soybean has been an important agricultural product in its home regions of China, Japan and South East Asia, diverse fermentations have been developed in an effort to increase the palatability of this rather bland food. At this remove in time it is not clear as to how the processes for producing soy sauce, miso, tempeh and other fermented products were initially developed. Consequently, it is also not clear as to how aware the pioneers developing these processes were of the changes in digestibility and toxicity which their processes effected in the beans. Despite this the improvements in flavour and nutritional value do go together in any properly conducted fermentation. It must also be emphasized that the remarks which are addressed to the soybean will apply to any other legume seed which is fermented before consumption by humans. Some reference will be made later in this chapter to fermentative upgrading of groundnut residues after the oil has been expressed from them, and Chapter 23 on African Fermented Foods refers to numerous fermentations of the seeds from tree legumes, and, even more remarkably, to fermenting the highly toxic castor oil seed (*Ricinus* spp.) to an apparently safe food flavouring. Tempeh is, while not unique, in the minority of fermented legume products which make a significant contribution to the total food intake of its consumers. Traditionally it has been a food of the poorer people in its Indonesian homeland, but it is attracting increasing interest in developed economies, in part because of the growth in vegetarian foods, and such interest tends to come from the more affluent sections of society, so tempeh's image in the USA, where it can now be purchased in mainstream supermarkets, is rather different. The discovery that it is a useful source of vitamin B_{12}, which is not present in vegetables, and thus presents a problem for vegans, and even for the stricter vegetarians, has strengthened interest in this food among Western consumers, and added new impetus to assessing its significance to the traditional consumer.

15.3 Tempeh production: an outline

The production process has been well reviewed in the literature of recent years. Among these reviews are National Academy of Sciences (1979), Pederson (1979), Ko (1982), Sharma & Sarbhoy (1984), Winaro (1986), Platt (1987), Teutonico (1987), Nout & Rombouts (1990), Hachmeister & Fung (1993), Golbitz (1995), and the most comprehensive technical review of all, Steinkraus (1995). Further reading, of a more general nature, although containing much technical material, is found is the publications of Shurtleff & Aoyagi (1980, 1985a,b; later editions of some of their books are now available). What follows is a general outline based on these sources, supplemented by my own observations in Indonesia. The process as carried out in Indonesia ranges from very basic village technology to fairly well-controlled production on a small industrial scale. On of its strengths within the developing Indonesian economy lies in the dispersed and small scale of production, which means that local suppliers are addressing local markets with fresh material, removing the need for elaborate provision for packaging and refrigeration. As the economy develops the process lends itself to development into more centralized production and distribution, and there are signs that this is already happening.

Tempeh production is regarded as a single-stage fermentation in which cooked soybeans are inoculated with a mould, fermented overnight in closed packages, then cooked for consumption as part of a main meal or a snack. The mould is generally assigned to *Rhizopus oligosporus*, although other members of this genus, such as *R. stolonifer*, *R. oryzae* and *R. arrhizus* have been isolated from good tempeh fermentations. The spore preparations which can now be purchased for controlled inoculation of industrial-scale fermentations are all *R. oligosporus*, so far as can be ascertained, and further discussion of the process here will be based on this mould. The selection of *Rhizopus* as the genus for this fermentation is one of those fortunate accidents which seem to characterize indigenous fermentations. When cooked, dehulled soybeans are wrapped in fresh banana leaves and then left overnight in the ambient conditions characteristic of Java (30–35°C, high humidity), they become covered with a dense web of fungal hyphae which binds them into a cake that can be sliced without disintegrating. This seems to be a practically pure culture of *Rhizopus*. The restricted access to oxygen within the package formed by the banana leaf wrapping will exert a selection pressure in favour of the desired mould, but I find it difficult to believe that this is the entire story. Banana leaves are a fairly natural choice for wrapping foods in the tropics, being fresh, readily available and large, and are used in this way throughout the plants' range; in West Africa for example, they are commonly used for wrapping cooked maize flour paste for fermentation by indigenous lactic acid bacteria. It is reported that certain other leaves (teak, hibiscus) can also be used in tempeh production, but I have no experience of this, whereas I have had several demonstrations of *Rhizopus* growth with banana leaves. I feel that there is scope for interesting research here, investigating if the banana leaf is a particular habitat for the mould, or if it exerts some selective action additional to that of the low availability of oxygen within the package.

The nutrition of the moulds which are involved in tempeh production is a key to optimizing the processes, particularly as the process is applied to substrates other than the traditional soybean. Graffham *et al.* (1995) have recently reported on this subject.

In fact, the microbiology of tempeh is rather more complex than the usual outline suggests. To understand this it is necessary to examine the process in a little more detail. The soybeans are mature seeds stored dry. For processing they need to be saturated with water, so they are mixed with an excess of water and left overnight at ambient temperature. In tropical conditions the seeds quickly take up water and begin to respire, thus generating further heat. The seeds also release a complex mixture of organic compounds into the water, and this supports the growth of microorganisms present on the seeds, in the water, and in the environment. Seed and microbial respiration render the mixture anoxic, and acid-forming bacteria such as lactic acid bacteria and spore-forming genera (*Bacillus* spp.) come to dominate the flora. By the following morning the mixture of beans and water is turbid, frothy and sour-smelling. These conditions will have killed the seeds, and they will have been significantly acidified. Some reports claim that the major part of the vitamin B_{12} in tempeh is formed during this stage, absorbed by the beans and carried through the rest of the process. The significance of acidification to the rest of the process is also debated.

The beans are then removed from the soak water, parboiled in fresh water, cooled, dehulled, the cotyledons separated from the skins, and the cotyledons cooked in fresh water until sufficiently softened. This last cooking stage requires careful judgement as the cotyledons must be soft enough to permit rapid growth of the mould, but firm enough to retain their integrity during subsequent processing. They are again cooled, typically by being spread out on trays, then ambient air is blown over them with occasional mixing by hand until sufficiently cool for the inoculum of *Rhizopus* to be added. The inoculum may be a culture from a banana leaf, a locally produced dry preparation, or spores produced under laboratory conditions and mixed with an inert powder (probably starch) to give a standardized spore concentration. The last type of inoculum is now available routinely in Indonesia and is favoured by the larger and more modern producers. After inoculation the material is packaged for fermentation. Increasingly, the traditional wrapping in banana leaves is being displaced by packing into plastic pouches. The latter process, although still normally labour-intensive as it is done by hand, is a considerable improvement on the slow wrapping in leaves, and can easily be mechanized if demand warrants it. For tempeh production a delicate balance must be struck in terms of access to the air. Under strictly anaerobic conditions mould growth would be poor or non-existent, but in the presence of too much oxygen the mould sporulates and gives a dark product with an unacceptable taste. When the correct balance is achieved the mould forms a tight mass of pale cream hyphae, binding the beans into a firm cake with a mild nutty taste after cooking. Plastic film of food grade is insufficiently permeable to oxygen as normally supplied, so specially prepared pouches with fine perforations spaced at about 1-cm intervals are now available; these give good growth of the mould without sporulation (although tiny black areas may occasionally develop around the holes). Since the film is transparent, mould growth is easily monitored by visual inspection. Where small packages are used, they may be printed with product details and then are retailed directly.

From this description it will be apparent that there is ample opportunity for other microbes to gain access to the tempeh. In industrial production dominance of the selected *Rhizopus* cultivar over other organisms is secured by adequate inoculation and reasonably careful working, together with the constraint imposed by the low oxygen concentration. To a Western observer the oxygen-deficient conditions and the fact that the fermentation is not acidic raises a fear of clostridial infection and food poisoning, but in practice this does not seem to occur. It fact there is a widely-held belief that the mould exerts an anti-bacterial action, and that tempeh protects against food poisoning. However, one bacterial species, *Klebsiella pneumoniae*, is reported to grow in all tempeh fermentations conducted under normal conditions, and the extensive literature on this subject is reviewed by Steinkraus (1995), whose group first reported its presence in the fermentation, and who demonstrated its involvement in vitamin B_{12} production.

The finished product has a fairly short shelf life, even if stored under refrigeration, and at ambient Indonesian temperatures it must be cooked and consumed fairly soon after purchase. The deterioration in unopened packages is chiefly attributable to continued activity of the mould. If the package is opened, the mould will sporulate, giving a very unacceptable material, and this exposure to the environment also carries the risk of contamination by other organisms. Nout *et al.* (1985) report on the effects of process variables and storage conditions on the product's quality and shelf life. A later report from this group (Nout *et al.*, 1987) focuses on the growth of *Bacillus cereus* on tempeh; the ability of this organism to survive the processing in spore form is not unexpected, and its capacity to cause food poisoning has been demonstrated in other foods. Samson *et al.* (1987) describe the microbiological quality of commercial tempeh on sale in the Netherlands, where, as the former colonists, the first Western interest in the product developed. Ashenafi & Busse (1989) describe the inhibition of various food-poisoning bacteria during tempeh production, caused by *Lactobacillus plantarum*.

15.4 Other ingredients

Legume seeds in general are rather bland and unexciting components of the diet, but they represent a good source of inexpensive protein acceptable to all people without any religious or other cultural problems restricting their use. Consequently it is unsurprising that there have been many investigations into the possibilities of using different substrates for the fermentation. The US Department of Agriculture group of Wang and Hesseltine developed tempeh formulations using mixtures of soy beans and wheat, and I found their preparations delicious. David & Verma (1981) used formulations with *Vicia faba*, and Paradeslopez *et al.* (1987) refer to their substrate as 'common beans'. In a subsequent study, Paradeslopez & Harry (1989) examined the effect of fermentation on various chemical components of both fresh and hardened beans, and in 1990, Paradeslopez *et al.* described the sensory evaluation of the resulting tempeh. Reduction in the flatulence produced by consumption of peas was an objective of the study by Nowak & Steinkraus (1988) who used the growth of, and gas production by *Clostridium perfringens* as a guide to flatulence production potential. Lupine tempeh was produced by Agosin *et al.* (1989) who described its chemical and nutritional properties, and another study (on *Lupinus angustifolius* seeds) was reported by Fudiyansyah *et al.* (1995).

Okara's suitability as a substrate was examined by Matsuo (1990a,b), and later reports (Matsuo, 1995a,b) describe anti-oxidant activity of the product in trials with rats fed oxidized oil, the utilization of the fibre in this tempeh by the rats' intestinal microflora, and (Matsuo, 1996) the ability of the rats to digest the protein in the tempeh. Even more surprising substrates, bitter and sweet apricot seeds, were fermented by Tuncel *et al.* (1990) to upgrade this by-product of the apricot fruit industry. The cyanogenic glycoside amygdalin is present in all apricot seeds, although at higher concentrations in the bitter varieties. It is also present in seeds of plums and other stone fruits, so a successful outcome to the study could have wider applications. While this study did not result in a process which yielded a product from the bitter seeds completely free of cyanide, or one reliably resistant to *Bacillus cereus* infection in either case, they were able to demonstrate that the fermentation is capable of yielding a product with an agreeable taste, and there were indications that further optimization of the process would result in an acceptably safe product.

Chickpea was the fermentation substrate in a study by Paradeslopez *et al.* (1991), and this legume, horsebean and pea were compared with soybeans in a study concerning the suitability of various indigenous Ethiopian beans for tempeh fermentation by Ashenafi & Busse (1991a,f). All of the legumes performed well in the various trials, supporting adequate growth of the mould, exhibiting changes during the fermentations broadly similar to those found with soybeans, and giving organoleptically acceptable products. Ashenafi & Busse (1991b) found that, as with soybean tempeh, their products from peas had short shelf lives (less than 1 week), even during storage at 4°C. On the other hand, chickpea tempeh had not spoiled even after 2 weeks of storage. Acidification of seeds during soaking had no effect on the shelf life of either type of tempeh, but inoculation of the cooked peas with *Lactobacillus plantarum* improved the shelf life of the resulting tempeh, and chickpea tempeh similarly treated had a shelf life in excess of 3 weeks. They went on to show (Ashenafi & Busse, 1991c) that the rise in pH of tempeh during *Rhizopus* fermentation increased the growth rate of

Salmonella infantis and *Escherichia coli* which had been introduced to experimental tempeh fermentations of horsebean, pea and chickpea, but that inoculating the cooked beans with *Lb. plantarum* at the same time resulted in decrease or complete inhibition in growth of the contaminant bacteria. The introduction of *Lb. plantarum* evidently did not produce unacceptable changes in the tempehs, and the authors suggest that this would be a satisfactory way to increase the safety of commercial tempeh production from their legumes. They further demonstrated the value of this approach in fermentations where naturally developing bacterial contaminations were studied (Ashenafi & Busse, 1991d) and in the case of *Listeria monocytogenes* (Ashenafi, 1991).

The African yam bean (*Sphenostylis stenocarpa*) was examined as a substrate for tempeh production by Njoku *et al.* (1991). A more general study of fermented tempeh-like products, considering them as a useful way of utilizing legume seeds, was reported by Piskorz *et al.* (1994). An application to a non-leguminous substrate, defatted rapeseed meal, and its effects on the elimination of antinutritional substances from that important by-product of the edible oil industry, was reported by Bau *et al.* (1994).

The nutritive quality of a flour prepared from a soybean–maize tempeh (30/70) was evaluated by Tchango (1995). Velvet beans (*Mucuna pruriens*) were the subject of a study by Higasa *et al.* (1996) who reported on changes in free amino acids during the fermentation. It seems unlikely that the country in which a legume is grown will make really significant changes in its chemical composition, or in the composition of tempeh made from it, but Vanderriet *et al.* (1987) report on the analysis of tempeh made with soybeans grown in South Africa.

This survey strongly suggests that the potential of tempeh-type fermentations is great, not least in their applications for upgrading underutilized materials and the by-products from food industries.

15.5 Applications of tempeh in the Western diet

The appearance of tempeh in Western foodstores has been previously alluded to, with particular reference to Holland and the USA. Wang, whose work with Hesseltine did so much to interest Western scientists in this product, has made several presentations on its potential to supply protein in the Western diet (Wang, 1983, 1984, 1985, 1986), along with that of tofu (a curd prepared by precipitation from an aqueous extract from soybeans). I have seen excellent quality tempeh for sale in California and Holland, but I am aware that some products, particularly from the more unscientific end of the 'wholefoods' and 'natural foods' spectrum, are of dubious microbial status. The Dutch authorities have also expressed some concerns about this matter, and in 1987 Samson *et al.* presented an authoritative report on the microbial quality of commercial tempeh in the Netherlands. It is very important that producers wishing to offer this product to the public be fully aware of its nature, and that they realize that products of this type are **not** intrinsically safe (see also Ashenafi, 1994). There is much anecdotal evidence, confirmed by some scientific studies, which indicate that the food has beneficial effects on the gastric microflora. Several of the general and review articles cited above carry informed discussions of this matter, and an African study has examined its effects on the nutrition of malnourished children, effects on their diarrhoea, and on the speed of their recovery from their unhealthy condition (Kalavi *et al.*, 1996). However, normally made tempeh has a somewhat alkaline reaction, and several of the studies cited in the previous section relate to the potential for growth of a range of undesirable organisms on the product, unless measures are taken to ensure scrupulous hygiene during its preparation.

15.6 Microbiology of the process

As previously established, the microbiology of the process appears to be rather more complex than some accounts may suggest. The step from reliance on cultures of uncertain microbiological status to the provision of standardized inocula of known potency may well have provoked arguments about quality similar to those found in the wine industry between advocates of pure-culture technique and those arguing that the best wine depends upon the interaction between various microbes. In practice, there can be little doubt of the benefits to most producers of the standardized inocula. For scientific study of the process it is necessary to have even more precise control of the microbiology; Ruiz-Teran & Owens (1996a,b) illustrate this in their study of chemical and enzymatic changes which occurred during fungal fermentations in the absence of bacteria. On the other hand it must be appreciated that in all practical situations where tempeh is being produced on a commercial scale it is impossible to effect a strictly axenic culture process. Moreover, not only are bacteria present, as will shortly be seen, they are responsible for at least one of the most desirable attributes of the food. Mulyowidarso *et al.* (1990) reported on some of the bacteria associated with the fermentation.

Kronenberg (1984) proposed that the process efficiency could be increased by using pre-germinated inoculum. Penaloza *et al.* (1991) showed that the growth of tempeh mould could be stimulated by supplementation with potassium ions in both solid substrate and liquid fermentations. Dereu *et al.* (1995a) discuss the influence which the acidity and initial temperature of the soybean substrate have on the germination of mould spores, and Davey *et al.* (1991) describe a method for monitoring the growth of the mould in normal tempeh fermentations by measuring changes in dielectric permittivity, which is a considerable advance on the difficult and often unreliable methods which attempted to measure components of the mycelium by chemical means. The extent to which the finished tempeh retains its integrity during subsequent stages, up to and including cooking, is a function of the efficiency with which the hyphal threads bind to each other, and also to the substrate. Arrifin *et al.* (1994) report on the assessment of this binding.

Mention of the effects of acidity on spore germination brings us back to the question of the events which occur during the soaking of the dried beans, a matter which seems to have received more attention for tempeh than soy sauce. Tanaka *et al.* (1985) made an early examination of the safety of tempeh made from unacidified soy beans. Fleet's group, who later championed the view that vitamin B_{12} formation during the soaking stage is an important contributor to the vitamin's presence in the finished product, reported on the microbial ecology of the soaking stage in a paper by Mulyowidarso *et al.* (1989). Tuncel & Goktan (1990) examined the effects of different soaking processes on the growth of various bacteria (including *Klebsiella pneumoniae*, which is regarded as a desirable one because of its part in providing vitamin B_{12}) during the subsequent tempeh fermentation. Ashenafi & Busse (1991e) examined development of microflora during the soaking of horsebean, pea and chickpea as well as soybean, and showed that coliform bacteria and yeasts were eliminated from soak water which had become acidic. Mulyowidarso *et al.* (1991a) studied the identities of acids formed during the soaking, and also the exchange of acids between soybeans and soak water. They also examined the changes in carbohydrate composition of the soak water during a 24-hour soak, using as controls beans that were soaked in aqueous solutions of antibiotics in order to suppress microbial growth (Mulyowidarso *et al.*, 1991b).

Although the beans are probably killed fairly early on in the soaking process due to the acidity and lack of oxygen, it seems reasonable to assume that some changes in preparation for germination will have taken place before death. In the case of beer brewing the changes (notably enzyme formation) which take place during a fairly extended period of germination of the barley comprise the essential malting stage. Suparmo & Markakis (1987) have tried the preparation of tempeh from germinated soybeans; results obtained in our laboratory by M.H. Gordon (unpublished) support the view that extensive changes in protein properties, etc. develop during even a fairly short period of germination, and that these could have significant effects on subsequent fermentations. Matsumoto & Imai (1990b) reported on the processing of soybeans for tempeh production, while Ruiz-Teran & Owens (1996b) investigated sterilizing soybean cotyledons for tempeh production by boiling them in a lactic acid solution.

15.7 Effects of fermentation on substrate composition

One of the chief benefits of prolonged cooking and subsequent fermentation of legumes (and of other seeds) is the reduction in concentrations of anti-nutritional factors, a matter which has been touched on above. Sutardi & Buckle (1985) have shown that phytic acid levels decline during the progression from raw soybeans to the final fried tempeh, while Roozen & Degroot (1985) used electrophoresis to follow levels of trypsin inhibitors during the fermentation. Liu & Markakis (1991), in a study of a trypsin inhibitor activity which could be extracted from soybeans with 85% aqueous ethanol, found that this activity actually increased during tempeh fermentation, but as it showed many similarities (for example its sensitivity to suppression by calcium ions) to fatty acid inhibition of trypsin activity, it should not be confused with the typical proteinaceous inhibitors of trypsin. Reddy & Pierson (1994) took a broader overview of the reductions in anti-nutritional and toxic compounds which fermentations can achieve, while Bau *et al.* (1994) demonstrated that fermentation with a tempeh mould could remove anti-nutritional substances from defatted rapeseed meal. Degradation of oligosaccarides, particularly those which cannot be digested by the human consumer, is of particular benefit to some fermentations, and has been investigated for tempeh moulds by Rehms & Barz (1995). Matsumoto & Imai (1990a) also report on changes in the chemical composition of tempeh during the fermentation stage of the process.

Among the many changes which can be expected to take place, lipid metabolism might be expected to be important, especially as the soybean is so oil-rich. However, the only recent references are reports by Dereu *et al.* (1994) on changes in the lipids during fermentation, and by Hering *et al.* (1991) who state that there is little quantitative change in lipid content during the fermentation, despite earlier suggestions that lipid metabolism was a significant source of energy for the mould.

Bioenrichment, the addition of factors such as vitamins during fermentations, is the subject of a special chapter in this book by Keith Steinkraus, a pioneer in the development of this concept. Indeed, some of his most significant work was done with tempeh fermentation. However, it is worthwhile to mention a few salient points here. Denter & Bisping (1994) report on the formation of B-vitamins by bacteria during the soaking process. Vitamin B_{12} activities in miso and tempeh are compared by Truesdell *et al.* (1987), while Okada (1989) reports on the isolation of vitamin B_{12}-forming bacteria, and Keuth & Bisping (1994) refer to *Citrobacter freundii* and *Klebsiella pneumoniae* as bacteria participating in the B_{12} production. Krusong *et al.* (1991) reported on a modified tempeh process which gave an increased level of vitamin B_{12} in the product. In a subsequent report Krusong *et al.*, 1993 added *Propionibacterium shermanii* to the list of those bacteria which can form vitamin B_{12} in tempeh.

15.8 Anti-oxidants

There are clear indications that tempeh possesses significant anti-oxidant activity, and this has been allied to claims about its health-promoting properties. Murakami *et al.* (1984) link this to the liberation of isoflavones by the fermentation. Ikeda *et al.* (1995) also report on changes in isoflavones during fermentations of defatted soybeans, while Wang & Murphy (1996) describe a mass–balance study of isoflavones during soybean processing. Stahl & Sims (1986) identify tempeh oil as an antioxidant, and Murata (1988) refers to the antioxidative stability of the whole tempeh, while Esaki *et al.* (1994) described the anti-oxidative activity of a range of fermented soybean products and later (1996) reported on the isolation of a new anti-oxidant from tempeh. Matsuo (1995b) showed that tempeh made from okara had anti-oxidant activity in rats which had been treated with oxidized oil. Klus & Barz (1995) reported on the hydroxylation of the isoflavones daidrein and glycitein (found in soybeans) by bacteria present in tempeh.

15.9 Enzymology

The enzymology of soy sauce and miso has been extensively studied, particularly in respect of the extracellular enzymes which are released by the mould during the koji stage; this is the key to the process, as these enzymes are hydrolases which digest and solubilize the proteins and polysaccharides in the substrate beans and wheat. *Rhizopus* spp. produce a range of extracellular enzymes and were among the first moulds to be employed in industrial enzyme production. It is therefore rather surprising that so few papers on tempeh enzymology were revealed in a computer search of the literature extending back to 1981.

Inhibition of enzymes can also be significant, beneficial or counterproductive, depending on the enzyme being inhibited and the process in which it is participating. Hermosilla *et al.* (1993) reported on the isolation and characterization of hydroxymethylglutaryl coenzyme A reductase inhibitors from various fermented soybean extracts.

On the theme of hydrolytic enzymes, Elegado & Fujio (1994) described the purification and properties of an endopolygalacturonase from *Rhizopus*; the relevance to the removal of flatulence-inducing oligosaccarides of the kind endemic in legumes will be apparent. The degradation of the specific oligosaccarides found in soybeans (stachyose, raffinose, melibiose) and also sucrose, by various *Rhizopus* used in tempeh production has been examined by Rehms & Barz (1995). Another endemic problem in legumes and indeed many seeds, is phytic acid, which binds nutritionally important divalent metals, rendering them unavailable for absorption by the consumer, and also ties up phosphate in a form unavailable to the consumer. Sutardi & Buckle (1988) have characterized the extracellular and intracellular phytases found in *R. oligosporus* used in tempeh production. Two reports (Baumann & Bisping, 1995; Dereu *et al.*, 1995b) deal with the hydrolysis of proteins during the tempeh fermentation, a topic which must have some bearing upon the texture of the finished product, and a very important one on its flavour, which is variously described as meaty, savoury and nutty, and is certainly much more pronounced and attractive than that of the original cooked beans. Proteolysis will also provide some of the starting materials from which volatile components of tempeh's flavour are derived; a study by Suprivanto *et al.* (1991) of the flavour volatiles in tempeh demonstrated that the treatments applied to the beans before fermentation had clear effects on the aroma profiles of the finished products.

15.10 Tempeh bongkrek and other problems

It is perhaps a little unfair to characterize tempeh bongkrek as a problem *per se*. Essentially, it is a fermentation practised by some of the poorest in Indonesian society. Coconut flesh is pressed to expel the oil, and this will also force out the sweet sap, leaving a press-cake which must be rather uninteresting as a food, and not particularly nutritious either. It was therefore subjected to a tempeh fermentation in an effort to improve it. Unfortunately, the fermentation is sometimes infection by *Pseudomonas cocovenenans* (Ko *et al.*, 1979; Arbianto, 1979) which forms two poisons, toxoflavin and the much more toxic bongkrek acid. The frequency of tempeh bongkrek poisoning outbreaks varied from year to year, but on average about 10% to 20% of those affected died. The Indonesian government has attempted to ban production of this food, but such bans are rather difficult to enforce in the sort of rural subsistence economy which prevails in the areas where it is found, and deaths continue to be reported. Buckle & Kartadarma (1990) have reported on the inhibition of toxins production in tempeh bongkrek containing *P. cocovenenans*, but the food is normally produced and consumed in conditions where there is very little scope for proper scientific control of the process.

Nout *et al.* (1993) reported on the production of biogenic amines and ethyl carbamate in soybean tempeh, and the effects of processing conditions on their production, while Thiebaud *et al.* (1995) reported on the production of airborne mutagens during the frying of tempeh, and also beef and pork. On the other hand Kiyosawa *et al.* (1995) demonstrated that alcohol-soluble components extracted from tempeh suppressed the responses of *Salmonella typhimurium* cultures to mutagens.

15.11 An overview of tempeh

No food is perfect, and tempeh is no exception despite the claims made by some of its Western advocates. Its great attractions lie in making bland soybeans (and other legumes, alone or in combination with cereals) into an attractive and very palatable food. This can be consumed in quantities sufficient to make a real contribution to both calorie and protein intake without the flatulence and other problems associated with consuming quantities of unfermented legumes, no matter how carefully the latter are cooked. Vitamin B_{12} production is another important advantage of tempeh production, and the relative contributions of bacteria active during the initial soaking of the beans and those present in the actual tempeh fermentation need to be assessed. *Klebsiella pneumoneae* does have what the advertising trade would call an 'image problem'. With a name like that, no matter what microbiologists may say about its safety under normal circumstances, its elimination from tempeh intended for developed markets would probably be highly desirable if adequate vitamin B_{12} levels can be obtained without its participation. Tempeh is on offer in some of the most advanced markets in the world, and I think that this is a desirable development which should be fostered and encouraged.

Finally, this discussion has dealt entirely with products for, and considerations concerning, the consumption of tempeh by humans, but there have been suggestions from time to time that it might have a place in the pet-food market. Some vegetarians and vegans own carnivorous pets (usually cats or dogs, of course) and even among omnivore pet owners there can be some disquiet about the foods which they provide for their animals. Even if there is not, many owners prefer the convenience of dried foods to meat (even in cans). Flatulence is just as prevalent a problem in animals as it is in humans, and a dog experiencing it in an enclosed space can make life rather suffocating for those around it. Consequently leading pet-food producers have occasionally approached scientists recognized as having interests in food fermentations with enquiries about producing foods suitable for this market. Unfortunately, none of the enquiries directed to me has so far resulted in a programme to develop such foods, but I remain optimistic that such developments will eventually take place.

15.12 Oncom; ontjom (fermented peanut press cake)

This interesting product seems to have attracted much less scientific attention than tempeh, probably because it is used on a smaller scale, both geographically and in tonnage terms. Basically it comprises a fermentation with species of *Neurospora* of the residues after oil has been extracted from peanuts (groundnuts, *Arachis hypogae*). Steinkraus (1995) gives an authoritative account of the production and application of this food, and there are no references in the literature subsequent to this review, so far as I have been able to ascertain. I will therefore confine myself to a brief outline of the subject. The essential raw material is the peanut residue, but this can be blended with other materials such as the residues from converting cassava (*Manioc* spp.) into tapioca and from extracting soybean 'milk' from the bean. The peanut press cake is a rather dry material, so it is soaked in water for 3–4 hours, then drained, crumbled to open it up, mixed with the other ingredient(s), steamed, formed into flat cakes, inoculated with a dried preparation of the mould (essentially a spore preparation), covered with the ubiquitous banana leaves after transfer to woven bamboo trays and fermented for 36–48 hours at ambient temperature and humidity. Under these conditions the mould grows through the cakes. *Neurospora* has pink spores and these impart their colour to the finished product. Older literature refers to the most important mould as *N. sitophila*, but this has now been described as *N. intermedia* by some workers. Some forms of ontjom use the tempeh mould *Rhizopus* spp. which gives a black ontjom, again because of spore formation. I have not sampled this food, but I am assured by Prof. Steinkraus and others that it is delicious when fried or otherwise cooked. Bacteria which are associated with the mould's growth supplement this food with vitamin B_{12} also.

15.13 Sufu

Again, the reader is referred to Steinkraus (1995) for details of this product. When raw soybeans are soaked in water, then ground with water and the liquid extract filtered off, a milky fluid results, which is colloquially called 'soy milk', although the EU frowns upon this name. The protein in this extract can be precipitated with calcium and/or magnesium salts to give the curd called 'tofu' (or 'dofu'). For sufu production this is cut into small cubes, dried for 10 minutes in an oven at 100°C, then inoculated with *Actinomucor elegans* and incubated until mould mycelium completely covers the tofu. The cubes are then placed in brine, which may contain rice wine, red rice, soy sauce and various other flavourings, possibly after the moulded cubes have been dry-salted for 3–4 days. The steeping in brine lasts for about 3 months, after which the product is ready for consumption. No reference to microbial growth during the brining stage has been encountered, although it seems reasonable to surmise that halotolerant bacteria, such as those which develop in the moromi (salt mash) stage of soy sauce production, would be able to grow, even in the 20% brine solution referred to by Steinkraus. All reports agree that his concentration of salt kills the sufu mould. Steinkraus' style in writing about this group of products suggests that he liked them, and a Japanese friend likened sufu to Camembert cheese, but some Western people have indicated that they found sufu very disagreeable.

15.14 Red rice; ang-kak

This product is definitely not a high-protein material, being based on rice, but I feel that the book would be incomplete without reference to it. It is another ancient Oriental product which attracts occasional flurries of interest in the West, this time on the basis that the group of red, yellow and purple pigments which it contains are probably safe, having been consumed for centuries without drawing adverse comment. Moreover, in a climate of opinion where the food colours developed for the European and North American markets receive increasing criticism (often on very dubious grounds) these 'natural' Oriental products should have something to offer. In practice, I do not know of any case where these colourings have found applications in Western foods. Again, I refer the interested reader to Steinkraus (1995) for a detailed account of production and properties of the material.

In outline, polished rice is steamed to cook and moisten it. It is then inoculated with a culture of the mould *Monascus purpureus* and incubated for up to 3 weeks at tropical ambient temperature. The mould completely penetrates the rice grains, and at the end of the fermentation they are red throughout and easily crumbled to a red dust. The pigments are soluble to a limited extent in water, but seem to be freely soluble in aqueous alcohol. There are reports that mutant strains of the mould which produce a great excess of pigment ('overproducers') can be fairly easily obtained, and there is some suggestion that these may be the key to future industrial production of the pigments, possibly in deep, stirred tank culture. Pigment production does not seem to depend upon solid-state growth conditions or the presence of rice, although the literature is not particularly explicit upon these points. Henning *et al.* (1992) comment upon the suitability of these pigments for colouring foods, particularly meat products. They are apparently also good for colouring imitation meat products, so the paper describes a method for detecting the *Monascus* pigments in such foods where their use is not yet permitted in Europe.

When searching the literature for recent information on red rice, I found that most of the references were not to ang-kak (and synonyms) but to *Oryza sativa* var. *indica* which is a rice cultivar having a natural red colour.

15.15 Conclusion

There are of course many foods not covered elsewhere in this text, and which could be discussed here. The interested reader is referred to the text by Steinkraus (1995) and to the glossaries by Campbell-Platt (Platt, 1987) and Wang & Hesseltine (1986). I regard the three groups dealt with here as indicative of the potential which these relatively unfamiliar (outside the regions where they were developed) products possess.

References

Agosin, E., Diaz, D., Aravena, R. & Yanez, E. (1989) Chemical and nutritional characteristics of lupine tempeh. *Journal of Food Science*, **54**, 102.

Arbianto. P. (1979) Bongkrek food poisoning in Java, in *Proceedings of the 5th International Conference on Global Impacts of Applied Microbiology* (ed. P. Matangkasombut), pp. 371–4.

Ariffin, R., Apostolopoulos, C., Graffam, A., MacDougall, D. & Owens, J.D. (1994) Asessment of hyphal binding in tempe. *Letters in Applied Microbiology*, **18**, 32–4.

Ashenafi, M. (1991) Growth of *Listeria monocytogenes* in fermenting tempeh made of various beans and its inhibition by *Lactobacillus plantarum. Food Microbiology*, **8**, 303–10.

Ashenafi, M. (1994) Microbiological evaluation of tofu and tempeh during processing and storage. *Plant Foods for Human Nutrition*, **45**, 183–9.

Ashenafi, M. & Busse, M. (1989) Inhibitory effect of *Lactobacillus plantarum* on *Salmonella infantis, Enterobacter aerogenes* and *Escherichia coli* during tempeh fermentation. *Journal of Food Protection*, **52**, 169–72.

Ashenafi, M. & Busse, M. (1991a) Microbial development during tempeh fermentation from various beans, and effect of *Lactobacillus plantarum* on the natural microflora. *International Journal of Food Science* and *Technology*, **26**, 501–6.

Ashenafi, M. & Busse, M. (1991b) Growth potential of *Salmonella infantis* and *Escherichia coli* in fermenting tempeh made from horsebean, pea and chickpea, and their inhibition by *Lactobacillus plantarum. Journal of the Science of Food and Agriculture*, **55**, 607–15.

Ashenafi, M. & Busse, M. (1991c) Development of microorganisms during cold-storage of pea and chickpea tempeh, and effect of *Lactobacillus plantarum* on storage microflora. *Journal of the Science of Food and Agriculture*, **56**, 71–8.

Ashenafi, M. & Busse, M. (1991d) Growth of *Bacillus cereus* in fermenting tempeh made from various beans, and its inhibition by *Lactobacillus plantarum. Journal of Applied Bacteriology*, **70**, 329–33.

Ashenafi, M. & Busse, M. (1991e) The microflora of soak water during tempeh production from various beans. *Journal of Applied Bacteriology*, **70**, 334–8.

Ashenafi, M. & Busse, M. (1991f) Production of tempeh from various indigenous Ethiopian beans. *World Journal of Microbiology and Biotechnology*, **7**, 72–9.

Bau, H.M., Villaume, C., Lin, C.F., Evrard, J. Quemener, B., Nicholas, J.P. & Mejean, L. (1994) Effect of a solid state fermentation using *Rhizopus oligosporus* sp. T-3 on elimination of antinutritional substances and modification of biochemical constituents of defatted rapeseed meal. *Journal of the Science of Food and Agriculture*, **65**, 315–22.

Baumann, U. & Bisping, B. (1995) Proteolysis during tempe fermentation. *Food Microbiology*, **12**, 39–47.

Buckle, K.A. & Kartadarma, E. (1990) Inhibition of bongkrek acid and toxoflavin production in tempe bongkrek containing *Pseudomonas cocovenenans. Journal of Applied Bacteriology*, **68**, 571–6.

Davey, C.L., Penazola, W., Kell, D.B. & Hedger, J.N. (1991) Real-time monitoring of the accretion of *Rhizopus oligosporus* biomass during the solid-substrate tempe feremention. *World Journal of Microbiology and Biotechnology*, **7**, 248–59.

David, I.M. & Verma, J. (1981) Modification of tempeh with the addition of bakla (*Vicia faba* Linn). *Journal of Food Technology*, **16**, 39–50.

Denter, J. & Bisping, B. (1994) Formation of B-vitamins by bacteria during the soaking process of soybeans for tempe fermentation. *International Journal of Food Microbiology*, **22**, 23–31.

Dereu, J.C., Ramdaras, D., Rombouts, F.M. & Nout, M.J.R. (1994) Changes in soya bean lipids during tempe fermentation. *Food Chemistry*, **50**, 171–5.

Dereu, J.C., Rombouts, F.M. & Nout, M.J.R. (1995a) Influence of acidity and initial substrate-temperature on germination of *Rhizopus oligosporus* sporangiospores during tempe manufacture. *Journal of Applied Bacteriology*, **78**, 200–8.

Dereu, J.C., Tenwolde, R.W., DeGroot, J., Nout, M.J.R., Rombouts, F.M. & Gruppen, M. (1995b) Protein hydrolysis during soybean tempe fermentation with *Rhizopus oligosporus. Journal of Agricultural and Food Chemistry*, **43**, 2235–9.

Elegado, F.B. & Fujio, Y. (1994) Purification and some properties of endopolygalacturonase from *Rhizopus* sp. LKN. *World Journal of Microbiology and Biotechnology*, **10**, 256–9.

Esaki, H., Onozaki, H. & Osawa, T. (1994) Antioxidative activity of fermented soybean products. *American Chemical Society Symposium Series*, **546**, 353–60.

Esaki, H., Onozaki, H., Kawakishi, S. & Osawa, T. (1996) New antioxidant isolated from tempeh. *Journal of Agricultural and Food Chemistry*, **44**, 696–700.

Fudiyansyah, N., Petterson, D.S., Bell, R.R. & Fairbrother, A.H. (1995) A nutritional, chemical and sensory evaluation of lupin (*L. angustifolius*) tempe. *International Journal of Food Science and Technology*, **30**, 297–305.

Golbitz, P. (1995) Traditional soyfoods – processing and products. *Journal of Nutrition*, **125(**Suppl. 3), S570–2.

Graffham, A.J., Gordon, M.H., Westby, A. & Owens, J.D. (1995) Nutrition of tempe molds. *Letters in Applied Microbiology*, **21**, 223–7.

Hachmeister, K.A. & Fung, D.Y.C. (1993) Tempeh, a mold-modified indigenous fermented food made from soybeans and/or cereal grains. *Critical Reviews in Microbiology*, **19**, 137–88.

Henning, W., Wedekind, R. & Bertling, L. (1992) Detection of *Monascus* pigments in meat-products. *Fleischwirtschaft*, **72**, 1428.

Hering, L., Bisping, B. & Rehm, H.J. (1991) Patterns and formation of fatty acids at tempe fermentation by several strains of *Rhizopus* sp. *Fette Wissenschaft Technologie*, **93**, 303–8.

Hermosilla, J.A.G., Jha, H.C., Egge, H., Mahmud, M., Hermana, S. & Rao, G.S. (1993) Isolation and characterization of hydroxymethylglutaryl coenzyme-A reductase inhibitors from fermented soybean extracts. *Journal of Clinical Biochemistry and Nutrition*, **15**, 163–74.

Higasa, S., Negishi, Y., Aoyagi, Y. & Sugahara, T. (1996) Changes in free amino acids of tempe during preparation with velvet beans (*Mucuna pruriens*). *Journal of the Japanese Society for Food Science and Technology (Nippon Shokuhin Kogaku Kaishi)*, **43**, 188–93.

Ikeda, R., Ohta, N. & Watanabe, T. (1995) Changes of Isoflavones at various stages of fermentation in defatted soybeans. *Journal of the Japanese Society for Food Science and Technology (Nippon Shokuhin Kogaku Kaishi)*, **42**, 322–7.

Kalavi, F.N.M., Muroki, N.M., Omwega, A.M. & Mwadime, R.K.N. (1996) Effect of tempe-yellow maize porridge and milk-yellow maize porridge on growth rate, diarrhoea and duration of rehabilitation of malnourished children. *East African Medical Journal*, **73**, 427–31.

Keuth, S. & Bisping, B. (1994) Vitamin B12 production by *Citrobacter freundii* or *Klebsiella pneumoniae* during tempeh fermentation, and proof of enterotoxin absence by P.C.R. *Applied and Environmental Microbiology*, **60**, 1495–9.

Kiyosawa, I., Matsuyama, J., Arai, C. & Setoguchi, T. (1995) Suppressive effects of the methanol extracts from soybean products on S.O.S. response of *Salmonella typhimurium* induced by mutagens, and their contents of isoflavones. *Journal of the Japanese Society for Food Science and Technology (Nippon Shokuhin Kogaku Kaishi)*, **42**, 835–42.

Klus, K. & Barz, W. (1995) Formation of polyhydroxylated isoflavones from the soybean seed isoflavones daidzein and glycitein by bacteria isolated from tempe. *Archives of Microbiology*, **164**, 428–34.

Ko, S.D. (1982) Indigenous fermented foods, in *Economic Microbiology, Vol. 7, Fermented Foods* (ed. A.H. Rose), Academic Press Inc., London. pp. 15–38.

Ko, S.D., Kelholt, A.J. & Kampelmacher, E.H. (1979) Inhibition of toxin production in tempe bongkrek, in *Proceedings of the 5th International Conference on Global Impacts of Applied Microbiology* (ed. P. Matangkasombut), pp. 375–88.

Kronenberg, H.J. (1984) Reduction of incubation time for tempeh fermentation by use of pregerminated inoculum. *Economic Botany*, **38**, 433–8.

Krusong, W., Yongsmith, B. & Sanchez, P.C. (1991) Note: Increased vitamin B-12 level in a modified tempeh process. *The Phillipine Agriculturist*, **74**, 89–94.

Krusong, W., Yongsmith, B. & Sanchez, P.C. (1993) Vitamin B12 production by *Propionibacterium shermanii* in tempeh. *Proceedings of the 3rd Annual Meeting of the Thai Society for Biotechnology on Current Advances in Biotechnology: Industry and Environment, 1991*, Publishers, The Thai Society for Biotechnology, Bangkok, pp. 113–19.

Kuo, J.C., Wang, S.Y., Peng, A.C. & Ockerman, H.W. (1989) Effect of tempeh on properties of hams. *Journal of Food Science*, **54**, 1186.

Liu, K.H. & Markakis, P. (1991) Aqueous ethanol extraction of soybean trypsin-inhibitors and characterization of a calcium-sensitive fraction. *Journal of Food Biochemistry*, **15**, 159–68.

Matsumoto, I. & Imai, S. (1990a) Changes of chemical composition of tempeh during fermentation. *Journal of the Japanese Society for Food Science and Technology (Nippon Shokuhin Kogaku Kaishi)*, **37**, 130–8.

Matsumoto, I. & Imai, S. (1990b) Processing of soybeans for manufacture of tempe. *Journal of the Japanese Society for Food Science and Technology (Nippon Shokuhin Kogaku Kaishi)*, **37**, 479–504.

Matsuo, M. (1990a) Effects of rice bran on properties of okara tempe. *Journal of the Japanese Society for Bioscience, Biotechnology and Agrochemistry (Nippon Nogeikagaku Kaishi)*, **64**, 1237–9.

Matsuo, M. (1990b) Suitability of okara tempe as foodstuff. *Journal of the Japanese Society for Bioscience, Biotechnology and Agrochemistry (Nippon Nogeikagaku Kaishi)*, **64**, 1235–6.

Matsuo, M. (1995a) Utilization of dietary fiber in a new foodstuff, okara-tempe, by microflora in rats. *Journal of the Japan Society for Bioscience, Biotechnology and Agrochemistry (Nippon Nogeikagaku Kaishi)*, **69**, 163–7.

Matsuo, M. (1995b) In-vivo antioxidant action of a new foodstuff, okara-tempe, in rats fed oxidised oil. *Journal of the Japanese Society for Bioscience, Biotechnology and Agrochemistry (Nippon Nogeikagaku Kaishi)*, **69**, 169–71.

Matsuo, M. (1996) Digestibility of okara-tempe protein in rats. *Journal of the Japanese Society for Food Science and Technology (Nippon Shokuhin Kogaku Kaishi)*, **43**, 1059–62.

Mulyowidarso, R.K., Fleet, G.H. & Buckle, K.A. (1989) The microbial ecology of soybean soaking for tempe production. *International Journal of Food Microbiology*, **8**, 35–46.

Mulyowidarso, R.K., Fleet, G.H. & Buckle, K.A. (1990) Association of bacteria with the fungal fermentation of soybean tempeh. *Journal of Applied Bacteriology*, **68**, 43–7.

Mulyowidarso, R.K., Fleet, G.H. & Buckle, K.A. (1991a) Changes in the concentration of organic acids during the soaking of soybeans for tempe production. *International Journal of Food Science and Technology*, **26**, 607–14.

Mulyowidarso, R.K., Fleet, G.H. & Buckle K.A. (1991b) Changes in the concentration of carbohydrates during the soaking of soybeans for tempe production. *International Journal of Food Science and Technology*, **26**, 595–606.

Murakami, H., Asakawa, T., Terao, J. & Matsushita, S. (1984) Antioxidative stability of tempeh and liberation of isoflavones by fermentation. *Agricultural and Biological Chemistry*, **48**, 2971–5.

Murata, K. (1988) Antioxidative stability of tempeh. *Journal of the American Oil Chemists Society*, **65**, 799–800.

National Academy of Sciences (1979) Indonesian tempeh, in *Microbial Processes: Promising Technologies for Developing Countries*, N.A.S., Washington, DC, pp. 32–6.

Njoku, H.O., Ofuya, C.O. & Ogbuli, J.N. (1991) Production of tempeh from the African yam bean *(Sphenostylis stenocarpa Hams)*. *Food Microbiology*, **8**, 209–14.

Nout, M.J.R. & Rombouts, F.M. (1990) Recent developments in tempe research. *Journal of Applied Bacteriology*, **69**, 609–33.

Nout, M.J.R., Bonantsvanlaarhoven, T.M.G., Dedreu, R. & Gerats, I.A.G.M. (1985) The influence of some process variables and storage conditions on the quality and shelf life of soybean tempeh. *Antonie van Leeuwenhoek Journal of Microbiology*, **51**, 532–4.

Nout, M.J.R., Beernink, G. & Bonantsvanlaarhoven, T.M.G. (1987) Growth of *Bacillus cereus* in soybean tempeh. *International Journal of Food Microbiology*, **4**, 293–301.

Nout, M.J.R., Ruikes, M.M.W., Bouwmeester, H.M. & Beljaars P.R. (1993) Effect of processing conditions on the formation of biogenic amines and ethyl carbamate in soybean tempe. *Journal of Food Safety*, **13**, 293–303.

Nowak, J. & Steinkraus, K.H. (1988) Effect of tempeh fermentation of peas on their potential flatulence productivity as measured by gas-production and growth of *Clostridium perfringens*. *Nutrition Reports International*, **38**, 1163–71.
Okada, N. (1989) Role of microorganisms in tempeh manufacture – isolation of vitamin B12-producing bacteria. *Japan Agricultural Research Quarterly*, **22**, 310–16.
Paradeslopez, O. & Harry, G.I. (1989) Changes in selected chemical and antinutritional components during tempeh preparation using fresh and hardened common beans. *Journal of Food Science*, **54**, 968–70.
Paradeslopez, O., Harry, G.I. & Montesrivera, R. (1987) Development of a fermentation procedure to produce a tempe-related food using common beans as substrate. *Biotechnology Letters*, **9**, 333–8.
Paradeslopez, O., Harry, G.I. & Gonzalezcastaneda, J. (1990) Sensory evaluation of tempeh produced by fermentation of common beans. *Journal of Food Science*, **55**, 123–6.
Paradeslopez, O., Gonzalezcastaneda, J. & Carabeztrejo, A. (1991) Influence of solid substrate fermentation on the chemical composition of chickpea. *Journal of Fermentation and Bioengineering*, **71**, 58–62.
Pederson, C.S. (1979) *Microbiology of Food Fermentations, 2nd edn*, Avi Publishing Co. Inc., Westport, Connecticut.
Penaloza, W., Davey, C.L., Hedger, J.N. & Kell, D.B. (1991) Stimulation by potassium ions of the growth of *Rhizopus oligosporus* during liquid-substrate and solid-substrate fermentations. *World Journal of Microbiology and Biotechnology*, **7**, 260–8.
Piskorz, B.J., Nowak, J. & Szebiotko, K. (1994) Fermented tempeh-like products as useful way of utilizing legume seeds. *Acta Biotechnologica*, **14**, 105–10.
Platt, G. Campbell (1987) *Fermented Foods of the World*, Butterworths Press, London.
Reddy, N.R. & Pierson, M.D. (1994) Reduction in antinutritional and toxic components in plant foods (A) by fermentation. *Food Research International*, **27**, 281–90.
Rehms, H. & Barz, W. (1995) Degradation of stachyose, raffinose, melibiose and sucrose by different tempe-producing *Rhizopus* fungi. *Applied Microbiology and Biotechnology*, **44**, 47–52.
Roozen, J.P. & Degroot, J. (1985) Electrophoresis and assay of trypsin-inhibitors in different stages of tempe production. *Journal of Food Biochemistry*, **9**, 37–48.
Ruiz-Teran, F. & Owens, J.D. (1996a) Chemical and enzymatic changes during the fermentation of bacteria-free soya bean tempe. *Journal of the Science of Food and Agriculture*, **71**, 523–30.
Ruiz-Teran, F. & Owens, J.D. (1996b) Sterilization of soybean cotyledons by boiling in lactic acid solution. *Letters in Applied Microbiology*, **22**, 30–3.
Samson, R.A., Vankooij, J.A. & Deboer, E. (1987) Microbiological quality of commercial tempeh in the Netherlands. *Journal of Food Protection*, **50**, 92–4.
Sharma, R. & Sarbhoy, A.K. (1984) Tempeh – a fermented food from soybean. *Current Science*, **53**, 325–6.
Shurtleff, W. & Aoyagi, A. (1980) *Tempeh Production. The Book of Tempeh, Volume II*, New Age Foods, Lafayette, California.
Shurtleff, W. & Aoyagi, A. (1985a) *The Book of Tempeh, 2nd edn*, Harper Colophon Books, New York.
Shurtleff, W. & Aoyagi, A. (1985b) *History of Tempeh*, The Soyfoods Center, Lafayette, California.
Stahl, H.D. & Sims, R.J. (1986) Tempeh oil – antioxidant. *Journal of the American Oil Chemists Society*, **63**, 555–6.
Steinkraus, K.H. (1995) *Handbood of Indigenous Fermented Foods: 2nd edn*, Marcel Dekker Inc., New York, pp. 11–110.
Suparmo & Markakis, P. (1987) Tempeh prepared from fermented soybeans. *Journal of Food Science*, **52**, 1736–7.
Suprivanto, Wada, K., Hayakawa, I. & Fujio, Y. (1991) Principal component analysis of aroma profiles of tempeh produced by using various starters. *Journal of the Society of Fermentation Technology (Hakkokogaku Kaishi)*, **69**, 331–6.
Sutardi & Buckle, K.A. (1985) Reduction in phytic acid levels in soybeans during tempeh production, storage and frying. *Journal of Food Science*, **50**, 260.

Sutardi & Buckle, K.A. (1988) Characterization of extracellular and intracellular phytases from *Rhizopus oligosporus* used in tempeh production. *International Journal of Food Microbiology*, **6**, 67–79.

Tanaka, N., Kovats, S.K., Guggisberg, J.A., Meske, L.M. & Doyle, M.P. (1985) Evaluation of the microbiological safety of tempeh made from unacidified soybeans. *Journal of Food Protection*, **48**, 438–41.

Tchango, J.T. (1995) The nutritive quality of maize-soybean (70/30) tempe flour. *Plant Foods for Human Nutrition*, **47**, 319–26.

Teutonico, R.A. (1987) Impact of biotechnology on the nutritional quality of foods, in *Food Biotechnology* ed. D. Knorr, Marcel Dekker Inc., New York, pp. 95–135.

Thiebaud, H.P., Knize, M.G., Kuzmicky, P.A., Hsieh, D.P. & Felton, J.S. (1995) Airborne mutagens produced by frying beef, pork and a soy-based food. *Food and Chemical Toxicology*, **33**, 821–8.

Truesdell, D.D., Green, N.R. & Acosta, P.B. (1987) Vitamin-B12 activity in miso and tempeh. *Journal of Food Science*, **52**, 493–4.

Tuncel, G. & Goktan, D. (1990) Effect of different methods of soaking soya beans on the growth of *Bacillus cereus, Klebsiella pneumoniae* and *Staphylococcus aureus* in tempeh. *Journal of the Science of Food and Agriculture*, **53**, 287–96.

Tuncel, G., Nout, M.J.R., Brimer, L. & Goktan, D. (1990) Toxicological, nutritional and microbiological evaluation of tempe fermentation with *Rhizopus oligosporus* of bitter and sweet apricot seeds. *International Journal of Food Microbiology*, **11**, 337–44.

Vanderriet, W.B., Wight, A.W., Cilliers, J.J.L. & Datel, J.M. (1987) Food chemical-analysis of tempeh prepared from South African-grown soybeans. *Food Chemistry*, **25**, 197–206.

Wang, H.L. (1983) Tofu and tempeh as potential protein-sources in the Western diet. *Journal of the American Oil Chemists Society*, **60**, 718.

Wang, H.L. (1984) Tofu and tempeh as potential protein-sources in the Western diet. *Journal of the American Oil Chemists Society*, **61**, 528–34.

Wang, H.L. (1985) Uses of whole soybeans in the West wuth emphasis on tofu and tempeh. *Abstracts of Papers of the American Chemical Society*, **190**, 127.

Wang, H.L. (1986) Uses of soybeans as foods in the West with emphasis on tofu and tempeh. *American Chemical Society Symposium Series*, **312**, 45–60.

Wang, H.L. & Hesseltine, C.W. (1986) Glossary of indigenous fermented foods, in *Indigenous Fermented Food of Non-Western Origin* (eds C.W. Hesseltine & H.L. Wang), Mycological Memoir No. 11, New York Botanical Garden and the Mycological Society of America, pp. 317–42.

Wang, H.J. & Murphy, P.A. (1996) Mass–balance study of isoflavones during soybean processing. *Journal of Agricultural and Food Chemistry*, **44**, 2377–83.

Winarno, F.G. (1986) Production and utilization of tempeh. *Journal of the American Oil Chemists Society*, **63**, 472.

16 Food flavours from yeast

H. STAM, M. HOOGLAND AND C. LAANE

16.1 Introduction

In the food industry yeast is a well-known microorganism (Lásztity, 1991; Thornton, 1992). From ancient times yeast has served many purposes. It is involved in baking, the production of (alcoholic) beverages, flavours and other food ingredients such as enzymes, colours, emulsifiers, stabilizers and thickeners. In addition, yeast and yeast-derived products increasingly find applications in the non-food sector, which includes animal feed, aquaculture, fermentation substrates, cosmetics and pharmaceutical products (Gordon, 1991). Figure 16.1 gives per segment an estimate of the current world market size for yeast and its derivatives. Of all the hundreds of yeast species known to-date, *Saccharomyces cerevisiae* forms the versatile workhorse from which more than 90% of the current yeast and yeast-based products originate. This yeast has a long history of safe use, is relatively easy to cultivate on a large scale, and has long been the prime focus of both fundamental and applied research.

The highest growth is expected in the higher-added value area of yeast-derived products, since the traditional active-yeast baking and beverages market is maturing. Furthermore, a shift towards non-food applications and yeast species other than *Saccharomyces cerevisiae* is envisaged. Typical examples include *Candida*, *Kluyveromyces* and *Schizosaccharomyces*, which are on the verge of being exploited to a greater extent in specific areas currently dominated by *Saccharomyces cerevisiae.* Important players in the above described world market are Bio-Springer, Gist Brocades, Quest International, DHW/Ohly, CPC and Universal foods.

In this chapter the emphasis will be on yeast and yeast-derived products which, among others, are tailored to deliver taste to a final food product. These products include active yeast for the *in situ* generation of a desired flavour profile in fermented food products (e.g. wine, beer, bread, soy sauce, etc.), as well as inactive yeast-based products, which can be added as ingredients to enhance or modulate the taste of a final food product (e.g. yeast extracts, aroma chemicals, etc.). Various applications of yeast in the flavour industry are presented in Figure 16.2. In this review the versatility of yeast as a flavouring tool in a variety of food applications will be described on the basis of our current understanding of market needs and technological possibilities.

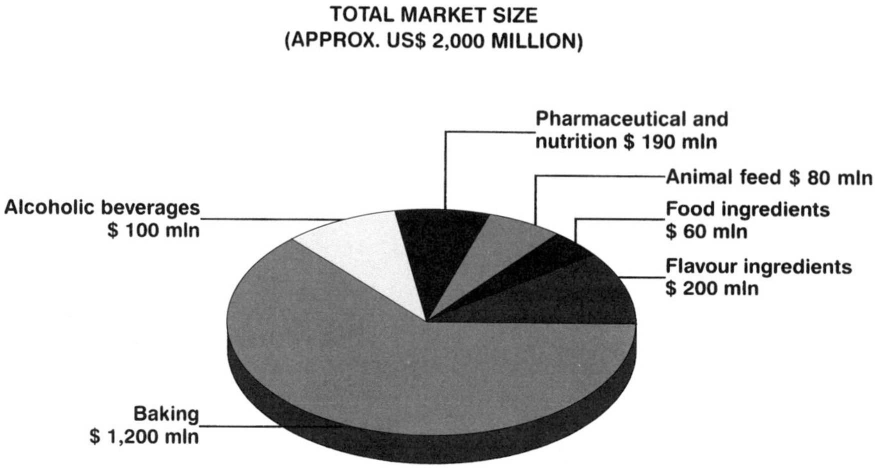

Figure 16.1 The current market for yeast and its derivatives.

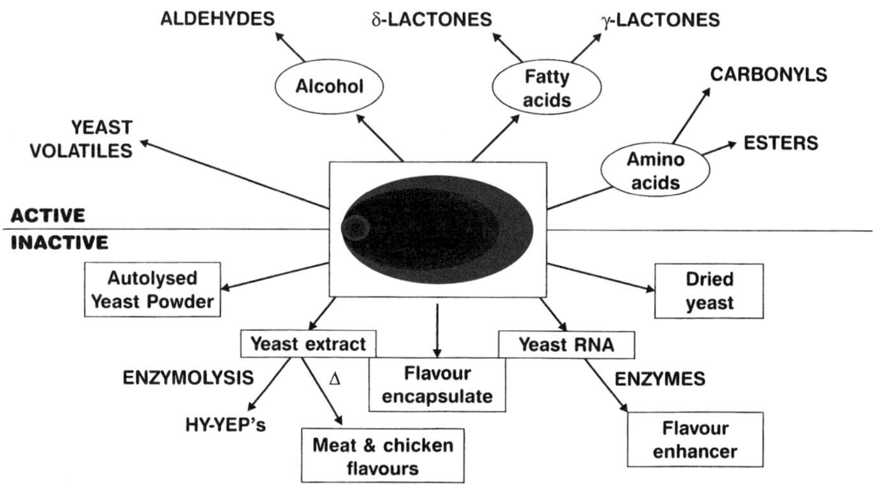

Figure 16.2 Applications of yeast in the flavour industry.

16.2 Active yeast-derived flavours

In many food systems live yeast is used to perform multiple tasks. Well-known examples include the production of (alcoholic) beverages such as wine, beer, spirits and liquor, the production of a whole variety of bakery products such as bread, crackers and biscuits, as well as the manufacturing of fermented food products such as soy sauce, ketjap, kefir, etc. In all these processes yeast contributes to the flavour of the final product by synthesis of a variety of chemical compounds. Some typical examples of the most important yeast-derived flavours are presented in Figure 16.3.

16.2.1 Fermented beverages

In the case of fermented beverages such as beer, wine and distilled beverages, the prime tasks of the yeast, in most cases *Saccharomyces cerevisiae*, is to produce alcohol and sensory active substances which contribute to the taste, aroma and mouthfeel of the final product. Depending on fermentation conditions and the raw materials used, the yeast is able to convert precursor molecules present in the fermentation broth into a complex mixture of flavouring substances. Besides transformation of raw material-derived precursor molecules, *de novo* synthesis of substances contributing to taste and aroma also takes place in the yeast's primary and secondary metabolism. Some major flavour compounds and their corresponding precursors are shown in Table 16.1. A list of the typical levels of some of the flavour components of beer, together with their flavour thresholds is listed by Moll (1991).

In the later stages of maturing processes, in particular in wine, also components released during the autolysis of yeast contribute to the final flavour (see Figure 16.3). These yeast-derived flavours contribute, together with the flavours already present in the raw materials, to a large extent, the final beer aroma. For example, in standard lager beer, over 200 different flavour molecules have been identified to make up the final flavour.

Quantitatively, by far the largest component of all alcohols present in beer is ethanol. At levels ranging typically from 3–5% (v/v), ethanol is present at two to four times its flavour threshold value, and therefore makes a significant contribution towards the flavour impact of beer. Glycerol,

Table 16.1 Major flavour molecules and corresponding precursors in yeast-fermented beverages

Flavour molecules	Precursor molecules	Major source
Higher alcohols, e.g. phenylethanol, iso-amylalcohol	Amino acids e.g. phenylalanine, 2-ketoglutaric acid	Raw materials + amino acid biosynthetic pathways
Organic acids, e.g. acetic acid, succinic acid, malic acid, long-chain fatty acids	Sugars, fatty acids	Yeast primary metabolism, fatty acid biosynthetic pathways
Esters, e.g. ethylacetate	Alcohols and acetyl CoA	Yeast primary metabolism
Carbonyl compounds: diacetyl, acetaldehyde	Amino acids	Raw materials + amino acid biosynthetic pathways
Sulphur compounds: DMS, H_2S, cysteine, methionine	Amino acids, *S*-methylmethionine, dimethylsulphoxide	Raw materials, amino acid biosynthetic pathways, yeast autolysis

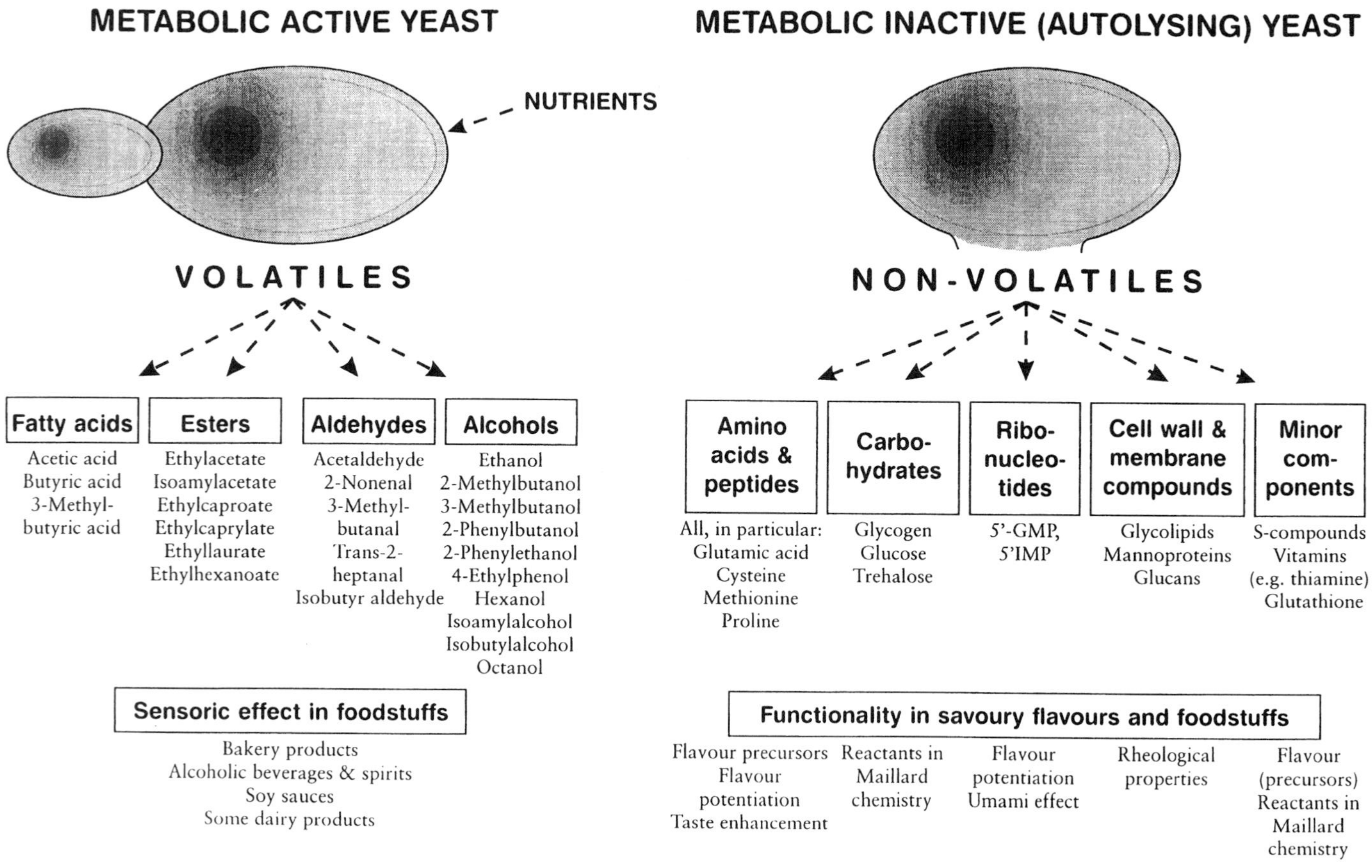

Figure 16.3 Role of yeast in flavour formation in foodstuffs.

although quantitatively important at levels of 1300–2000 mg/l, is present at concentrations below its flavour threshold value.

The group of compounds referred to as the higher, or fusel alcohols are present collectively at levels of around 100 mg/l. They are formed during protein synthesis by the yeast via trans-amination of amino acids, followed by the decarboxylation and reduction of the resultant α-keto acid. Higher alcohols are present in beer generally at levels below their flavour threshold value, although one example, iso-amyl alcohol (from leucine metabolism) can exceed its threshold level.

One of the simplest acids, carbonic acid – the product of dissolved carbon dioxide – has under normal circumstances probably the greatest influence on beer flavour. At levels between 3000–5500 mg/l, it is present at concentrations which are three to five times higher than the flavour threshold of 1000 mg/l.

Typically, the other organic acids, of importance to beer flavour, comprise a mixture of fatty acids of varying chain length. Quantitatively, acetic acid is the most abundant fatty acid found in beer, although normally it does not exceed its flavour threshold value. However, octanoic and decanoic acid – both of which give rise to a soapy flavour character, often associated with lagers – can exceed threshold levels. The lower chain length fatty acids (<C12) are produced during fermentation, whereas the higher chain length (>C12) tend to be consumed by yeast during beer fermentation. Of the other organic acids found in beer only succinic, pyruvic, malic and citric acid are of quantitative interest. Individually, however, these acids have little flavour impact in beer.

The volatile esters are an important group of flavour components as they are often present at levels above their flavour threshold. The formation of esters in beer appears to be linked to yeast lipid metabolism (Engan, 1981).

During fermentation, esters are believed to be formed primarily from the enzyme-mediated reaction between organic acids and alcohols. This is because the levels of esters found in fermenting beer are too high to be explained by the direct, enzyme-free reaction between alcohols and acids (Nordström, 1965). The latter reaction mechanism, however, may play a role in ester formation during the aging process of beer.

Not surprisingly, as ethanol and acetic acid represent quantitatively the highest levels of alcohol and acid respectively, ethyl acetate is the most abundant ester in beer. At typical levels of 20–40 mg/l, it often exceeds its flavour threshold of around 30 mg/l and contributes a fruity, solvent-like character to beer. Other esters, including ethyl hexanoate, ethyl octanoate, amyl acetate and iso-amyl acetate can also exceed their flavour thresholds and similarly contribute to the fruity flavour of beer.

Generally, aliphatic carbonyl compounds have high flavour potential, when compared with their corresponding carboxylic acid or alcohol, while unsaturated carbonyls have an even greater flavour potency. During fermentation, however, yeasts tend to prevent the accumulation of carbonyls by reducing them to alcohols (Peppard & Halsey, 1981).

Although acetaldehyde is quantitatively the most important carbonyl compound in beer, it is not normally present above its threshold level. The presence of acetaldehyde or 'green apple' flavour is often an indication of a microbial infection, either by yeast or bacteria (Rainbow, 1981).

Probably the best understood carbonyl flavour compounds associated with yeast fermentations are the vicinal diketones, which include diacetyl and, to a lesser extent, 2,3-pentan-dione (Wainwright, 1973). Diacetyl is produced by the oxidative decarboxylation of α-acetolactate and is a normal product of brewery fermentation (see Figure 16.4). With a flavour threshold of around 70–80 ppb, diacetyl gives rise to a buttery, creamy, toffee flavour. In certain beers, particularly English ales, it is considered beneficial at levels up to 200 ppb, but in blander-flavoured lagers it is regarded as an off-note. Traditionally, one of the main purposes of the secondary fermentation or lagering stage was to allow yeast sufficient time to reduce diacetyl to the less flavour-active acetoin and 2,3-butanediol. In modern breweries, because of their knowledge of the mechanism of formation of diacetyl, brewers can control the level of diacetyl in beer by a variety of ways, although low-temperature storage (lagering) is still widely practised.

Sulphur dioxide is a normal product of yeast fermentation, and levels are more dependent on fermentation conditions rather than on the yeast strain used. Although quantitatively the most important sulphur-containing compound in beer, it is seldom present in levels above its flavour threshold.

Hydrogen sulphide (H_2S) has a very low flavour threshold (5 ppb) and levels in beer are mainly dependent upon yeast strain. In some beers, it can exceed its flavour threshold level, where it imparts a rotten egg-like odour to beer.

The organic sulphur compounds are important contributors to flavour in beer and compounds of this type have relatively low flavour thresholds (<100 ppb). The flavours are typically associated with sweet corn-like flavours, e.g. dimethylsulphide (DMS) and cooked vegetable, e.g. mercaptans. In some beers they are considered beneficial and in others not. Often they can be prime indicators of the distinctive character of a given beer type, although there is relatively little evidence which relates the organic sulphur profile of a beer to a particular yeast strain and/or fermentation condition.

During fermentation undesirable flavours, so-called 'off-notes', may be formed by the yeast, or may also be removed (upgrading effect). Well known in this respect is the unbalanced formation of diacetyl and other carbonyl compounds in beers and sulphur components (e.g. DMS, H_2S and mercaptans) in wines. Off-notes may also be carried over as such from poor raw materials or from infecting microorganisms.

Certain carbonyls, particularly unsaturated aldehydes of C6–C12 chainlength, are very flavour-active, with typical threshold levels of around 0.05–1 ppb. They are associated with the formation of a stale, papery off-flavour as beer is stored or aged. In the presence of live yeast, the accumulation of these stale flavour compounds to levels above their threshold values does not take place. This is primarily because brewing yeasts have a tendency to reduce aldehydes to less flavour-active alcohols (Peppard & Halsey, 1981). For the most part, however, beer is filtered to remove yeast and other solids to give a bright product. In the absence of yeast, very low levels of dissolved oxygen can ultimately give rise to the formation of stale, carbonyl off-flavour compounds during storage. Trans-2-nonenal, with a flavour threshold of around 0.1 ppb, is a well-researched example of a carbonyl compound associated with stale off-flavour formation in beer (Drost *et al.*, 1990).

The production of certain volatile phenols, particularly 4-vinylguaiacol (4VG, flavour threshold 300 ppb) from ferulic acid, is regarded as a serious off-flavour problem in most breweries. The formation of 4VG is usually the result of an infection by a strain of wild yeast, also termed phenolic off-flavour (pof) yeasts (McMurrough *et al.*, 1996). These yeasts possess the gene which codes for ferulic acid decarboxylase and can therefore, unlike most brewing yeasts, produce 4VG from ferulic acid. In some beers, however, notably wheat (Weizenbier) and Lambic beers, the formation of 4VG by pof yeasts is an important character impact component of the flavour profile. Table 16.2 gives on overview of the most important yeast-derived beer off-flavours known to-date (Baigrie, 1994).

In conclusion, alcoholic beverages are a complex mixture of compounds, many of which are formed as a result of yeast and the fermentation process. In general, ethanol, carbonation, some esters and organic sulphur compounds are ubiquitous flavour components of all alcoholic beverages, especially beers which are derived from yeast and/or fermentation.

(a) Metabolic pathway engineering and flavour formation. Recent attempts to increase flavour profiles of beers, or to decrease off-flavour

Table 16.2 The main off-flavours found in beer

Flavouring compound	Flavour description	Cause
Fatty acids (C6–C12)	Caprylic, fatty	Raw materials
3-Methyl-2-butene-1-thiol	Sunstruck	Light
Diacetyl	Buttery	Brewing process
Ferulic acid derivatives	Phenolic	Contaminating bacteria
Dimethyl sulphide (DMS)	Sulphury	Brewing process
DTMS	Onion	Brewing process
1-Naphtol and derivatives	Stale	Energy
Aldehydes, sulphur, oxidation products		

formation comprise the usage of genetic engineering (reviewed by Hammond, 1995). One of the first successful examples is the construction of brewer's yeast with a reduced capacity to form diacetyl. Diacetyl formation occurs from α-acetolactate, an intermediate of the valine biosynthetic pathway (Figure 16.4). The various genes of the valine biosynthetic pathway (*ILV* genes) have been isolated. Using antisense techniques with a part of the *ILV2* gene, Vakeria *et al.* (1991) were able to reduce the level of α-acetolactate mRNA and hence the amount of enzyme formed. Diacetyl formation was indeed decreased, but the resulting yeast appeared not very effective in wort fermentation.

Another approach was to increase the metabolic flux to valine by constructing yeast strains containing multiple copies of the *ILV5* or *ILV3* genes. Multicopy transformants containing the *ILV5* gene produced less diacetyl due to an increased acetolactate–reductoisomerase activity (Villanueba *et al.*, 1990; Goossens *et al.*, 1993). Multiple copies of the *ILV3* gene resulted in an increased corresponding enzyme activity, though without any influence on diacetyl production (Goossens *et al.*, 1987). Obviously, the conversion of α-acetolactate to dihydroxyisovalerate catalysed by the

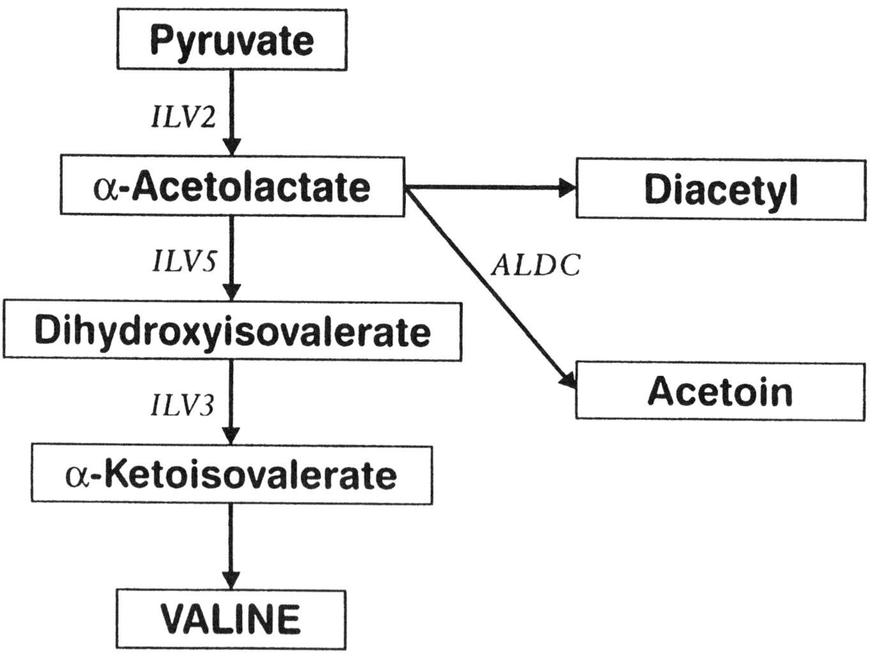

Figure 16.4 Valine biosynthetic pathway in *Saccharomyces cerevisiae* and site of activity of α-acteolactate decarboxylase. *ILV2*, acetolactate synthase; *ILV3*, dihydroxyacid dehydratase; *ILV5*, reductoisomerase; *ALDC*, alphaacetolactate decarboxylase. (From Hammond, 1995.)

ILV5 gene product is a reaction with a high metabolic flux control coefficient (Fell & Thomas, 1995) in the valine biosynthetic pathway.

Sulphur metabolism is very important in the formation of flavour and off-flavours in beer (Figure 16.5). Like diacetyl, H_2S is a molecule which causes an off-flavour and has to be removed during maturation of beers. Genes have been isolated which play a role in the conversion of H_2S to e.g. cystathionine (Figure 16.5). Overexpression of two of these genes, *NHS5* and *MET25*, in brewer's yeast resulted in a decreased H_2S formation in pilot scale fermentations (Tezuka *et al.*, 1992).

Sulphur dioxide (SO_2), however, is a compound which is important as an antioxidant and in sequestering carbonyl compounds which act as stale flavours (see section 16.2.1). Various metabolic pathway engineering studies have been reported in order to increase the flux to a concentration of SO_2, which is still below its threshold level. Overexpression of the *MET3* and *MET14* genes resulted in higher levels of SO_2 (Korch *et al.*, 1991). Disruption of the *MET10* gene resulted both in a reduced level of H_2S in combinantion with an increased SO_2 production.

Iso-amyl acetate is produced from iso-amyl alcohol during the production of sake and is an important flavour of sake. The production of iso-amyl

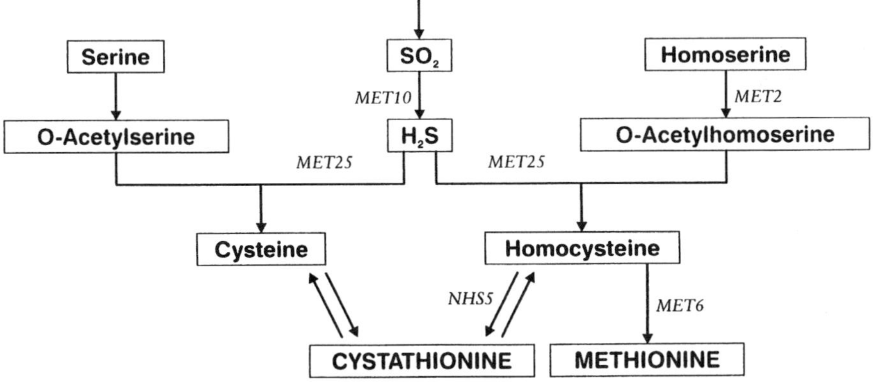

Figure 16.5 Genes of the sulphate metabolic pathway in *Saccharomyces cerevisiae*. *MET2*, homserine acetyltransferase; *MET3*, ATP sulphurylase; *MET6*, homocysteine methyltransferase; *MET10*, sulphite reductase; *MET14*, APS kinase; *MET25*, O-acetylhomoserine, O-acetyl sulphydrase; *NHS5*, cysthathionine beta synthase. (From Hammond, 1995.)

acohol is connected to the leucine biosynthetic pathway. Overexpression of the *LEU4* gene resulted in increased iso-amyl alcohol and iso-amyl acetate production, leading to sake with an increased flavour content (Hirata *et al.*, 1992).

Recently, the gene for alcohol acetyltransferase, responsible for the formation of acetyl esters from acetylCoA and alcohols, has been isolated and used to construct multicopy transformants. The transformants showed indeed an increased production of the desirable acetate esters (Minetoki *et al.*, 1993).

From these examples it is clear that metabolic pathway engineering can be used in order to increase flavour formation, or decrease off-flavour production. Although profound knowledge on biochemistry, regulation of biosynthetic pathways and metabolic flux analysis is needed in order to clone or disrupt the right genes, it is not the limiting factor for the usage of genetically engineered yeast strains in the brewing process. Until now, only pilot plant fermentations have been carried out using genetically engineered yeast strains. Legislation requires that the commercial usage of these strains will result in products which fall in the category 'Novel Foods'. Apart from legislation, public acceptance might be another problem. According to Hammond (1995), it is important that the consumers are being informed about the benefits and possibilities of modern biotechnology in order to form a profound opinion on these issues.

16.2.2 Baking

In case of bakery products yeasts are mainly selected for 'gassing power', *i.e.* their capacity to generate CO_2, which to a large extent determines the textural properties of the baked product. In most cases, *Saccharomyces cerevisiae* is used for this purpose. However, in some bread types other yeasts are also being applied. For example, *Saccharomyces rosei* (osmophilic yeast) is used in high-sugar-containing doughs for the manufacture of sweet rolls and Danish pastries. In sourdough applications, acid-resistant yeasts, e.g. *Candida krusei* are also being used (Spicher & Stephan, 1982). Due to the increasing degree of process rationalization within the bakery industry, more and more attention is paid to selecting yeasts which are providing a flavour profile reminiscent of baked goods produced in a more traditional/artisanal way.

Baking experiments and model studies with yeast/sucrose homogenates revealed that yeast provides the most important precursor pool for enzymatically or thermally generated flavours (Schieberle, 1990a). Yeast generates high amounts of free amino acids, either via biosynthesis during metabolism or from breakdown of flour proteins, as well as monosaccharides and intermediates of the glycolytic pathway (Grosch & Schieberle, 1991; Schieberle 1996).

When yeast is added to dough, a four-fold increase in free amino acids was observed (El-Dash & Johnson, 1970). Although during fermentation the amount of free amino acids in dough is reduced significantly, the overall amount in dough after fermentation is about twice the amount originally present at the start of fermentation. Besides the effect on non-enzymatic browning, increased amino acid levels have a pronounced effect on bread flavour formation.

Bread flavour can be divided into crust and crumb flavour, which are distinctly different. This is due to different temperatures, enzyme activities and water contents in crumb and crust. The direct way of improving the flavour quality of bakery products in general is through bioconversion by the yeast of small molecules present in the raw materials or by *de novo* synthesis of flavouring substances from glucose. For example, linoleic acid present in wheat is converted by the yeast to 2-4-heptenal, which contributes to the characteristic sweet taste of bread crumb.

The crumb flavour is to a large extent caused by enzymatically catalysed reactions. Crumb flavour is usually associated with yeast primary and secondary metabolites, also interacting with metabolites from *Lactobacillus* species. As mentioned before, the levels of many amino acids decrease significantly during fermentation of dough, and aldehydes (Table 16.3) are formed as important odour active products (El-Dash & Johnson, 1970; Grosch & Schieberle, 1991).

The aldehydes, which can also be formed during the biosynthesis of amino acids (Piendl, 1969), are further reduced in the dough by yeast enzymes to the corresponding alcohols, e.g. 3-methylbutanol and 2-phenylethanol, which are among the most important flavour substances in crumb flavour (Table 16.3). The contribution of these two alcohols is almost independent of the heating process, and increases significantly, when the fermentation time of the dough is elongated (Schieberle, 1990a; Schieberle & Grosch, 1991). Another important flavour substance in bread crumb is furaneol, which is generated enzymatically by yeast in relatively high concentrations (Schieberle, 1990a).

Another approach to improve flavour quality is by selecting yeasts which are able to accumulate/generate molecules known to be effective in the formation of crust flavour, in particular in the flavour-generating Maillard reaction during baking. In the crust, many of the important aroma compounds are also derived from baker's yeast, but their formation is strongly dependent on exposure to high temperatures (Schieberle, 1990a; Grosch & Schieberle, 1991). For example, 2-acetyl-1-pyrroline is the most important flavour impact compound in the crust, and is formed only during baking (Schieberle, 1990a). Schieberle (1992a) demonstrated that dihydroxyacetone and ornithine are the main precursors of this very potent flavour substance in bread. Dihydroxyacetone phosphate is a glycolytic intermediate generated by the yeast in the dough. Other important flavour

Table 16.3 Major important flavour substances in bread and their origin

Flavour substance	Crumb	Crust	Precursor	References[c]
2-Acetyl-1-pyrroline	+[a]	+++	Ornithine, proline, citrulline, dihydroxy-acetonephosphate, glyceraldehyde-3-phosphate	1,2,3
2-Acetyl-1,4,5,6,-tetrahydropyridine	+	+++	Proline, fructose	1,2
3-Methylbutanol	+++[b]	–	Leucine	1,4,5,6
2-Phenylethanol	+++[b]	–	Phenylalanine	1,4,5,6
Methional	++	+++	Methionine	1,4,7,8
Phenylacetaldehyde	+	++	Phenylalanine	1,4,7,8
3-Methylbutanal	+	+++	Leucine	4,7,8
4-Hydroxy-2,5-dimethyl-3(2H)-furanone	++	+++	Glyceraldehyde-3-phosphate, dihydroxy-acetonephosphate, fructose-1,6-diphosphate, fructose-6-phosphate	1,9,10
Methylpropanal	+	+++	Valine	4,7,11
2,3-butanedione	++	++	Pyruvate, acetyl-CoA, monosaccharides	4,8

[a], + low, ++ moderate, +++ high contribution to flavour.
[b], High contribution to crumb flavour, when dough is long fermented or when a liquid preferment is used. In short-fermented dough, their contribution is somehow lower.
[c] References: 1. Schieberle (1990a); 2. Schieberle (1992a); 3. Schieberle (1990b); 4. Grosch & Schieberle (1991); 5. Schieberle & Grosch (1991); 6. Gassenmeier & Schieberle (1995); 7. Belitz & Grosch (1992); 8. Schieberle & Grosch (1994); 9. Schieberle (1992b); 10. Preiniger *et al.* (1994); 11. Rychlik & Grosch (1996).

compounds in bread crust are 2-acetyl-1,4,5,6-tetrahydropyridine (ACTP) and 4-hydroxy-2,5-dimethyl-3(2H)-furanone (HDF), which are formed during the Maillard reaction of proline and fructose (ACTP), respectively from fructose phosphates (HDF). In addition the potent Strecker aldehydes, methional, 3-methylbutanal and methylpropanal contribute significantly to the flavour of bread crust, and to a somewhat lower extent also to the flavour of bread crumb (Grosch & Schieberle, 1991).

Consequently, bread flavour formation can be improved by using naturally selected or recombinant yeast strains capable of providing a greater flavour precursor pool, or producing enhanced levels of flavour-active compounds. Flavour delivery to baked goods can also occur via autolysis of the yeast cells during the baking process. During this process, release of yeast cell constituents, i.e. glycolytic intermediates such as dihydroxyacetone phosphate and (free) amino acids, may have a significant contribution to the overall aroma and textural aspects as well.

Enzymes secreted by yeast which are able to form precursors for the Maillard reaction are also important for flavour formation. For example, cysteine and proline liberated by proteolysis from dough proteins can serve, like glutathione, as a precursor in thermally generated flavours. Glutathione has a dual function in baking. Besides delivering taste via Maillard reactions, it affects the redox status of the dough and hence influences the textural properties via the amount of thiol-crosslinks between the gluten network.

Since many yeast metabolites, amino acids and enzymes are involved in the formation of bread flavour (precursor) molecules, molecular biology can be used to optimize yeast for flavour formation. Until now, not many examples are known in this area, but since most of the biosynthetic pathways to these molecules are known and the techniques and knowledge of molecular biology and metabolic pathway engineering are increasing, this will be the subject of much research in the future. However, the influence of legislation and public acceptance on the commercial usage of genetically modified organisms in such a tradional 'art' as baking can not be neglected.

16.2.3 Other fermented foods

Another example of active flavour formation by yeasts is found in, among others, japanese soy sauce (Shoyu) manufacturing, in particular during the brine fermentation or 'Moromi' stage of the process (Roling, 1995). A consortium of halotolerant yeasts such as *Zygosaccharomyces rouxii*, *Candida versatilis* and *C. etchelsii* proliferates and develops a typical alcoholic fermentation, which finally results in the characteristic 'fermented' soy sauce flavour consisting of well over 100 compounds. Important flavour molecules produced here are 4-ethylguaiacol, 4-ethylphenol and 2-phenylethanol (see Figure 16.3).

In kefir, a fermented milk product, two yeast species, *Saccharomyces kefir* and *Candida kefir*, in symbiosis with lactic acid bacteria, produce acetaldehyde, diacetyl and acetone resulting in a mildly alcoholic, yeasty sour flavour profile (Oberman, 1982). Recently, new savoury products (Aromild) with strong flavour-potentiating properties (high 5′-IMP and 5′-GMP) was put into the market by Kohjin Co. Ltd. This new product is based on the combination of soy sauce and yeast extract derived from 'high RNA' *Candida* yeast.

In vegetable fermentations, e.g. cucumbers, many yeast species are found (Vaughn, 1982), but their role in flavour formation is yet unclear.

16.3 Inactive yeast-derived flavours

Inactive yeast-derived flavours constitutes those flavour preparations which are sold in an inactive form, typically in a paste or powder form. Their main purpose is to deliver taste to a final food product. Active yeast-derived flavouring products include, as described in section 16.2, live yeast which is typically sold as a paste and for certain applications as a powder. Such products are usually not sold for their flavouring purposes but for their capacity to generate CO_2 or ethanol. However, this situation is bound to change.

Todate, many inactive yeast-derived flavouring preparations are on the market. These range from a wide variety of yeast extracts, dried yeast preparations to yeast-derived volatiles, aroma chemicals and flavour delivery systems. Recently, a novel range of products, 'HyBake' (manufactured by Quest International) has been added to this large portfolio, i.e. a dry fermented food ingredient where, among others, yeast is used to deliver taste, texture and process tolerance to a final food product. The world market for inactive yeast-derived flavours currently amounts to about \$260M, but is rapidly expanding (see Figure 16.1). Figure 16.6 gives an overview of the application areas for each type of yeast-derived product.

16.3.1 Yeast extracts

(a) Yeast extract market. Yeast extracts form an important component in savoury flavours. The world market for savoury ingredients, taste enhancers, secondary flavourings and nutrients from yeast etc. is estimated at \$1000M. Several market parties estimate the annual growth of this segment at 10–12%. This growth is mainly caused by the growth in Latin America and Asia and the growth of ready meals and savoury snacks, as well as by the shift towards more natural, healthy taste enhancers in Europe and Northern America. Another growing area for yeast extracts is the non-food

	CULINARY	DAIRY	BEVERAGES	BAKERY	SNACKS
Yeast extracts	X X			X	X X
Flavour enhancers	X X		X		X X
Dried yeast	X				X
Yeast volatiles			X X		
Aroma chemicals	X	X X	X X		
Autolysed yeast	X			X	X

Figure 16.6 Main application areas for inactive yeast-derived flavours.

sector. Key sectors include animal feed, aquaculture, fermentation substrates and cosmetics (Gordon, 1991). The total market volume for yeast extracts is estimated at 47 000 tons in 1995, of which more than half is consumed in Europe (Table 16.4).

The usage of yeast extracts as food ingredients started at the beginning of this century when spent brewer's yeast in England and Germany became available as raw material for the development of yeast extracts. This surplus yeast is relatively inexpensive and is often used to produce valuable, mainly savoury ingredients. Primary grown yeast, i.e. yeast grown specifically for use in the food industry (baker's yeast and yeast extract) are relatively more expensive, but can add more value to the final product.

Yeast extracts are mainly used as feed stock in the fermentation industry and as flavourings in the food industry. In fermentation, i.e microbial nutrition, yeast extract is mainly used as a versatile source of peptides, free amino acids and vitamins, and is particularly useful to support the growth of fastidious microorganisms for starter culture production and manufacturing of diagnostics.

In the food industry, yeast extract is applied mainly in the savoury area in a variety of applications such as soups and bouillons, snacks, noodle seasonings, bakery products, etc. In the beverage industry there are applications in beer and distilleries. On a global basis yeast extracts forms a business worth in excess of $250 M.

Yeast extract as savoury ingredient can offer cheesy, meaty or chicken-like notes in for example soups, gravy sauces and vegetarian products where a meaty-type savoury flavour is desired. Other well-known products responsible for savoury characteristics are soy sauce and Hydrolysed Vegetable Proteins (HVP). Soy sauce fermentation is a classical example of food fermentation and has been produced for thousands of years, especially in the far east. HVP is a product obtained after acid hydrolysis of soy beans, wheat or other plant substrates. In the past few years there has been some concern about the use of HVP (Thornton 1996). Hydrolysis of plant protein in the presence of HCl might result in the formation of chloropropanols,

Table 16.4 Market size and expectations of savoury ingredients. Ingredient consumption (1995) (×1000 tons)

	Europe	North America	Asia	Rest of World	Total
HVP	36	27	25	10	98
Meat and bone extracts	4	5	30	2	41
MSG	45	45	700	30	820
Nucleotides	1	1	2	1	5
Yeast extract	26	11	7	3	47
Soy sauce	3	5	4200	1	4209

HVP, hydrolysed vegetable protein; MSG, monosodium glutamate.

which are known to be carcinogenic compounds. Alternatives for HVP are protein hydrolysates obtained by enzymatic treatment, yeast extracts or a combination of these.

(b) Yeast extract production. As described above, yeast specialties such as yeast autolysates and extracts are usually produced from spent brewer's and baker's yeast (both *Saccharomyces cerevisiae*), although other yeast species such as *Candida utilis* (previously designated as *Torula* yeast), *C. marxianus* and *Kluyveromyces lactis* are being grown for the same purpose. Brewer's yeast (*S. cerevisiae*) is the spent (exhausted) yeast released from wort fermentation in the beer manufacturing process. Since brewer's yeast is a side product, its production is not a dedicated process as in baker's yeast production, and the quality of the yeast is largely dictated by the raw materials used and requirements of the final beer product. Typically, brewer's yeast is recycled several times before it is available to the food industry or discarded as animal feed.

Unlike brewer's yeast, baker's yeast is produced as a main rather than a by-product (so-called primary-grown yeast). Thus, baker's yeast is grown in a highly aerobic fed-batch fermentation process, usually on a standardized (50% TIS = Total Inverted Sugar) blend of cane and beet molasses, in six subsequent stages (D1/D6, Figure 16.7) to a final yeast biomass concentration of about 50–70 kg/m^3. Although fed-batch cultivation is the preferred

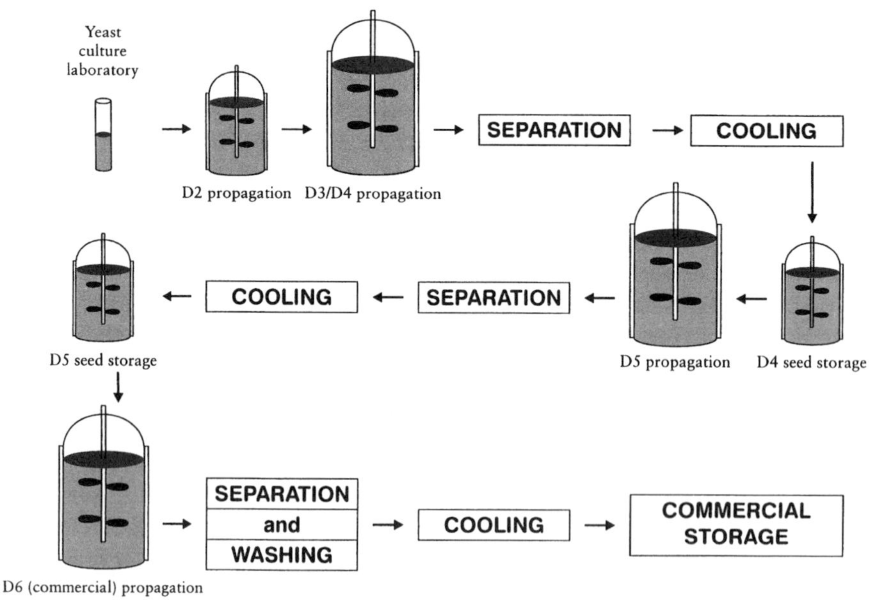

Figure 16.7 Industrial yeast propagation.

method, continuous propagation systems have recently been reported and may be used in the future (Voigt & Walla, 1995).

After final propagation in D6, which usually takes about 13–16 hours, the yeast is separated off and washed simultaneously by counter-current centrifugation and collected as a yeast cream having a biomass concentration of approximated 190 kg/m^3. Yields obtained on molasses are typically in the order of 0.38–0.45 g yeast dry matter (Ydm)/g TIS dependent on the quality of the molasses blend and efficiency of the yeast propagation process. Efficiency of yeast propagation is largely determined by maintaining respiratory growth during propagation as much as possible. Subsequently, the yeast cream is stored at 4°C for a short period until it is further processed into 'actives' (i.e. pressed or instant dry yeast for bakery and some other applications (distillery, wine)), or 'inactives' (i.e. yeast autolysate or yeast extract paste or powder for use in the food and fermentation industry).

Yeast autolysate and yeast extract production for food applications is initiated by quickly increasing the temperature to 40–60°C, the (optional) addition of salt or other plasmolysing agents and with the addition of hydrolytic enzymes (i.e. proteases, nucleases).

During yeast hydrolysis, intracellular yeast enzymes and added enzymes (proteases, nucleases) are acting together, resulting in the fast release of soluble and colloidal matter consisting mainly of proteins, peptides, carbohydrates and nucleic and cell wall material, which are subsequently hydrolysed into respectively free amino acids, peptides, sugars and nucleic acids. Recently, temperature sensitive autolysing yeast strains have been selected after mutagenesis and responsible genes have been isolated. These strains showed increased autolysis at 37°C (Asahi Breweries, 1995). An important process parameter in autolysis is the ratio of alpha-amino nitrogen to total nitrogen (AN/TN) and/or the ratio of free amino acids to total amino acids (FAA/TAA), both in combination with the degree of solubilization which is usually indicated by a Brix value of the cleared extract. At the end of the hydrolysis phase, cell debris (mainly cell walls and other insoluble matter) is removed by centrifugation and the remaining liquid fraction is further processed into yeast extract powder or paste (>70% dry matter), involving process steps such as evaporation, (ultra)filtration and spray drying. In the autolysate process there is no separation of insoluble matter and the whole matrix including cell walls and membranes is (roller) dried (Figure 16.8).

As will be described in later, the desirable characteristics of yeast to be used as starting material for yeast extracts differ considerably from those required for living yeast cells. In the latter case, the primary function of the active yeast cell is either ethanol formation (in brewing and distilling), or CO_2 production causing a leavening of the dough before baking. Moreover, it is desirable that the active yeast cells have a prolonged shelf life which is

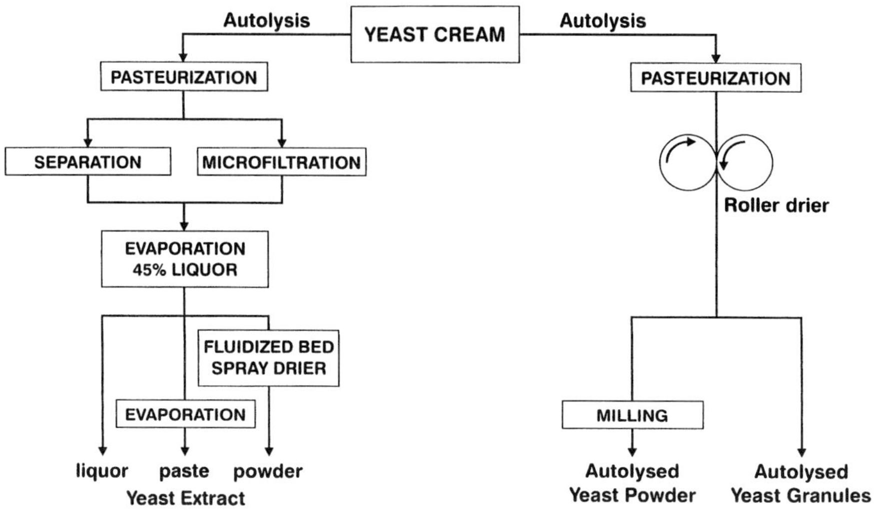

Figure 16.8 Industrial yeast extract production.

stimulated by a strong cell wall and the presence of a storage compound, e.g. trehalose. In the case of yeast extract, the specific molecular composition of the cell determines the quality of the derived flavour characteristics. Besides the yield on substrate after propagation, the yield after autolysis is an important parameter, which implies that cells should be in such a physiological state after propagation, that the autolysis process can proceed effectively. To that end, thin cell walls might be desired. It is obvious that in the ideal situation the conditions for yeast propagation should be adapted with respect to the final application of the biomass produced. Since the capacity of CO_2 production and ethanol formation is not needed for the generation of yeast extracts, the use of food-grade, Crabtree-negative yeast strains could be an advantage. In Crabtree-negative strains ethanol formation is absent, or limited, resulting in an increased yield on substrate. Alternative strains broaden the range of substrates that can be used (Vij & Gandhi, 1993). Some growth parameters of interesting yeast strains are presented in Table 16.5.

(c) Yeast extracts flavours and functionality. A few reports have been published about the molecular composition of yeast extracts and the volatile components of yeast autolysates. The first investigations describe the flavour compounds in various fractions of autolysed yeast powder and a total of 48 compounds was identified comprising various carboxylic acids, benzoids, furanoids, thiophenoids, pyrazines, pyrolle and thiazole (Davidek *et al.*, 1979; Hasjlova *et al.*, 1980). In another study, 37 sulphur-containing components were identified in a bakery-autolysed yeast extract which was

Table 16.5 Some growth parameters of different yeasts

Parameter	Brewer's yeast	Baker's yeast	Lactic yeast	Fodder yeast
Yeast strain	*S. cerevisiae*	*S. cerevisiae*	*Kluyveromyces* sp.	*C. utilis*
Metabolism	Fermentative	Respiratory	Respiratory	Respiratory (Crabtree-negative)
Raw materials	Barley malt, wheat	Beet- and cane molasses	Whey, deproteinized milk	Sulphite spent liquor, ethanol, alkanes
Growth temperature	Lager: 12–15°C Ale: 15–22°C	30–35°C	30–40°C	30–40°C
pH	5.5	4.4–5.5 controlled	4.5–6.5 controlled	3.5–5.5
Time	Lager: 100–120 h Ale: 72–140 h	13–16 h	20–30 h	20–40 h
Biomass (start)	3 kg/m^3	10–20 kg/m^3	10–20 kg/m^3	10–20 kg/m^3
Biomass (end)	15–18 kg/m^3	50–60 kg/m^3	60–100 kg/m^3	60–150 kg/m^3

processed together with a sugar compound in a Maillard type reaction (Golovnya *et al.*, 1983). Analysis of the volatile components of a yeast extract, comprising also spices and vegetable extracts, by GC and GC/MS resulted in the identification of 125 compounds. It was concluded that these compounds were derived from sugar/amino acid interaction, lipid degradation and from thermally degraded thiamine, all of which were probably derived from the yeast metabolism (Ames & Macleod, 1985). Werkhoff *et al.* (1991) identified 115 sulphur-containing flavour compounds from yeast extracts after distillation/extraction and concluded that these components are responsible for the main impact of flavour. The compounds identified were aliphatic sulphur-containing components, hemidithioacetals, sulphur-substituted furans, thiophenes, thiazoles, five- and six-membered ring polysulphides, C3–C12-alkyl-substituted perhydro-1,3,5-dithiazines (thialdines), bicyclic perhydro-1,3,5-dithiazines, 1,4,6, trimethylperhydro-1,3,5-oxathiazine and trimethylperhydro-1,3,5-thiadiazine. A few newly identified furan components in yeast extract are presented in Figure 16.9 (Werkhoff *et al.*, 1991).

Flavour dilution analysis, a technique to identify the odour significance of aroma components, showed that in unheated yeast extract the most significant compounds were butanoic acid, methylpropanoic acid, 2- and 3-methylbutanoic acid, 2,5-dimethyl-4-hydroxy3(2)furanone, 4-vinyl-2-methoxyphenol and phenylacetaldehyde. After heating, 2-acetyl-1-pyrroline, methional, γ-decalactone and γ-dodecalactone appeared (Schieberle, 1990a).

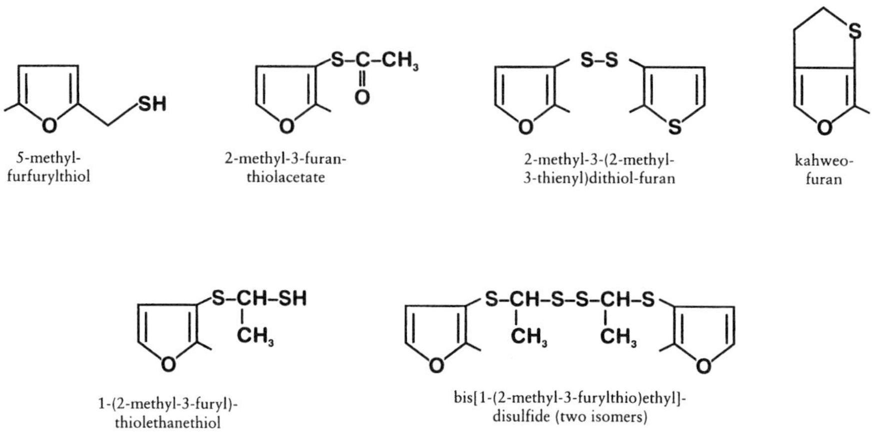

Figure 16.9 Some newly identified furan components in yeast extracts. (From Werhoff *et al.* (1991).)

Finally, Ames & Elmore (1992) compared three different yeast extracts with different aroma properties and identified 200 components, including 47 sulphur compounds. Most of these were derived form sugar and/or amino acid interactions, or thiamine degradation. Differences in manufacturing these yeast extracts became manifest in different compositions.

It is obvious that the combination of many compounds is responsible for the flavour character of yeast extracts. Especially sulphur-containing molecules are responsible for the meaty character. In this respect, gluthathione may also act as an important source of cysteine in process flavour reactions (Wierzbicka *et al.*, 1989). Many attempts have been made to select or genetically engineer *Saccharomyces cerevisiae* to contain more than the average concentration of 0.5–2% on dry weight of glutathione (Hamada *et al.*, 1986; Lang-Hinrichs & Stahl, 1988; Ohtake *et al.*, 1988, 1989; Shimizu *et al.*, 1991; Alfafara *et al.*, 1992; Sakato & Tanaka, 1992; Udeh & Achremowicz, 1994). By genetic engineering, sometimes in combination with downstream processing, concentrations of 3–8% glutathione have been reached (Hamada *et al.*, 1986; Yeast extract YH provided by Kohjin Co, Ltd). In general, levels of glutathione are increased by overexpressing the two enzymes involved in the biosynthesis of glutathione (i.e. γ-glutamylcysteine synthetase and glutathione synthetase), or inhibiting the enzyme involved in the breakdown of glutathione (i.e. γ-glutamyl transpeptidase; see Figure 16.10).

Besides genetic engineering, advanced control systems (e.g. a fuzzy control of ethanol production) can be used to adjust the feeding rate during propagation resulting in optimal concentrations of glutathione (Shimizu *et al.*, 1991, 1992). Cysteine and ammonium slphate were found to increase

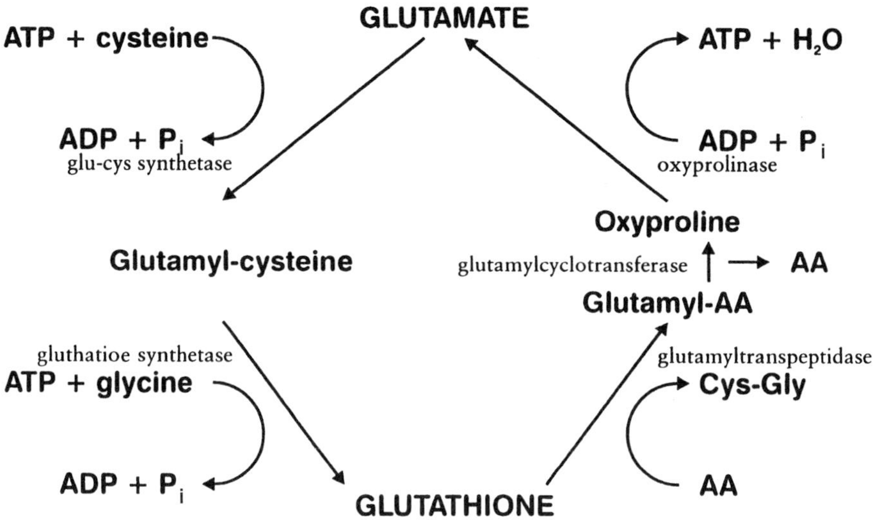

Figure 16.10 The glutathione cycle in yeast.

glutatahione biosynthesis in the yeast. L-glutamic acid and glycine had little influence on glutathione production (Udeh & Achremowicz, 1994).

It is also clear that many of these compounds are formed when a heating step is involved during/after the yeast extract manufacturing, due to the fact that Maillard-type reactions take place between the metabolites synthesized by the yeast cell. More specifically, sugar components, amino acids, sulphur compounds and thiamine are important precursors for savoury flavour character impact compounds in yeast extracts. From a flavourist point of view, however, the relationship between molecular compostion of yeast extract and its functionality in savoury applications (e.g. in soups, meat products and snacks) is often difficult to assess. Thus, only through a very systematic method of classifying sensory properties using a range of specific flavour descriptors, is it possible to 'map' the yeast extract flavour profile and divide the descriptors into categories, each category having its specific application area. A mapping profile and their preferred application areas for a range of commercial yeast extracts (Quest International) are presented in respectively Figure 16.11 and Table 16.6. Typical compositions of yeast extract products are presented in Table 16.7.

The amino acids act as precursors in a Maillard-type of reaction, or have as such a certain, mainly bitter, taste (Table 16.8: Nagodawithana, 1995).

Despite this knowledge, very few reports have appeared in the literature describing attempts to increase the level of these compounds in yeast during propagation, or in yeast extracts during autolysis. Some Japanese authors describe the isolation of *Kluyveromyces lactis* strains enriched in L-

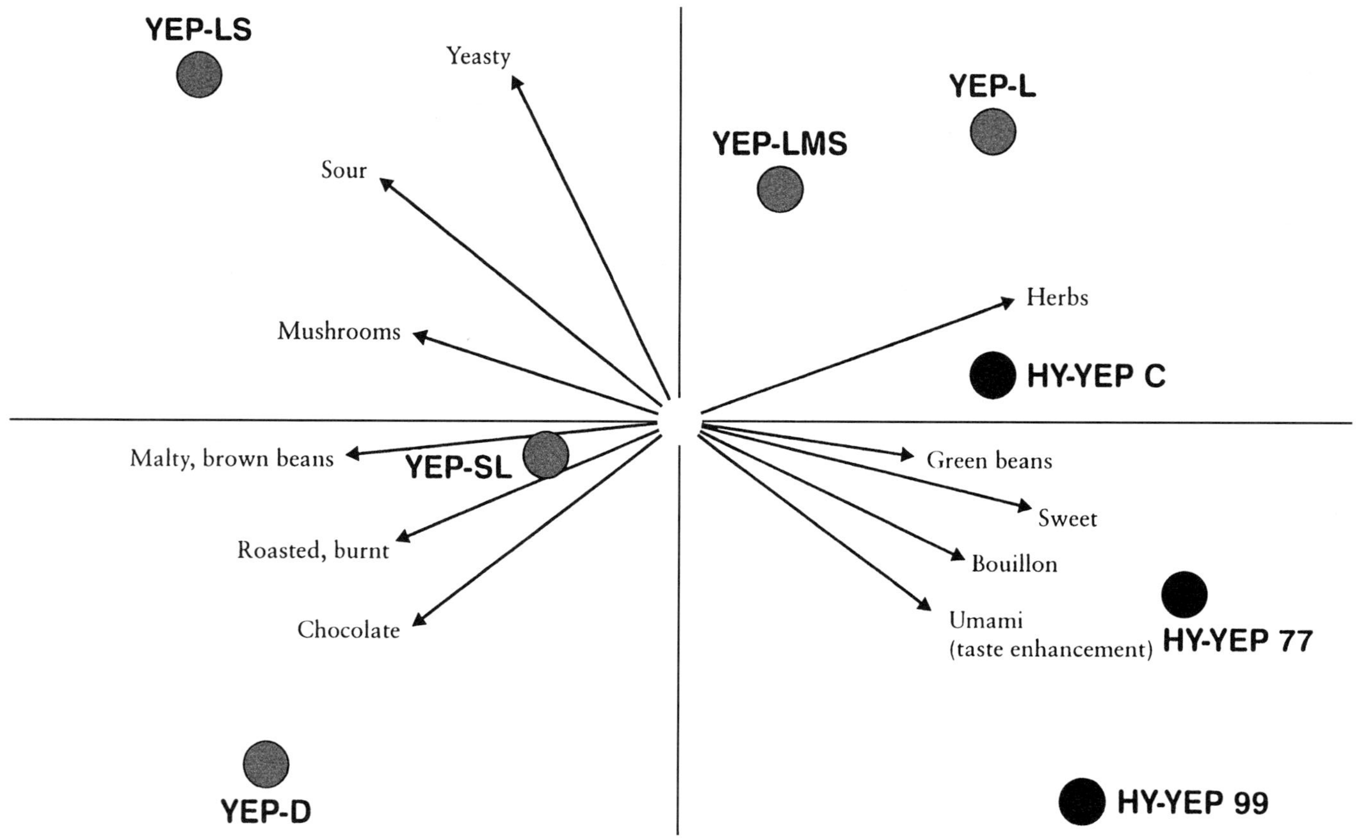

Figure 16.11 Mapping profile of commercial yeast extracts. YEP and HY-YEP represent classes of yeast extract products.

Table 16.6 Yeast speciality range

All yeast specialties are kosher-certified

	Product	Taste	Soups dry		Soups wet		Sauces		Bouillon cubes		Meat products		Prepared meals	Snacks – dust-on	Snacks – slurry	Snacks – pre-extrusion	Snacks – baked goods
			0.2–0.3%	0.3–0.8%	0.3–1.0%	0.3–0.6%	0.5–1.0%	0.6–1.2%	10–30%	15–35%	1.0–1.5%	1.5–2.0%	0.5–1.0%	1.5–2.0%	1.0–1.5%	1.5–2.0%	1.0–1.5%	1.0–2.0%
1	**AYG**	Slightly toasted, yeasty, savoury	●			●			●		●		●		●	●	●	●
	AYG-LS		●			●			●		●		●		●	●	●	●
	AYP		●			●			●		●		●		●	●	●	●
	AYP-LS		●			●			●		●		●		●	●	●	●
	AYP-X2	Roasted, meaty, savoury	●			●			●		●		●		●	●	●	●
2	**YEP S Paste**	Toasted, meaty, sweet			●			●		●		●		●				
	YEP LS Paste	Dark, savoury, slightly meaty			●			●		●		●		●				
	YEP 80 Paste	Toasted, meaty, sweet			●			●		●		●		●				
	YEP SL 73 Paste	Mild bouillon, savoury			●			●		●		●		●				
	YEP L Powder	Light, meaty, brothy		●			●		●		●		●		●	●	●	●
	YEP D Powder	Dark, meaty, chocolate		●			●		●		●		●		●	●	●	●
	YEP LS Powder	Dark, roasted, low sodium		●			●		●		●		●		●	●	●	●
	YEP LMS Powder	Light, mild brothy, reduced sodium, clear soluble		●			●		●		●		●		●	●	●	●
3	**Hy-YEP C Powder**	Bouillon, mild umami		●			●		●		●		●		●	●	●	●
	Hy-YEP 77 Paste	Bouillon, sweet, strong umami		●			●		●		●		●		●	●	●	●
	Hy-YEP 77 Powder	Strong umami, clean		●			●		●		●		●		●	●	●	●
	Hy-YEP 99 Powder	Extra strong umami, low sodium	●			●			●		●		●		●	●	●	●

1 Autolysed yeast powder/granules (AYP/AYG).
2 YEP yeast extracts.
3 Hy-YEP taste-enhancing extracts.

Table 16.7 Composition of a typical yeast extract

Main components (g/100 g)	
Protein	67.9
Fat	0.1
Carbohydrates	17.6
Total sugar	4.3
Dietary fibre	4.14
Moisture	4.00
Ash	8.37
Vitamins (mg/100 g)	
Riboflavin	5.61
Thiamine	7.36
Niacin	7.68
Pantothenic acid	21.9
Biotine	0.05
Pyridoxine	2.97
Folic acid	2.81
Vitamin B_{12}	<0.001
Minerals (mg/100 g)	
Aluminium	5.96
Barium	0.2
Calcium	29.0
Chromium	0.2
Copper	0.12
Iron	3.33
Magnesium	40.2
Manganese	0.15
Phosphorus	1006.0
Potassium	3520.0
Sodium	815.0
Zinc	17.4
Amino acids (mg/g)	
Alanine	43.5
Arginine	32.0
Aspartic acid	64.1
Glutamic acid	106.6
Glycine	28.3
Histidine	13.7
Isoleucine	30.1
Cystine	6.8
Leucine	43.7
Lysine	49.8
Methionine	9.3
Phenylalanine	26.4
Proline	22.3
Threonine	29.6
Serine	28.5
Tryptophan	6.38
Tyrosine	22.9
Valine	33.8

From Nagodawithana (1995).

Table 16.8 Flavour characteristics of amino acids

Taste	Amino acid
Bitter (intense)	Tryptophan, phenylalanine, arginine, isoleucine
Bitter (less intense)	Leucine, histidine[a], valine, ornithine, methionine, lysine[a]
Sweet (intense)	Hydroxyproline, proline, alanine[a]
Sweet (less intense)	Lysine[a], valine, alanine[a], threonine, serine[a], glycine, glutamic acid[a]
Sour (intense)	Aspartic acid, glutamic acid[a], asparagine
Sour (less intense)	Histidine[a], serine

[a] Concentration-dependent.
From Nagodawithana (1995).

methionine after growth on whey (Kitamoto & Nakahara, 1994). This work was done with the intention of increasing the level of sulphur-containing amino acids in single-cell protein production. The level of specific metabolites could be increased using different propagation conditions, alternative substrates or strains, classical mutagenesis, or alternative processing. On the other hand, molecular biology has evolved to such an extent that genetic engineering of yeast, in order to increase fluxes to specific metabolites by changing regulatory mechanisms, is becoming a reality. Cloning of genes involved in glutathione metabolism has been described earlier. The entire sequence of the yeast genome has recently become fully available, opening up the possibilities of profound genetic and physiological studies followed by metabolic reprogramming.

However, another new science area called Metabolic Control Analysis (MCA) showed that the old dogma of a single rate-limiting step in metabolic pathways should be abandoned, and that the flux control over a specific pathway is shared by many enzymes, limiting the possibility of an easy enhancement in fluxes by the overexpression of one or two genes.

(d) Yeast extract and flavour enhancement. Apart from flavour and flavour precursor molecules, yeast extracts can be rich in components which enhance the overall flavour impact. These components have no or little aroma attributes by themselves, but are able to enhance the flavour contribution of other molecules, resulting in a yeast extract with improved flavour or mouthfeel. Well-known flavour enhancers are succinic acid, monosodium glutamic acid (MSG), disodium-5′-inositate and disodium-5′-guanylate. These compounds accentuate meaty flavours and find ample applications in soups and bouillons (Nagodawithana, 1992). Disodium-5′-inositate and disodium-5′-guanylate can be obtained from RNA via the enzymic hydrolysis of 5′-inosine monophosphate (5′-IMP) and 5′-guanosine monophosphate (5′-GMP). Yeasts can be rich in RNA, e.g. *Candida utilis* might contain 10–15% RNA and baker's yeast 8–11% (Schoenberg, 1993).

Cells growing at a high specific growth rate require a lot of RNA for protein synthesis and hence the level of RNA under these conditions is increased. At the end of industrial propagations the specific growth rate normally drops rapidly, causing a decrease in RNA content. This again stresses the importance of adapting cultivation conditions to the desired characteristics of the final product. Phosphodiesterase degrades RNA into the 5′-nucleotides. Addition of this enzyme from fungal (*Penicillium citrinum*) or plant origin (malt rootlets, cereal germs) to yeasts enriched in RNA will yield 5′-GMP, 5′-UMP, 5′-AMP and 5′-CMP during autolysis. Only 5′-GMP has flavour-enhancing properties, but 5′-AMP can be converted to the flavour enhancer 5′-IMP by adenylic deaminase. Various yeast companies such as Quest International, Red Star, Ohly and Gist Brocades have introduced yeast extracts containing 1.5–6% of total 5′-GMP and 5′-IMP (Nagodawithana, 1992). Low levels of glutamate perform a synergistic effect with 5′-IMP and 5′-GMP.

MSG has a suspicious label due to the possibility of the 'Chinese restaurant syndrome' (reviewed by LaBell, 1993). However, the Food and Drug Administration in the USA (FDA) considers MSG as a safe substance. Although it is doubtful whether MSG really causes allergenic reactions, there are no reports known of yeasts enriched in glutamic acid causing such reactions. Apparently, the glutamic acid is in a form not harmful for human beings and yeast extracts enriched in glutamic acid could be a useful replacer for MSG. Yeast extracts should be labeled as yeast extract or natural flavours. In case of high glutamate content the FDA suggest that the phrase 'contains glutamate' should be added.

16.3.2 Yeast-derived volatiles

During fermentation, yeasts produce substantial amounts of CO_2 and to a lesser extent other organic volatiles which, in most cases, are released into the environment. This typical 'yeasty' smell can be experienced in the vicinity of large-scale yeast factories. In most cases these volatiles are a mixture of acetaldehyde, ethylbutyrate, lactones, diacetyl, and fusel alcohols. Particularly during beer fermentation, interesting components are released by the CO_2 exhaust. Several attempts have been made to trap these volatiles from the exhaust system of the fermenter. However, their concentration is so low that an effective trapping and collection via scrubbing is not very cost-effective.

On the other hand, an absorption/desorption-technology called ASPROC has been developed, by which flavour recovery can be achieved from CO_2 streams generated by beer fermentation. The vapours are demoistured and conveyed over a porous powdery polymer. Only the so-called 'positive' flavours are absorbed. There is no affinity for CO_2, ethanol, sulphury components and aldehydes. A high affinity for esters, hop components and higher alcohols is claimed. Desorption takes place by stripping with ethanol. Examples of flavours successfully isolated are iso-amyl acetate, ethyl hexanoate, ethyl octanoate, and ethyl decanoate, the concentration of these products ranging from 600 to 1000 ppm. The alcohol volume used per elution on a pilot plant column amounts to 3 litres for stripping and cleaning. However, there is only a very limited interest for these esters. In principle, however, this approach could yield concentrates with a typical yeasty note, or even single topnotes when the concentrates are further processed, for example, by distillation.

16.3.3 Yeast-derived aroma chemicals

Research aimed at producing yeast-derived aroma chemicals was initiated in the early 1980s when consumers began to prefer natural aroma chemicals over their synthetic equivalent. Due to its biocatalytic versatility, its Generally Regarded As Safe (GRAS)-status and its ease of production, *Saccharomyces* species were often chosen as the preferred microorganism to produce such natural flavours, although other yeast species are used as well. Table 16.9 gives a list of natural aroma chemicals currently produced on a commercial scale by various types of yeast.

The production of natural aroma chemicals is based on the principle that natural precursor, *viz.* compounds which are naturally present in nature, are converted by enzymes and/or microorganisms into the desired flavour compound.

(a) Acetaldehyde. Acetaldehyde is used in flavours for fruit, butter rum, wine, etc. (Arctander, 1969; Table 16.9). Different methylotrophic yeast species (e.g. *Hansenula*, *Pichia* and *Candida*) contain an alcohol oxidase which is capable of converting ethanol into acetaldehyde. This enzyme is induced during growth on methanol. At the end of the logarithmic growth

Table 16.9 Natural aroma chemicals produced by yeast

Flavouring compound	Precursor	Flavour description[a]
Acetaldehyde	Ethanol	Pungent, coffee/wine-like in high dilutions
γ-Decalactone	Ricinoleic acid ex castor oil	Fruity, peach
δ-Decalactone	Linoleic acid 11-hydroxypalmitic acid Massoi lactone	Sweet coconut–peach-milk-like
2-Phenylethanol	Phenylalanine	Rose–honey-like

[a] Flavour descriptions from Arctander (1969).

phase cells are harvested and incubated with ethanol. In this way, concentration of about 1.5% acetaldehyde can be achieved, which can be raised to the desired level. The substrate specificity of the alcohol oxidase varies from yeast to yeast. Hence, this procedure can also be used to convert other, longer-chain alcohols such as hexanol, etc.

(b) γ-Decalactone. Lactones are odorous molecules often involved in fruit flavours (Pagot & Belin, 1996). For a long time, they could only be obtained by extraction from fruits or by chemical synthesis. One of the key compounds is γ-decalactone, which smells like peach, and can be produced by certain fungi and yeasts from certain hydroxylated fatty acids found in nature. Especially Ricinoleic acid [12(R)-hydroxy-9(7)-octadecenoid acid] is an especially good precursor, since it is readily available from castor oil (Maume & Cheetham, 1991; Hagedorn & Kaphammer, 1994). During growth, this fatty acid is partly converted by β-oxidation to 4-hydroxydecanoic acid, which lactonizes spontaneously to the R(+)-isomer. This isomer happens to be the one found in nature and thus is indistinguishable from fruit-derived γ-decalactone. In contrast, chemically synthesized γ-decalactone is a racemic mixture of the R(+) and S(−)-isomers. By using a second organic phase during fermentation, the lactone is extracted continuously from the broth and as a consequence high production yields have been achieved.

(c) δ-Decalactone. Like γ-decalactone, the delta-form is used in fruit flavours but is also widely applied as a butter flavour. Delta-decalactone has a creamy sweet coconut, peach-like taste in concentrations below 2 ppm (Arctander, 1969). Several biocatalytic routes to natural δ-decalactone have been explored. An overview of the different enzymic/microbial routes to δ-decalactone, is shown in Figure 16.12.

One of the approaches involves the hydroperoxydation of linoleic acid by the soybean enzyme lipoxygenase, followed by reduction to the hydroxy-fatty acid derivative, which in turn is converted by the β-oxidizing system of the yeast to yield 5-hydroxydecanoic acid (Hou, 1995). At low pH this intermediate lactonizes to the desired S(−)-isomer of δ-decalactone. An alternative approach is to reduce the double bond in Massoi lactone, isolated from Massoi bark oil, by yeast (Schaft *et al.*, 1992). This reductive route yields the R(+) isomer. Mixtures of both isomers can be isolated from coconut oil (75% R(+) and 25% S(−1)). A third possibility includes the biocatalytic conversion of 11-hydroxyhexadecanoic acid by a selected *Saccharomyces* species to the intermediate 5-hydroxydecanoic acid, which in turn can be lactonized at low pH. The precursor, 11-hydroxyhexadecanoic acid, is present in relatively large amounts in the resin of Mexican Jalap roots as a glucosidically bound hydroxyfatty acid. Upon hydrolysis this precursor can be liberated, purified and used as such

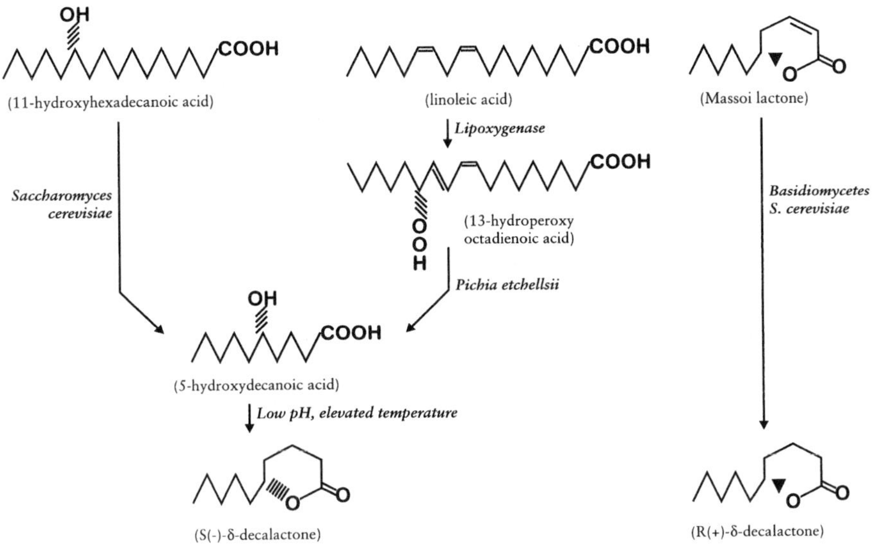

Figure 16.12 Schematic overview of the different enzymatic/microbial routes to natural δ-decalactone.

in the fermentation process. The different processing steps (Boog *et al.*, 1991) are illustrated schematically in Figure 16.13.

(d) 2-Phenylethanol and phenylacetate. As described in section 16.2.1, aroma compounds are important for the quality of many alcoholic beverages. Many of these compounds are formed by the yeast cells during fermentation. In such cases, *S. cerevisiae* is not only responsible for ethanol production, but also plays an important role in flavour formation. Phenyl ethanol is a (minor) constituent of the so-called fusel oil, a product formed during the manufacturing of ethanol (spirit alcohol). Higher alcohols (fusel alcohols) are the most abundant volatile flavour components in alcoholic beverages (Engan, 1981). They include propan-1-ol, 2-methylpropan-1-ol (isobutanol), (S)-(−)-2-methylbutan-1-ol (optically active amyl alcohol), 3-methylbutan-1-ol (iso-amyl alcohol) and 2-phenylethanol (phenethyl alcohol). Fusel alcohols are formed by all yeast species (Guymon *et al.*, 1961).

2-Phenylethanol and 2-phenylacetate have been characterized as rose-type flavours and are important aroma compounds in such alcoholic beverages as saké and wine. The levels in saké are 20–70 ppm for phenyl ethanol and 5–8 ppm for phenyl acetate, respectively (Yoshizawa, 1966). 2-Phenylethanol is produced by the yeast during alcoholic fermentation either by conversion of phenylalanine present in the medium or by *de novo* synthesis.

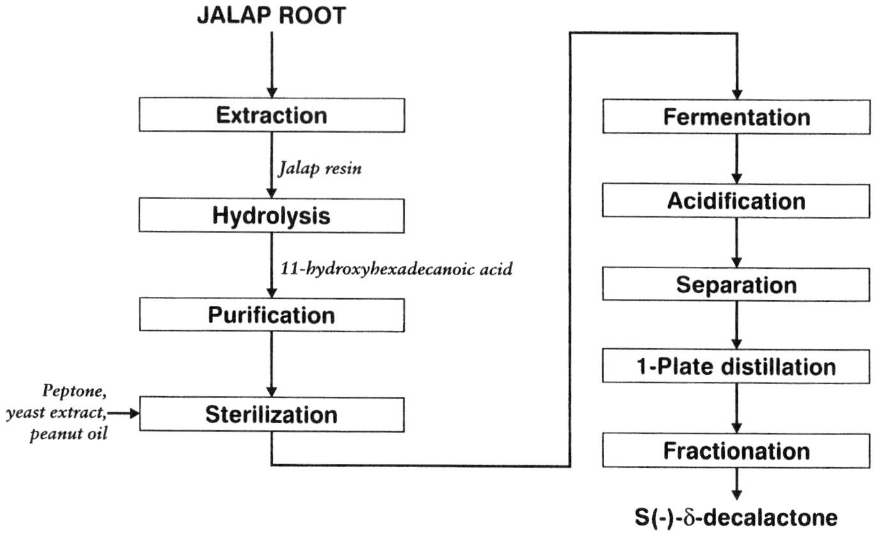

Figure 16.13 Steps involved in δ-decalactone production process from Jalap root.

Many reports describe attempts to increase the production of these compounds either by *in situ* production of selected/mutagenized yeast strains, or by cultivating yeast strains under controlled conditions in the fermenter. In the latter case the compound is isolated after fermentation and can be added as natural aroma ingredient or, after compounding with other natural ingredients, as natural flavour.

2-Phenylethanol is synthesized by decarboxylation of phenylpyruvate which is a precursor of phenylalanine. Regulation of aromatic amino acid biosynthesis is presented in Figure 16.14 (Akita *et al.*, 1990).

In *S. cerevisiae*, the first step of aromatic amino acid biosynthesis is catalysed by 3-deoxy-D-arabino-heptolusonate-7-phosphate synthases (DAHP synthase). The enzyme occurs in two isomeric forms, one being sensitive to feedback inhibition by phenylalanine (encoded by *ARO3* gene), the other by tyrosine (encoded by the *ARO4* gene). Both enzymes are under general amino acid biosynthesis control.

Several mutants have been described able to overproduce 2-phenylethanol. Akita *et al.* (1990) isolated mutants of the saké yeast *S. cerevisiae* resistant to p-fluorophenylalanine (FPA). These mutants produced increased amount of both 2-phenylethanol (200 ppm) and phenylacetate, because of impaired regulation of phenylalanine biosynthesis. Some of these mutants showed increased tryptophol production, suggesting mutations in a common regulation mechanism. Saké brewed with the mutants contained large amounts of 2-phenylethanol and phenyl

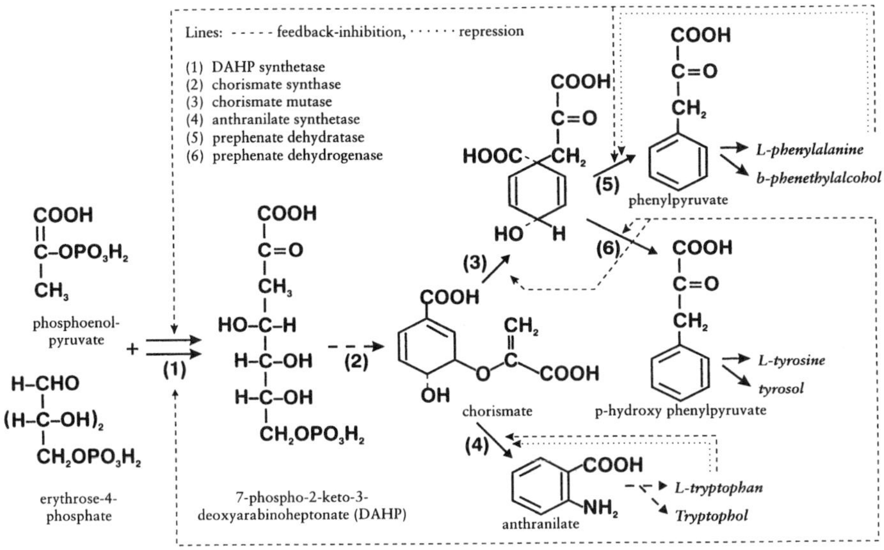

Figure 16.14 Regulation of aromatic amino acid biosynthesis in *S. cerevisiae*.

acetate and possessed a fine aromatic character and good quality (Akita *et al.*, 1990).

Fukuda *et al.* (1990a,b) isolated *p*-fluoro-dl-phenylalanine (PFP) and *o*-fluoro-dl-phenylalanine (OFP) mutants of *S. cerevisiae* capable of over-production of 2-phenyl ethanol. Two classes of mutants were found. In the OFP-resistant mutants the feedback inhibition of the tyrosine-dependent DHAP synthase was released, resulting in an increased flux to chorismate. In the FPA-resistant mutants, prephenate dehydrogenase (PDH) activity, a key enzyme of the flow to tyrosine, was decreased together with an increased activity of DAHP synthase, resulting in an increased flux to phenyl pyruvate. Similar mutants were isolated from haploid laboratory strains by the same researchers (Fukuda *et al.*, 1991a) and it was shown that the decreased activity of PDG was due to a mutation in the *TYR1* gene encoding PDG. The mutant gene was cloned and used to construct transformants which showed expression of the altered gene together with overproduction of 2-phenylethanol. Haploid mutants with increased DAHP synthase activity were also characterized genetically. The mutation occurred at the *ARO4* locus and transformant containing the mutated *ARO4* gene showed expression of a tyrosine feedback-insensitive DHAP synthase again, together with increased production of 2-phenylethanol and tyrosine (Fukuda *et al.*, 1991b). The transformants were used to produce saké with improved flavour characteristics (Fukuda *et al.*, 1992).

Aoki & Uchida (1990) isolated FPA mutants (after NTG mutagenesis) of the yeast *Zygosaccharomyces rouxii* used in the production of soy sauce. 2-Phenylethanol is also an important flavour component in soy sauce. The mutants showed a 38-fold increase in 2-phenylethanol production. Probably the biochemical pathway to 2-phenylethanol is the same as in *S. cerevisiae*. The authors suggest that a decrease in prephenate dehydrogenase activity, as described above, reduced the intracellular concentration of tyrosine and derepressed prephenate biosynthesis. Most of this prephenate will then be converted to phenyl pyruvate, the precursor of 2-phenylethanol.

Another approach to increase 2-phenylethanol production is conversion of the precursor phenylalanine. Albertazzi *et al.* (1994) described 2-phenylethanol and 2-phenylacetate production by various yeast strains in media containing 2 g/l phenylalanine. High levels of phenylacetate were produced by *Pichia pastoris* and *Kloeckera saturnus* (700–1700 mg/l). *Hansenula anomala* and *S. cerevisiae* were shown to be effective 2-phenylethanol producers.

In the catabolic conversion of phenylalanine to phenylethanol, three enzymes are involved (Figure 16.15): an amino acid aminotransferase, first postulated by Sentheshanmuganathan (1960) and Sentheshanmuganathan & Elsden (1958), who indicated that this enzyme is a transferase rather than a deaminase. The second enzyme is pyruvate decarboxylase, which decarboxylates the resulting keto acid to the aldehyde. The third enzyme is an alcohol dehydrogenase, which catalyses the NADH-dependent reduction of the aldehyde to the corresponding primary alcohol.

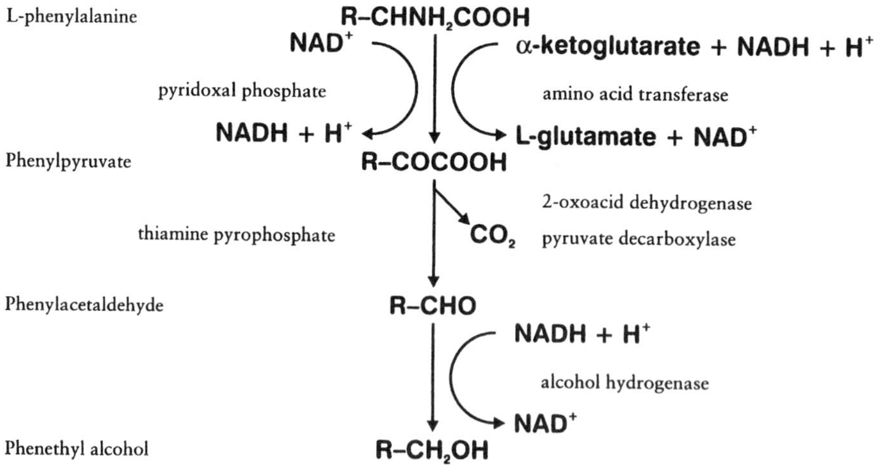

Figure 16.15 Modified Ehrich–Neubauer pathway of phenylethyl alcohol formation from L-phenylalanine in yeast.

It has been shown (for wine yeasts) that the isoenzyme involved in branched-chain aldehyde reduction is the classical fermentative isoenzyme ADH-1 (Singh & Kunkee, 1976).

Phenylethanol and other fusel alcohols are formed when yeast is grown using amino acids as nitrogen source, as in wort, or with ammonium salts, but the presence of glucose is essential. In wort, the carbon skeletons arise from the wort amino acids; when using ammonium salts they are produced *de novo* from glucose. Fusel alcohol formation is promoted by aerobiosis (Guymon *et al.*, 1961; Kunkee & Singh, 1975; Kuriyama & Kobayashi, 1993). Until recently, it was assumed that the Ehrlich–Neubauer route was the major, if not the only, route for metabolic transformation of (branched-chain) amino acids in *S. cerevisiae.* Recent work however (Dickinson & Dawes, 1992) has shown that the branched-chain 2-keto-acid dehydrogenase multi-enzyme found in bacteria is also present in *S. cerevisiae.* The presence of this enzyme may explain why not all added amino acid could be recovered into identified metabolic products (Derrick & Large, 1993). In the case of phenylalanine, the 2-keto acid dehydrogenase must play a major role, since pyruvate decarboxylase from yeast cannot convert phenylpyruvate (Bergmeyer, 1983).

16.4 Perspectives and future developments

The trend towards natural food ingredients combined with the positive image, the detailed knowledge and the versatility of food-grade yeasts, clearly has sparked renewed interest in the use of these microorganisms for the food industry. Yeast delivers a range of unique products which cannot be mimicked by chemicals. To date, the prefered yeast is *Saccharomyces cerevisiae*, but it can be expected that future yeast-derived ingredients will increasingly apply other yeast species. Likely candidates are *Kluyveromyces* and *Candida* species, which grow rapidly on simple substrates (see Table 16.5).

It can be expected that yeasts will be trained to accumulate or generate increased levels of flavour substances, precursors or flavour enhancers in order to meet the diversifying market needs. The tools to achieve this goal are at hand and include classical mutagenesis, metabolic reprogramming, raw materials and strain selection combined with careful process control.

The lessons learned from traditional fermented food systems are bound to lead to a range of new, multifunctional food ingredients in which yeasts play a key role, besides other microorganisms and enzymes, in creating the desired functionality. In these products, yeast has multiple functions. Besides delivering taste, it will affect the final product with respect to texture, appearance, process tolerance and nutritional value.

The prospects for novel yeast-derived products are excellent now that the rapidly expanding technological possibility can serve the market needs. As a result, rapid growth is expected for yeast-derived products in both the food and non-food sector.

Acknowledgements

We thank Ronald Boog, Paul Cornelius, Dave Evans, Josef Kerler, Cor Vos, Maarten van Oort, Dave Powell, Diederik Schmedding, Glynn Shore, and Hugo Weenen from Quest International for their valuable input and comments. Furthermore, we are grateful to Laura Ausma and Saskia Overhage for typing the manuscript and to JeanineTeeling for creating the figures.

References

Akita, O., Ida, T., Obata, T. & Hara, S. (1990) Mutants of *Saccharomyces cerevisiae* producing a large quantity of β-phenethyl alcohol and β-phenethyl acetate. *Journal of Fermentation and Bioengineering*, **69**(2), 125–8.

Albertazzi, E., Cardillo, R., Servi, S. & Zucchi, G. (1994) Biogeneration of 2-phenylethanol and 2-phenylethylacetate, important aroma components. *Biotechnology Letters,* **16**(5), 491–6.

Alfafara, C., Miura, K., Shimizu, H., Shioya, S. & Suga, K.I. (1992) Cysteine addition strategy for maximum glutathione production in fed-batch culture of *Saccharomyces cerevisiae. Applied Microbiology and Biotechnology*, **37**, 141–6.

Ames, M.J. & Elmore, J.S. (1992) Aroma components of yeast extracts. *Flavour and Fragrance Journal*, **7**, 89–103.

Ames, M.J. & Macleod, G. (1985) Volatile components of a yeast extract composition. *Journal of Food Science*, **50**, 125–8.

Aoki, T. & Uchida, K. (1990) Enhanced formation of 2-phenylethanol in *Zygosaccharomyces rouxii* due to prephenate dehydrogenase deficiency. *Agricultural and Biological Chemistry*, **54**(1), 273–4.

Arctander (1969) *Perfume and Flavor Chemicals*, Montclair, NJ, USA.

Asahi Breweries (1995) Temperature sensitive autolysing *Saccharomyces cerevisiae* is prepared by mutating wild type *Saccharomyces cerevisiae* strain and separating out autolysing variants useful for production of seasoning proteins. Patent JO6253825.

Baigrie, B. (1994) Taints and off-flavours in foods. *International Food Ingredients*, **4**, 35–9.

Belitz, H.-D. & Grosch, W. (1992) *Lehrbuch der Lebensmittelchemie,* **4**. Auflage, Berlin, 304–61.

Bergmeyer, H.O. (1983) Samples, Reagents, Assessment of results. *Methods of Enzymatic Analysis*, II, Third edition.

Boog, A.L.G.M., Peters, A.L.J. & Roos, R. (1991) Process for producing delta lactones. EP409320.

Davidek, J., Hasjlova, J., Kubelka, V. & Velisek, J. (1979) Flavor significant compounds in yeast autolysates, Gistex X-II powders. Part 1. Acid Fraction. *Die Nahrung*, **23**, 673–80.

Derrick, S. & Large, P.J.J. (1993) Activities of the enzymes of the Ehrlich pathway and formation of branched-chain alcohols in *Saccharoymyces cerevisiae* and *Candida utilus*. Grown in continuous culture on valine or ammonium as sole nitrogen source. *Journal of General Microbiology*, **139**, 2783–92.

Dickinson, J.R. & Dawes, I.W. (1992) The catabolism of branched-chain amino acids occurs via 2-oxoacid dehydrogenase in *Saccharomyces cerevisiae. Journal of General Microbiology*, **138**, 2029–33.

Drost, B.W., Van Den Berg, R., Freijee, F.J.M., Van Der Velde, E.G. & Hollemans, M. (1990) Flavour Stability. *Journal of the American Society of Brewing Chemists,* **48**, 124–31.

El-Dash, A. & Johnson, J.A. (1970) Influence of yeast fermentation and baking on the content of free amino acids and primary amino groups and their effect on bread aroma stimuli. *Cereal Chemistry*, **47**, 247–59.

Engan, S. (1981) Beer composition: volatile substances, in *Brewing Sciences*, Vol. **2** (ed. R.A. Pollock), Academic Press, London, pp. 93–165.

Fell, D.A. & Thomas, S. (1995) Physiological control of metabolic flux: the requirement for multisite modulation. *Journal of Biochemistry*, **311**, 35–9.

Fukada, K., Wantanabe, M., Asano, K., Ueda, H. & Ohta, S. (1990a) Breeding of brewing yeast producing a large amount of 2-phenylethyl alcohol and 2-phenylethyl acetate. *Agricultural Biological Chemistry*, **54**(1), 269–71.

Fukuda, K., Watanabe, M. & Asano, K. (1990b) Altered regulation of aromatic amino acid biosynthesis in 2-phenylethyl-alcohol-over-producing mutants of saké yeast *Saccharomyces cerevisiae. Agricultural Biological Chemistry*, **54**(2), 3151–6.

Fukuda, K., Wantanabe, M., Asano, K., Ouchi, K. & Takasawa, S. (1991a) A mutated ARO4 gene for feedback-resistant DAHP synthase which causes both *o*-fluoro-DL-phenylalanine resistance and 2-phenethyl alcohol overproduction in *Saccharomyces cerevisae. Current Genetics*, **20**, 453–6.

Fukuda, K., Wantanabe, M., Asano, K., Ouchi, K. & Takasawa, S. (1991b) Isolation and genetic study of *p*-fluoro-DL-phenylalanine-resistant mutants overproducing 2-phenethyl alcohol in *Saccharomyces cerevisiae. Current Genetics*, **20**, 449–52.

Fukuda, K., Wantanabe, M., Asano, K., Ouchi, K. & Takasawa, S. (1992) Molecular breeding of a sake yeast with a mutated ARO4 gene which causes both resistance to ortho-fluoro-DL-Phenylalanine and increased production of beta-phenyl alcohol. *Journal Fermentation and Bioengineering*, **73**(5), 366–9.

Gassenmeier, K. & Schieberle, P. (1995) Potent aromatic compounds in the crumb of wheat bread (French-type) – influence of pre-ferments and studies on the formation of key odorants during dough processing. *Zeitschrift fur Lebensmittel – Unterschung und Forschung*, **201**(3), 241–8.

Golovnya, R.V., Misharina, T.A., Garbuzov, G. & Medvedyev, F.A. (1983) Volatile sulfur containing compounds in stimulated meat flavour and their comparison with the constituents of natural aroma. *Die Nahrung*, **27**, 237–49.

Goossens, E., Debourg, A., Villanueba, K.D. & Masschelein, C.A. (1993) Decreased diacetyl production in lager brewing yeast by integration of the *ILV5* gene. *Proceedings of the European Brewery Convention Congress*, Oslo, pp. 251–8.

Goossens, E., Dillemans, M., Debourg, A. & Masschelein, C.A. (1987) Control of diacetyl formation by the intensification of the anabolic flux of acetohydroxyacid intermediates. *Proceedings of the European Brewery Convention Congress*, Madrid, 553–60.

Gordon, I.R. (1991) Yeast extracts and HVP: W. European markets and opportunities 1990–1995. Food Ingredients Europe. Conference proceedings, Porte de Versailles, Paris, pp. 211–14.

Grosch, W. & Schieberle, P. (1991) Bread, in *Volatile Compounds in Food and Beverages* (ed. H. Maarse), Marcel Dekker, Inc., New York, pp. 41–77

Guymon, J.F., Ingraham, J.L. & Crowell, E.A. (1961) Influence of aeration upon the formation of higher alcohols by yeasts. *American, Journal of Enology and Viticulture*, **12**, 60–6.

Hagedorn, S. & Kaphammer, B. (1994) Microbial biocatalysis in the generation of flavor and fragrance chemicals. *Annual Review of Microbiol*ology, **48**, 773–800.

Hamada, S., Tanaka, H.H. & Sakato, K. (1986) Process for producing glutathione. US patent 4582801.

Hammond, J.R.M. (1995) Genetically-modified brewing yeasts for the 21st century. Progress to date. *Yeast*, **11**, 1613–27.

Hasjlova, J., Velisek, J., Davidek, J. & Kubelka, V. (1980) Flavour significant compounds in yeast autolysates, Gistex X-II powders. Part II. Neutral and basic fractions. *Die Nahrung*, **24**, 875–81.

Hirata, D., Aoki, S., Wantanabe, K., Tsukioka, M. & Suzuki, T. (1992) Stable overproduction of isoamyl alcohol by *Saccharomyces cerevisiae* with chromosome-integrated multicopy *LEU4* genes. *Bioscience, Biotechnology and Biochemistry*, **56**, 1682–3.

Hou, C.T. (1995) Microbial oxidation of unsaturated fatty acids. *Advances in Applied Microbiology*, **41**, 1–23.

Kitamoto, H.K. & Nakahara, T. (1994) Isolation of an L-methionine-enriched mutant of *Kluyveromyces lactis* grown on whey permeate. *Process Biochemistry*, **29**, 127–31.

Korch, C., Mountain, H.A., Gyllang, H., Winge, M. & Brehmer, P. (1991) A mechanism for sulphite production in beer and how to increase sulphite levels by recombinant genetics. *Proceedings of the European Brewer Convention Congress*, Lisbon, pp. 201–8.

Kunkee, R.E. & Singh, R. (1975) Dehydrogenase activity for higher alcohols in cell free extracts of *Saccharomyces cerevisiae*. *Journal of the Institute of Brewing*, **81**, 214–7.

Kuriyama, H. & Kobaysashi, H. (1993) Effects of oxygen supply on yeast growth and metabolism in continuous fermentation. *Journal of Fermentation and Bioengineering*, **75**, 364–7.

Labell, F. (1993) Glutamates under the gun. *Food Processing*, May, 39–45.

Lang-Hinrichs, C. & Stahl, U. (1988) Verfahren zur Synthese von Glutathion in Hefen. EP 0300168.

Lásztity, R. (1991) Yeast biomass, a potential source of food ingredients. *International Food Ingredients*, **3**, 4–8.

Maume, K.A. & Cheetham, P.S.J. (1991) The production of γ-decalactone by fermentation of castor oil. *Biocatalysis*, **5**, 79–97.

Mcmurrough, I., Madigan, D., Donnelly, D., Hurley, J., Doyle, A., Hennigan, G., McNulty, N. & Smyth, M.R. (1996) Control of ferulic acid and 4-vinyl guaiacol in brewing. *Journal of the Institute of Brewing*, **102**, 327–32.

Minetoki, T., Bogaki, T., Iwamutsu, A., Fujii, T. & Hamachi, M. (1993) The purification, properties and internal peptide sequences of alcohol acetyltransferase from *Saccharomyces cerevisae* Kyokai No. 7. *Bioscience, Biotechnology and Biochemistry*, **57**, 2094–8.

Moll, M. (1991) *Beers & Coolers* (translated by Wainwright, T.), Intercept Ltd, Andover.

Nagodawithana, T. (1992) Yeast-derived flavors and flavor enhancers and their probably mode of action. *Food Technology*, November, 138–44.

Nagodawithana, T. (1995) Protein hydrolysates: past and present, in *Savoury Flavours* (ed. Esteeky Associates Inc.), pp. 225–63.

Nordström, K. (1965) Possible control of volatile ester formation in brewing. *EBC Proceedings, 10th Congress, Stockholm*, Elsevier, Amsterdam, pp. 195–208.

Obermann, H. & Libudzisz, Z. (1997) Fermented milks, in *Microbiology of Fermented Foods*, 2nd edition, (ed. B.J.B. Wood), Chapman & Hall, London.

Ohtake, Y., Watanabe, K., Tezuka, H., Ogata, T., Yabuuchi, S., Murata, K. & Kimura, A. (1988) The expression of the γ-glutamyl cysteine synthetase gene in *Escherichia coli B* in *Saccharomyces cerevisiae*. *Agricultural and Biological Chemistry*, **52**, 2753–62.

Ohtake, Y., Watanabe, K., Tezuka, H., Ogata, T., Yabuuchi, S., Murata, K. & Kimura, A. (1989) Expression of the glutathione synthetase gene of *Escherichia coli B* in *Saccharomyces cerevisiae*. *Journaly of Fermentation and Bioengineering*, **68**, 390–4.

Pagot, Y. & Belin J.-M. (1996) Fatty acid cellular metabolism and lactone production by the yeast *Pichia guilliermondii*. *Applied Microbiology and Biotechnology*, **45**, 349–54.

Peppard, T.L. & Halsey, S.A. (1981) Malt flavour – Transformation of carbonyl compounds by yeast during fermentation. *Journal of the Institute of Brewing*, **87**, 3860–90.

Piendl, A. (1969) Brauereitechnologie und Molekularbiologie. *Brauwissenschaft*, **22**, 175.

Preininger, M., Rychlik, M. & Grosch, W. (1994) Potent odorants of the neutral volatile fraction of Swiss Cheese (Emmentaler). *Developments in Food Science*, **35**, 267–70.

Rainbow, C. (1981) Beer spoilage microorganisms, in *Brewing Science*, 2nd edn, (ed. R.A. Pollock), Academic Press, London, pp. 491–550.

Roling, W. (1995) *Traditional Indonesian soy sauce (kecap) production: microbiology of the brine fermentation*, PhD Thesis, University of Amsterdam, The Netherlands.

Rychlik, M. & Grosch, W. (1996) Identification and quantification of potent odorants formed by toasting of wheat bread. *Food Science and Technology*, **29**, 515–25.

Sakato, K. & Tanaka, H. (1992) Advanced control of glutathione fermentation process. *Biotechnology and Bioengineering*, **40**, 904–12.

Schaft, P.H., Van Der Burg, N., Ter Bosch, S., Van Den & A.M. Cohen (1992) Microbiol production of natural δ-decalactone and δ-dodecalactone from the corresponding α, β-unsaturated lactones in Massoi bark oil. *Applied Microbiology and Biotechnology*, **36**, 712–16.

Schieberle, P. (1990) Studies on bakers Yeast as source of maillard-type bread flavour compounds, in *The Maillard Reaction in Food Processing, Human Nutrition and Physiology* (eds P.A. Finot, H.U. Aeschbacher, R.F. Hurrell), *Birkhauser Verlag*, Basel, pp. 187–96.

Schieberle, P. (1992a) Bildung wichtiger R_staromastoffe in Lebensmitteln aus Getreide, in *Getreide, Mehl und Brot*, **11**, 338–42.

Schieberle, P. (1992b) Formation of furaneol in heat-processed food, in *flavour Precursors, Thermal and Enzymatic Conversions* (eds R. Teranishi, G.R. Takeoka, M. Göntert), American Chemical Society, Washington, DC, pp. 164–74.

Schieberle, P. (1996) The role of free amino acids present in yeast as precursors of the odorants 2-acetyl-1-pyrroline and 2-acetyltetrahydro- pyridine in wheat bread crust. *Zeitschrift fur Lebensmittel – Unterschung und Forschung*, **191**, 206–9.

Schieberle, P. & Grosch, W. (1991) Potent odorants of the wheat bread crumb. *Zeitschrift fur Lebensmittel – Unterschung und Forschung*, **192**, 130–5.

Schieberle, P. & Grosch, W. (1994) Potent odorants of rye bread crust – differences from the crumb and from wheat bread crust. *Zeitschrift fur Lebensmittel – Unterschung und Forschung*, **198**, 292–6.

Schoenberg, E. (1993) The technology and application of yeast derived flavour enhancers. *European Food and Drink Review*, 55–7.

Sentheshanmuganathan, S. (1960) The mechanism of the formation of higher alcohols from amino acids by *Sacchoromyces cerevisae. Biochem. J.*, **74**, 568–76.

Sentheshanmuganathan, S. & Elsden, S.R. (1958) The mechanism of the formation of tyrosol by *Saccharomyces cerevisiae. Biochem. J.*, **69**, 210–18.

Shimizu, H., Araki, K., Shioya, S. & Suga, K. (1991) Optimal production of glutathione by controlling the specific growth rate of yeast in fed-batch culture. *Biotechnology and Bioengineering*, **38**(2), 196–205.

Shimizu, H., Miura, K., Alfafara, C.G., Shioya, S., Suga, K. & Suzui, K. (1992) Fuzzy control of ethanol concentration and its application to glutathione production in yeast fed-batch culture. *IFAC Symposium Series,* (Volume 10, Modeling and Control of Biotechnical Processes), pp. 433–6.

Singh, R. & Kunkee, R.E. (1976) Alcohol dehydrogenase activities of wine yeast in relation to higher alcohol formation. *Applied and Environmental Microbiology*, **32**(5), 666–70.

Spicher, G. & Stephan, H. (1982) Die Hefe des Sauerteiges, in *Handbuch sauerteig. Biologie, Biochemie, Technologie* (ed. D. Behr), pp. 71–81.

Tezuka, H., Mori, T., Okumura, Y., Kitabatake, K. & Tsumura, Y. (1992) Cloning of a gene suppressing hydrogen sulphide production by *Saccharomyces cerevisae* and its expression in a brewing yeast. *Journal of the American Society of Brewing Chemists,* **50**, 130–3.

Thornton, J. (1992) Brewers' yeast extract. A survey of the types available to the food processor. *International Food Ingredients*, **2**, 40–3.

Thornton, J. (1996) Developments in savoury flavour systems. *International Food Ingredients*, **2**, 19–21.

Udeh, K.O. & Achremowicz, B. (1994) Production of yeast biomass with elevated content of glutathione. *Polish Journal of Food Nutritional Science*, **3/44**(No. 1), 93–100.

Vakeria, D., Box, W.G. & Hinchliffe, E. (1991) The control of gene expression in brewing yeast with anti-sense RNA. *Proceedings of the European Brewery Convention Congress*, Lisbon, pp. 305–12.

Vaughn, R.H. (1982) The microbiology of vegetable fermentations, in *Microbiology of Fermented Foods*, Vol. 1 (ed. B.J.B. Wood), Elsevier Applied Science Publishers, pp. 49–111.

Villanueba, K.D., Goossens, E. & masschelein, C.A. (1990) Sub-threshold vicinal diketone levels in lager brewing yeast fermentations by means of *ILV5* gene amplification. *Journal of the American Society of Brewing Chemists*, **48**, 111–14.

Voight, J.C. & Walla, G. (1995) A novel yeast propagation system. *Proceedings of the 5th Scientific and Technical Convention*, Victoria Falls, March 1995, pp. 173–8.

VIJ, S. & Gandhi, D.N. (1993) Whey – An alternative substrate for the production of bakers yeast. *Indian Food Industry*, **12**(5), 41–3.

Wainwright, T. (1973) Diacetyl – A review. Part I – Analytical and biochemical considerations. Part II – Brewing experience. *Journal of the Institute of Brewing*, **79**, 451–70.

Werkhoff, P., Bretschneider, W., Emberger, R., Güntert, M., Hopp, R. & Köpsel, M. (1991) Recent developments in the sulfur flavour chemistry of yeast extracts. *Chemie Mikrobiologie Technologie der Lebensmittel,* **13**, 30–57.
Wierzbicka, G.T., Hagen, T.M. & Jones, D.P. (1989) Glutathione in food. *Journal of Food Composition and Analysis*, **2**, 327.
Yoshizawa, K. (1966) On various factors affecting formation of isobutanol and isoamylalcohol during alcoholic fermentation. *Agricultural and Biological Chemistry*, **30**, 634–41.

17 Biology and technology of mushroom culture

R.J. SCRASE AND T.J. ELLIOTT

17.1 Mushrooms: an overview of the product

The mushroom industry world-wide currently has a turnover of £3,000 million. This industry is by far the most economically important industry based on a solid-state fermentation.

While the cultivation of *Agaricus bisporus* is by far the largest component of the industry in the USA and Europe, shiitake (*Lentinus edodes*), paddy straw mushrooms (*Volvariella volvaceae*) and oyster mushrooms (*Pleurotus spp.*) are produced in significant quantities world-wide (Table 17.1). Shiitake production figures for the US in 1991–1992 were 12300 tonnes with a further 14000 tonnes of dried Shitake (equivalent to a fresh weight of 126000 tonnes) being imported.

Many other species, e.g. Shaggy Ink-Cap (*Coprinus comatus*) and the woodland Blewit (*Lepista nuda*), have already been domesticated, but the majority of mushroom-producing species have yet to receive serious or even passing attention. Consequently, the mushroom grower has the opportunity, not only to grow and eat certain 'wild' mushrooms, but also to contribute to an understanding of the parameters controlling the growth and reproduction, and hence domestication, of new species. The techniques for cultivating mushrooms, for many species, follow the same basic pattern, and once this pattern has been learned, it can be applied to novel species. From among many thousands of basidiomycete fungi, a few hundred have been induced to produce a fruitbody in artificial or semi-natural conditions at one time or another. The fact that only a dozen or so of these organisms are under regular commercial production (Chang & Hayes, 1978; Wood, 1989) suggests that they represent a relatively untapped potential, and that there is a need for a better understanding of the processes underlying fruiting.

It is important to appreciate that basidiomycete fruitbodies are important not only as food but also as potential sources of medicine and animal feeds, and that basidiomycetes are nearly unique in possessing the growth form and enzymic capability to degrade fully the plant polymers lignin, hemicellulose and cellulose. These facts make these organisms ideal candidates for the bioconversion of lignocellulosic wastes produced by agriculture and plant-processing industries (Zadrazil & Reiniger, 1988). The process of

Table 17.1 World production of edible cultivated mushrooms

Species	Common name	Production (tonnes × 1000, fresh wt)
Agaricus bisporus	Button mushroom	2000 (200)
Pleurotus spp.	Oysters	850 (800)
Lentinus edodes	Shii-take	400 (210)
Auricularia spp.	Woody ear	365 (360)
Volvariella volvacea	Straw mushroom	130 (10)
Tremella spp.	Jelly fungus	104 (104)
Hericium erinaceus	Monkey head	90 (90)
Flammulina velutipes	Winter mushroom	60 (40)
Pholiola nameko	Viscid mushroom	53 (32)

Estimated from various sources.
Figures in brackets are for China.

bioconversion can reduce the quantity of waste, reduce the waste's potential for pollution, and provide products with increased value.

An increased understanding of the biology of basidiomycetes will also present new possibilities for forestry and nature conservation management. The ubiquitous roles of basidiomycetes as nutrient recyclers, pathogens, or mycorrhizal associates in all terrestrial ecosystems demand their inclusion in the consideration of strategies for forestry production and the conservation of biodiversity.

Being able to grow new species of basidiomycete also creates new aesthetic possibilities in the garden and home.

Considering the importance of being able to fruit basidiomycetes, and the fact that while several hundred species have been brought to fruition, only a handful are under commercial cultivation, one can consider the obstacles to the application of what is already known about basidiomycete production. Perhaps the major obstacle to the widespread use of basidiomycetes by humans has been the limited range of approaches to that use. In the area of mushrooms for food, past conservatism of food tastes in the Anglo-Saxon world are presently being overcome by the large food distributors using product proliferation as a route to continued profits growth. This trend is being slowed for mushrooms by an Agaricus mushroom-growing industry which is locked into a particular production system, that does not happen to suit the production of many other species. Despite this, and uncoordinated marketing, *Pleurotus* and *Lentinus* species are now found regularly on the

UK supermarket shelf. This was not the case 10 years ago. The limited use of mushrooms for medicine in the west can be contrasted with their use in China and other far eastern countries (Chang & Miles, 1989). In the latter case, basidiocarps have a role in herbal medicine. This role is currently under intense investigation by the pharmaceutical industry. Pharmaceutical companies are constrained by their own production systems, and have so far limited themselves to making use of the chemicals produced by the fungal mycelia in liquid culture, although this is changing. While suitable modification of the external environment can release much of the mycelium's developmental plasticity in liquid culture (Deacon, 1984) it may well be worth producing fruitbodies for drug extraction. Work at Xenova has demonstrated conclusively that different compounds are produced within the fruitbody compared with the supporting mycelium. Some of these have been demonstrated to be pharmaceutically active (Penn, 1994).

In the area of forestry, the forester's pragmatic view of the role of Basidiomycetes as either destructive pathogen or beneficial mycorrhizal symbiont distorts the perception of their possible value in other roles. There are possibilities such as that of 'inoculating' forests with the naturally occurring (or other) competitors of pathogens to control pathogenic activity (Dowson *et al.*, 1988). Alternatively, artificial woodland communities could be created specifically to avoid the level of damage that monoculture is prone to from pathogens and changing environental conditions. Deliberately introduced Basidiomycetes have been used in the reconstruction and decontamination of damaged ecosystems: methods have included the introduction of basidiomycete symbionts onto tree-seedlings, and the re-introduction of saprotrophs to new woodland. Breeding basidiomycetes to introduce genetic diversity, and mediating conservation practice in the light of fungal ecology, and the needs of the fungal component of the community are roles that have still to be developed; however, the potential benefits of understanding the ecological role of Basidiomycetes and being able to manipulate them are clear.

To summarize, basidiomycete cultivation provides humans with the opportunity for converting waste into food and medicines for themselves and their animals. Basidiomycete cultivation can also present the opportunity for habitat management that in turn can allow more effective nature conservation and contribute to the development of sustainable forestry production systems.

17.1.1 Basidiomycetes life-cycle

Basidiomycetes reproduce by producing spores. They can reproduce asexually (mitospores) but manipulation, and breeding usually depends on the use of the sexually produced basidiospores. Sexual reproduction is fundamentally similar to that in other organisms. It involves the fusion of two haploid nuclei to form a diploid nucleus, which undergoes meiosis. The way in which parental nuclei are brought together and the regulatory mechanisms that control outbreeding are now discussed.

Single spores will germinate to produce a *monokaryotic* mycelium. Compatible adjacent monokaryotic mycelia can fuse to produce a *dikaryotic* mycelium. Compatibility in Basidiomycetes is complex. There are often two gene loci, A and B, with either two alleles at each (tetrapolar compatability) or multiple alleles at each (multipolar compatability) controlling compatability. Successful mating depends on the partners differing at each locus, so AB will successfully mate with ab, but not with Ab, etc.

After two compatible strains fuse, nuclei are exchanged between the two mycelial networks. The nuclei do not undergo meiosis at this point; instead, the nuclei divide and migrate so that eventually every hyphal compartment within the mycelium contains two nuclei, one of each type. The fungus continues to grow in this form, termed a *dikaryon*, and now has the capability to produce fruitbodies. Dikaryotic mycelium typically exhibits clamp-connections and these can be used to confirm dikaryotization in most, but not all basidiomycetes. Some species will produce mushrooms from homokaryotic mycelia, but this is unusual and the mushrooms are rarely fertile.

At a later stage, when the mycelium has sufficient food reserves, it can be induced to form fruitbodies. The fruitbody itself is composed of dikaryotic hyphae, but in the later stages of development it forms basidia which cover the hymenial surfaces such as those found on the gills or surrounding the pores. The nuclei fuse in the individual basidia to form diploid nuclei. This is followed by meiosis, and the four diploid nuclei move into the developing basidiospores which are then released to continue the cycle. *Agaricus bisporus* is atypical in only producing two spores from each basidium (Figure 17.1).

17.1.2 Role of Basidiomycetes in ecosystems

The purpose of this section is to give a very brief overview to create a context for the following sections that give specific information about mushroom production. Basidiomycetes are ubiquitous in terrestrial ecosystems. As heterotrophs, their near-unique ability to produce enzymes that degrade lignocellulose coupled with a body-form composed of a network of microscopic tubes that can intimately penetrate a substrate, determines the functions the group fulfill in ecosystems.

(a) Necrotrophic basidiomycete fungi. These kill living tissues before utilizing them, and some species with this habit cause considerable economic loss, particularly in mono-cultural forestry systems. The honey fungus, *Armillaria mellea* is a notorious example. Interestingly, this species is

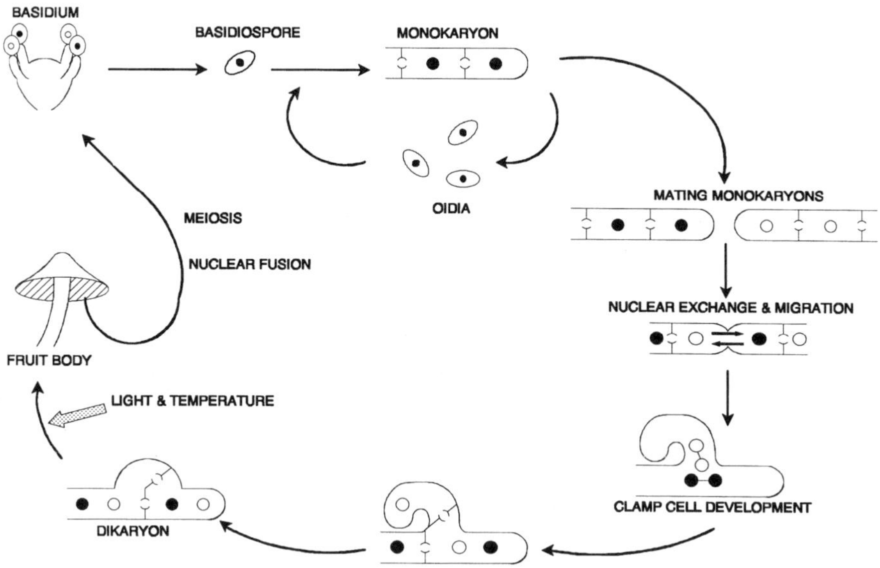

Figure 17.1 Life cycle of Basidomycetes. (Courtesy of C. Kingsnorth, HRI.)

rare in undisturbed woodland, probably due to the presence of competitors in leaf-litter that are not present in parks and gardens.

(b) Saprotrophic basidiomycete fungi. These utilize dead material, the degree of niche-specificity varying as much as in any other group. The Oyster mushroom, *Pleurotus ostreatus* grows naturally on dead or dying deciduous wood, woodland blewits, *Lepista nuda* are found on leaf-litter, and Shaggy Ink-Caps, *Coprinus comatus* grow in nitrogen-enriched grassland. In many cases, the fundamental niche of a species is wider than the realized niche. For example, *Lepista nuda* is a leaf-litter decomposer in the wild but can be cultivated on mushroom compost or straw, demonstrating a wider fundamental niche than its realized niche: the niche *L. nuda* occupies as a result of interaction with its ecosystem.

The *Hygrocybe* genus (waxcaps) are so closely associated with unfertilized and undisturbed natural and semi-natural grassland that they are being used to classify this endangered habitat (Boertmann, 1966).

(c) Mycorrhizas. These are symbiotic associations familiar to everyone who has ever collected mushrooms. The ocurrence of many mushroom species beneath particular trees is due to an intimate connection between fungal hyphae and the plant's roots where the root is encased in a fungal sheath. In this example of a biotrophic heterotrophy the fungus mediates the uptake of minerals and water by the plant and provides protection from

pests and diseases; in return, it obtains nutrients from the plant's living cells. These associations are so vital for the establishment and growth of trees, particularly on poor and polluted soils, that tree seedlings are frequently inoculated with mycorrhizal fungus, and will fail to grow in the absence of suitable fungi. Mycorrhizal basidiomycetes are typically very difficult to grow in culture and have only been commercially produced by growing them in association with their host trees. This is the basis of truffle-culture and may form the basis of chanterelle culture (Danell, 1997). The current trade in chanterelles picked from the wild is worth about £1 billion (Watling, 1997). These heterotrophic behaviours are not mutually exclusive, but form a continuum.

In conclusion, basidiomycetes are a primary conduit for the flux of carbon from biological material to the atmosphere. They are also an important component of soils, and regulate the fluxes of mineral nutrients into plants.

17.1.3 Overview of the mushroom production process

Mushroom production depends on the following stages:

- culture (spawn) production;
- substrate (compost) production;
- inoculation of substrate;
- colonization of the substrate;
- initiating fruitbody formation; and
- cropping.

(a) Spawn production. Having isolated a pure culture, it is a relatively straightforward operation to increase the mycelial mass to make a mushroom 'spawn' for inoculating into a bulk substrate. While it is possible to purchase pure cultures of a very wide range of basidiomycetes, preparing cultures is a relatively straightforward microbiological procedure. Cultures can be obtained by germinating spores or growing mycelium from excised basidiocarp material. To collect spores, the cap is severed from the stem of a fresh mushroom and placed gills down on a piece of clean paper or a clean glass surface such as a microscope slide. A drop or two of water is added to the cap surface to aid spore release. A container is placed over the mushroom to lessen evaporation and disturbance from air-currents. After a few hours a spore deposit should be visible. The spores should be stored in a labelled, airtight container until use. If the cap is suspended a short distance above the surface, contamination by invertebrates is less likely.

Basidiospores have a notoriety for being difficult to germinate (Madelin, 1966). Part of the reason is historical in that early workers discarded their cultures after only a few days, guaranteeing failure with organisms that may take 4 weeks or more to germinate. However many species have resisted all attempts to make their spores germinate. Starting with a fruitbody, there are four basic techniques for germinating basidiospores, listed here in ascending order of complexity.

1. Allow spores to fall onto a nutrient agar.*
2. Collect fresh spores, mix them with sterile distilled water, plate out a series of dilutions onto agar.
3. Add spores to a plate pre-inoculated with mycelia of the same species.
4. Add spores to a plate containing colonies of *Rhodotorula mucilaginos* var. *sanguinea* (a yeast).

***Note:** all fungi require several specific nutrients for growth and reproduction. For the growth of mycelium in small quantities, these are commonly provided by mixing a gelling agent (agar) with a natural product such as malt extract.

Variations and additions to these techniques can be found in the references, but from the experience of previous workers it is possible to make some generalizations about the ease with which different types of fungi may be isolated. The spores of coprophilous and lignicolous species usually germinate readily, as do Agaricaceous forest-inhabiting litter decomposers. Litter-decomposing gasteromycetes need to be co-cultured with *Rhodotorula* (Stevens, 1976). Mycorrhiza forming species are generally difficult to germinate, e.g. *Cantherellus cibarius* (Straatsma *et al.*, 1986).

Initiating a culture from tissue taken from a basidiocarp is usually the easiest way to create a culture. The indeterminate nature of mycelial development allows a piece of the stipe or cap to de-differentiate when placed onto a suitable substrate and it begins to grow as somatic mycelium. This may have all the properties of the 'parent' mycelium and eventually be used to initiate a new reproductive cycle. The method is very straightforward:

1. Use recently picked young basidiocarps and brush off any surface detritus.
2. Break the fruitbody in half.
3. Use a sterile needle or scalpel to remove pieces of tissue about 2 × 2 × 2 mm. Take about 10 samples from different areas, as the best area varies. Include tissue from the stipe of agarics.
4. Partially immerse the tissue samples into agar.
5. Incubate in the dark at 20–25°C.

Essentially, sterile tissue is exposed in a way that maintains its sterility, so that samples can be obtained. Some fruitbodies lack pileus tissue between the cuticle and hymenium. In this case, use the stipe. Frequently, especially in the case of soft, watery specimens, e.g. *Hygrocybe*, and the lower stipe of most others, the removed tissue will be contaminated with bacteria and other fungi, making the isolation of a pure culture difficult. Bacteriocidal compounds added to the agar can partly control the problem. New mycelium should grow away from the inoculum within 3–7 days. Following sub-culturing and microscopic examination to ensure a pure culture, the culture can be stored by growing mycelium over the surface of an agar slope in a small bottle. The surface can then be submerged in sterile mineral oil, and the bottle refrigerated at 2°C. Alternatively, samples can be stored in liquid nitrogen. Cultures stored under oil can survive for many years; all of 30 wood-inhabiting species stored under mineral oil for 27 years produced a viable culture (Perrin, 1979). The easiest system is to store cultures in water as described by Jones *et al.* (1991). However, cultures can senesce and die, so it is necessary to remove and regrow stored cultures on a regular basis. How often this is done depends on the practitioner: Watling (1971) suggests every 6 months while Stamets (1983) is content with an annual overhaul.

This technique works for polypores, agarics, puffballs and some boletes (Stevens, 1974), but the mycorrhizal boletes have been found to require special attention. Sometimes the hyphae do not grow out of the block if placed on solid agar, but when the tissue is floated on the edge of a small piece of sterilized paper on a nutrient solution, hyphal filaments can be observed (Norkrans, 1949). It is important that the tissue is no smaller than 5 mm^2 due to hyphal death during cutting preventing new growth. Re-culturing should take place every few months and cultures should be stored at 10°C (Stevens, 1974). If the basidiocarp is very small, it can be surface-sterilized (see below) and used as the inoculum.

Cultures can also be obtained from tissue derived from other organs.

The methods below (Watling, 1971) and for mycorrhiza essentially involve surface-sterilization and plating onto agar media. Sclerotia are surface-sterilized with 10% sodium hypochlorite solution for 10–20 minutes and then sliced to obtain internal tissue as the inoculum. Rhizomorphs and mycelial cords from the base of basidiomes may be treated in a similar way. Wood can be used as a source of inoculum by removing a sample from the growth front, the edge of decay, and soaking it in 10% sodium hypochlorite for 10–20 minutes, rinsing, and placing a fragment on agar. Mycorrhiza can be used; the majority of basidiomycetes are ectomycorrhizal; in their case wash mycorrhizal roots in sterile water three times, cut and plate out onto agar. Mycorrhizal species often require thiamine and low-molecular weight carbohydrate sources so it is advisable to use a mycorrhizal medium, e.g. Modess's medium.

Mushroom mycelium can be grown on sterilized cereal grain, and grain colonized with mycelium (grain-spawn) is typically used as a bulk inoculum into composts. Grain makes an ideal inoculum because it gives a large number of inoculation sites, each with a high inoculum potential. The high inoculum potential derives from the nutrient base the outgrowing fungi can utilize. The large number of inoculation sites help to ensure that the compost is rapidly permeated. This is as important for the exclusion of competitors as it is for the rapid production of a yield of fruitbodies. Rye and millet are most often used because of their low price and ability to separate into individual kernels. Buy organic grain to avoid the unfortunate effects of fungicides!

These are typical grain-spawn recipes: 240 ml grain to 170–200 ml water; 600 ml grain to 400–460 ml water (adjust to 50% moisture by weighing a sample, drying it, and re-weighing). It is necessary to add calcium for the growth of *Agaricus bisporus* and *Coprinus comatus* (and probably others) on grain: add chalk (calcium carbonate) and gypsum (calcium sulphate) in the ratio 1:4 to give 1–3% by weight.

Sawdust/bran spawn made by adding 4 parts sawdust: 1 part cereal bran, adjust to 60% moisture is used as an alternative for wood-rotters.

The above must be sterilized before being inoculated (a two-stage 'cook' is best). The spawn is autoclaved at 121°C for 30 minutes. After 24 hours the process is repeated. This kills any heat-resistant organisms that have survived and germinated after the first period of autoclaving. Glass 'Kilner-jars' and a domestic pressure cooker are sufficient for small-scale spawn production. The grain can be inoculated by placing mycelium-coated agar on the grain, closing the container, and then shaking it to distribute fragments of mycelium. Alternatively, mycelium can be suspended in sterile water and the mixture added to the grain. The inoculated containers should be stored at 20–25°C in the dark. Mycelium should be visible after a few days, at which point a further shake will ensure an even distribution of mycelium. Grain-spawn is commercially available for many species. The spawn can be refrigerated, but not frozen, until it is used to inoculate compost.

(b) Compost production. The production of sufficient mycelial biomass to support a useful mushroom yield has traditionally taken place within quantities of naturally ocurring materials. These materials can be relatively unprocessed, e.g. freshly cut wood (logs), or transformed by procedures known as 'composting'.

The compost (this term is used loosely as the substrates mentioned have not necessarily been through a composting process of alteration by microorganisms before fungal inoculation) not only provides the nutritional requirements of the fungus but also the physical and chemical conditions that give the fungus a competitive edge over other microorganisms.

All saprotrophic Basidiomycetes require carbon compounds. In nature these are usually obtained from plant polymers. The enzymes that achieve the breakdown of these polymers can together be effective against a wide range of cellular architectures, and so have an action that is non-specific enough to allow the grower to replace the natural diet with more readily available alternative, e.g. substituting cereal straw for wood. Nitrogen compounds also have to be provided; sources containing the nitrate ion are usually acceptable. Essential elements and vitamins are not usually growth-limiting except for calcium which is often added to regulate pH and to modify the structure of the compost.

Almost any organic material can be composted, but it must be mixed to give a carbon/nitrogen ratio of between 80:1 and 10:1 (Charlesworth, 1995). Chemical analysis of wood shows a C/N ratio of 500–1250:1 (Stamets & Chilton, 1983). Other natural substrata have nitrogen contents that are growth-limiting, so nitrogen supplementation increases mushroom yields in many compost systems. One example is *Lepista nuda*, the woodland Blewit; this fungus utilizes leaf-litter in nature but has been grown semi-commercially on *Agaricus bisporus* mushroom compost (Frische & Loone, 1989).

The fact that *Lepista nuda* can fruit on this range of composts demonstrates a fundamental ecological niche for the Blewit that is wider than its perceived ('realized') niche in the wild. This contrast between the perceived niche of *Lepista nuda* and its fundamental niche may well be found in other species, in which case the example of *Lepista nuda* is a message of hope to the cultivator because it raises the possibility that the composts listed below may serve as a suitable substrate for many different saprotrophic species.

Those fungi which are presently under large-scale cultivation are grown on a spectrum of composts; from non-sterile fresh lignocellulosic material such as straw used for *Pleurotus* spp. (Zadrazil & Reiniger, 1988) and logs used for *Lentinus edodes* (Przybylowicz & Donoghue, 1988) through *Volveriella volvaceae* (Chang & Hayes, 1978) on straw composted for a few days, to *Agaricus bisporus* on straw supplemented with nitrogen (manure) composted in two phases over 3 weeks. The processes of compost preparation will only be outlined below; fuller treatments are available in the references (Stamets & Chilton, 1983; Fermor *et al.*, 1985).

Natural bulk substrates. The use of non-composted and semi-composted materials is common among commercial growers of *Pleurotus*, *Volvariella*, *Flammulina* and *Stropharia*. These materials can be sterilized, pasteurized, or used untreated in their natural state. For example, I have grown *Pleurotus ostreatus* outdoors on a 'bale' of unsterilized wheat straw, soaked in cold tap-water, with good results. The *Pleurotus* mycelium grew in all but the outer 5 cm of the bale. When growing mushrooms indoors however, problems with fungal and other contaminants make it prudent to pasteurize or sterilize the compost.

Steineck (1984) listed many species he had grown on straw or wood-chips buried in damp and shady places in his garden and these materials can form the basis of both garden and commercial-scale cultivation.

CEREAL STRAW Straw, clean and unspoiled by any previous decomposition, is prepared by chopping the straw into 2–8 cm lengths and soaking in water for about 3 days to achieve 75% moisture. Baled cereal straw is ideal as it needs no further chopping. This is sufficient for competive species such as *Pleurotus ostreatus* but to reduce losses from competition the straw can be pasteurized by using hot water or sterilized by steam. Straw is usually inoculated with grain spawn at the rate of 1–2% by weight (Stamets & Chilton, 1983).

DECIDUOUS LEAVES Leaves were used by Matruchot (1917) to grow *Lepista nuda* in caves left by limestone mining. In the process of replicating his experiments in a local disused mine it was found discovered that standard mushroom compost could be used as a substrate instead. The yields from mushroom compost are higher, but, alas, are generally used in a less romantic setting. Leaf mould has a long history of use in horticulture, traditionally by being stacked outside in piles 60–90 cm deep for 2 years to allow it to decompose (Larkcom, 1976). The resulting fibrous material is used as a mulch or as the water-holding component of seed, cutting and potting composts.

The need to replace peat (Pryce, 1991) may create a niche for leaf-mould. Preliminary experiments using leaves inoculated with *Lepista nuda* produced horticulturally useful leaf-mould in 6 months (Scrase, 1993). Leaves are often available in the autumn from civic authorities and should not be collected from the wild.

WOOD-LOGS The most notable commercial species grown on wood is *Lentinus edodes* or Shiitake. Logs (traditionally oak) 7–15 cm in diameter and 90 cm long are cut between autumn and spring when sap content is highest. Care is taken to leave the bark intact as this has consequences for subsequent fruiting. Holes 2–3 cm deep are drilled 20 cm apart into the log; these holes are then plugged with sawdust spawn or mycelium-colonized dowels (Przybylowicz & Donoghue, 1990).

WOOD-CHIPS Freshly cut chips of alder (*Alnus glutinous*), maple (*Acer* spp.) and fir (*Abies* spp) can support mycelial growth. Because alder is high in sugar content, without resins, and abundant, it has often been selected (Stamets & Chilton, 1983).

In spring, freshly cut branches are chipped, mixed with the inoculum, which itself can be grown on wood-chips, and made into a ridge bed on the ground. Pieces 2–8 cm long give better results than sawdust in non-sterile conditions. The beds size commonly used in 60 cm wide by 15 cm deep. Incidentally, the beds of wood-chips and bark used to cushion falling children in parks frequently support massive fruitings of mushrooms, e.g. *Hypholoma fasciculare* (sulphur tuft) and the false morel *Gyromitra esculenta* (personal observation).

SAWDUST The particulate nature of sawdust makes nutrients more readily available; consequently, competing organisms can rapidly invade these materials. Therefore the sawdust-based substrate is usually sterilized or pasteurized before inoculation. Generally, the sawdust type is matched to the fungus, but alternatives to the usual host species can be substituted.

Even softwood sawdust can be substituted for hardwood if it is aged for about a year to allow the reduction in the amount of resins and phenolics. More commonly, growers use a blend. The time taken to degrade the substratum is related to particle size and in Shiitake the optimum has been found to be 2–3 mm (Nisikado *et al.*, 1942). Smaller particles reduce gas flow and restrict growth. Sawdust is usually supplemented to increase the amount of nitrogen and easily assimilated carbohydrates. While high levels of supplementation result in higher yields it also encourages the growth of competitors. The recipes shown below are a result of finding a balance between these two opposing considerations (Stamet & Chilton, 1983).

Standard formulae	Low-contamination formulae
sawdust 80%	sawdust 90%
cereal bran 20%	cereal bran 10%

The optimum moisture level is between 55% and 70% by dry weight before heat treatment. This can be established by oven-drying a sample; however, many growers estimate the proper moisture content by squeezing the mixture in one hand: drops of free water should be squeezed out.

Air exchange is provided by using low-density polyethylene bags or by inserting cotton wool, foam, etc. into the neck of the container holding the substrate.

OTHER LIGNOCELLULOSIC WASTES A tremendous range of materials has been used, especially for the growth of *Pleurotus* spp., e.g. sugar-cane bagasse, pomace (apple waste), sugar beet pulp, coffee, tea, cotton wastes, etc. Unless the waste contains a toxic component, it is likely that by adjusting moisture levels to around 70%, providing aeration, and optimising carbon/nitrogen ratios, any plant waste can be degraded by basidiomycete fungi. At the XIV the Congress on the Science and Cultivation of Edible Fungi (Oxford, 1995), delegates reported on several instances of using basidiomycetes to transform waste materials, e.g. *Auricularia polytricha* has been grown on coffee wood in Mexico, and *Pleurotus ostreatus* used to turn waste-paper into a horticulatural mulching material in the Seychelles.

Agaricus bisporus compost. Fermor *et al.* (1985) give a very full account of the processes involved in preparing compost for *Agaricus bisporus* while Genders (1969) gives a more colloqial description; the following however is drawn primarily from Stamets & Chilton (1983). The whole problem of producing *Agaricus* compost can be avoided by obtaining mushroom compost from a commercial compost producer, complete except for the inoculation of mushroom spawn. The commercial composting process is divided into two stages, commonly called Phase 1 and Phase 2, but the process called 'Long composting' is more appropriate for the amateur grower, as there is no need for the environmentally controlled rooms used for phase 2. processing. Both systems will be descibed here.

The raw materials for the compost are straw, manure, calcium sulphate (gypsum), water, and nutritional supplements. Wheat straw is preferred because its resilience maintains an open, well-ventilated compost structure. The straw is the primary source of carbohydrates and in addition it provides lignin which becomes converted to the nitrogen-rich lignin–humus complex, which provides a source of protein to the mushroom mycelium. Calcium sulphate improves compost structure by aggregating colloidal particles and it also provides a source of metabolic calcium.

Horse-manure (usually ready mixed with bedding-straw) is the traditional source of nitrogen, while so-called 'synthetic' composts are made with chicken manure, but other sources of organic and inorganic nitrogen can be used. Elephant manure has been found satisfactory! (Genders, 1969). The aim is to creat an initial mixture with a nitrogen content of 1.5–1.7%, with a water content of 70% (±3%), and a carbon : nitrogen ratio of approximately 30 : 1. Many materials can be used as supplements to modify the C: N ratio, e.g. urea, cereal bran, or sugar beet pulp. The following recipes are based on using fresh materials, but are given as dry weight ratios. Horse manure has proportions of manure and straw 100 to gypsum 5, while synthetic compost is made up from chicken manure in these proportions: wheat straw 52, chicken manure 45, gypsum 3.5.

To make mushroom compost, fresh manure and straw is mixed, soaked and then stacked. To prevent the development of anaerobic conditions within the stack the vertical cross-section is generally not more than 2 × 2 metres. This allows the microorganisms present in the raw materials to begin to degrade the stack. Their activity heats up the stack which in turn makes conditions ideal for further decomposition by further microorganisms. The stack core is depleted of oxygen in 48 to 96 hours and it is then rebuilt. This procedure is repeated three times during the first phase to allow the decomposition to continue, typically during the third, fifth and seventh day after initial mixing. The centre of the stack should reach 65–80°C. At the end of this first phase the compost should be deep brown, be flecked with whitish colonies of actinomycetes, smell of ammonia, have a pH of 8.0–8.5, and release liquid when firmly squeezed.

In the second phase, the compost is filled into trays, beds or bags, and pasteurized (58–60°C). This kills nematodes, fly eggs and larvae, mites and competitor fungal mycelium and spores. The compost is 'fermented' for a further 5–8 days. During this time the internal temperature of the compost is held at about 50°C by controlling ventilation. This is optimal for the actinomycetes that incorporate ammonia at this stage. The presence, or rather absence, of ammonia is usually detected by smell, but cresyl orange indicator or gas detection tubes can be used. After the ammonia concentration has dropped to 10–20 μg/litre, the temperature is allowed to drop to 25°C and the compost is inoculated.

In the process of 'long composting,' the whole composting procedure takes place outdoors. The stack is turned four times, on days 6, 10–12, 13–15, and 15–17. The anaerobic centre zone is moved to the outside on each occasion. The finished compost is pasteurized for 4 hours. If this is not possible, the outer shell of the stack is discarded and the compost showing the white flecking indicating strong actinomycete activity is used.

Inoculation and spawn running. Mushroom spawn is mixed into the compost, usually as 1–2% by wet weight. Conditions then need to be maintained for the strain in question. The primary requirement is to maintain the core compost temperature at the optimum temperature. This is 25°C for many commercial strains of *Agaricus*, *Lepiota*, *Lepista* and *Coprinus*. The compost usually contains sufficient oxygen and water to support the development of the fungal mycelium for the duration of the growth period. The success of the mycelial growth or 'spawn-run' can literally be seen by the change in colour of the compost. The whitish colour of mycelium is also evident. Similar changes can also be seen with other compost types, leaf mould and straw becomes paler as the fungal enzymes literally bleach their surroundings. However, the apparently successful utilization of the substrate is no guarantee of success when it comes to mushroom production. On several occasions we have had straw bales smelling sweetly of fungal decomposition, all but the outermost layer of straw charged with mycelium, ready to produce . . . nothing!

(c) Induction and maintenance of fruitbody production (cropping). Fruiting can frequently be initiated by inducing the mycelium to go through a developmental switch from mycelial extension to the production of mushroom primordia ('pinning'). Conditions are then maintained that facilitate the development of these primordia into mushrooms ('fruiting').

The initial switch is usually achieved by radically altering the environmental conditions surrounding the mycelium, typically reducing temperature and increasing oxygen. The subsequent mushroom development then requires higher humidity (Flegg & Wood, 1985). While the understanding of the mechanisms of basidiomycete morphogenisis have yet to be synthesized into a complete story, some of the components can be found in the B.M.S. symposium: Developmental Biology of Higher Fungi (Moore *et al.*, 1985), but this area is undoubtedly a major problem in the scientific understanding of mushroom growing. We cannot definitely say why on some occasions mushrooms fruit from a mycelium, and on other occasions they do not – it is another part of life's rich complexity. Observations of fungi in their natural environment may guide the process of cultivation, but it should be remembered that cultivation often removes the constraints imposed by the stress of interactive living. Species that exhibit a ruderal strategy by exploiting recently disturbed habitats may be good candidates for cultivation, as a ruderal strategy implies the rapid utilization of resources and then an early and complete commitment to reproduction. In contrast, the low productivity of stress-tolerant species may not in fact be due to an inability to reproduce with the high yields required for commercial production, but merely be due to the constraints to nutrient uptake imposed by the substrate, e.g. the beefsteak mushroom *Fistulina hepatica* growing on the heartwood of oak. In the wild, mushrooms can be predicted to an extent when certain temperature and moisture criteria are met. As fruiting has traditionally been initiated by changing external environmental conditions, each of these will be considered in turn.

Carbon dioxide. The carbon dioxide concentration in the gaseous environment is a crucial factor in the release of fruiting. Excess CO_2 concentrations can inhibit fruiting completely, even if fruiting has been initiated and the mycelium has formed hyphal knots. At lower concentrations the fruitbody is malformed, typically with an elongated stipe and reduced pileus in Agaricales. *Pleurotus* spp. can produce coralloid branching in place of the pileus at CO_2 concentrations between 600 and 5000 ppm (Chang & Hayes, 1978). Growth parameters listed by one of the authors (Scrase, 1993) show maximum values allowable for fruiting of between 600 and 5000 ppm of CO_2 depending on the species. The aptness of CO_2 concentrations as a fruiting trigger is obvious. The wood-inhabiting species require the lower concentrations; presumably this avoids initiating fruitbody formation within the wood. The coprophiles tolerate higher concentrations, perhaps enabling them to avoid desiccation or the loss of the physical base from which to fruit. An exception to this rule is *Polyporus brumalis*, which can fruit at CO_2 concentrations of more than 5% (Plunkett, 1956). This may be explained by the fact that *P. brumalis* fruits at low humidities and needs to initiate fruiting deep within its substratum to grow a fruit body that can withstand external conditions. Although CO_2 concentration has such a dramatic effect on basidiomycete morphogenesis, no conclusions can be drawn about the metabolic pathway through which the activity is mediated. In *S. commune*, colonies grown under high concentrations of CO_2 have larger diameters and dry weights, and the hyphae have wider and longer compartments (Wessles, 1992). Carbon dioxide fixation has been shown in the microfungi, but the position in basidiomycetes is less clear. Pyruvate carboxylase and phosphoenolpyuvate carboxylase have been isolated from extracts of *S. commune* hyphae (Tashibana *et al.*, 1967) and pyruvate carboxylase, malic enzyme and phosphoenolpyruvate carboxylase have been isolated from *A.*

bisporus (Bachofen & Rast, 1968). The primary product of these enzymes is oxaloacetate, an intermediate of the tricarboxylic acid cycle which feeds mainly into amino and nucleic acid production. We are left with the question, does CO_2 fixation contribute directly to increased growth? Lack of an accumulation of carbohydrate in hyphae and a lack of isocitrate lyase (a central enzyme in the glyoxalate cycle) in *S. commune* grown at high CO_2 (Cotter *et al.*, 1970) suggests that CO_2 does not increase carbohydrate metabolism, but operates through other mechanisms. It is possible that the reduction of CO_2 fixation may alter the C:N ratio of available nutrients, which has been shown to act as a fruiting trigger. Carbon dioxide fixation does not provide the mycelium with significant quantities of carbon, so a more likely scenario is an effect on the redox potential or oxygen uptake. Experiments tracing the incorporation of traceable forms of carbon into fungal mycelium are needed.

From a practical view-point, ventilation of the fruiting chamber or room is increased to reduce carbon dioxide below the required threshold. Because ventilation has consequences for the control of temperature and humidity, as well as pest management, varieties that are relatively unaffected by carbon dioxide could be comercially useful.

Light. Light has a role in the initiation of fruiting, and then in the normal development of stipe and pileus in most species. Ross (1984) states that the mycelium has to 'attain competence before it can receive and respond to external stimuli'. He also poses the following questions: 'What stimulus is required, at what age is the mycelium able to respond to the stimulus, and what parts of the mycelium are able to respond? What is the molecular apparatus that receives the stimulus and how is that reception translated into genomic activity?' In the case of *Coprinus congregatus*, the stimulus is ultra-violet and blue light of wavelengths 320–400 nm and 400–520 nm, respectively. This suggests a flavin as a receptor but none has been isolated (Wessels, 1992). The timing of the stimulus is important; the duration of light and dark necessary for normal development depends on species and strain (Manachere, 1976). The time it takes to 'attain competence' depends on the quality of nutrition (Ross, 1984). Light may be influential at two developmental stages. Light is initially 'received' by young, actively growing mycelium. This same mycelium is then stimulated again by light to produce primordia. The first period of light creates an increase in branching angle and septation, leading to the creation of a network of thick-walled cells; these can act as a nutrient sink and hence create the possibility for primordia to expand. Additionally, there is a reduction in the activity of a wall-bound phenol oxidase, associated with melanization (which may be caused by the inactivation of the enzyme). A second period of light was necessary for primordia initiation in *C. congregatus*. In *Coprinus cinereus* the effect of light can be bypassed by low temperature treatment and nutrient starvation (Wessels, 1991). The effect of light on *Schizophyllum commune* can be completely inhibited by high CO_2 concentration. Light has been found to raise cAMP levels. Protein phosphorylation induced by cAMP has been linked to activation of glycogen phosphorylase and inhibition of glycogen synthase; this could mobilize reserves of glycogen for fruitbody growth. It has also been shown that cAMP protein phosphorylation is regulated by mating-type genes (Wessels, 1992). Adding cAMP may 'kick-start' fruiting in some species or varieties, but can cause abnormal morphognesis in others. In conclusion, Ross's questions have still to be answered, but there are clear possibilities for using light to control the timing and duration of mushroom production in light-sensitive species. The existence of light-independent mutants raises the possibility of breeding out the need for light during mushroom growth.

Temperature. A reduction in temperature often releases fruiting in temperate species. Usually, the change needs to exceed 5°C, although some species or varieties have a threshold temperature above which they fail to fruit. It is possible that the effect of a drop in temperature on the relative humidity of compost air-spaces and ambient air or the general temperature effect of lowering metabolic rates and hence CO_2 concentrations could be the mechanism for change. Tropical species are usually triggered by rainfall rather than temperature. *Pleurotus ostreatus* is often given an extreme cold shock, 25°C down to 2–5°C, to induce fruiting, although this is not strictly necessary. This extreme change may damage the mycelium in some way. The stress proteins produced in other eukaryotic cells after temperature change are brought to mind. There is the possibility of breeding out the need for a temperature change before fruiting, although the more obvious option is to use non-temperate strains of the species in question.

Water. The optimum water content for the mycelial substratum is between 35–60% for wood, and 60–80% for other substrates. These figures reflect the need for oxygen balanced against the requirement for water. The mycelium has a limited ability to waterproof itself as it leaves a solid substratum and enters the air. Aerial mycelium, aerial rhizomorphs and mushrooms all lose water that must be replaced from the submerged mycelium. In mushrooms, water may be drawn in osmotically by the stipe producting trehalose or mannitol and the pileus producing urea or mannitol, depending on the species (Gooday, 1984; Hammond, 1984). Other osmotically active compounds may well be found, since having different compounds for stipe and pileus could provide for independent control of their expansion. Additionally, evaporation may create a flow analogous to plant transpiration. If evaporation exceeds this flow, mushroom growth ceases, and so the ideal relative humidity (RH) for the fruiting of many species is between 75–95%. The fact that fruiting can be inhibited by humidities above 95% RH may indicate the importance of evaporation for normal development. However, very high humidities cause condensation on the surface of the substratum and this may be the cause of the real problem by cutting off oxygen to the primordia. Surface water also encourages the growth of bacterial diseases. The compartmentation of niches within wood (Cooke & Rayner, 1984) according to water potential underlines the fact that different species have different optima. Although the fundamental niche for this characteristic is likely to be wider than the realized niche, water content is crucial. During preparation of inocula of the wood decomposer *Coriolus versicolor* for insertion into logs, wooden dowels were sterilized within Kilner jars and themselves inoculated with mycelium on agar. After mixing, the jars were incubated. Growth was best at a zone in the middle of the jar; waterlogged dowels at the base of the jar were lifeless and the dryness of the dowels at the top restricted growth. This system of setting up a gradient for moisture is now being used to determine the optimum conditions for growth in other species.

Mechanical damage. An anecdote from a witness to Shiitake growing in Papua New Guinea gave rise to this section. It was stated that incubated logs were beaten all over to provoke fruiting. This may have a similar effect to the sequence of allowing the logs to dry, and then soaking them, which is used to induce quick, heavy crops in Shiitake. The practice of raking casing soil in *Agaricus* cultivation has a similar effect. The effect could be due to a cessation of growth at the colony margin, if the young hyphae produce a translocatable growth inhibitor or prevent primordia growth by being a more competitive sink for resources. Alternatively, the release of mycelial contents may have a stimulating effect. Whatever the cause, this is an option for the frustrated grower!

External structural organization. Tamblyn & DeCosta (1958) had great success in producing fruitbodies of wood-inhabiting basidiomycetes. Their method was to place an autoclaved wooden top onto an incubated sawdust and bran mixture. The top was hollowed to receive water. It could be that this method provided sufficient gradients of water and air such that ideal conditions for fruiting would be found somewhere in/on the top. It is possible, however, that the very structure of the wood orientated mycelium into primordia. Whatever, fruiting bodies usually formed. *P. ostreatus* initials grown on straw often arise inside the lumen of the straw. If there is anything in this it raises the possibility of using natural or artificial three-dimensional structures to case mushroom composts.

Casings. When a compost is colonized by fungal mycelium it may be covered by a layer of material to induce fruiting. In *Agaricus bisporus*, the layer – commonly composed of moist peat and chalk – is called the 'casing'.

The function of casing layers is to induce fruitbody initiation. The physical and chemical properties must be capable of retaining water, allowing gas flow, having a low salt content, a low level of nutrients and, for *Agaricus bisporus*, be slightly alkaline. A casing often enhances fruiting even if it is not strictly necessary, e.g. *Coprinus comatus*. There is a relationship between the bacteria in the casing and fruitbody initiation which may be due to stimulation by the bacteria, or may be due to the bacteria removing substances inhibitory to fruiting. This latter theory is supported by the facts that activated charcoal can be used to replace the activity of bacteria and that different bacteria; *Pseudomonas putida, Bacillus megaterium, Arthrobacter terregens* and *Rhizobium melioti* have similar effects on the fungus (Flegg & Wood, 1985). The casing also tops the cropping surface from drying out and provides a humid microclimate. The low level of nutrients causes a change in mycelial branching angle that can lead to formation of cords and then primordia. When conduction trials on new species it makes sense to cover the colonized compost with a casing with a gradient to give different depths of cover in the hope that somewhere on the slope will have the ideal microenvironment for fruiting.

Mushroom growth maintenance. Once the primordia have formed, the control of environmental conditions may become less critical, and the optimum conditions for the expansion of the fruitbody may be different. In *A. bisporus* for example, the temperature can be raised from 16°C to 25°C once the mushroom is >6 mm in diameter, without aborting growth. This fact is used to control flushing duration. The pattern from the literature is that optimal mushroom growth is at lower humidities than fruitbody initiation. This could simply be because of the decreased surface area/volume ratio in the growing mushroom making evaporation a relatively smaller problem. The movement of resources from vegetative mycelium to fruitbody has consequences for the survival of that mycelium. One strategy used by Shiitake growers is to limit fruitbody yield in a single flush, in order to allow the mycelium to be vigorous enough to defend and maintain its territory, and flush repeatedly, providing greater long-term yields.

17.2 Production of *Agaricus bisporus*

In the previous sections we have provided a generic overview of the processes involved in the production of all types of mushrooms. In this section the production of the principal cultivated mushroom, *Agaricus bisporus*, is described in more detail. For more detailed information still the reader is referred to the books *Modern Mushroom Growing* (Vedder, 1978) and *The Cultivation of Mushrooms* (van Greinsven, 1988). For a general account of the biology and technology of *A. bisporus* which covers the areas described in this section and in section 17.3 see Flegg *et al.*, 1985. For more specific information, the Proceedings of the International Congress of the Science and Cultivation of Edible Fungi published since 1950 and known as the Mushroom Science series are a good source, e.g. Elliott (1995). Throughout the rest of this section the use of the term 'mushroom', should be taken to mean *A. bisporus* unless otherwise indicated. The earliest known description of mushroom cultivation is that of Bonnefons in 1650 (Bonnefons, 1650), but the modern industry has really developed in the last 100 years and particularly in the period after World War II. In the developed world, mushroom production is intensive, highly mechanized and carried out in purpose-built structures in which the cropping environment, in particular temperature and humidity, can be carefully controlled.

In the developing world, much lower levels of technology are utilized with the inevitable, but non-problematic consequence in this situation of reduced yields and quality. Table 17.2 shows levels of mushroom production from the major producing countries but, as indicated earlier, mushrooms are grown in most countries of the world from Iceland to the Yemen.

Mushrooms, as stated, can be grown not only in purpose-built structures – often insulated plastic tunnels – but also in a wide variety of other buildings converted for the purpose and in caves, tunnels and mine-workings.

Table 17.2 Production of the white button mushroom, *Agaricus bisporus*, worldwide

Country	Production (tonnes)
USA[a]	300 000
The Netherlands[a]	225 000
China	200 000
France[a]	170 000
UK[a]	134 000
(GB)	(94 000)
Italy[a]	101 000
Poland	70 000
Germany[a]	60 000
Spain[a]	60 000
Taiwan	60 000
Canada[a]	55 000
Belgium–Luxembourg[a]	43 000
Ireland[a]	40 000
Australia[a]	38 000
South Africa[a]	18 000

Compiled from various sources.
[a] Data from *Der Champignon* (1996), **6**, 270, and from the International Society for Mushroom Science. Figures given are for 'cut' mushrooms, i.e. with 'trimmed' stalks; approximately 20% should be added for uncut weight.

The commercial cultivation of the mushroom involves three essential elements: substrate (compost), preparation inoculum (or to use the industry's jargon 'spawn') preparation, and thirdly the production of the mushrooms themselves. These essential elements can be further subdivided in process terms and these are described below and summarized in Table 17.3.

17.2.1 Making compost

The quality and nutrient value of the substrate, i.e. compost, is the major determinant of yield. Composts result from the activities of a variety of microbes on a diverse range of raw materials. Historically, all mushroom farms made their own compost, but a recent trend has seen the growth of dedicated compost makers who supply their product to growers. Composts are made from cereal straws and animal manures and almost any combination of straws and manures can be made into a mushroom-producing compost. From country to country, economics and easy availability determine compost formulation.

In the UK and elsewhere in the developed world there has been a move away from traditional horse manure/stable bedding-based composts to mixtures based on wheat straw and the litter from intensively reared chickens. Typical formulations are shown in Table 17.4.

A properly prepared compost favours the growth of mushroom mycelium at the expense of other competitor fungi.

In the early history of mushroom cultivation compost was prepared in a single step out of doors. The stack was turned periodically until microbial activity subsided – indicated by the compost no longer heating up, freedom from evolving ammonia and a relatively pleasant odour! This single-phase system was used until c. 1915 when it was discovered serendipitously that allowing compost to heat up in stacked beds was beneficial to yield and reduced pests and pathogens. Modern composting practice follows these principles and is a two-phase process. Phase I is preceded by a so-called prewetting stage in which the straw is wetted to increase its moisture content. Phase I is carried out in windrow stacks with varied but limited protection from the environment, while in phase II the compost is kept in a controlled environment at 52°C for c. 7 days and at some point during this process the temperature is raised to 60°C for several hours. This pasteurization period kills insects pests, fungal pathogens and viruses, but leaves the populations of thermotolerant and thermophilic microorganisms unharmed. This microbial biomass, which can be up to 2% of compost dry weight also contributes to the nutrition of the mushroom (Fermor & Wood, 1991).

The current emphasis in compost R&D is to reduce odour and run-off pollution. This is being achieved by a range of approaches which all have the same basic objectives: (i) to be able to control the environment

Table 17.3 Production cycle of *Agaricus bisporus*

	Activity			
Stage	Human	Microbes	Temperature	Duration
Compost preparation			**Uncontrolled environment**	
Pre-wet	Wetting straw and other ingredients	Mesophiles active	Ambient, some heating up	c. 7 days
Phase I	Making stacks. Mixing ingredients, wetting, adding proprietary activators	Mesophiles and thermophiles active	From ambient up to 70°C in centre of stack	c. 14 days
			Controlled environment	
Phase II	Transfer to controlled environment	Thermotolerant fungi/actinomycetes predominant	Held at 52°C except for pasteurization at 60°C	c. 7 days (2–3 h for pasteurization)
Cropping				
Spawning	Mixing spawn with compost	–	Ambient	<1 day
Spawn-run	–	Mushroom mycelium colonizes the compost	25–28°C	c. 14 days
Casing	Covering compost with 'casing' peat/chalk mixture	–	Ambient	<1 day
Pre-cropping	–	Mushroom mycelium grows into the casing	25°C	c. 7 days
Initiation of mushroom fruitbodies	Introduction of fresh air, ruffling	Mushroom mycelium begins transition from vegetative stage to fruiting stage	16–18°C	c. 14 days
Cropping	Harvesting of mushrooms. Watering	Mushrooms produced in flushes at c. 7-day intervals	16–18°C	c. 28 days
End of crop	'Cooking-out'/cleaning of cropping rooms. Disposal of 'spent' compost	Mushroom and other microorganisms killed	'Steam' sterilization, 60–70°C	c. 1 day
Total duration				c. 94 days

Table 17.4 Formulations to produce 1 tonne fresh weight of mushroom compost

Ingredients (excluding added water)	Dry matter (%)	Fresh weight (kg)	Dry weight (kg)
Horse-manure/stable bedding compost			
Stable bedding	37	882	326
Chicken manure	60	88	53
Gypsum	80	27	22
Chicken litter/straw compost			
Wheat straw	85	322	274
Chicken manure	60	356	214
Gypsum	80	27	22

Modified from Derikx *et al.* (1991).

Table 17.5 Microorganisms abundant in compost during phase I and phase II

Bacteria	*Bacillus subtilis*
	Bacillus coagulans
	Fluorobacterium spp.
	Pseudomonas spp.
Actinomycetes	*Streptomyces thermovulgaris*
	Thermoactinomyces vulgaris
Fungi	*Aspergillus fumigatus*
	Chaetomium thermophile
	Cladosporium herbarium
	Humicola grisen-thermoidea
	Humicola languginosa
	Mucor spp.
	Mucor pusillus
	Penicillium spp.
	Trichoderma viride

throughout composting; and (ii) to speed up the process. These approaches have been variously described, but commonly, as indoor-composting or controlled environment composting.

17.2.2 Compost microbiology

Composting is the result of microbial activity and a number of authors have isolated and enumerated the microbes present in compost. Table 17.5 (modified from Derikx *et al.*, 1991) lists those that are abundant or frequently found during both phase I and phase II.

During the composting process there is a microbiological succession. Initially, mesophiles are predominant, utilizing the most readily available carbohydrate and nitrogen sources. As the compost temperature rises as a result of this intense microbial activity there is a progressive reduction in species types with thermotolerant and thermophilic microorganisms becoming pre-eminent. The mesophilic bacteria are replaced by thermophilic spore formers; mesophilic fungi are succeeded by thermotolerant species which grow and sporulate up to 45°C. In the later stages of composting, when readily available nutrients are limited, thermophilic actinomycetes increase in number producing sporulating colonies on the compost surface and imparting a white colouration known as 'fire fang' by mushroom growers. These actinomycetes produce cellulases which can attack long-chain carbohydrates. About 10% of the celluloses/hemicelluloses present are utilized by the compost microflora during the composting process (Smith, 1980). However, this fraction still represents 30–50% of the remaining dry matter, on the completion of composting.

Compost and the composting process represent a complex ecosystem and while the broad principles are understood in microbial terms, the interactions and relative importance of specific microorganisms in the process is less clear. During phase II, the presence and growth of specific thermophilic fungi, in particular *Scytalidium thermophilum*, is considered paramount by some research groups (Houdeau *et al.*, 1991; Straatsma *et al.*, 1994). *S. thermophilum* is believed to play a role in selectivity; there appears to be a relationship between mushroom yield and density of *Scytalidium* in the compost; and this fungus has been shown to stimulate the extension rate of mushroom mycelium *in vitro* (Straatsma *et al.*, 1991). Microorganisms; bacteria, fungi and actinomycetes, their interactions and successional relationships, are the key to the production of high-yielding mushroom composts. They change the chemical composition, effect the physical nature of the ingredients, contribute to the nutrition of the mushroom, and produce a substrate which favours the growth of mushroom mycelium. And they can manage all of this with little intervention from humans! So, to quote from Stutzenberger (1990), 'salute when you pass the compost pile wherein reigns the noble thermophile . . .'.

17.2.3 Spawn and spawn running

Once a compost has been prepared it is ready for colonization by the mushroom mycelium. The mycelium is introduced into the compost in the form of colonized cereal grains called spawn. These grains provide a convenient inoculum which can be readily mixed through the compost and this process is usually mechanized. For mushroom production, 0.5–0.8% spawn by weight of compost in used. Mushroom spawn production is in the hands of a sophisticated and technologically advanced industry. Rye and millet grains are the cereals of choice, but almost any type of grain can be used. The grain is wetted to raise its moisture content to c. 40% by weight. The wetted grain is mixed with gypsum ($CaSO_4$, 2% by weight) and powdered chalk ($CaCO_3$ 0.5% by weight) and autoclaved to eliminate potential contaminants, in particular, thermophilic bacteria which can cause 'souring' of the wetted grain. The autoclaved grain is inoculated with mushroom mycelium and incubated to allow colonisation. A range of spawn types are available to the grower.

After the spawn is mixed with compost it begins to grow and ramify through it – a process described as 'spawn running'. A successful spawn run requires a temperature of around 24–25°C, which is the optimum for vegetative growth of mushroom mycelium and a high humidity (c. 90% RH) to prevent the compost drying out. Mushroom compost can be spawn run in individual containers or in bulk and it has also been found beneficial during spawn-running to have a concentration of CO_2 of around 2%.

The colonization of the compost is completed in 11–14 days, depending on spawn type.

17.2.4 *Supplementation*

Growth and yield can be improved by the addition of nutritional supplements, normally at the time of spawning. A variety of proprietary products are available and they are essentially high-protein, slow-release materials; their use has become widespread since their development in the mid-1970s.

Contraindications for use are where control of compost temperatures are problematic, where thorough mixing of supplement with compost is difficult, and where the quality of the compost may be suspect.

17.2.5 *Casing*

For colonized compost to produce mushrooms it must be covered with a so-called casing layer. In the early days of mushroom production, loam soil was used but this has how been superseded by the use of mixtures of peat and chalk. This peat casing is used to cover the compost to a depth of 3.5–5 cm, has an alkaline or neutral pH, provides physical support for the developing mushroom, and has good water-holding capacity which has been shown to be essential if high yields are to be obtained.

The role of the casing in stimulating fruiting is not completely resolved. It is nutrient-poor in comparison with the compost, which contributes to promoting the transition to the fruiting phase and initiation is also encouraged by reducing CO_2 to ambient levels and by lower temperature (16–18°C for fruiting). The bacterial microflora naturally present in the casing also has a vital role to play. Use of a sterile casing layer results in a distinct lack of mushrooms. Experimentally casing a compost with sterilized charcoal does result in the production of mushrooms and this has led to the hypothesis that the bacterial microflora may metabolize or bind a self-inhibitor of fruiting produced by the mushroom mycelium. A Petri-dish assay of bacteria present in the casing has shown that strains of *Pseudomonas putida* can stimulate the production of fruitbody initials.

Other materials have been evaluated for use as casings, e.g. coir, vermiculite, but none has yet been shown adequately to replace peat-chalk casing. The experimental casings either caused a drop in yield and/or encouraged the development of pests and diseases. A likely way forward will be the ability to substitute other materials for some of the peat in casing mixes.

(a) Spawned casing. A recent change in cropping systems has been the use of spawn in the casing layer. This ensures uniform colonization of the casing by the mushroom mycelium and contributes to an improved synchrony in and distribution of developing fruitbodies. The technique was devised by Maccanna (1985) who mixed colonized compost with the casing. This practice has now been superseded with proprietary casing spawn products in which the mycelium is grown on a relatively inert carrier, as a viable alternative to grain spawn, which cannot in practice be used in the casing layer. The vast majority of intensive growers will now use spawned casing.

17.2.6 Cropping

Between 2–3 weeks after the colonized compost has been covered with the casing layer the first mushrooms appear. During this period the growth of the mycelium from the compost into the casing occurs, in grower parlance, 'the case run'. A technique called 'ruffling' is sometimes used when the mycelium has about one-third colonized the casing. 'Ruffling' is achieved, in effect, by raking the casing. This can be done by hand or by specialized machines and has consequences similar to spawning the casing; an evenness in colonization and a better control of initiation.

When the mycelium reaches the surface of the casing, fruiting is triggered by the introduction of fresh air and a lowering of the temperature as described earlier. Mushroom production occurs at regular intervals (c. 7 days) in well-defined 'flushes' or 'breaks'. In modern commercial production the trend has been to reduce the number of flushes harvested and many growers now take only three flushes compared with five to six flushes 10–15 years ago. Flushes tend to reduce in size, with most mushrooms coming in the first and second flushes, which provides a rationale for harvesting only a few flushes. In practice, the number of flushes harvested is driven by economics; being able to throughput more crops with larger flushes is more profitable than extended and fewer crops. Temperature changes can be used during cropping to delay or speed up flushes to match the demands of the market.

The casing will be watered as necessary during cropping, always between flushes, to prevent drying out and associated problems such as the development of cap spotting bacterial diseases.

17.2.7 Harvesting, storage and packaging

A prerequisite for profitability from modern production systems is to be able to harvest and market mushrooms of the highest quality. Harvesting, the picking of the mushrooms, is still done by hand and picking costs represent 30–40% of production costs in intensive systems. For the major producers there is significant interest in the development of mechanized picking aids and ultimately in robotic harvesting.

Mushrooms are susceptible to discoloration through bruising and have a relatively short storage life (1–3 days at 18°C). Loss of quality post-harvest results from discoloration, continuing development, loss in weight, and changes in texture and flavour. Various post-harvest technologies are used to maintain mushroom quality Noble & Burton (1993). Reducing the temperature of the mushrooms is the most widely used technique for extending storage life (Burton, 1989).

The optimum storage temperature is c. 2°C (freezing will cause irreversible tissue damage). Ice bank cooling of the bulk produce has the benefit of reducing the weight losses associated with traditional cooling technologies.

Modern packaging designs must protect the mushroom from bruising. Experimentally, it has been shown that modified atmosphere packaging can extend the storage life of mushrooms but this system is not widely used for economic reasons.

17.2.8 End of crop

The final stage in the cropping process is the treatment of the compost and casing and the structures from which the mushrooms have been harvested. This treatment should prevent the carry over of any form of contamination to the next crop (see section 17.3). Traditionally, steam has been used to 'cook-out' *in situ* and is very effective; a temperature of 71°C maintained for 2 hours has been shown to kill mushroom and all pests and pathogens. Steam, however, is damaging to equipment and buildings. A second method based on methyl bromide fumigation is also effective but little used due to its toxic nature; the risks associated with its use have led to it being banned in some parts of the world.

There is a move away from *in situ* 'cook-outs', steam or chemical, towards less drastic methods involving disinfecting the emptied containers used for cropping (if not disposable, i.e. bags) and washing down the cropping house. These methods must be balanced against the risk of the carryover of pests and diseases.

The 'spent' compost/casing mixture can be used as a soil improver and, if sufficiently weathered, as a casing but, in general it remains a waste product requiring disposal.

17.2.9 Mushroom growing systems

Almost any container that will hold colonized compost can be used for mushroom cropping The 'containers' most frequently used are polythene bags, trays (generally wooden) and shelves (aluminium). In these systems there is a realistic relationship between volume of compost and the cased area available for cropping. A more recent development is a containerless production system in which the spawned compost is pressed into blocks which have in themselves sufficient structural integrity.

The production process, as we have seen, has distinct elements; phase II composting, spawn-running and the cropping of the mushrooms. These elements have differing environmental requirements and usually are carried out in different purpose-built rooms with the compost being moved between them; a system described as three-zone.

Variations on this theme are seen in the industry with two-zone and even one-zone systems where all operations are carried out *in situ* in the same room. The container used for production usually has an influence on the number of zones in the production process.

(a) Trays. Tray production is generally a three-zone process. Mechanization is invariably used, with the trays transported using forklift trucks and by the use of conveyor tray-handling lines for tray-filling, spawning and casing. For cropping, trays are normally stacked four or five high. There is no standard size for trays; they are rectangular with a cropping area usually between $1\,m^2$ and $3\,m^2$ and a depth of 20 cm. Wood is still the material of choice, although metal and plastic trays have been developed; wooden trays are robust, flexible and repairable. In mechanized lines, if jams occur, wooden trays break before the line does!

(b) Shelves. Cropping on shelves is historically a single-zone system in which compost is filled into shelves, three or four high. Their width is determined by the reach of the pickers, although this system perhaps has most potential for the use of mechanical harvesting. Their length is to a large extent determined by the power of machinery needed to fill the shelves and empty the compost after cropping. The shelf system has been extensively developed in Holland and for large-scale production is probably the system of choice.

(c) Bags. Bag production is the simplest production system with regard to technology. Bags can vary in dimensions, but are usually man-handled at some stage in the production process which limits their size and weight; a common size of bag would have a cropping area of $0.2\,m^2$ and contain 25 kg of compost. They are usually cropped in a single layer on the floor. The evolution of the bag system has been correlated with the development of centralized specialized compost making and the bag grower typically will receive spawn run compost and will only have responsibility for cropping.

17.3 Pest and diseases of *Agaricus* and their control

As with any other intensive cropping system, the production of *A. bisporus* can be subject to a wide range of pest and disease problems including viruses, bacteria, fungi and flies. These pests and diseases are not necessarily restricted to *A. bisporus*, but because of its pre-eminence much of the published literature relates to this species. For many of the other species pest and disease problems will become increasingly apparent as production intensity increases. In this section the major pest and disease problems of *A. bisporus* are described. For those wishing to know more then the book *Mushrooms: Pest and Disease Control* (Fletcher *et al.*, 1989) is an invaluable source, as is *The Pests of Protected Cultivation* (Hussey *et al.*, 1969).

A *sine qua non* for preventing the outbreak of pests and diseases, for limiting their spread, and for ensuring the efficacy of control measures when problems do arise, is hygiene. Hygiene is, without doubt, the single most important control measure for pests and diseases and the concept of hygiene should inform the whole process of mushroom production from the design and layout of buildings right through to the handling and storage of the harvested mushrooms. With regard to design, the key element is to ensure that the opportunity for transmission of pests and diseases from 'dirty' areas to 'clean' areas is minimized. Strategies for pest and disease control should therefore be hygiene-based with the primary aim of preventing the problems in the first place. The overall watchword for P&D control is 'integration'; all control measures from hygiene to the application of specific control chemicals should be considered in the round. There is interest in biological control agents as alternatives to chemicals and, as the pressures to reduce the use of chemicals increase, the need for effective biologicals may become acute.

Pests, pathogens and related problems can affect mushroom cropping either directly by causing spoilage of the mushrooms themselves or indirectly by reducing yield. For the broad spectrum of problems, spoilage of the crop is generally the major component of crop loss. The threat from pests and disease is ever-present. Research is directed at understanding the epidemiology of pathogens and developing methods of control based on more effective use of chemicals, or their replacement by biologicals. Problems alter with time, often as a consequence of changes in methods of production. For example, the bulk processing of compost when introduced resulted in competitor moulds such as *Penicillium chermesinum* causing real problems; this fungus although known to be found occasionally in compost was hitherto trivial. Similarly, in bulk spawn running problems with viral infections can be exacerbated.

In the section below the order of presentation of the major pests and diseases broadly reflects both their current and historical importance as crop pathogens.

17.3.1 Fungi

The mushroom is subject to attack by a wide range of fungal pathogens. The three species causing major problems are: *Verticillium fungicola* var. *fungicola*, 'dry bubble'; *Mycogone perniciosa*, 'wet bubble'; and *Dactylium dendroides* (Syn: *Hypomyces rosellus*) 'cobweb'. The descriptions in inverted commas are the grower names for these diseases and indicate the symptoms caused. *Verticillium* is the most troublesome and if unchecked can totally destroy a crop. Early infection results in the mushrooms failing to develop, remaining as small undifferentiated masses of tissue; later infection distorts the developing mushroom. Current control exploits good hygiene and fungicides containing prochloraz-manganese.

Mycogone is commonly found but is not normally a serious problem. It causes the formation of sclerodermoid masses on which amber to brown drops of liquid develop, the 'wet' of 'wet bubble'. This fungus can be controlled by benomyl-based fungicides as well as by prochloraz-manganese. *Dactylium* is rarely epidemic, but is troublesome to eliminate and seems to be an increasing problem. It grows as a coarse 'cobweb' of mycelium over the developing mushrooms. It can be controlled by good hygiene and the use of fungicides. Other fungi are troublesome in the compost and prevent mushroom mycelial growth. They may be both competitors for the resource and actual pathogens on the mushroom mycelium *Dielhomyces microsporus*, the false truffle, can colonize compost at the expense of the mushroom. The most effective control is an adequate pasteurization phase and the avoidance of elevated temperatures during spawn-running (27–28°C) which favour its growth.

Members of the genus *Trichoderma*, known in the industry as green moulds, can be very troublesome competitors in compost, can act as microparasites to the mushroom hyphae, and can cause cap-spotting on the mushrooms themselves. The epidemiology of these green moulds is not completely understood. Effective pasteurization aids control but reinfection via spores can occur. This disease develops more readily around the mushroom-colonized grain of the spawn inoculum; an effective control is to treat the spawn rather than the compost with benomyl fungicides. This targets the fungicide to the point of infection and reduces significantly the amount of fungicide needed to treat a crop. *Trichoderma* species are also a problem in Shiitake cultivation, developing around the Shiitake inoculum and limiting growth.

17.3.2 Bacteria

The main bacterial disease of mushroom is bacterial or brown blotch. The surface of the mushroom caps becomes covered with brown blotches and they are rendered unmarketable. The causal organism is *Pseudomonas tolassii*. A related species, *P. gingeri*, causes ginger-coloured blotches. Biological antagonists have been developed for blotch disease, but the most effective control is via the environment through control of humidity and avoiding the 'dew-point' and the presence of free water on the mushroom caps. The use of chlorine as a bactericide in the water used on the crop also aids control.

Other bacteria can cause problems from time to time. Mushroom gills can be attacked by *P. agarci* causing the symptom 'drippy gill' which gives the disease its name. Bacterial soft rots can cause problems in the cultivation of another *Agaricus* species, *A. bitorquis*, with the higher temperatures required for its growth.

17.3.3 Flies

Mushroom flies belonging to the insect group Diptera are a perennial problem in mushroom growing. Three groups are involved, sciarids, phorids and cecids. Sciarids, mainly the species *Lycoriella auripila*, can affect the stages of mushroom production post-pasteurization. Their larvae will tunnel in mushroom stipes but the most important damage is done to the 'pin-head' stage in mushroom development; damaged 'pin-heads' do not become mushrooms. The adult flies are a nuisance to the pickers but more importantly can act as dispersal agents for a range of pathogens. Stopping flies gaining access to phase II and spawn-running rooms is an important element in control strategies. Casing treatment with insecticides such as diflubenzuron will kill larvae. Adult flies can be knocked down with pyrethroid sprays. The phorid fly *Megaselia halterata* is a damaging pest because its larvae feed solely on the mycelium of the mushroom. Infestations can therefore reduce yields, although sciarid larvae, larva for larva cause more damage, i.e. at equivalent larva densities, sciarids cause more crop loss. The insecticide, diazonon is incorporated in the compost to kill phorid larvae. The adult flies cannot get airborne when temperatures are below 11–12°C, so in some climates this fly will tend not to be a pest at certain times of the year.

The third fly pest, the cecid, is rarely seen as an adult. Its larvae are paedogenetic; each cecid larva can give birth to 12–20 'daughter' larvae in the absence of adult flies. Following infection, enormous populations can rapidly develop. Stringent hygiene coupled with effective pasteurization provides control.

17.3.4 Viruses

Mushroom disease caused by a virus was first described in 1950 (Sinden & Hauser, 1950) but the fact that a virus was the causal organism was not demonstrated until much later (Gandy & Hollings, 1962) when it was shown that diseased mushrooms contained isometric virus particles. Virus infections could have quite catastrophic effects on cropping; symptoms of virus disease are highly variable but the classic symptom is regarded as the 'drumstick', a graphic description of the affected mushrooms in which the stipe is elongated and the cap fails to develop. In terms of diagnosis, the only consistent and meaningful symptom is otherwise unexplained loss of yield. Viruses can be transmitted between mushroom strains by hyphal anastomosis, but infected mushroom spores are a more important source of potential infection. There is currently no effective treatment to control virus once developed. Rigorous hygiene is the key to managing virus as a problem and a change of mushroom strain may help to break the chain of infection.

The presence of a small amount of virus may be non-consequential for cropping. What controls the transition for low to high levels of virus, to the point at which the level of virus can affect yield, is still not understood. Molecular approaches through recombinant DNA technologies may offer the prospect of control by limiting viral replication.

17.3.5 Mites

Mites are prevalent during phase I of composting, but are killed during the pasteurisation component of phase II. Mites are seen in the crop after spawning but most do not affect cropping. The mushroom mite, *Taisonemus myceliophagus* is the only mite which has been clearly shown to cause damage. The mushroom stipes become discoloured at the base and the mycelial attachment to the casing layer may be severed. The presence of these mites is encouraged by the growth of various contaminant fungi and they are an indicator that all is not well with the compost. A more dramatic indicator of poor compost quality is swarms of the red pepper mite (*Pygmephorus* spp.). These mites do not feed on mushroom mycelium but on contaminating moulds in the compost. Predatory mites which feed on mushroom fly pest larvae are also found and attempts have been made to develop them as biological control agents.

17.3.6 Nematodes

Nematodes (eel-worms), which are found associated with compost and casing, are saprophagous, feeding on decaying organic matter, and mycophagous, feeding on mycelium. They are not normally a primary problem but a secondary indicator that all is not well with the crop. Enormous infestations can develop in poor composts but, in practice a well-controlled pasteurization during phase I will prevent a problem occurring. A beneficial use of a different group of nematodes, the insect parasites, has been their development and exploitation as a commercial product for the control of mushroom fly-pests (Richardson & Grewal, 1991).

17.3.7 Disorders

A number of other conditions occur during cropping in which neither pests nor diseases have been incriminated. Examples are 'hard gill' in which the gill tissue does not form properly and is brittle or may in extreme cases be absent, and waterlogging in which the mushroom tissue becomes spongy and saturated with water. These conditions, for want of a better explanation, are attributed to abiotic factors such as temperature, humidity, CO_2 concentration and watering regimes, acting alone or in combination.

A very specific condition known as rose-comb in which distorted gill-tissue develops on the mushroom cap and in which the affected mushrooms are grotesque in appearance, has been shown to be caused by contamination with hydrocarbons such as diesel oil or creosotes. Elimination of the contaminant eliminates the rose-comb.

17.4 Species grown commercially in an *Agaricus* growth system or requiring 'casing'

The following species can be grown on standard *Agaricus* compost. Several have a requirement for light to induce fruiting.

17.4.1 Coprinus comatus

Coprinus comatus is a coprophilic species, often found in huge quantities after artificial enrichment of grassland by slurry or manure. The species grows very readily on mushroom compost, with yields approaching those for agaricus. The fruitbodies release their spores by a controlled 'self-destruction' in which the fruitbody deliquesces into a black 'ink'.

Consequently, the shelf life for this species is insufficient for mass marketing. The species is ideal however, for home growth and consumption. Fruitbodies can be preserved by cooking as this destroys the enzymes that cause deliquescence.

Basic growth data

Spawn run:	25°C, RH 90%+, at 10 days
Casing:	3–5 cm peat/chalk, pH 7.0–7.5
Post-casing:	25°C, RH 90%+, at 10 days
Pinning:	18°C, RH 95%+, CO_2 < 1000 ppm, daylight
Cropping:	18°C, RH 85–90%, CO_2 < 1000 ppm, daylight
	Flushing interval: 7–10 days

17.4.2 Lepiota spp.

Parasol mushrooms are considered good to eat. Reports of successful fruiting in the literature have not led to a commercial variety as yet, however. The fruitbody bruises easily, and the shelf life is only 3–6 days for specimens from the wild. These factors will constrain commercial exploitation. Use *Agaricus* growth procedures, adjusting the temperature regime according to the source of the strain.

17.4.3 Lepista nuda

Blewits (*L. nuda* or *saeva*) have dense flesh and a long shelf-life. They have a strong flavour and make a good addition to soups and stews. A commercial strain bred at the mushroom research station at Horst, Nederlands has not met its early promise and this and other strains give erratic, low yields. Despite this, blewits are being grown commercially. It seems to grow better on a less rich compost, this can be achieved by adding straw or leaves into the *Agaricus* compost before spawning.

Basic growth data

Spawn run:	23°C, RH 90%+, at 25 days
Casing:	3–5 cm peat/chalk, pH 7.0
Post-casing:	20°C, RH 90%+, at 10 days
Pinning:	15°C, RH 95%, CO_2 < 1000 ppm, daylight
Cropping:	15°C, RH 85–90%, CO_2 < 1000 ppm, daylight
	Flushing interval: 14+ days

17.4.4 Psilocybin spp.

Psilocybin species are generally coprophiles and will grow on mushroom compost. Growth parameters vary, depending on the species. The data listed below is for *Psilocybin cubensis*, a sub-tropical and tropical species.

Basic growth data

Spawn run:	30°C, RH 90%+, at 10 days
Casing:	3–5 cm peat/chalk, pH 7.0
Post-casing:	30°C, RH 90%+, at 7 days
Pinning:	25°C, RH 95%+, CO_2 < 5000 ppm, daylight, 7 days
Cropping:	25°C, RH 85–90%, CO_2 < 5000 ppm, daylight
	Flushing interval: 5–8 days.

17.4.5 *Stropharia rugosa-annulata*

This is a large mushroom, commercially grown on wheat straw, but also grown on wood-chips mixed with horse-manure. It is included in this section because of its requirement for a casing layer.

Basic growth data

Spawn run:	25°C, RH 90%+, at 14–28 days
Casing:	3–5 cm peat/chalk, pH 5.7–6.0
Post-casing:	25°C, RH 90%+, at 10 days
Pinning:	18°C, RH 95%.001, CO_2 < 1000 ppm, daylight, mist with water daily
Cropping:	18°C, RH 85–90%, CO_2 < 1000 ppm, daylight
	Flushing interval: 10–15 days

17.4.6 *Morchella*

The high value of morels stimulates an international trade in wild-picked mushrooms. As ascomycetes, the reproductive cycle is different to basidiomycetes, but attempts at cultivation have been based on experience with basidiomycetes. Morel Mountain in the US successfully fruit a variety of *Morchella esculenta*. Although their methodology is based on two patents (Ower, 1982), nobody else has succeeded in reliably fruiting morels, suggesting that they are fortunate with their strain, or that some vital details are missing from the patents. The patents describe growing mycelia on sterilized grain, and casing the grain with peat–bark–fibre-based mix. Resting bodies called sclerotia form in this casing. These are then removed and soaked to stimulate morel formation. One of the authors (Scrase, R.J.) succeeded in growing sclerotia, but has failed to grow morels. Stamets (1993) has achieved primordia and is worth reading for his morel anecdotes.

17.5 'Wood-rotting' species

So many wood-rotting species have been fruited that it is fairly safe to apply the techniques used to fruit Shiitake to other wood-rotters. The list below gives some examples:

Gilled mushrooms	Polypores	Others
Agrocybe aegerita	*Ganoderma lucidum*	*Hericium erinaceus*
Flammulina velutipes	*Grifola frondosa*	*Auricularia polytricha*
Hypholoma capnoides	*Polporus umbellatus*	
Hypzizygus tessulatus	*Polyporus sulphureus*	
Pholiota nameko	*Coriolus versicolor*	

17.5.1 Shiitake

Growing Shiitake outdoors on logs is a tradition stretching back a thouand years. Inoculating logs with cultured mycelium dates from 1904 (Mimura, 1904). The logs produce mushrooms 6–12 months later. Most shiitake mushrooms found on the supermarket shelves have grown from bags containing sterilized sawdust enriched with cereal bran. The bags are inoculated by an injection of a suspension of mycelial fragments. This leads to a rapid colonization of the substrate and fruitbodies can appear in as little as 2 weeks, although 4 weeks is more usual.

(a) Shiitake growing on hardwood logs. The following is an overview; the reader is referred to the *Shiitake Growers Handbook* (Przybylowicz, 1988) for full details.

Shiitake utilize sapwood, so freshly cut logs with little or no heartwood are chosen. The total yield depends on the density of the tree which explains why oak (*Quercus* spp.) is preferred.

Log moisture is crucial; shiitake grow best between 35% and 55% log moisture. Avoiding damage to the bark is important, as the bark provides a protective layer against diseases and limits moisture loss. The optimum size log depends on the handling system, lengths between 0.9–1.2 m with a diameter of 15–18 cm are typical. The logs are cut during the dormant season, this ensures a maximium of available carbohydrate.

The logs are inoculated by introducing mushroom mycelium on wooden plugs or in sawdust into holes drilled into the logs. Grain spawn should be avoided as it attracts the attention of seed-eating birds and mammals. The hole entrance is then sealed with wax to exclude contaminants and retain water.

Strains are available to suit different weather conditions and to grow fruibodies with particular characteristics. Logs are incubated at ambient temperatures, but protected from temperature extremes by shading. The logs are typically stacked in alternate layers. Irrigation may be necessary to maintain log moisture between 35% and 55%.

Fruiting is induced by lowering the temperature; this can occur naturally, but the grower may soak logs and increase the ventilation to achieve the same result.

Basic growth data

Spawn run:	6–24 months
Pinning induction:	cool, daylight, 1–3 days
Pinning:	cool, daylight, 3–10 days
Fruiting:	14–21 days
Flushing interval:	20–180 days

Note: precise conditions depend on strain and environmental conditions.

(b) Shiitake growing on sawdust. Shiitake can be grown on a wide variety of particulate materials composed of cellulose and lignin. Because the nutrients are readily available to other organisms, the substrate must be sterilized. Hardwood sawdust is mixed with bran or other sources of nitrogen. Bags full of sterilized substrate are inoculated and then stacked in environmentally controlled rooms.

Basic growth data

Spawn run:	1–4 months, RH 95%+
Induction:	1–3 days, light, $CO_2 < 1000$ ppm, RH 95%+
Pinning:	3–10 days, light, $CO_2 < 1000$ ppm, RH 95%+
Cropping:	14–21 days, $CO_2 < 1000$ ppm, light, RH 60–80%
Flushing interval:	10–30 days.

17.6 Mycorrhizal species

Although many mycorrhizal fungi may be grown in pure culture, the growth is very slow, and rarely leads to fructification, so for practical purposes the mycelium must be grown with its plant partner. The economic importance of mycorrhiza to forestry and land reclamation has led to fairly well-defined techniques for establishing mycorrhiza.

Surface-sterilized seeds are germinated in the presence of hyphal fragments or spores of the fungal partner. Later, the established mycorrhizal seedling is planted in appropriate conditions to maintain the relationship. Not all mycorrhizal establishment takes place in the laboratory however; tree nurseries' seed beds often contain appropriate fungi, and forestry workers deliberately transfer inoculum into seedling beds. Wild mycorrhiza have been manipulated by the addition of fertilizer to boost yields, with variable results.

17.6.1 Establishing mycorrhiza

The following is taken from *Methods and Principles of Mycorrhizal Research* (Schenk, 1982). The isolation, maintenance and growth in pure culture of mycorrhizal fungi has already been dealt with.

(a) Pure-culture ectomycorrhiza synthesis. Melin (1936) first established the techniques to create ectomycorrhiza. He used flasks containing sterile sand moistened with a nutrient solution into which an aseptically germinated seedling and single fungus culture were introduced. Subsequently sand has been replaced by vermiculite and peat (10:1 by volume) moistened with MMN nutrient solution. The fungus is grown in liquid culture. Broken glass in the flasks fragments the mycelium when shaken to accelerate the growth of mycelial volume and to provide more inoculation centres. Healthy seed is used and prepared as follows (Douglas fir):

1. Rinse in running water for 60 minutes.
2. Wash in mild detergent: 2 drops of Tween in 500 ml water, shake 90 minutes.
3. Rinse in running water for 90 minutes.
4. Soak in 30% hydrogen peroxide for 60 minutes.
5. Rinse in 2 litres of sterile water.

The seeds are aseptically planted in nutrient agar and then incubated for 7–10 days to check for contamination. Stratification can be provided in a refrigerator. The routine is varied depending on the plant species. Hydrogen peroxide will not penetrate all seed coats. Germinants are ready when the radicles are 1–2 cm long and the cotyledons are still within the seed coat. Fungi are inoculated at this time for vigorously growing trees such as *Pinus*, but slower growing types such as *Alnus* are inoculated after about a month. The growth vessels are illuminated and the temperature regulated. Synthesis take 4–6 months. The seedling can then be removed for planting. Low levels of phosphate in the planting site help to maintain the mycorrhiza, but do not guarantee that the chosen fungus is not replaced by other species. From the foresters' point of view, the fungal symbiont is irrelevant as long as the seedling becomes established.

Soil inoculation. Soil is removed from plantations with established mycorrhiza and either mixed with the seedling growth medium (5–10% by volume) or mixed with water (1 kg/20 l) and watered in. This method has obvious problems of non-specificity, pests and weeds, and transport.

Inoculation from pre-infected seedlings. Seedlings prepared as listed above are planted into nursery beds and as roots touch, the fungus spreads.

Ectomycorrhiza. These are harvested and mixed into the seed bed at the rate of 1 kg/m^3.

Spore inoculum. Spores can arrive naturally, the probability depending on the vicinity of mature woodland and the season. While most Basidiomycetes produce their spores when host roots are still actively growing, the practice of planting out seedlings in spring has necessitated finding ways of collecting and storing spores. Species of *Pisolithus, Scleroderma* and *Rhizopogon* are most often used in North America because of their wide host tolerance. Fruitbodies are collected from the wild, crushed, and air-dried at 25°C until they have a moisture level of 13–15%. Spores are either dusted onto the soil and watered in or mixed with a carrier of paper and mixed into the nursery soil. Approximately 1 mg of spores are used per seedling. Spores have the advantage of lightness and needing no extensive treatment compared with mycelium, but the demand for inoculum material exceeds supply. Presumably, intraspecific competition reduces the effectiveness of spores as an inoculating method but this disadvantage is outweighed by the practical advantages of using spores. Additionally there is little control over viability and differences between strains affecting the ability to form mycorrhiza. Lastly, ectomycorrhiza take 3–4 weeks longer to form when spores are used compared with other methods. This delay in establishing a mycorrhiza may be a problem at the critical time of seedling establishment.

Vegetative inoculum. Moser (1958) first made the bridge between establishing mycorrhiza in the laboratory and deliberately trying to inoculate trees. He wished to combine *Pinus cembra* with an alpine strain of *Suillus plorans* to improve the survival rate of seedlings planted near the timber line. His method was to produce mycelium in 10-litre flasks and then use this to inoculate 51 flasks containing peatmoss and more nutrients. When the moss was permeated with mycelium it was used as a plant growth medium. Interestingly, he noted that some contaminated flasks produced better mycorrhiza than the pure cultures! Moser grew several *Suillus* species, *Paxillus involutus, Amanita muscaria* and *Lactarius porninsis* among others. This method has been used subsequently to grow many other species combinations. A recent variation on this theme is in the form of a commercial product known as MYCORHIZ. Liquid culture of *Pisolithus tinctorius* grown on MMN is produced in shake flasks, 14-l fermentors and large industrial fermenters grown with high aeration rates (one volume change/minute) and continuous agitation at 28–32°C for 7–14 days. Sterile moss/vermiculite/MMN (28 : 1 : 14 by volume) is inoculated with 5–20% by volume with actively growing culture. After 1–4 weeks the colonized substratum is washed to remove excess nutrients and dried to a moisture level of 25%. The shelf life at 5°C is 5–6 weeks, so the product is only just commercially viable.

17.6.2 Commercialization of ectomycorrhizal species: truffles and Chanterelles

Plantations of trees co-inoculated with the truffle (*Tuber melanosporum*) mycelium have now been established for sufficiently long to demonstrate that this is an economically viable prospect, in France at least. Precise information about yields is hard to come by because the profits derived from them are taxable! (Shaw, 1996). Truffles generally appear after the host tree is over 10 years old, with yields increasing in line with the biomass of the tree. The recent announcment of a fruiting of *Chanterelle cantherellus* (Danell, 1997) on an artificially estabished mycorrhiza suggests this species will be the next to be commercialized.

17.7 Future prospects

Looking into our crystal ball, we make the following suggestions.

1. The list of domesticated species will inevitably become longer, with further impetus to research probably coming from the need to produce pharmaceutically useful species.
2. *Chanterelle* mycorrhiza will act as a model system, allowing the role of a number of associated microorganisms in fruiting to be further elucidated, and lastly, genetic engineers will re-introduce genes into Chanterelles, allowing the mycelium to extend and branch in a manner similar the most prolific free-living saprotroph. Consequently, the engineered strain will grow on ordinary mushroom compost!
3. Lastly, peat/chalk casings will be replaced by an eco-friendly re-usable paper based matting ready inoculated with just the right mix of microorganisms to give perfect fruitbody initiation every time.

Acknowledgement

Some of the material within this chapter has appeared previously in *The Mycologist.*

References

Boertmann, D. (1996) *The genus* Hygrocybe, Fungi of Northern Europe 1.
Bonnefons, N. de (1650) *Le Jardinier Français*, Paris.
Buczacki, S. (1989) *Fungi of Britain and Europe*, Collins.
Burton, K.S. (1989) The quality and storage life of *Agaricus bisporus. Proceedings of the Twelfth International Congress on the Science and Cultivation of Edible Fungi*, Braunschweig, Germany, Vol. 1, pp. 287–93.
Charlesworth, K. (1995) Life, the universe and (almost) everything 240: Muck and Magic, *New Scientist*, 1979, 53.
Chang, S.T. & Hayes, W.A. (eds) (1978) *The Biology and Cultivation of Edible Mushrooms.* Academic Press, New York.
Danell, E. & Camacho, F.J. (1997) Successful cultivation of the golden chanterelle, *Nature*, **385**, 303.
Deacon, J.W. (1984) *Introduction to Modern Mycology*, Blackwell Scientific Publications, Oxford.
Derikx, P.J.L., Op Den Camp, H.J.M., Van Der Drift, C., Van Griensven, L.J.L.D. & Vogels, G.D. (1990) Biomass activity during the production of compost used as a substrate in mushroom cultivation. *Applied and Environmental Microbiology*, **56**, 3029–34.

Dowson, C.G., Rayner, A.D.M. & Boddy, L. (1988) Inoculation of mycelial cord-forming basidiomycetes into woodland soils. *New Phytologist*, **109**, 335–41.
Elliott, T.J. (1995) *Mushroom Science 14*, Proceedings of the 14th International Congress on the Science and Cultivation of Edible Fungi (published in two volumes, pp. 926), A.A. Balkema, Rotterdam.
Fermor, T.R. & Wood, D.A. (1991) Mushroom compost microbial biomass: a review. *Mushroom Science*, **13**(1), 191–9.
Fermor, T.R., Randle, P.E. & Smith, J.F. (1985) Compost as a substrate and its preparation, in *The Biology and Technology of the Cultivated Mushroom* (eds P.B. Flegg, D.M. Spencer & D.A. Wood), John Wiley & Sons, Chichester, New York, pp. 81–110.
Flegg, P.B., Spenser, D.M. & Wood, D.A. (Eds) (1985) *The Biology and Technology of the Cultivated Mushroom*. Wiley, Chichester.
Flegg, P.B. & Wood, D.A. (1985) Growth and fruiting, in *The Biology and Technology of the Cultivated Mushroom* (eds P.B. Flegg, D.M. Spencer & D.A. Wood), John Wiley & Sons, chichester, New York, pp. 141–78.
Fletcher, J.T., White, P.F., Gaze, R.H. (1989) *Mushrooms: Pest and Disease Control, 2nd. edn.* Intercept.
Fries, N. (1966) Chemical factors in the germination of spores of Basidiomycetes, in *The Fungus Spore* (ed. M.F. Madelin), Butterworths, London, pp. 189–200.
Fries, N. (1973) Effects of volatile organic compounds on the growth and development of fungi. *Transactions of the British Mycological Society*, **60**, 1–21.
Fries, N. (1978) Basidiospore germination in some mycorrhiza-forming Hymenomyceyes. *Transactions of the British Mycological Society*, **70**, 319–24.
Frische, G. & van Loone, P. (1989) Breeding experiments with the wood blewit (Lepista nuda), in Mushroom Science XII (part 1), *Proceedings of the Twelth International Congress on the Science and Cultivation of Edible Fungi*, Germany (1987).
Gandy, D.G. & Hollings, M. (1962) Die-back disease of mushrooms: a disease associated with a virus. *Report of the Glasshouse Crops Research Institute*, 1961, 103–8.
Griensven, L.J.L.D. van (1988) *The Cultivation of Mushrooms*, Darlington Mushroom Laboratories Ltd, Rustington, Sussex, UK.
Griffin, D.M. (1972) *Ecology of Soil Fungi*, Chapman & Hall, London.
Houdeau, G., Olivier, J.M. & Chabbert, B. (1991) Improvements in indoor short composting. *Mushroom Science*, **13**(1), 215–20.
Hussey, N.W., Read, W.H. & Hesling, J.J. (1969) *The Pests of Protected Cultivation*, Arnold, London.
Jennings, D.H. & Lysek, G. (1996) *Fungal Biology*, Bios.
Jones, R.J., Sizmur, K.J. & Wildman, H.G. (1991) A miniaturised system for storage of fungal cultures in water. *Mycologist*, **5**, 184–6.
Kehl, H. (1943) Zur Keimungsphysiologie der Champignonsporen. *Gartenbauwissenschaft*, **17**, 156–70.
Larkcom, J. (1976) *Vegetables from Small Gardens*, Faber & Faber, London.
Maccanna, C. (1985) *Commercial Mushroom Production*, The Agricultural Institute, Dublin.
Madelin, M.F. (ed.) (1966) *The Fungus Spore*, Butterworths, London.
Matruchot, L. (1917) Variations Experimentales du Tricholoma nudum. *Revue Gen. Bot.*, 503.
McTeague, D.M., Hutchinson, S.A. & Reed, R.I. (1959) Spore germination in *Agaricus campestris. Nature*, **183**, 1736.
Meredith, D.S. (1960) Further observations on fungi inhabiting pine stumps. *Annals of Botany*, **24**, 63–78.
Melin, E. (1936) Methoden der experimentellen Untersung Mykotropher. *Pflanzen. Handb. Biol. Arbeitsmeth.*, Abt. II, 1015–108.
Melin, E. (1962) Physiological aspects of mycorrhiza of forest trees, in *Tree Growth* (ed. E. Kozlowski), Ronald Press Co., pp. 247–63.
Mills, G.L. & Eilers, F.I. (1973) Factors influencing the germination of basidiospores of *Coprinus radiatus. Journal of General Microbiology*, **77**, 393–401.
Moore, D., Casselton, L.A., Wood, D.A. & Frankland, J.C. (eds) (1985) *Developmental Biology Of Higher Fungi*, BMS Symposium 10, Cambridge University Press.
Mushroom growing for everyone, Faber & Faber, London.

Nisikado, Y., Kimura, K. & Miyawaki, T. (1942) The mycelial growth in pure culture on the sawdust medium prepared of various kinds of tree. *Ber. Ohara Inst. Landw. Forsch.*, **9**, 39–60.
Noble, R. & Burton, K.S. (1993) Postharvest storage and handling of mushrooms: Physiology and technology. *Postharvest News and Information*, **4**, 125–9.
Norkrans, B. (1949) Some mycorrhizal forming Tricholoma species. *Svensk. bot. Tidskr.*, **43**, 485–90.
Nuss, I. (1982) Die bedeutung der proterosporen: schussfolgerungen aus unter-suchungen an Ganoderma. *Pl. Syst. Evol.*, **141**, 53–79.
Penn, J. (1994) Alternatives to the usual cocktails for fungal fermentations. *Discovering Drugs from Nature*, IBC.
Perrin, P. (1979) Long-term storage of cultures of wood-inhabiting fungi under mineral oil. *Mycologia*, **71**, 867.
Pryce, S. (1991) *The Peat Alternatives Manual*, Friends of the Earth, London.
Przbylowicz, P. & Donoghuue, J. (1990) *Shiitake Growers Handbook*, Kendall/Hunt, Iowa.
Richardson, P.N. & Grewal, P.S. (1991) Comparative assessment of biological (Nematodes: *Steinernema feltiae*) and chemical methods of control of the mushroom fly, *Lycoriella auripila* (Diptera: Sciaridae). *Biocontrol Science and Technology*, **1**, 217–28.
Risbeth, C. (1950) Observations on the biology of *Fomes annosus. Annals. of Botany*, **14**, 365–83.
Scrase, R.J. (1993) *The Cultivation of Basidiomycetes*, MPhil. Thesis, Bath University.
Scrase, R.J. (1995) Cultivating mushrooms – the Potential. *Mycologist*, **9**(1), 18–19,
Scrase, R.J. (1995) Cultivating mushrooms – from Pure Culture to Spawn Production. *Mycologist*, **9**(2), 53–6.
Scrase, R.J. (1996) Cultivating mushrooms: making composted and non-composted substrates. *Mycologist*, **10**(2), 52–5.
Shaw, P.J.A., Lankey, K. & Jourdan, A. (1996) Factors affecting yield of *Tuber melanosporum. Mycological Research*, **100**(10), 1176–8.
Sinden, J.W. & Hauser, E. (1950) Report on two new mushroom disease. *Mushroom Science*, **1**, 96–100.
Smith, J.F. (1990) Composted substrates used in the cultivation of white button mushroom, *Agaricus bisporus. Aspects of Applied Biology*, **24**, 131–43.
Stamets, P. & Chilton, J.S. (1983) *The Mushroom Cultivator*, Agarikon Press, Olympia, Washington.
Stamets, P. (1993) *Growing Gourmet and Medicinal Mushrooms*, 10 Speed Press.
Steineck, H. (1984) *Mushrooms in the Garden*, Mad River Press, California.
Stevens, R.B. (1974) *Mycology Guidebook*, University of Washington Press, Seattle, London.
Straatsma, G., van Griensven, L. & Bruinsma, J. (1986) Root influence on in vitro growth of hyphae of the mycorrhizal mushroom *Cantherellus cibarius* replaced by carbon dioxide. *Physiological Plant*, **67**, 521–8.
Straatsma, G., Gerrits, J.P.G., Gerrits, T.M., Op Den Camp, H.J.M. & Van Greinsven, L.J.L.D. (1991) Growth kinetics of *Agaricus bisporus* mycelium on solid substrates (mushroom compost). *Journal of General Microbiology*, **137**, 1471–7.
Straatsma, G., Olijnsma, T.W., Gerrits, J.P.G., Amsing, J.G.M., Op Den Camp, H.J.M. & Van Griensven, L.J.L.D. (1994) Inoculation of *Scytalidium thermophilum* in button mushroom compost and its effect on yield. *Microbiology*, **60**, 3049–54.
Stutzenberger, F. (1990) Thermostable fungal α-glucosidases. *Letters in Applied Microbiology*, **11**, 173–8.
Vedder, P.J.C. (1978) *Modern Mushroom Growing*, Educaboek Culernberg.
Watling, R. (1971) Basidiomycetes, in *Methods in Microbiology* (ed. C. Booth), Academic Press, London, New York, p. 231.
Watling, R. (1982) *How to Identify Mushrooms to Genus V: Cultural and Developmental Features*, Mad River Press, Eureka, CA.
Watling, R. (1997) The business of fructification. *Nature*, **385**, 299.
Wood, D.A. (1989) Mushroom Biotechnology. *International Industrial Biotechnolgy*, **9**(1), 5–8.
Zadrazil, F. & Reiniger, P. (eds) (1988) *Treatment of Lignocellulosics with White Rot Fungi*, Elsevier, London.

18 Algae as food

M.A. BOROWITZKA

18.1 Introduction

Algae have been used as human food for thousands of years in all parts of the world. The most commonly consumed macroalgae include the red algae *Porphyra* (nori, kim, laver), *Asparagopsis taxiformis* (limu), *Gracilaria, Chondrus crispus* (Irish moss) and *Palmaria palmata* (dulse), the kelps *Laminaria* (kombu), *Undaria* (wakame) and *Macrocystis*, and the green algae *Caulerpa racemosa*, *Codium* and *Ulva* (see Tseng, 1981; Druehl, 1988; Mumford & Miura, 1988 for reviews). These algae are either harvested from wild populations or are farmed. These algae usually are eaten either fresh, dried or pickled (Abbott, 1988). Several macroalgae are also the source of hydrocolloids such as agar-agar and carrageenan which are widely used in the food industry as stabilisers, thickeners and gelling agents.

Some microalgae, especially the blue-green algae or cyanobacteria have also been used as a food source or nutritional supplements for a long time (see Ciferri, 1983). More recently, the range of microalgal species used as food supplements has increased greatly and their application is becoming more widespread. This chapter will consider only the microalgae and their application as food or nutritional supplements.

Although there are more than 40000 species of eukaryotic and prokaryotic (cyanobacteria) algae, only a few species are used directly as food or as food supplements at present (Table 18.1).

18.2 *Spirulina* – food and health food

The cyanobacterium (blue-green alga), *Spirulina (S. platensis, S. maxima)* has been a traditional food in Chad, where 9–13 g per meal is consumed in 10–60% of meals (Jassby, 1988b). There are also reports that the Aztecs around Lake Texcoco regularly consumed *Spirulina* (Ciferri, 1983). Before 1981, food-grade *Spirulina* was mainly produced for a small health food market in Japan (about $20\,t\,y^{-1}$), however since then the demand for this alga has increased throughout the world and present production of food-grade *Spirulina* is at least $900\,t\,y^{-1}$ and demand exceeds supply. The major producers are in Thailand and the USA, with smaller producers in China, India and Taiwan.

Table 18.1 Microalgae currently used (or proposed) for human consumption

Alga	Source	Reference
Cyanobacteria (blue-green algae)		
Spirulina platensis *S. maxima*	Cultured (USA, Thailand, China, Taiwan, India, etc.)	Jassby (1988b); Richmond (1988); Belay *et al.* (1994)
Nostoc commune	Field-collected	Martinez (1988)
Aphanizomenon flos-aquae	Field-collected (Lake Klamath, USA)	
Chlorophyta		
Chlorella spp.	Cultured (Taiwan, Japan)	Soong (1980)
Dunaliella salina (for beta-carotene)	Cultured (Australia, Israel, USA)	Borowitzka & Borowitzka (1989a,b)
Scenedesmus spp.	Cultured (Czech Republic and experimental)	Becker (1988); Grobbelaar *et al.* (1995)
Haematococcus pluvialis (for astaxanthin)	Cultured (experimental only)	Borowitzka (1992)

Spirulina platensis grows best at a high pH of about pH 9–11 and at high bicarbonate concentrations. At this high pH, competition by other algae is greatly reduced and *Spirulina* can therefore be grown in open ponds. The most commonly used pond type are paddle-wheel driven raceway ponds, i.e. long channels arranged in single or in multiple loops (Shimamatsu, 1987; Seshadri & Jeeji Bai, 1992). These rectangular ponds are either constructed of concrete or are lined with PVC or a similar material. The bends in the channels result in hydraulic losses and some problems with solids deposition. This is generally ameliorated by using a single-loop (raceway) configuration and installing baffles at the corners. For example, Shimamatsu (1987) describes a very efficient design using closely spaced turning vanes installed along the diagonals from the centre wall to the corners of the pond. These vanes are spaced less than 50 cm apart, producing two right-angle changes in flow direction. An alternative design has been tested by Dodd (1986) in Singapore. In this design the wall at the end of the pond without the paddle-wheel is eccentrically placed and baffles are located in the water flow. This creates a curved zone of accelerating flow followed by a flow expansion zone after the directional change has been made. The rate of constriction of the curved zone is sufficient to avoid eddies or velocities below those causing deposition on the back side of the centre wall and baffles.

The water depth in the ponds generally varies from 30 to 50 cm depending upon season and algal density. This depth is a compromise between pond size, flow velocity and optimal light absorption by the algal culture. Laboratory studies have indicated that flow velocities of $60\,cm\,s^{-1}$ are optimal (Richmond, 1988); however, in production systems it has been found that velocities of $30\,cm\,s^{-1}$ are the maximum which can be used since higher velocities result in fragmentation and increased coiling of trichomes which eventually leads to decreased harvesting efficiencies (Belay *et al.*, 1994). The costs of pond maintenance and energy requirements for mixing also increase significantly at the higher flow velocities (c.f. Oswald, 1988).

The basic culture medium generally used is Zarrouk medium (Zarrouk, 1966); however, each production plant has simplified and optimized this medium to reduce costs and improve productivity. The medium is usually recycled after harvesting. Since *Spirulina* grows best at high pH and high bicarbonate concentration, CO_2 is often added to the culture. Where CO_2 is added, this is done by bubbling the gas through the culture until it is trapped in a floating plastic bubble which keeps the gas in contact with the medium to increase the time available for the CO_2 to be absorbed by the medium.

In such an open system the maintenance of a predominantly unialgal culture throughout the growing season is of critical importance. Continuous recycling of the medium can lead to contamination by other algae (Richmond *et al.*, 1990), as well as build-up of organic matter which appears to inhibit *Spirulina* growth and which may also lead to excessive foaming (Belay *et al.*, 1994). The major algal contaminant which has been reported in *Spirulina* production plants is the green alga *Oocystis* sp., which has been found to occur mainly in the colder months of the year when *Spirulina* growth is reduced; however, processes have been developed to control green algal contamination (Belay *et al.*, 1994). Zooplankton contamination can also be controlled through careful pond management.

Spirulina is among the best-studied algae in large-scale outdoor culture and serves as a good example of the major factors which limit productivity in outdoor algal cultures. Temperature has a large effect on productivity and *Spirulina* has a temperature optimum between 35 and 37°C (Richmond, 1988). Maintaining the culture temperature near optimum by heating of the culture in winter and heating in the morning in summer can enhance productivity significantly (Richmond *et al.*, 1990). Irradiance is also important, and Richmond *et al.* (1990) have demonstrated that the decline in productivity in outdoor cultures in the afternoon can be largely attributed to photoinhibition; it is estimated that photoinhibition can cause a loss as high as 30% of the potential output rate. The effect of photoinhibition however, can be ameliorated somewhat by selecting strains which are resistant to photoinhibition (Vonshak, 1987; Vonshak & Guy, 1992). The potential productivity of the system is also affected by the time of day at which harvesting occurs. Harvesting in the evening is most effective since it appears to reduce the night-time loss of biomass due to respiration by the algal biomass. Richmond *et al.* (1990) report that in two commercial operations (Ein-Yahav Algae, Israel, and Siam Algae, Thailand) there was an increase in net productivity of 15–20% when harvesting was carried out mainly in the late afternoon rather than the morning.

Since *Spirulina* is filamentous, harvesting is relatively easy. Initial concentration is usually done using either an inclined gravity screen or a vibrating screen filter. Further dewatering may be achieved by a vacuum belt filter resulting in a paste of about 15% solids. The mesh size of the filters used and the flow rates need to be optimized to achieve a good rate of biomass removal without causing too much breakage of the trichomes. This is made more complicated by the fact that there is a constant change in the size and morphology of the trichomes over the year and also with increasing frequency of harvest (Belay *et al.*, 1994). The main factors affecting the degree of coiling of the trichomes appear to be nutrient concentration and light intensity.

Since the final product is destined for human consumption there is a requirement for some basic quality standards. Such standards for *Spirulina* are shown in Table 18.2. Monitoring of product quality is essential in the manufacture of a product for human consumption, since bacterial contamination is impossible to eliminate in open-air cultures. The bacterial count is however, reduced during processing. For example, dewatering and spray-drying of *Spirulina* reduces the standard plate count (SPC) to 10^4–$10^6\,g^{-1}$,

Table 18.2 Guidelines for quality of *Spirulina* powder

	Jassby (1988a)	Indian Standard[a]
Microbial contaminants		
Coliform (or *E. coli*)	Negative	Negative
Salmonella	Negative	Negative
Shigella	Negative	Negative
Staphylococcus (coagulase + ve)	Negative	Negative
Mould	$<100\,g\,l^{-1}$	
Heavy metal contaminants		
Lead	$<2.0\,\mu g\,g^{-1}$	$<2.5\,\mu g\,g^{-1}$
Mercury	$<0.1\,\mu g\,g^{-1}$	$<0.1\,\mu g\,g^{-1}$
Cadmium	$<0.1\,\mu g\,g^{-1}$	$<1.0\,\mu g\,g^{-1}$
Arsenic	$<2.0\,\mu g\,g^{-1}$	$<1.1\,\mu g\,g^{-1}$
Other contaminants		
Unidentified insect fragments	<150 per 50 g	
Rodent hairs	<0.5 per 50 g	
Appearance and texture		
Moisture	<7%	<9%
Total ash (on dry basis)		<9%
Acid-insoluble ash (on dry basis)		<0.5%
Protein (on dry basis)		>55%
Phaeophorbide (total)	$<1.2\,mg\,g^{-1}$	
Appearance	Uniform powder	
Colour	Depends on species	
Odour	Fresh, not decayed	
Taste	Mild, not salty or bitter	

[a] Indian Standard IS 12895: 1990 (Amendment No. 1, June 1991).

Table 18.3 Typical analyses of commonly eaten algae (in $mg\,kg^{-1}$ unless otherwise stated)

	Spirulina[a]	Chlorella[b]
Vitamins		
Provitamin A	2 330 000 $IU\,kg^{-1}$	55 500 $IU\,kg^{-1}$
Beta-carotene	1400	1808
Vitamin E*	100	<10 $IU\,kg^{-1}$
Thiamin B_1	35	15
Riboflavin B_2	40	48
Niacin B_3	140	238
Vitamin B_6	8	17
Vitamin B_{12}	3.2	1259
Inositol	640	1650
Folic acid	0.1	269
Biotin	0.05	1916
Pantothenic acid	1	13
Minerals		
Calcium	10 000	
Phosphorus	8000	
Magnesium	4000	
Iron	1500	
Zinc	30	
Copper	12	
Manganese	50	
Chromium	2.8	
Sodium	9000	
Potassium	14 000	
Amino acids ($g\,kg^{-1}$)		
Alanine	47	48
Arginine	43	36
Aspartic acid	61	52
Cysteine	6	4
Glutamic acid	91	63
Glycine	32	34
Histidine	10	13
Isoleucine	35	26
Leucine	54	53
Lysine	29	35
Methionine	14	15
Phenylalanine	28	31
Proline	27	29
Serine	32	28
Threonine	32	27
Tryptophan	9	6
Tyrosine	30	21
Valine	40	36
Essential fatty acids		
Linoleic	8	
Gamma-linolenic	10	
Pigments		
Phycocyanin	150	
Chlorophyll	11	20
Carotenoids	3.7	

[a] Belay *et al.* (1994).
[b] Hansakul (1995).

and subsequent heat treatment at 65°C for 30 sec results in a SPC of 10^2–$10^4\,g^{-1}$ (Jassby, 1988a). The other potential health threat is phaeophorbides, which are naturally occurring degradation products of chlorophylls which have lost their magnesium and phytol residue. Phaeophorbide formation is the result of acidity which removes the magnesium atom from the chlorophyll and the activity of the chlorophyllase enzyme which catalyses the loss of the phytol residue. High levels of phaeophorbides in microalgae products (especially *Chlorella*) have been implicated in an outbreak of photosensitive dermatitis in Tokyo in 1977 (Tamura *et al.*, 1979; Miki *et al.*, 1980). This has led to the Ministry of Public Welfare in Japan recommending a maximum level of $0.8\,mg\,g^{-1}$ of phaeophorbide, and a level of $1.2\,mg\,g^{-1}$ of total phaeophorbide (the amount that potentially can form) for microalgal products. In order to reduce the level of phaeophorbide, the Ministry further recommends heating the dried powder for 3 min at 100°C to inactivate the chlorophyllase and to avoid contact with alcohol and acetone which enhance chlorophyllase activity (Tamura *et al.*, 1979).

Extensive toxicity and nutritional studies have been undertaken with *Spirulina* (e.g. Bizzi *et al.*, 1980; Yoshino *et al.*, 1980; Chamorro & Salazar, 1990; Mitchell *et al.*, 1990; Kapoor & Mehta, 1992, 1993) and these have confirmed the safety of this alga. *Spirulina* is rich in proteins, vitamins, essential amino acids (with the exception of methionine and cystine, which may be limiting; Shubert, 1988), minerals and essential fatty acids such as gamma-linolenic acid, and many pre-clinical and clinical studies suggest a wide range of therapeutic effects ranging from cholesterol reduction to enhancement of the immune system (see Belay *et al.*, 1993 for review). The composition of *Spirulina* is given in Table 18.3.

18.3 *Dunaliella salina* – production of beta-carotene

The halophilic green flagellate, *Dunaliella salina* (sometimes also referred to as *D. bardawil*), is the best natural source of the carotenoid beta-carotene. This alga is cultivated commercially in Australia, the USA and Israel, and the beta-carotene is either extracted for use in nutritional supplements or as a natural pigmenter, or the dried algae are tableted and sold as health food. *Dunaliella salina* is extremely well suited for commercial large-scale culture for the following reasons:

1. It grows at extremely high salinities up to NaCl saturation (i.e. >30% NaCl) with an optimum salinity for growth at about 22% NaCl. This means that there are very few competing species or predators and therefore simple open-pond culture is feasible.
2. It can accumulate a high concentration of beta-carotene (up to 14% of dry weight); this is a much higher content of beta-carotene than any other natural source.

3. The beta-carotene is very valuable and thus while the production process can be quite expensive, a profit can still be made.

Dunaliella salina is grown in either very large (5 to >200 Ha), shallow, unstirred ponds or in shallow paddle-wheel raceway ponds which are usually no greater than 1000 m^2 in area. The extensive culture method is used by the two largest producers, Betatene Ltd and Western Biotechnology Ltd, both of whom are located in Australia. The more intensive raceway system is used by the smaller producers in Israel and the USA. As in any industrial process, the choice if culture system is dictated by economics. Thus, in Australia where land costs are very low, the climate is ideal, and seawater and/or natural brines can be used, the expansive method of culture is economically possible as long as a very efficient and cost-effective harvesting system exists. Higher land costs (as in Israel) and the cost of salt for the medium (as in California, USA) mean that a more intensive culture process is required.

Most of the producers use a semi-continuous culture process, harvesting part of the ponds at required intervals and recycling the brine. Detailed descriptions of the process used by some of the producers can be found in Borowitzka & Borowitzka (1989a) and Schlipalius (1991). In the very large open ponds there is no control over temperature or most other environmental factors. The main parameter which can be controlled to some degree by the operator is salinity. The salinity chosen for culture is a compromise between several factors, i.e. the growth rate is higher at lower salinities (below about 22% (w/v) NaCl) and the beta-carotene content is highest at the highest salinity (about 33% NaCl) (Borowitzka, *et al.*, 1985). However, at lower salinities there is the possibility of contamination by protozoa and even the brine shrimp, *Artemia*, which can decimate a culture in a very short period of time. Non-carotenogenic species of *Dunaliella* such as *D. viridis*, also grow better at low salinities (Moulton *et al.*, 1987). Rainfall will also reduce the salinity and this means that when rain is likely the operating salinity needs to be higher so that the rain does not reduce the salinity to a level where the contaminating invertebrates can proliferate. In large open ponds excessive dilution by rain can be avoided by maintaining a layer of salt on the pond floor. The need for high irradiance and high temperatures, as well as the need to minimize the effects of rain, means that all *Dunaliella* plants are located in semiarid regions. For example, the *D. salina* production plant of Western Biotechnology Ltd is located at Hutt Lagoon in Western Australia. Hutt Lagoon is a natural salt lake and the algal ponds are constructed directly on the lake bed. The algal culture used is the natural population found in the lake and the selective growth of *D. salina* is encouraged through careful management of salinity and the addition of nutrients. Evaporative losses are made up by the addition of seawater pumped from the nearby ocean. The climate at Hutt Lagoon is extremely favorable to *Dunaliella* growth with few cloudy days, high ambient temperatures and with rain occurring only over a 3-month period.

Harvesting of *D. salina* is difficult due to the fact that the cell densities in the cultures are usually no more than 0.1 g dry weight l^{-1}, the cells are of about the same density as the brine, and the cells are very fragile as they do not have a cell wall. In addition, if they are ruptured, the beta-carotene is rapidly oxidized. Each of the producers has their own proprietary harvesting method and these include centrifugation, flocculation or a method which exploits the hydrophobic nature of the cell membrane (Ruane, 1974; Curtain & Snook, 1983; Curtain *et al.*, 1987; Mohn, 1988, 1992).

Once harvested, the biomass may be spray- or drum dried, or the beta-carotene may be extracted directly using hot oil or other solvents (e.g. Rüegg, 1984; Nonomura, 1987; Qin *et al.*, 1994). The extracted beta-carotene can be concentrated, crystallized and a range of different formulations produced, depending on the final application. The commonly prepared products are beta-carotene solutions or suspensions in vegetable oil for use as a natural food colouring or for the preparation of soft-shell capsules, or water dispersable beadlets mainly for use in the preparation of tablets.

Dunaliella beta-carotene is composed mainly of two isomers, all-*trans* and 9-*cis* in approximately equal amounts (Ben-Amotz *et al.*, 1982, 1988). The ratio of the 9-*cis* to the all-*trans* increases with increasing irradiance to which the algae are exposed. On the other hand, the synthetic beta-carotene is only in the form of the all-*trans* isomer. Studies on the biological activity of these two isomers have shown that the 9-*cis* isomer is apparently absorbed by the body more rapidly than the all-*trans* isomer (Mokady *et al.*, 1990; Levin & Mokady, 1994; Levin *et al.*, 1994). This suggests that the 9-*cis* isomer may be more biologically active; however, to date all large-scale trials on the effects of beta-carotene on human health have been carried out only with the synthetic all-*trans* isomer. Toxicological studies with *Dunaliella* have also failed to show any adverse effects (Mokady, 1992).

18.4 *Chlorella*

The green alga *Chlorella* (mainly *C. vulgaris*, *C. ellipsoidea* and *C. pyrenoidosa*) has long been cultured for use as a nutritional supplement, especially in Japan and Taiwan. The cultivation process is basically a series of batch cultures, increasing in volume by a factor of approximately 5 to 10 at each step. The algae are grown initially in small culture flasks indoors and these cultures are then used as inocula for outdoor ponds. The most commonly used outdoor culture systems are circular ponds with a centrally pivoted arm which has blades mounted along its length and which thus mixes the culture as it rotates. The largest ponds are approximately 500 m^2 in area, with a depth of about 20 cm. This pond type was the first large-scale microalgal culture system and is rather inefficient. Mixing in these ponds is very uneven, especially in the larger ponds where the outer part of the mixing arm travels much faster than the inner part. In the largest ponds additional rotating blades are mounted along the inner part of the rotating arm so as to increase the degree of mixing at the centre of the pond. Paddle wheel ponds are also used by some producers.

Chlorella grows not only photoautotrophically, but also can be grown mixotrophically or heterotrophically in the dark with acetate or glucose as a carbon source (Kawaguchi, 1980). At present, *Chlorella* is grown mainly for the health food market; however, Running *et al.* (1994) have developed a process for the production of L-ascorbic acid (vitamin C) using heterotrophically grown *C. pyrenoidosa* growing on glucose. The process has been improved through the production of mutants and concentrations of 1–2 $g\,l^{-1}$ of L-ascorbic acid have been reported.

Harvesting of *Chlorella* is by centrifugation. More recently, at least one producer has expoited the fact that some strains of *Chlorella* auto-flocculate. In this system, the mixing of the culture was stopped at night and the algae were allowed to auto-flocculate and settle on the floor of the pond. In the morning the settled algal biomass was collected before mixing was again commenced. This preliminary harvesting step greatly decreased the volume of culture which had to be centrifuged, resulting in a marked reduction in the cost of harvesting.

After harvesting the biomass is spray- or drum-dried and the powder sold directly or used to make tablets. *Chlorella* has a very thick and resistant cell wall, which must be broken to make the alga digestible. This is usually done either by milling or by sonication before drying. In most cases it is also necessary to pasteurize the algal powder to produce a product suitable for human consumption. The typical composition of *Chlorella* powder is given in Table 18.3.

18.5 Other algae

Several other algae are under development as food sources or as sources of nutritional supplements. In the 1970s, extensive work was done in India with the green alga *Scenedesmus* as a source of single-cell protein (Becker & Venkatamaran, 1980, 1982). Although this study achieved its aims of producing algal biomass for human and animal consumption, the cost of production was generally too great for this alga to be an alternative protein source for developing countries. *Scenedesmus* is however, still grown in the Czech Republic at Trebon. The culture system used here consists of a sloping flat area. The algae are pumped to the top of the slope and flow down it, with a culture depth of about 5 mm. Shallow baffles are placed at right angles to the culture flow at regular intervals to enhance turbulence. Once the culture has reached the bottom of the slope it is collected in a tank where carbon dioxide can be added. The culture is then recirculated to the top of the slope. At night, the whole culture volume is held in the tanks to minimize cooling, thus allowing the culture to reach its optimum temperature earlier in the morning as well as saving energy. The cell densities reached by this culture system are much higher than in more conventional paddle-wheel pond systems, probably because of the greater availability of light to the shallow algal culture. The *Scenedesmus* produced is sold as a health food and extracts are also prepared for use in human and animal health (J. Doucha and I. Setlik, personal communication).

More recently, the work on microalgae for the production of long-chain omega-3 and omega-6 polyunsaturated fatty acids (PUFAs) is of particular interest. A number of algal species, mainly marine diatoms, dinoflagellates and cryptophytes, produce significant quantities of lipid high in these fatty acids (Borowitzka, 1988; Bajpai & Bajpai, 1992, 1993).

Studies by Martec Biosciences in the USA found that a heterotrophically growing strain of the diatom *Nitzschia* could achieve cell densities of 45–48 $g\,l^{-1}$ in 64 hours in a stirred tank reactor. This diatom can contain up to 50% of dry weight as lipid, with containing 4–5% eicosapentaenoic acid (EPA) (Kyle & Gladue, 1991). This is equivalent to a productivity of EPA of 0.25 $g\,l^{-1}\,day^{-1}$. Other diatoms, such as the non-photosynthetic *Nitzschia alba*, also contain high amounts of EPA (Tornabene *et al.*, 1974) and the productivity of EPA can be enhanced significantly by optimising culture conditions (e.g. Chrismadha & Borowitzka, 1994a). Heterotrophic culture overcomes the major limitation to algal culture, namely light, and much higher cell densities are reached compared with phototrophic cultures. Unfortunately, the number of algal species which can be grown heterotrophically is limited and the proximate composition of the algae is also different in heterotrophically grown cells (Chen & Johns, 1991).

The most promising organism identified so far for the production of docosahexaenoic acid (DHA) is the dinoflagellate, *Crypthecodinium cohnii* (Takeuchi & Uehara, 1993). In heterotrophic culture, cell densities of up to 40 $g\,l^{-1}$ have been achieved, with a cell lipid content of 15–30%. The DHA content of this lipid is reported to be 25–30% (Kyle *et al.*, 1991; Kyle, 1992; Anon, 1995). This is equivalent to an approximate DHA productivity of 1.2 $g\,l^{-1}$. Other promising DHA-producing organisms are the thraustochytrids (organisms with a close affinity to the red and brown algae), *Schizochytridium* spp. (Barclay *et al.*, 1994) and *Thraustochytrium* spp. (Bajpai *et al.*, 1991a,b). Barclay *et al.* (1994) report DHA productivities of 1 $g\,l^{-1}$ of long-chain omega-3 fatty acids with this organism. This productivity is much greater than the best productivities reported for fungi such as *Mortierella* (e.g. Shinmen *et al.*, 1992).

In the near future several microalgae may also find application as sources of bioactive molecules in human health care (see Borowitzka, 1995 for review). There are numerous reports in the literature of the isolation of bioactive molecules from microalgae; however, the most exciting of these are the range of anti-cancer and anti-viral compounds produced by a number of blue-green algae and dinoflagellates. Many of these compounds are too complex for chemical synthesis so that they will probably have to be produced by algal culture. Their large-scale culture will require cost-effective reliable closed culture systems (see below).

18.6 New culture systems

One major limitation to the increased use of microalgae for food and nutritional supplements is the cost of culture and the need for new closed culture systems. The existing outdoor culture systems are suitable only for a small number of species which grow in highly selective environments. Most other species cannot be maintained in monoalgal culture in open systems. Furthermore, contamination, especially by heavy metals, is a problem in many cases.

The main reason for the relatively high cost of algal culture is the algal requirement for light. This limitation can be overcome for some species which can be grown heterotrophically; however, the number of species able to grow heterotrophically is limited and heterotrophically grown algae often have a different chemical composition compared with phototrophically grown algae. Over the past decade or so there has been great interest in the development of closed algal systems. Some of the systems developed include:

1. Bag culture which is widely used for the culture of algae for aquaculture (Daintith, 1993).
2. Alveolar panels and other flat plate reactors of various designs (Tredici *et al.*, 1991; Torzillo *et al.*, 1993; Silva & Cortinas, 1994).
3. Stirred tank reactors with internal illumination (Pohl *et al.*, 1988; Wohlgeschaffen *et al.*, 1992; Hilaly *et al.*, 1994).
4. Tower reactors with internal fibre-optic illumination (Mori *et al.*, 1987; Burgess *et al.*, 1993).
5. Suspended narrow bags or tubes (Groeneweg, 1978; Cohen & Arad, 1989).
6. Tubular reactors (Pirt *et al.*, 1983; Torzillo *et al.*, 1986; Chaumont *et al.*, 1993; Richmond *et al.*, 1993; Chrismadha & Borowitzka, 1994b).

Of these, the tubular photobioreactors – in particular the helical tubular photobioreactors such as the 'Biocoil' – appear the most promising for commercial-scale production. Most of the tubular photobioreactors now under development use clear tubes made of glass, PVC, silicon, teflon or similar materials, with a tube diameter of between 500 and 250 mm. This narrow tube diameter minimizes the length of the light path and also results in reduced adhesion of the algae to the inner surface of the tube due to high turbulence. The tubes are arranged in various configurations including long parallel arrays laying on the ground or arranged vertically with 180° bends at the end, similar arrays floating in water for support and temperature control, or wound helically around an upright cylindrical support (Chaumont *et al.*, 1988; Torzillo *et al.*, 1993; Chrismadha & Borowitzka, 1994b; Lee *et al.*, 1995). The latter system, the 'Biocoil' (Robinson *et al.*, 1988), appears to be the most promising design. The algal culture is circulated within the tubes by means of pumps (e.g. diaphragm, lobe, peristaltic or centrifugal pumps) or by an airlift with flow rates of about $30\,cm\,s^{-1}$. Temperature control is either by a heat exchanger incorporated into the system, or by evaporative cooling by spraying water over the surface of the tubes. One problem of these systems is the accumulation of photosynthetically produced oxygen in the tubes, which inhibits algal photosynthesis. These reactors must therefore include a gas exchange system which removes some of the oxygen and adds carbon dioxide. The addition of carbon dioxide is usually on a pH-stat system. The design criteria for such systems are reviewed in detail by Borowitzka (1996) and the general design criteria for closed systems are summarised in Table 18.4.

Although not all algae are suited for culture in such closed systems, a very wide range of species have been cultured on the laboratory and pilot scale in the 'Biocoil', including *Spirulina platensis, Chlorella* spp., *Porphyridium cruentum*, *Phaeodactylum tricornutum, Tetraselmis* spp., *Isochrysis* spp.,

Table 18.4 Fundamental design criteria for closed microalgal cuture systems

Criterion	Reason	Solution
Light utilization efficiency	Capture the maximum amount of light for maximum growth	Minimize the length of the light path
Good mixing (turbulence)	Minimize settling of cells, which can lead to culture instability; minimize unstirred layer around cells in order to optimize nutrient uptake	Use narrow tubes or flat plate reactor with appropriate pumping or mixing system
Temperature control	Maximize algal productivity by maintaining the culture at the optimum temperature	Control temperature using a heat exchanger and/or by evaporative cooling by spraying water on to the reactor surface
Efficient gas transfer	Photosynthetic O_2 must be removed and CO_2 must be added to maximize productivity	

Chaetoceros muelleri, *C. calcitrans*, *Pavlova lutheri*, *Chroomonas salina*, *Haematococcus pluvialis* and *Anabaena* sp.

These closed systems not only allow a wider range of algae to be cultivated on a large scale than is possible with open systems, but they also have the advantage of reducing possible contamination from other organisms or airborne contaminants such as heavy metals, so that a better food quality product can be produced. Culture conditions can also be controlled better so that the chemical composition of the algae can be optimized and a product of reliable composition and nutritional value can be produced. The development of economically viable closed algal culture systems will allow the production of a wider range of algal species for human consumption and for the production of specific chemicals such as the long-chain polyunsaturated fatty acids.

18.5.1 Scale-up

As in all microbial culture systems, scale-up presents its own problems. The culture of microalgae at the laboratory scale presents few problems and high productivities can be reached. When scaling-up, however, the usual experience is that there is a reduction in productivity. This is especially true of the various open-culture systems. In most commercial algal plants, culture scale-up proceeds in steps of 5 to 10 times the previous volume. This is necessary since several critical factors change at each stage of scale-up; e.g. the surface area to volume ratio (important for both gas exchange and light availability) and the overall mixing regime. Scale-up usually also means moving from artificial illumination indoors, to outdoors into daylight. The increased irradiance outdoors is generally beneficial, being about 10-fold greater than the maximum irradiance which can be achieved indoors; however, temperature control is generally impossible outdoors. Furthermore, open outdoor cultures are exposed to potential dilution by rain and this can present significant management problems to the operator. Our experience with open systems has been that each increase in scale has led to new problems and had to be optimized before proceeding to the next step. Closed culture systems, especially the tubular photobioreactors, do not suffer as much from this problem. For example, scaling up a culture of the gren alga, *Tetraelmis chuii*, from a 60 l 'Biocoil'-type reactor to a 680 l reactor resulted in identical growth rates and cell densities. It remains to be seen whether other closed systems have similar scale-up properties.

18.7 Conclusion

Large-scale microalgal culture is still in its infancy; however, recent developments in new culture systems and new algae and algal products are very encouraging. The next few years are likely to see significant expansion in microalgal culture and the appearance of new algae and algal products in the marketplace. Improved understanding of the physiology and biochemistry of the algae will also aid in optimizing algal culture systems and in the production of consistent quality algal products for human nutrition and health care. Genetic engineering methods being developed for microalgae will also permit the creation of strains which have enhanced productivities of desired compounds, thus making microalgal culture more economically viable.

References

Abbott, I.A. (1988) Food and food production from seaweeds, in *Algae and Human Affairs* (eds C.A. Lembi & J.R. Waaland), Cambridge University Press, Cambridge, pp. 135–47.

Anon. (1995) Culturing method providing algae containig docosahexaenoic acid. Japan Patent Number 7,075,557.

Bajpai, P. & Bajpai, P.K. (1992) Arachidonic acid production by microorganisms. *Biotechnology and Applied Biochemistry*, **15**, 1–10.

Bajpai, P. & Bajpai, P.K. (1993) Eicosapentaenoic acid (EPA) production from microorganisms – a review. *Journal of Biotechnology*, **30**, 161–83.

Bajpai, P.K., Bajpai, P. & Ward, O.P. (1991a) Optimization of production of docosahexaenoic acid (DHA) by *Thraustochytrium aureum* ATCC 34304. *Journal of the American Oil Chemists Society*, **68**, 509–14.

Bajpai, P., Bajpai, P.K. & Ward, O.P. (1991b) Production of docosahexaenoic acid by *Thraustochytrium aureum. Applied Microbiology and Biotechnology*, **35**, 706–10.

Barclay, W.R., Meager, K.M. & Abril, J.R. (1994) Heterotrophic production of long chain omega-3 fatty acids utilizing algae and algae-like microorganisms. *Journal of Applied Phycology*, **6**, 123–9.

Becker, E.W. (1988) Micro-algae for human and animal consumption, in *Micro-algal Biotechnology* (eds M.A. Borowitzka & L.J. Borowitzka), Cambridge University Press, Cambridge, pp. 222–56.

Becker, E.W. & Venkatamaran, L.V. (1980) Production and processing of algae in pilot plant scale. Experiences of the Indo-German project, in *Algae Biomass* (eds G. Shelef & C.J. Soeder), Elsevier/North Holland Biomedical Press, Amsterdam, pp. 35–50.

Becker, E.W. & Venkataraman, L.V. (1982) *Biotechnology and Exploitation of Algae – The Indian Approach.* German Agency for Technical Co-op., Eschborn.

Belay, A., Ota, Y., Miyakawa, K. & Shimamatsu, H. (1993) Current knowledge on potential health benefits of *Spirulina. Journal of Applied Phycology*, **5**, 235–41.

Belay, A., Ota, Y., Miyakawa, K. & Shimamatsu, H. (1994) Production of high quality *Spirulina* at Earthrise Farms, in *Algal Biotechnology in the Asia-Pacific Region* (eds S.M. Phang, K. Lee, M.A. Borowitzka & B. Whitton), Institute of Advanced Studies, University of Malaya, Kuala Lumpur, pp. 92–102.

Ben-Amotz, A., Katz, A. & Avron, M. (1982) Accumulation of β-carotene in halotolerant algae: purification and characterisation of β-carotene-rich globules from *Dunaliella bardawil* (Chlorophyceae). *Journal of Phycology*, **18**, 529–37.

Ben-Amotz, A., Lers, A. & Avron, M. (1988) Stereoisomers of β-carotene and phytoene in the alga *Dunaliella bardawil. Plant Physiology*, **86**, 1286–91.

Bizzi, A., Chiesara, E., Clementi, F., Della Torre, P., Marabini, L., Rizzi, R. & Villa, A. (1980) Trattamenti prolungati nel ratto con diete contenenti proteine di *Spirulina.* Aspetti biochimici, morfologici e tossicologici, in *Prospettive della coltura di* Spirulina *in Italia* (ed. R. Materassi), Accademia dei Georgofili, Firenze, pp. 205–28.

Borowitzka, M.A. (1988) Microalgae as sources of essential fatty acids. *Australian Journal of Biotechnology*, **1**, 58–62.
Borowitzka, M.A. (1992) Comparing carotenogenesis in *Dunaliella* and *Haematococcus*: Implications for commercial production strategies, in *Profiles on Biotechnology* (eds T.G. Villa & J. Abalde), Universidade de Santiago de Compostela, Santiago de Compostela, pp. 301–10.
Borowitzka, M.A. (1995) Microalgae as sources of pharmaceuticals and other biologically active compounds. *Journal of Applied Phycology*, **7**, 3–15.
Borowitzka, M.A. (1996) Closed algal photobioreactors: design considerations for large-scale systems. *Journal of Marine Biotechnology*, **4**, 185–91.
Borowitzka, L.J. & Borowitzka, M.A. (1989a) Industrial production: methods and economics, in *Algal and Cyanobacterial Biotechnology* (eds R.C. Cresswell, T.A.V. Rees & N. Shah), Longman Scientific, London, pp. 294–316.
Borowitzka, L.J. & Borowitzka, M.A. (1989b) β-Carotene (Provitamin A) production with algae, in *Biotechnology of Vitamins, Pigments and Growth Factors* (ed. E.J. Vandamme), Elsevier Applied Science, London, pp. 15–26.
Borowitzka, L.J., Moulton, T.P. & Borowitzka, M.A. (1985) Salinity and the commercial production of beta-carotene from *Dunaliella salina. Nova Hedw., Beih.*, **81**, 217–22.
Burgess, J.G., Iwamoto, K., Miura, Y., Takano, H. & Matsunaga, T. (1993) An optical fibre photobioreactor for enhanced production of the marine unicellular alga *Isochrysis* aff. *galbana* T-Iso (UTEX-LB-2307) rich in docosahexaenoic Acid. *Applied Microbiology and Biotechnology*, **39**, 456–9.
Chamorro, G. & Salazar, M. (1990) Teratogenic study of *Spirulina* in mice. *Archives of Latinoamerican Nutrition*, **40**, 86–94.
Chaumont, D., Thepenier, C., Gudin, C. & Junjas, C. (1988) Scaling up a tubular photoreactor for continuous culture of *Porphyridium cruentum* from laboratory to pilot plant (1981–1987), in *Algal Biotechnology* (eds T. Stadler, J. Mollion, M.C. Verdus, Y. Karamanos, H. Morvan & D. Christiaen), Elsevier Applied Science, London, pp. 199–208.
Chaumont, D., Dos Santos, P.F., Gudin, C., Assise, D. & Chaintron, G. (1993) Dispositif de production intensive et contrôlée de micro-organismes photosynthétiques fragiles. France Patent Number 2685344.
Chen, F. & Johns, M.R. (1991) Effect of C/N ratio and aeration on the fatty acid composition of heterotophic *Chlorella sorokiniana. Journal of Applied Phycology*, **3**, 203–9.
Chrismadha, T. & Borowitzka, M.A. (1994a) Effect of cell density and irradiance on growth, proximate composition and eicosapentaenoic acid production of *Phaeodactylum tricornutum* grown in a tubular photobioreactor. *Journal of Applied Phycology*, **6**, 67–74.
Chrismadha, T. & Borowitzka, M.A. (1994b) Growth and lipid production of *Phaeodactylum tricornutum* in a tubular photobioreactor, in *Algal Biotechnology in the Asia-Pacific Region* (eds S.M. Phang, Y.K. Lee, M.A. Borowitzka & B.A. Whitton), Institute of Advanced Studies, University of Malaya, Kuala Lumpur, pp. 122–9.
Ciferri, O. (1983) *Spirulina*, the edible microorganism. *Microbiology Reviews*, **47**, 551–78.
Cohen, E. & Arad, S. (1989) A closed system for outdoor cultivation of *Porphyridium. Biomass*, **18**, 59–67.
Curtain, C.C. & Snook, H. (1983) Method for harvesting algae. International Patent Number WO83/01257.
Curtain, C.C., West, S.M. & Schlipalius, L. (1987) Manufacture of β-carotene from the salt lake alga *Dunaliella salina*; the scientific and technical background. *Australian Journal of Biotechnology*, **1**, 51–7.
Daintith, M. (1993) *Live Feeds for Marine Aquaculture: a Training Guide*, Aquaculture Sourcebook, Launceston.
Dodd, J.C. (1986) Elements of pond design and construction, in *CRC Handbook of Microalgal Mass Culture* (ed. A. Richmond), CRC Press, Boca Raton, pp. 265–83.
Druehl, L.D. (1988) Cultivated edible kelp, in *Algae and Human Affairs* (eds C.A. Lembi & J.R. Waaland), Cambridge University Press, Cambridge, pp. 119–34.
Grobbelaar, J.U., Nedbal, L., Tichy, L. & Setlik, I. (1995) Variation in some photosynthetic characteristics of microalgae cultured in outdoor thin-layered sloping reactors. *Journal of Applied Phycology*, **7**, 175–84.

Groeneweg, J. (1978) Einsatz hängender Plastikschläuche zur Massenkultur fädiger Algen. *Archives of Hydrobiology Supplenant*, **51**, 349–54.
Hansakul, W. (1995) *Chlorella* nutrients and its beneficial properties, in *Mass Cultures of Microalgae* (eds V. Thirakhupt & V. Boonakijjinda), Proceedings of the Research Seminar and Workshop, Silpakorn University, Thailand. November 18–23, 1991. UNESCO, Thailand, pp. P9–P25.
Hilaly, A.K., Karim, M.N. & Guyre, D. (1994) Optimization of an industrial microalgae fermentation. *Biotechnology and Bioengineering*, **43**, 314–20.
Jassby, A. (1988a) Some public health aspects of microalgal products, in *Algae and Human Affairs* (eds C.A. Lembi & J.R. Waaland), Cambridge University Press, Cambridge, pp. 181–202.
Jassby, A. (1988b) *Spirulina*: a model for microalgae as human food, in *Algae and Human Affairs* (eds C.A. Lembi & J.R. Waaland), Cambridge University Press, Cambridge, pp. 149–79.
Kapoor, R. & Mehta, U. (1992) Development and sensory evaluation of *Spirulina* supplemented recipes, in Spirulina. *ETTA National Symposium* (eds C.V. Seshadri & Bai N. Jeeji), MCRC, Madras, pp. 134–9.
Kapoor, R. & Mehta, U. (1993) Effect of supplementation of blue green alga (*Spirulina*) on outcome of pregnancy in rats. *Plant Foods for Human Nutrition*, **43**, 29–35.
Kawaguchi, K. (1980) Microalgae production systems in Asia, in *Algae Biomass Production and Use* (eds G. Shelef & C.J. Soeder), Elsevier/North Holland Biomedical Press, Amsterdam, pp. 25–33.
Kyle, D.J. (1992) Microbial oil mixtures and uses thereof. World Patent Number 9,212,711.
Kyle, D.J. & Gladue, R.M. (1991) Eicosapentaenoic acids and methods for their production. World Patent Number 9,114,427.
Kyle, D.J., Reeb, S.E. & Sicotte, V.J. (1991) Docosahexaenoic acid, methods for its production and compounds containing the same. World Patent Number 9,111,918.
Lee, Y.K., Ding, S.Y., Low, C.S., Chang, Y.C., Forday, W.L. & Chew, P.C. (1995) Design and performance of an alpha-type tubular photobioreactor for mass cultivation of microalgae. *Journal of Applied Phycology*, **7**, 47–51.
Levin, G. & Mokady, S. (1994) Antioxidant activity of 9-cis compared to all-trans beta-carotene in vitro. *Free Radical Biology and Medicine*, **17**, 77–82.
Levin, G., Ben-Amotz, A. & Mokady, S. (1994) Liver accumulation of soluble all-trans or 9-cis beta-carotene in rats and chicks. *Comparative Biochemistry and Physiology, A*, **107**, 203–7.
Martinez, M.R. (1988) *Nostoc comune* Vauch., A nitrogen-fixing blue-green alga, as source of food in the Philippines. *Philippine Agriculture*, **71**, 295–307.
Miki, K., Tajima, O., Matsuura, E., Yamada, K. & Fukimbara, T. (1980) Isolation and identification of a photodynamic agent of *Chlorella. Japanese Agriculture and Chemical Society Journal*, **54**, 721–26.
Mitchell, G.V., Grundel, E., Jenkins, M. & Blakely, S.R. (1990) Effects of graded dietary levels of *Spirulina maxima* on vitamin-A and vitamin-E in male rats. *Journal of Nutrition*, **120**, 1235–40.
Mohn, F.H. (1988) Harvesting of micro-algal bimomass, in *Micro-Algal Biotechnology* (eds M.A. Borowitzka & L.J. Borowitzka), Cambridge University Press, Cambridge, pp. 395–414.
Mohn, F.H. (1992) Technique and cost distribution for harvesting the microalga *Dunaliella salina* and impact on β-carotene production, in *Proceedings, 1st European Workshop on Microalgal Biotechnology*, Institut für Getreideverarbeitung, Potsdam, pp. 7–12.
Mokady, S. (1992) Nutritional, toxicological and therapeutic aspects, in Dunaliella: *Physiology, Biochemistry, and Biotechology* (eds M. Avron & A. Ben-Amotz), CRC Press, Boca Raton, pp. 217–29.
Mokady, S., Avron, M. & Ben-Amotz, A. (1990) Accumulation in chick livers of 9-cis versus all-trans β-Carotene. *Journal of Nutrition*, **120**, 889–92.
Mori, K., Ohya, H., Matsumoto, K. & Furune, H. (1987) Sunlight supply and gas exchange systems in microalgal bioreactor, in *Controlled ecological life support system* (eds R.D. MacElroy & D.T. Smernoff), NASA, Moffett Field, California, pp. 45–50.

Moulton, T.P., Sommer, T.R., Burford, M.A. & Borowitzka, L.J. (1987) Competition between *Dunaliella* species at high salinity. *Hydrobiologia*, **151/152**, 107–16.
Mumford, T.F. & Miura, A. (1988) *Porphyra* as food: cultivation and economics, in *Algae and Human Affairs* (eds C.A. Lembi & J.R. Waaland), Cambridge University Press, Cambridge, pp. 87–117.
Nonomura, A.M. (1987) Process for producing a naturally derived carotene-oil composition by direct ectraction from algae. U.S. Patent Number 4680314.
Oswald, W.J. (1988) Large-scale algal culture systems (engineering aspects), in *Micro-Algal Biotechnology* (eds M.A. Borowitzka & L.J. Borowitzka), Cambridge University Press, Cambridge, pp. 357–94.
Pirt, S.J., Lee, Y.K., Walach, M.R., Pirt, M.M., Balyuzi, H.H.M. & Bazin, M.J. (1983) A tubular bioreactor for photosynthetic production of biomass from carbon dioxide: design and performance. *Journal of Chemical Technology and Biotechnology*, **33B**, 35–58.
Pohl, P., Kohlhase, M. & Martin, M. (1988) Photobioreactors for the axenic mass cultivation of microalgae, in *Algal Biotechnology* (eds T. Stadler, J. Mollion, M.C. Verdus, Y. Karamanos, H. Morvan & D. Christiaen), Elsevier Applied Science, London, pp. 209–17.
Qin, H., Yu, G. & Yuan, S. (1994) Collecting *Dunaliella* and extracting beta-carotene. China Patent Number 1,084,848.
Richmond, A. (1988) *Spirulina*, in *Micro-Algal Biotechnology* (eds M.A. Borowitzka & L.J. Borowitzka), Cambridge University Press, Cambridge, pp. 85–121.
Richmond, A., Lichtenberg, E., Stahl, B. & Vonshak, A. (1990) Quantitative assessment of the major limitations on productivity of *Spirulina platensis* in open raceways. *Journal of Applied Phycology*, **2**, 195–206.
Richmond, A., Boussiba, S., Vonshak, A. & Kopel, R. (1993) A new tubular reactor for mass production of microalgae outdoors. *Journal of Applied Phycology*, **5**, 327–32.
Robinson, L.F., Morrison, A.W. & Bamforth, M.R. (1988) Improvements relating to biosynthesis. European Patent Number 261,872.
Ruane, M. (1974) Recovery of algae from brine suspensions. Australia Patent Number 486 999.
Rüegg, R. (1984) Extraction process for beta-carotene. U.S. Patent Number 4439629.
Running, J.A., Huss, R.J. & Olson, P.T. (1994) Heterotrophic production of ascorbic acid by microalgae. *Journal of Applied Phycology*, **6**, 99–104.
Schlipalius, L. (1991) The extensive commercial cultivation of *Dunaliella salina. Bioresource Technology*, **38**, 241–3.
Seshadri, C.V. & Jeeji Bai, N. (eds) (1992) Spirulina. *ETTA National Symposium.* MCRC, Tharamani, Madras.
Shimamatsu, H. (1987) A pond for edible *Spirulina* production and its hydraulic studies. *Hydrobiologia*, **151/152**, 83–9.
Shinmen, Y., Kawashima, H., Shimizu, S. & Yamada, H. (1992) Concentration of eicosapentaenoic acid and docosahexaenoic acid in an arachidonic acid-producing fungus, *Mortierella alpina* 1S-4, grown with fish oil. *Applied Microbiology and Biotechnology*, **38**, 301–4.
Shubert, L.E. (1988) The use of *Spirulina* (Cyanophyceae) and *Chlorella* (Chlorophyceae) as food sources for animals and humans. *Progress in Phycological Research*, **6**, 237–54.
Silva, H.J. & Cortinas, T.I. (1994) Vertical thin-layer photoreactor for controlled cultivation of cyanobacteria. *World Journal of Microbiological Biotechnology*, **10**, 145–8.
Soong, P. (1980) Production and development of *Chlorella* and *Spirulina* in Taiwan, in *Algae Biomass* (eds G. Shelef & C.J. Soeder), Elsevier/North Holland Biomedical Press, Amsterdam, pp. 97–113.
Takeuchi, D. & Uehara, K. (1993) Docosahexaenoic acid production eliminating generation of fish malodour. World Patent Number 9320225.
Tamura, Y., Maki, T., Shimamura, Y., Nishigaki, S. & Naoi, Y. (1979) Causal substances of photosensitive dermatitis due to *Chlorella* ingestion. *Food Hygene Journal*, **22**, 183–8.
Tornabene, T.G., Kates, M. & Volcani, B.E. (1974) Sterols, aliphatic hydrocarbons, and fatty acids of a nonphotosynthetic diatom *Nitzschia alba. Lipids*, **9**, 279–84.
Torzillo, G., Pushparaj, B., Bocci, F., Balloni, W., Materassi, R. & Florenzano, G. (1986) Production of *Spirulina* biomass in closed phtobioroeactors. *Biomass*, **11**, 61–4.

Torzillo, G., Carlozzi, P., Pushparaj, B., Montaini, E. & Materassi, R. (1993) A 2-plane tubular photobioreactor for outdoor culture of *Spirulina*. *Biotechnology and Bioengineering*, **42**, 891–8.

Tredici, M.R., Carlozzi, P., Zittelli, G.C. & Materassi, R. (1991) A vertical alveolar panel (VAP) for outdoor mass cultivation of microalgae and cyanobacteria. *Bioresource Technology*, **38**, 153–9.

Tseng, C.K. (1981) Commercial cultivation, in *The Biology of Seaweeds* (eds C.S. Lobban & M.J. Wynne), Blackwell Scientific, Oxford, pp. 680–725.

Vonshak, A. (1987) Strain selection of *Spirulina* suitable for mass production. *Hydrobiologia*, **151/152**, 75–7.

Vonshak, A. & Guy, R. (1992) Photoadaptation, photoinhibition and productivity in the blue-green alga, *Spirulina platensis* grown outdoors. *Plant and Cell Environment*, **15**, 613–16.

Wohlgeschaffen, G.D., Rao, D.V.S. & Mann, K.H. (1992) Vat incubator with immersion core illumination – a new, inexpensive setup for mass phytoplankton culture. *Journal of Applied Phycology*, **4**, 25–9.

Yoshino, Y., Hirai, Y., Takahashi, H., Yamamoto, N. & Yamazaki, N. (1980) The chronic intoxication test on *Spirulina* product fed to Wistar rats. *Japanese Journal of Nutrition*, **38**, 221–6.

Zarrouk, C. (1966) Contribution à l'étude d'une cyanophycée. Influence de divers facteurs physiques et chimiques sur la croissance et la photosynthèse de *Spirulina maxima*, Doctoral Thesis University of Paris.

19 Bio-enrichment: production of vitamins in fermented foods

K.H. STEINKRAUS

19.1 Introduction

Less than a hundred years ago, vitamin deficiency diseases were common in the United States. Unfortunately hundreds of millions of people still suffer from a variety of vitamin-deficiency diseases in the developing world. Each year millions of children become permanently blind via xerophthalmia caused by a lack of vitamin A in the diet. Surrounded by quantities of green and leafy vegetables, no child should be lacking vitamin A and have to suffer life-time blindness. Lack of proper nutrition education is the problem.

Thiamine deficiency leading to polyneuritis and muscular weakness characteristic of beri-beri is common among those subsisting primarily on highly polished rice and systems have been devised to enrich polished rice with thiamine. The problem would not exist if people were willing to consume brown or less highly polished rice; however, those eating predominantly polished rice are very sensitive and do not like the flavour or texture of rice retaining portions of the bran. Also, the keeping quality of brown rice is poorer. Insects rapidly invade the more nutritious brown rice.

People subsisting primarily on maize are in danger of niacin deficiencies leading to pellagra characterized by dermatitis, diarrhoea and dementia. Ascorbic acid deficiency leads to bleeding gums, loss of teeth and scurvy. Riboflavin deficiency leads to fissures at the corners of the mouth and abnormalities of skin and mucous membranes. Vitamin D deficiency leads to rickets in children. Exposure of infants to sunlight would generally prevent this but mothers in certain parts of the world prefer to protect their infants from sunlight. Deficiency of vitamin B_{12} – cobalamin – leads to pernicious anemia. This is rare in people consuming milk and meats. With the increase in vegetarianism, the incidence of B_{12} deficiency is increasing in the United States. Except for those people who follow unusual diets, for example, vegetarian diets for religious or personal reasons or those who follow weight reduction diets or for the poor and particularly the aged poor, vitamin deficiency diseases in the developed world have been largely eliminated by enrichment or fortification of the basic foods.

19.2 Enrichment/fortification

The process of adding nutrients to processed foods to restore the nutrient to its original level is called 'restoration' (LaChance, 1978). An example is the addition of ascorbic acid to 'instant' dried potatoes.

'Enrichment' in the Western world is the process of adding nutrients to selected formulated or manufactured foods as a public health measure. Examples are the addition of vitamin D to milk, vitamins A and D to margarine, and riboflavin to bread. 'Fortification' is the process of adding nutrients to levels coinciding with the image or place of the food in the diet. Thus, fruit juices may be fortified with ascorbic acid. Adding vitamin A to margarine is fortification based upon its relation to butter (LaChance, 1978).

Present levels of enrichment in the United States are presented in Table 19.1.

The National Academy of Sciences, National Research Council recommends that cereal-grain products be fortified to vitamin levels indicated in Table 19.2.

Vitamins are available for enrichment of food products in the Western world as shown in Table 19.3.

Vitamin A, vitamin C, thiamine, riboflavin, niacin and pantothenate are all quite inexpensive by Western standards, costing less than 10 cents (US) per 1000 mg.

Table 19.1 Levels of enrichment of several cereal products in the United States

Food	Thiamine	Riboflavin (mg/100 g)	Niacin
White flour	0.44–0.55	0.26–0.33	3.5–4.4
Bread	0.24–0.40	0.16–0.35	2.2–3.3
Corn meal	0.44–0.66	0.26–0.40	3.5–5.3
Rice	0.44–0.88	0.26–0.53	3.5–7.0

From Bender (1973).

Table 19.2 Levels of vitamins recommended for fortification of cereal-grain products by the National Academy of Sciences

Vitamin	Recommended level (mg/100 g)
A	0.48
Thiamine	0.64
Riboflavin	0.40
Niacin	5.29
Pyridoxine	0.44
Folic acid	0.07

From NAS/NRC (1974).

Table 19.3 Availability and prices of selected vitamins

Vitamin	Price per gram[a] $ (US)
Vitamin A (Retinal)	0.21
Vitamin D (ergocalciferol)	8.40
Vitamin C (L-ascorbic acid)	0.033
Thiamine	0.136
Riboflavin	0.163
Niacin (niacinamide)	0.024
(nicotinic acid)	
Biotin	22.80
Vitamin B_{12}	21.50
Pyridoxine	0.435
Pantothenate	0.122
Folic acid	0.903

[a] Sigma Chemical Corp., St. Louis, 1995.

Table 19.4 Recommended Daily Allowances of selected vitamins for adult males

Vitamin	Recommended daily allowance	Approximate cost per adult/year $ (US)
Vitamin A	1000 µg	0.077
Vitamin D	10 µg	0.031
Vitamin C	60 mg	0.723
Thiamine	1.5 mg	0.074
Riboflavin	1.7 mg	0.101
Niacin	19 mg	0.166
Pyridoxine	2.0 mg	0.318
Pantothenic acid	4–7 mg	0.312
Vitamin B_{12}	2 µg	0.016
Folacin	200 µg	0.006
Biotin	30–100 µg	0.830
	Total	2.65/year

From NAS/NRC (1989).

The recommended daily allowances of selected vitamins for adult males are given in Table 19.4.

The prices of vitamins are low enough so that foods in the Western world can be enriched at a rather insignificant increase ($2.65/adult/year) in the cost of the food (Table 19.4). Eliminating biotin, the cost of vitamins for a year would be only $1.82 for an adult. At times, it has been suggested that foods such as bread in India be enriched to eliminate some of the vitamin malnutrition in that country. However, when the daily requirements of particular vitamins are multiplied by 1 000 000 000 consumers, the total cost becomes prohibitive.

19.3 Sources of vitamins

Thiamine was first isolated from rice polishings (bran). Niacin is synthesized during the germination of oats, barley, wheat and maize. It is also synthesized in commercial quantities by *Xanthomonas pruni* (Davis *et al.*, 1951). The only vitamins produced commercially in significant quantities by fermentation are riboflavin and vitamin B_{12}.

Present world consumption of riboflavin is about 1.25 million kg (Lago & Kaplin, 1981). Although riboflavin is produced by *Candida* yeasts, solvent-producing bacteria such as *Clostridium acetobutylicum*, *Escherichia coli*, *Aerobacter aerogenes*, *Azotobacter* and *Mycobacterium*, the principal species used in commercial production are *Ashbya gossypii* and *Eromothecium ashbyii*, two closely related *Ascomycetes*. *E. ashbyii* has yielded as much as 2.2 mg riboflavin per ml on whey, skim milk, sucrose medium (Hendrickz & de Vleeschauwer, 1955). Superior cultures of *Ashbya* that produce 11 times more riboflavin under commercial conditions have been produced by chemical mutation (Lago & Kaplin, 1981).

Total sales of vitamin B_{12} in 1980 were estimated to be about 5000 kg (Lago & Kaplin, 1981). Merck and Co. use a selected strain of *Pseudomonas denitrificans* which produces 30 times the amount of cyanocobalamin produced by the original culture. *Propionibacterium shermanii* is also utilized for commercial production of vitamin B_{12}. *Streptomyces olivaceus* produces 55–66 μg vitamin B_{12} per gram mycelium (Pfeifer *et al.*, 1954). *Bacillus megaterium* also produces yields of 0.5 μg/ml vitamin B_{12} and higher (Garibaldi *et al.*, 1953).

Most of the other vitamins are chemically synthesized.

Modern food technology uses enrichment or fortification to improve nutritive value of foods (Harris, 1959). It is possible to accomplish much the same enrichment by use of suitable fermentations. Platt (1946, 1964) coined the phrase 'biological ennoblement' for the latter. This paper deals with a number of foods in which there are striking examples of 'biological ennoblement'.

The vitamin content of selected yeasts is given in Table 19.5.

Table 19.5 Vitamin content of some selected yeasts[a]

Yeast type	Thiamine	Riboflavin	Nicotinic acid	Pyridoxine	Pantothenic acid	Folic acid	Biotin	Reference
Brewer's	10.4–25	2.5–8.0	30–62.7	2.3–4.0	7.2–8.6	1.9–3.0	0.11	Peterson (1948)
Saccharomyces cerevisiae	2.8–4.1	3.9–6.2	27.7–56.8	–	–	1.9–3.6	0.045–0.36	Peterson (1948)
Torulopsis utilis	3.5–3.8	5.4–6.2	51.1–60.0			1.0–1.5		Agarwal *et al.* (1947)
Candida arborea	3.1–3.3	5.2–6.9	49.2–58.0			1.5–1.6		Agarwal *et al.* (1947)

[a] All values expressed as mg/100 g dry yeast.

Fermentations which involve yeasts and particularly those in which the yeasts are consumed along with the substrate tend to be enriched in the B vitamins. These include the primitive beers and wines which will be used as the first examples of bio-enrichment.

19.3.1 Kaffir (sorghum) beer

Kaffir beer is an alcoholic beverage with a pleasantly sour taste and the consistency of a thin gruel. It is the traditional beverage of the Bantu people of South Africa. Alcohol content may vary from 1 to 8% (v/v). Kaffir beer is generally made from kaffircorn (*Sorghum caffrorum*) malt and unmalted kaffircorn meal. Maize or finger millet (*Eleusine coracana*) may be substituted for part or all of the kaffircorn depending upon the relative cost (Schwartz, 1956). Even cassava and plantains may be used (Platt, 1964).

The kaffircorn grain is steeped for 6–36 hours. It is then drained and placed in layers and germinated with periodic moistening for 4–6 days. Germination continues until the plumule is 2.5 cm (1 in) or longer in length, and then sun-dried. Over 80% of the malt is sold for home brewing of kaffir beer.

The essential steps in brewing are: mashing, souring, boiling, conversion, straining and alcoholic fermentation. Mashing is carried out in hot (50°C) water. Proportions of malted to unmalted grains vary but 1:4 is satisfactory. About 4.5 l 1 gallon of water is added for about every 1.8 kg 4 lb of grain. Souring begins immediately due to the presence of lactobacilli. A temperature of 50°C favours development of *Lactobacillus delbrueckii*. Souring is complete in 6–15 hours. Water is added and the mixture is boiled. It is then cooled to 40–60°C and additional malt is added. Conversion proceeds for 2 hours and then the mash is cooled to 25–30°C. Yeasts present on the malt are responsible for the natural fermentation. However, *Saccharomyces cerevisiae* isolated from kaffir beer is frequently inoculated in the more modern breweries. Kaffir beer is ready for consumption in 4–8 hours. It is drunk while still actively fermenting, a state which may continue for up to 40 hours. Ethanol content is generally from 2–4% (w/v). The beer also contains from 0.3 to 0.6% lactic acid and 4–10% solids.

Consumption of 3 litres (5 pints) of kaffir beer which requires approximately 425 g of grain is not at all unusual for a working man each day. The question has been raised, how does the consumption of the beer diet compare in nutrient intake with a diet in which the grains were consumed directly? Platt (1964) compares the two diets in Table 19.6.

It is seen that caloric content of the two diets is quite similar. Only 37 calories are lost in the diet containing the beer. However, most notable is the doubling of riboflavin and near-doubling of nicotinic acid content of the diet containing beer due to synthesis of vitamins during malting and fermentation. Pellagra, which is relatively common in people subsisting on maize diets, is never noted in those consuming usual amounts of Kaffir beer (Platt, 1964).

The infants in the villages are given the dregs which contain very little alcohol, are rich in vitamins and protein content and are very likely highly beneficial to infants' nutrition.

The effect of germination on vitamin content of kaffircorn and maize is presented in Table 19.7. Thiamine decreases; riboflavin shows a substantial increase; nicotinic acid remains nearly constant in Kaffircorn but increases by about 50% in maize. Thus, the microorganisms involved in the fermentation must be contributing additional thiamine, riboflavin and nicotinic acid to the product.

Novellie (1966a,b) has found that the content of thiamine, riboflavin, and niacin in kaffir beers has tended to decrease (Table 19.8). The proportion of sorghum to maize used in brewing has decreased concomitantly from 4.9:1 to 1.2:1 or less. Following analyses of brewing practices and changes in vitamin contents during fermentation, Novellie (1966b) concluded that the

Table 19.6 Comparison of diet with and without maize beer[a]

	Amount of food eaten	
Food item	Diet without beer	Diet with kaffir beer
Maize, wholemeal	350 g	137.5 g
Maize, 60% extraction	350 g	137.5 g
Maize beer	–	5 pints (2840 ml)
Vegetables	130 g	130 g
Sweet potatoes	470 g	470 g
Kidney beans	30 g	30 g
Calories	3016	2979
Vitamin B_1	2.00 mg	1.95 mg
Riboflavin	1.13 mg	2.32 mg
Nicotinic acid	11.70 mg	20.30 mg

[a] From Platt (1964).

Table 19.7 Effect of germination on the vitamin content of kaffircorn and maize[a]

			Vitamin content (μg/g; dry basis)		
Sample		Moisture %	Thiamine	Riboflavin	Nicotinic acid
Kaffircorn	Grain	11.6	3.34	1.29	35.1
	Malt	10.7	1.73	2.41	34.1
Maize	Grain	12.1	3.50	1.46	19.6
	Malt	8.6	2.32	2.02	29.1

[a] From Goldberg & Thorp (1946).

change in substrate ratio was the primary factor responsible for the loss in vitamins. In particular, one brewery that was late in changing to low sorghum/maize ratios was also late in producing low-vitamin beer (Table 19.9). Changes in brewing practices have probably played a minor role.

This illustrates very well that modification and modernization of traditional fermented foods should only be undertaken when the nutritional value can be maintained. Kaffir beer is a very important source of nutrition to the indigenous people. It would be a tragedy if modernization of kaffir brewing were accompanied by a higher incidence of malnutrition among the Bantus.

Aucamp *et al.* (1961) reported chemical analyses of municipal kaffir beers (Table 19.10). The content of ascorbic acid in kaffir beer is negligible. However, 2 litres of beer per day, which is a rather common level of consumption, would supply 140%, 72% and 36% of the US recommended allowance for adult males (Food and Nutrition Board, 1974) of thiamine, riboflavin and nicotinic acid, respectively (Table 19.11).

Kaffir beer nutritionally is a very important part of the South African village diet.

Table 19.8 Average vitamin values for kaffir beer produced in large-scale breweries, 1954, 1959 and 1963

	Average vitamin content (μg/100 ml)		
Year	Thiamine	Riboflavin	Nicotinic acid
1954	153	95	494
1959	80	34	194
1963	32	36	189

From Novellie (1966b). Reprinted from *Food Technology*, **20**, 1607–8 (1966). © by Institute of Food Technologists.

Table 19.9 Vitamin content of kaffir beer made at a municipal brewery

			Vitamin content (μg/100 ml)		
Year	Brewery	Sorghum/ maize ratio	Thiamine	Riboflavin	Nicotinic acid
1954	Old	High	226	171	600
1959	Old	High	176	45	289
1963	Old	6:1	150	87	480
1964	Old	3:1	121	48	328
	New	0.875:1	29	35	233

From Novellie (1966b). Reprinted from *Food Technology*, **20**, 1607–8 (1966). © by Institute of Food Technologists.

Table 19.10 Composition of municipal kaffir beers[a]

	Range	Mean	No. of analyses
pH	3.2–3.7	3.4	10
Alcohol (% w/v)	1.8–3.9	3.0	17
Solids (% w/v)			
Total	3.0–8.0	5.4	17
Insoluble	2.3–6.1	3.7	6
Nitrogen (% w/v)			
Total	0.059–0.137	0.093	16
Soluble	0.010–0.017	0.014	9
Thiamine (μg/100 ml)	20–230	93.0	21
Riboflavin (μg/100 ml)	27–170	56.0	21
Nicotinic acid (μg/100 ml)	130–660	315.0	21
Ascorbic acid (mg/100 ml)	0.01–0.15	0.04	7

[a] From Aucamp *et al.* (1961).

Table 19.11 Amounts of B vitamins supplied by 2.2 litres (0.5 gallon, US) of kaffir beer[a]

	Thiamine (mg)	Riboflavin (mg)	Nicotinic acid (mg)
Minimum	0.38	0.51	2.5
Maximum	4.4	3.2	12.5
Mean	1.8	1.1	6.0
Minimum daily requirement[b]	1.0	1.2	10.0
Recommended allowance[c]	1.5	1.7	19.0

[a] Data calculated from values in Table 19.10.
[b] From Harris (1959).
[c] Adult males, NAS/NRC (1980).

19.3.2 Palm toddies

Palm sap is a sweet, clear and colourless liquid containing about 10–12% sugar and neutral in reaction (Okafor, 1975b).

Palm wine is essentially a heavy milky-white, opalescent suspension of live yeasts and bacteria with a sweet taste and vigorous effervescence (Okafor, 1975a,b).

The yeast *Saccharomyces cerevisiae* seems to be always present, whereas *Lactobacillus plantarum*, *Leuconostoc mesenteroides* and other bacterial species vary (Okafor, 1972; Faparusi, 1973). Van Pee & Swings (1971) also reported the discovery in palm wines of a motile rod-shaped high alcohol-producing anaerobic bacterium closely related to *Zymomonas mobilis*.

An enormous amount of palm wine is consumed in Southern Nigeria. Four million people each drink approximately half a litre per day which contains approximately 300 calories from the sugar and alcohol, 0.5–2 g of protein and important amounts of water-soluble vitamins. During the first 24 hours, there is as much as 83 mg ascorbic acid/litre. This declines later. Thiamine increases from 25 μg to 150 μg/litre. Riboflavin content increases from 35 to 50 μg/litre in the first 24 hours. Pyridoxine increases from 4 to 18 μg/litre. Vitamin B deficiencies are particularly widespread among pregnant women and teenagers. Palm wine is an important addition to their diets (Bassir, 1968).

Van Pee & Swings (1971) reported that palm juice and wines also contain vitamin B_{12} as follows:

Fresh palm juice (26 samples)	17–180 pg/ml vitamin B_{12}
Palm wine (12-hour fermentation)	140–190 pg/ml vitamin B_{12}
Palm wine (24-hour fermentation)	190–280 pg/ml vitamin B_{12}

The vitamin B_{12} content is very important for people consuming primarily vegetarian foods. The fact that the fresh palm juice contains vitamin B_{12} suggests that there is bacterial growth in the fresh juice before it is 'fermented'.

The major nutrients in palm wine are sugars and B complex vitamins, and the cheapest source of B vitamins in Ghana is palm wine (Nyako, 1977). During the first 3–4 days, the B vitamins increase in concentration, after which no further increase takes place (Faparusi, 1977). As palm wine is undrinkable after 3 or 4 days, the vitamin content at 24 hours is more representative of the vitamins actually consumed. Bassir (1968) and Faparusi & Bassir (1972) have reported the following for 24-hour oil and raphia palm wines: thiamine 8–25 μg/litre, riboflavin 8–35 μg/litre, pyridoxine 4–30 μg/litre, ascorbic acid 7.4–9.0 mg/100 ml, a decrease from about 11 mg/100 ml in fresh sap. Vitamin B_{12} increases with time up to 190–500 ng/litre at 24 hours (Okafor, 1978).

The following vitamins were found in the microbial cells and fluid of fermenting coconut toddy (Leong, 1953): thiamine (23.9 μg/100 ml), riboflavin (10.8 μg/100 ml), and nicotinic acid (369 μg/100 ml).

Palm toddies play an important role in nutrition among the economically deprived in the tropics.

19.3.3 Pulque

Pulque is the national drink of Mexico where it was inherited from the Aztecs (Goncalves de Lima, 1975). It is produced by fermentation of the juices (called aguamiel) of certain cactus plants of the genus *Agave*. The Agave plants grow on very poor soil and consumption of pulque is of particular importance in the diets of the poor and low-income people.

Pulque is a cloudy, white, viscous alcoholic beverage produced by the natural fermentation of aguamiel.

In the traditional process, the floral stem of mature plants 8–10 years old is cut off and the juice accumulates in the cavity. The juices are collected daily by oral suction and the juice is then fermented in wooden, leather or fibre-glass tanks with capacity of about 700 litres. The inoculum is pulque from a previous fermentation. Fermentation time is 8–30 days, depending upon temperature and the season. Generally 30–70 litres of fresh juice are added to a tank and an equal amount is withdrawn. Thus, the process is semi-continuous.

The essential organisms are homofermentative and heterofermentative *Lactobacillus* sp., *Leuconostoc* sp. (closely related to *Leuconostoc dextranicum* and *Leuconostoc mesenteroides*), *Saccharomyces carbajali* (closely related to *Saccharomyces cerevisiae*), and *Pseudomonas lindneri*. The *Leuconostoc* sp. produce bacterial polysaccharides or dextrans which are responsible for the characteristic viscosity of pulque.

Because of problems controlling the natural fermentation, pure-culture techniques have been developed in recent yeats (Sanchez-Marroquin, 1953).

The pulque fermentation has been modernized (Sanchez-Marroquin, 1977). Mechanical presses are used to collect the juice which is then treated to remove impurities, filtered and pasteurized. Pure cultures are used for the fermentation. A separate fermentation with *Leuconostoc* sp. is utilized to provide the desired degree of viscosity in the product. Fermentations are carried out in fibre-glass tanks. Optimum fermentation temperature is 28°C and optimum pH is 5.0. The overall result is a higher quality pulque produced in 48–72 hours versus 8–30 days required for natural fermentations.

The vitamin contents of traditional and modern pilot plant produced pulque are given in Table 19.12. It is seen that modernising the production of pulque has not decreased the vitamin content. Pulque continues to be an important source of nutrition among the economically deprived.

Table 19.12 B vitamins in traditional and pilot plant pulque

Number of samples	Vitamins	Traditional[a,b] (tinacales)	Pilot plant[a,b]
10	Thiamine	5.2–29.0	10.2–38.0
10	Niacin	54.0–515.0	115.0–635.0
10	Riboflavin	18.0–33.1	19.4–35.6
10	Pantothenic acid	60.0–355.0	15.9–32.2
5	*p*-Aminobenzoic acid	12.3–20.7	15.9–32.2
5	Pyridoxine	14.2–33.4	19.1–36.4
5	Biotin	9.1–32.2	11.0–35.3

[a] Range of values obtained.
[b] μg/100 ml.
Source: Sanchez-Marroquin (1977).

19.3.4 South American chicha

Chicha is a fermented beverage which is produced traditionally among the Indians of the Andes by pre-chewing maize, yuca, mesquite, quinoa, or plantain. Salivary diastase is the amylolytic agent. Well-made chicha is an attractive beverage, clear and effervescent, resembling apple cider in flavour. It generally contains 5% or less ethanol with a range of 2–12% (Cutler & Cardenas, 1947).

It would be expected that the nutritive value of chicha would be relatively high. The alcohol supplies calories; the microorganisms, if consumed with the chicha, would be rich in B vitamins and protein. Escobar *et al.* (1977) reported that riboflavin doubled while thiamine and niacin remained fairly constant during fermentation. Their results were obtained using a pure culture of *S. cerevisiae*. It would be worthwhile to study the vitamin contents of traditional chichas.

19.3.5 Indonesian tape ketan

Indonesian tape ketan is a sweet-sour, alcoholic rice food in which an amylolytic mould *Amylomyces rouxii* and at least one yeast of the *Endomycopsis burtonii* type hydrolyse steamed rice starch to maltose and glucose and produce ethanol and organic acids which provide an attractive flavour and aroma. If yeasts of the genus *Hansenula* are present, the acids and ethanol are esterified producing highly aromatic esters. Fermentation time is 2–3 days at 30°C. With continued incubation, the product becomes more liquid.

Detailed studies have been made of the biochemical and nutritional changes that occur during tape fermentations (Ko, 1972; Cronk *et al.*, 1977).

The ethanol (up to 8.5% (v/v)) serves as a source of calories and probably helps prevent transmission of disease or toxin-producing microorganisms in tape. The protein content of rice, which generally is about 7–8%, rises to a maximum of 16% in tape (dry solids basis) due to losses of total solids and synthesis of protein by the microorganisms (Cronk *et al.*, 1977). Lysine, the first limiting amino acid in rice, is selectively synthesized and increases by 15% in tape ketan. *A. rouxii* by itself or in combination with *E. burtonii*, *Candida lactosa*, *C. melinii*, *C. parapsilosis*, *Hansenula anomala*, *H. malanga* or *H. subpelliculosa* raises the thiamine content from 0.04 mg/100 g to 0.11–0.13 mg/100 g, a 300% increase of great nutritional importance for people subsisting principally on polished rice (Cronk *et al.*, 1977).

19.3.6 Acid-leavened bread and pancakes

Foods of the Indian idli type, acidified and leavened through fermentation by heterofermentative lactic acid bacteria, constitute a very interesting group of cereal-based foods of considerable potential importance in the developing and also the developed world where they are largely unknown (Ramakrishnan, 1979; Steinkraus, 1983). Idli is closely related to sourdough bread of the Western world, but it does not depend upon wheat or rye as a source of protein to retain the carbon dioxide gas during leavening. Leavening is produced by bacterial rather than by yeast activity.

White polished rice and dehulled black gram are soaked separately for 5–10 hours during the day. In the evening, the rice is drained and coarse ground and the black gram is drained and finely ground. They are then combined with 1% salt and sufficient water to make a relatively thick batter. The batter is incubated overnight, during which time *Leuconostoc mesenteroides* produces acid and CO_2 leavening the batter, and *Streptococcus faecalis* produces additional acid. The batter is steamed in the morning yielding acidified, leavened bread-like cakes or the batter can be made thinner and fried as a pancake (then called dosai) (Mukherjee *et al.*, 1965). Proportions of rice to black gram can vary from about 4:1 to 1:2 and still yield a satisfactory texture and flavour. Soybean cotyledons can be used to replace the black gram dhal.

Rajalakshmi & Vanaja (1967) reported that thiamine and riboflavin increase during fermentation (Table 19.13). Fermentation results in an increase in both free and total niacin (Rajalakshmi *et al.*, 1964). In addition, Akolkar (1977) reported increases in thiamine, riboflavin and niacin during the idli fermentaion (Table 19.14). Rajalakshmi & Vanaja also reported improvements in weight gain per gram protein intake and increases in the liver contents of thiamine, riboflavin, xanthine oxidase, and succinic dehydrogenase in rats fed fermented idli. These results were consistent with the findings of higher thiamine and riboflavin levels in fermented idli and the observed increases in nitrogen retention and riboflavin content of idli-fed rats.

The changes in thiamine, riboflavin and niacin contents were higher with fermentation by pure cultures than with natural fermentation. This bonus in vitamins is largely retained during short-term autoclaving of idli batter

Table 19.13 Composition of fermented and unfermented idli

	Amount in 100 g dry material[a]		
Food	Protein (g)	Thiamine (mg)	Riboflavin (mg)
Unfermented idli	13.9 (13.7–14.9)	0.21 (0.20–0.25)	0.25 (0.24–0.29)
Fermented idli	14.0 (13.9–14.1)	0.58 (0.51–0.64)	0.54 (0.43–0.60)

[a]Values are means of 4–7 determinations with range shown in parentheses.
Adapted from Rajalakshmi & Vanaja (1967).

(Table 19.14). A loss of less than 15% has been associated with ordinary steaming.

19.3.7 Indonesian tempe

Centuries ago the Indonesians, without modern chemistry and microbiology, developed a meat analogue by a fermentation process in which soybeans are soaked, dehulled, partially cooked and inoculated with moulds belonging to the genus *Rhizopus*. Incubated in a warm place (30–33°C), the soybean cotyledons are knitted into a compact cake by the fibrous mould mycelium in 1–3 days. The product, called 'tempe kedele', can be sliced thinly and deep-fat fried or cut into chunks and used in place of meat in soup and other dishes. Tempe contains over 40% (w/w) protein (dry solids basis) and a flavour and texture which appeals to the consumers. It is a good substitute for meat and it has begun to serve the same role in diets of American vegetarians (Steinkraus *et al.*, 1960; Hesseltine, 1961; Saono *et al.*, 1977; Shurtleff & Aoyagi, 1979, 1980; Wang & Hesseltine, 1979).

The essential microorganism for the production of the typical tempe bean cake is a mould belonging to genus *Rhizopus*. A number of species can complete the essential step of knitting the soybean cotyledons into a firm cake. These include *Rhizopus oligosporus*, *R. stolonifer*, *R. oryzae* and *R. arrhizus* (Hesseltine *et al.*, 1963; Dwidjoseputro & Wolf, 1970). The best of the moulds so far discovered is *Rhizopus oligosporus*, Northern Regional Research Laboratories (NRRL) strain 2710. This strain was originally isolated by Keith H. Steinkraus at Cornell University from dried powdered

Table 19.14 Vitamin changes in naturally fermented, experimentally fermented and steamed soy idli batter[a]

		Fermented batter[a]			
	Raw mixture[b]	Control[c]	Uninoculated, unautoclaved	Inoculated[d]	Steamed[e] inoculated
Thiamine (mg)	0.45	0.56	0.89	1.03	1.00
	(0.44–0.47)	(0.54–0.58)	(0.87–0.97)	(1.02–1.05)	(1.00–1.02)
Riboflavin (mg)	0.46	0.53	1.47	1.85	1.55
	(0.44–0.48)	(0.50–0.56)	(1.45–1.50)	(1.80–1.94)	(1.45–1.63)
Niacin (mg)	3.2	3.34	4.68	5.61	4.85
	(3.19–3.23)	(3.28–3.36)	(4.66–4.70)	5.43–5.84)	(4.80–4.86)

[a] Values per 150 g raw ingredients; average of three trials; range given in parentheses.
[b] Rice meal (100 g) and soy dhal flour (50 g)
[c] Autoclaved batter incubated without inoculation.
[d] Autoclaved batter inoculated with mixture of *L. mesenteroides*, *L. fermenti*, *L. delbrueckii* and *Bacillus* spp.
[e] Autoclaved at 5 lb/in^2 for 5 min.
Source: Akolkar (1977).

tempe brought to the United States by Miss Yap Bwee Hwa in 1958 and identified by Dr C.W. Hesseltine in the Northern Regional Research Laboratories, Peoria, Illinois.

Steinkraus *et al.* (1960) reported that the only essential microorganism in the tempe fermentation was the mould. Subsequently, however, it was found that tempe contained vitamin B_{12} (Steinkraus *et al.*, 1961). When a search was made for the source of the vitamin B_{12}, it was found that all commercial tempes studied contained a Gram-negative rod along with the mould. This rod produced the vitamin B_{12} (Liem *et al.*, 1977). Tempe made with the pure mould contains no vitamin B_{12} activity. The Gram-negative rod was subsequently identified as a non-pathogenic strain of *Klebsiella pneumoniae* (Curtis *et al.*, 1977). Since vitamin B_{12} is so important in the diets of vegetarians, from the nutritional viewpoint, it must be considered that *Klebsiella pneumoniae* is essential in the tempe fermentation.

The subject of vitamin B_{12} production is complicated and confused because of the various molecular types of vitamin B_{12}. These include cyanocobalamin, with a cyanide (-CN) attached to the cobalt, hyroxycobalamin with a hydroxyl (-OH) group attached to the cobalt; methylcobalmin with a methyl (CH_3-) group attached to the cobalt; and adenosylcobalamin with a complex adenosyl group attached to the cobalt (Steinkraus, 1986). Hydroxycobalamin rather than cyanocobalamin is the form most suitable for therapeutic use (Matthews and Linnell, 1979). Only use of the phagocytic protozoan *Ochramonas malhamensis* as test organism measures 'true' vitamin B_{12} (Ford, 1953; Williams *et al.*, 1956). *Lactobacillus leichmanii*, commonly used as test organism since it is easier to work with, measures not only 'true' vitamin B_{12} but also pseudo vitamin B_{12} activity. That vitamin B_{12} activity is produced by *K. pneumoniae* and *Citrobacter freudii* growing in tempe is not a question, but the actual amounts and the relative proportions of each of the cobalamins in tempe has not yet been studied (Steinkraus, 1986).

Microbiological assays indicated that the vitamin content of tempe is higher than the starting soybeans in certain cases and lower in others (Steinkraus *et al.*, 1961; Roelofsen & Talens, 1964; Napavarn *et al.*, 1977). Riboflavin doubles, niacin increases 7-fold, and vitamin B_{12} activity increases 33-fold. Thiamine unfortunately decreases (Table 19.15) (Steinkraus *et al.*, 1961). Pantothenate has been reported to stay approximately the same (Steinkraus *et al.*, 1961) or to double or quadruple (Murata *et al.*, 1967, 1968). Murata *et al.* (1967) reported increases of riboflavin from 8- to 47-fold, pyridoxine from 4- to 14-fold, and niacin from 2- to 5-fold in tempes manufactured and sun-dried in Indonesia.

Keuth and Bisping (1993) studied the production of vitamins by pure cultures of tempe molds and bacteria during the tempe solid substrate fermentation. *Rhizopus oligosporus*, *R. arrhizus* and *R. stolonifer* produced

Table 19.15 Comparison of certain vitamins in soybeans and in tempe

	Concentration (per gram)	
Vitamin	Soybeans	Tempe
Riboflavin	3 μg	7.0 μg
Pantothenate	4.6 μg	3.3 μg
Thiamine	10.0 μg	4.0 μg
Niacin	9.0 μg	60.0 μg
B_{12}	0.15 ng	5.0 ng

From Steinkraus *et al.* (1961).

riboflavin, nicotinic acid, nicotinamide and pyridoxine (vitamin B_6). *R. oligosporus* was the best vitamin producer. Thiamine content decreased during fermentation. *Citrobacter freundii* yielded 99.5 ng B_{12} per gram wet weight tempe) after 34 hours. This is a considerably higher level of vitamin B_{12} than Liem *et al.* (1977) reported, i.e. 5 ng B_{12} per gram tempeh using *Klebsiella pneumoniae*. Bisping *et al.* (1993) isolated two strains of *C. freundii* in the soak water which were the best producers of vitamin B_{12}. In the presence of *Rhizopus*, vitamin B_{12} production was three times higher than in the presence of *C. freundii* alone (Bisping *et al.*, 1993). Okada *et al.* (1985) isolated a number of bacterial species that increased during storage of tempe and were capable of producing vitamin B_{12} Mulyowidarso *et al.* (1989) demonstrated that there is a strong microbial fermentation of complex ecology that develops during soaking of soybeans. Mulyowidarso *et al.* (1990) reported that bacteria grew to viable populations of 100 million to 1 billion cfu/g during the tempe fermentation. *Bacillus pumilus* and *Bacillus brevis* were the predominant bacterial species, reaching populations of 100 million cfu/g during a 48-hour fermentation. Species of *Streptococcus faecium*, *Lactobacillus casei*, *Klebsiella pneumoniae* and *Enterobacter cloacae* also contributed to the fermentation and achieved populations of 1 to 10 million cfu/g (Okada *et al.*, 1985).

Biotin and total folate compounds were respectively 2.3 and 4–5 times higher in tempe than in unfermented soybeans (Murata *et al.*, 1970). The increases in riboflavin, niacin, pyridoxine and vitamin B_{12} activity are of considerable importance nutritionally. In the Western world, the consumer gets vitamin B_{12} from milk or meats, while vegetarians usually have to find alternative sources of B_{12}, often a vitamin capsule. Tempe serves not only as a protein-rich meat substitute but also as a source of vitamin B_{12}.

It has been shown that *Klebsiella pneumoniae* isolated from Indonesian tempe can be inoculated into Indian idli and enrich it with vitamin B_{12} (Parekh, 1979).

19.4 Bio-enrichment with protein

In the Indonesian tape ketan fermentation referred to earlier, rice starch is hydrolysed to maltose and glucose and fermented to ethyl alcohol. The loss of starch solids results in a doubling of the protein content (from about 8% to 16% in rice) on a dry solids basis (Cronk *et al.*, 1977). Thus, this process provides a means by which the protein content of high-starch substrates can be increased for the benefit of consumers needing higher protein intakes. This is particularly important for people consuming principally cassava which has a protein content of about 1% (wet weight basis). If the tape ketella fermentation is applied to cassava as it is in Indonesia, the protein content can be increased to at least 3% (wet weight basis) – a very significant improvement in nutrition to the consumer. In addition, the flavour of the cassava becomes sweet/sour and alcoholic, flavours that consumers may prefer to the bland starting substrate.

19.4.1 Bio-enrichment with essential amino acids

The Indonesian tape ketan/tape ketella fermentation not only enriches the substrate with protein; the microorganisms also selectively enrich the rich substrate with lysine, the first essential limiting amino acid in rice (Cronk *et al.*, 1977). This improves the protein quality.

Several researchers have reported from 10.6 to 60.0% increases in methionine during the Indian idli fermentation (Rao, 1961; Rajalakshmi & Vanaja, 1967; Steinkraus *et al.*, 1967). An increase in methionine, a limiting essential amino acid in legumes greatly improves the protein value.

The above are examples of bio-enrichment through fermentations that are well-accepted by millions of consumers. There are numerous other food fermentations used in many parts of the world. Most of these have been inadequately studied to determine what are the synthetic capabilities of their essential edible microorganisms. With further study, other methods of bio-enrichment may become known and available for enhancing the vitamin content, essential amino acid contents and related nutritional qualities of man's foods.

19.5 Summary

While the affluent Western world can afford to enrich its basic foods with chemically available vitamins, most of the developing world must rely upon biological enrichment of foods through fermentation. Every culture studied to date has its own food fermentations which are accomplished at costs within the financial means of most of the people. While some of the indigenous food/beverage fermentations, for example those reported in this chapter, have been extensively studied, there are numerous fermentations that have not been sufficiently studied to determine the essential micro-organisms, their sequences, their synthetic capabilities, particularly of vitamins, essential amino acids, protein and other biochemical/nutritional changes occurring in the substrate.

There is a potential gold-mine of information awaiting discovery in the indigenous food/beverage fermentations that will enable the world's people to bio-enrich their basic foods at relatively low cost in the coming years when world population reaches 6 billion about the year 2000 and 8–12 billion in the 21st century.

References

Agarwal, P.N., Singh, K., King, P.S. & Peterson, W.H. (1947) Yield and vitamin content of food yeasts grown on different kinds of molasses. *Archives Biochemistry*, **14**, 105–15.

Akolkar, P.N. (1977) Studies on soyidli fermentation. PhD Thesis. M.S. University of Baroda, Baroda, India.

Aucamp, M.C., Crieff, J.T., Novellie, L., Papendick, B., Schwartz, H.M. & Steer, A.G. (1961) Kaffircorn malting and brewing studies. VIII. Nutritive value of some kaffircorn products. *Journal of the Science of Food and Agriculture*, **12**, 449–56.

Bassir, O. (1968) Some Nigerian wines. *West African Journal of Biology and Applied Chemistry*, **10**, 42–5.

Bender, A.E. (1973) *Nutrition and Dietetic Foods*, Chemical Publishing Co., New York.

Bisping, B., Hering, L., Baumann, U., Denter, J., Keuth, K. & Rehm, H.J. (1993) Tempe fermentation: some aspects of formation of gamma-linoleic acid, proteases and vitamins. *Biotechnology Advances*, **11**, 481–93.

Cronk, T.C., Steinkraus, K.H., Hackler, L.R. & Mattick, L.R. (1977) Indonesian tape ketan fermentation. *Applied and Environmental Microbiology*, **33**, 1067–73.

Curtis, P.R., Cullen, R.E. & Steinkraus, K.H. (1977) Identity of a bacterium producing vitamin B-12 activity in tempe. Symposium on Indigenous Fermented Foods, Bangkok, Thailand.

Cutler, H.C. & Cardenas, M. (1947) Chicha, a native South American beer. *Bot. Mus. Leafl., Harv. Univ.*, **13**(6), 33–60.

Davis, B.D., Henderson, L.M. & Powell, D. (1951) Production of Niacin by *Xanthomonas pruni*. *Journal of Biological Chemistry*, **189**, 543–9.

Dwidjoseputro, D. & Wolf, F.T. (1970) Microbiological studies of Indonesian fermented foodstuffs. *Mycopathological and Mycological Applications*, **41**, 211–22.

Escobar, A., Gardner, A. & Steinkraus, K.H. (1977) Studies on South American Chicha. Symposium on Indigenous Fermented Foods. Bangkok, Thailand, Nov. 21–27.

Faparusi, S.I. (1973) Origin of initial microflora of palm wine from oil palm trees (*Elaeis guineensis*). *Journal of Applied Bacteriology*, **36**, 559–65.

Faparusi, S.I. (1977) Nigerian palm wine-emu. Symposium on Indigenous Fermented Foods, Bangkok, Thailand.

Faparusi, S.I. & Bassir, O. (1972) Factors affecting the quality of Palm-wine 2. Period of storage. *African Journal of Biology and Applied Chemistry*, **15**, 24–8.

Ford, J.E. (1953) The microbiological assay of 'vitamin B12'. The specificity of the requirement of *Ochramonas malhamensis* for cyanocobalamin. *British Journal of Nutrition*, **7**, 299–306.

Garibaldi, J.A., Ijichi, K., Snell, N.S. & Lewis, J.C. (1953) *Bacillus megatherium* for biosynthesis of cobalamin. *Industrial Engineering and Chemistry*, **45**, 838–946.

Goldberg, L. & Thorp, J.M. (1946) A survey of vitamins in African foodstuffs. VI. Thiamin, riboflavin and nicotinic acid in sprouted and fermented cereal products. *South African Journal of Medical Science*, **11**, 177–85.

Goncalves de Lima, O. (1975) Pulque, Balche Pajauaru. Univ. Fed. Pernambuco, Recife, Brasil.

Harris, R.S. (1959) Supplementation of foods with vitamins. *Agricultural and Food Chemistry*, **7**(2), 88–102.

Hendrickz, H. & de Vleeschauwer, A. (1955) Production of riboflavin by *Eromothecium ashbyii. Meded. Landb Hoofesch Gent.* **20**, 229–35.

Hesseltine, C.W. (1961) Research at Northern Regionai Research Laboratory on fermented foods, in *Proceedings of Conference on Soybean Products for Protein in Human Foods.* USDA, Peoria, Illinois, pp. 67–74.

Hesseltine, C.W., de Camargo, R. & Rackis, J.J. (1963) A mould inhibitor in soybeans. *Nature*, **200**, 1226–7.

Keuth, S. & Bisping, B. (1993) Formation of vitamins by pure cultures of tempe moulds amd bacteria during the tempe solid substrate fermentation. *Journal of Applied Bacteriology*, **75**, 427–34.

Ko, S.D. (1972) Tape fermentation. *Journal of Applied Microbiology*, **23**, 976–8.

LaChance, P.A. (1978) Enrichment, restoration, fortification and nutrification, in *Encyclopedia of Food Science* (eds M.S. Peterson & A.H. Johnson), AVI Publishing Co., Westport, Conn., pp. 232–7.

Lago, B.D. & Kaplan, L. (1981) Vitamin fermentations: B_2 and B_{12}, in *Advances in Biotechnology. Vol. III. Fermentation Products* (eds M. Moo-Young, C. Vezina & K. Singh), Pergamon Press, Oxford.

Leong, P.C. (1953) The nutritive value of coconut toddy. *British Journal of Nutrition*, **7**, 253–9.

Liem, I.T.H., Steinkraus, K.H. & Cronk, T.C. (1977) Production of vitamin B-12 in tempeh, a fermented soybean food. *Applied and Environmental Microbiology*, **34**, 773–6.

Matthews, D.M. & Linnell, J.C. (1979) Vitamin B12: an area of darkness. *British Medical Journal*, **2**, 533–5.

Mukherjee, S.K. Albury, M.N., Pederson, C.S., van Veen, A.G. & Steinkraus, K.H. (1965) Role of *Leuconostoc mesenteroides* in leavening the batter of idli, a fermented food of India. *Applied Microbiology*, **13**, 227–31.

Mulyowidarso., R.K., Fleet, G.H. & Buckle, K.A. (1989) The microbal ecology of soybean soaking for tempe production. *International Journal of Food Microbiology*, **8**, 35–46.

Mulyowidarso, R.K., Fleet, G.H. & Buckle, K.A. (1990) Association of bacteria with the fungal fermentation of soybean tempe. *Journal of Applied Bacteriology*, **68**, 43–7.

Murata, K., Ikehata, H. & Miyamoto, T. (1967) Studies on the nutritional value of tempeh. *Journal of Food Science*, **32**, 580–6.

Murata, K., Miyamoto, T. & Taguchi, F. (1968) Biosynthesis of B vitamins with *Rhizopus oligosporus. Journal of Vitaminology (Kyoto)*, **14**, 191–7.

Murata, K., Miyamoto, T., Kokufu, E. & Sanke, Y. (1970) Studies on the nutritional value of tempeh. III. Changes in biotin and folic acid contents during tempeh fermentation. *Journal of Vitaminology (Kyoto)*, **16**, 281–4.

NAS/NRC. (1974) Proposed Fortification Policy for Cereal-Grain Products. Food and Nutrition Board, National Academy of Sciences National Research Council, Washington, D.C.

NAS/NRC. (1989) Recommended Dietary Allowances. Committee on Dietary Allowances. Food and Nutrition Board, Washington, D.C.

Napavarn, N., Chalothorn, T.A. & Somari, S. (1977) Factors affecting fermentations and vitamin B-12 content in tempeh and tempeh-like products. Symposium on Indigenous Fermented Foods, Bangkok, Thailand.

Novellie, L. (1966a) Bantu beer – popular drink in South Africa. *International Brewer and Distiller*, **1**, 27–31.

Novellie, L. (1966b) Biological ennoblement and kaffir beer. *Food Technology*, **20**, 1607–8.

Nyako, K.O. (1977) Palm wine – an alcoholic beverage of Ghana. Symposium on Indigenous Fermented Foods, Bangkok, Thailand.

Okada, N., Hadioetomo, R.S., Nikkuni, S. & Itoh, H. (1985) Isolation of bacteria producing V. B-12 from fermented soybean tempeh in Indonesia. *Report of the National Food Research Institute*, **46**, 15–20.

Okafor, N. (1972) Palm-wine yeasts from parts of Nigeria. *Journal of the Science, Food and Agriculture*, **23**, 1399–407.

Okafor, N. (1975a) Preliminary microbiological studies on the preservation of palm wines. *Journal of Applied Bacteriology*, **38**, 1–7.

Okafor, N. (1975b) Microbiology of Nigerian palm wine with particular reference to bacteria. *Journal of Applied Bacteriology*, **38**, 81–8.

Okafor, N. (1978) Microbiology and biochemistry of oil-palm wine. *Advances in Applied Microbiology*, **24**, 237–56.

Parekh, B.J. (1979) Personal communication. M.S. University, Baroda, India.

Peterson, W.H. (1948) Yeasts in feeding. Proceedings of the Symposium, Nov, 1948, p. 26; abstracted in *Wallenstein Laboratory Communications*, **13**(43), 401 (1950).

Pfeifer, V.F., Vojnovich, C. & Heger, E.N. (1954) Vitamin B_{12} by fermentation with *Streptomyces olivaceus. Industrial Engineering and Chemistry*, **46**, 843–9.

Platt, B.S. (1946) 'Nutrition in the British West Indies,' Colon. Rep. No. 195. HM Stationery Office, London.

Platt, B.S. (1964) Biological ennoblement: Improvement of the nutritive value of foods and dietary regimens by biological agencies. *Food Technology*, **18**, 662–70.

Rajalakshmi, R., Nanavaty, K. & Gumazha, A. (1964) Effect of cooking procedures on the free and total niacin content of certain foodstuffs. *Indian Journal of Nutrition and Dietetics*, **1**, 276–80.

Rajalakshmi, R. & Vanaja, K. (1967) Chemical and biological evaluation of the effects of fermentation on the nutritive value of foods prepared from rice and gram. *British Journal of Nutrition*, **21**, 467–73.

Ramakrishnan, C.V. (1979) Studies on Indian fermented foods. *Baroda Journal of Nutrition*, **6**, 1–57.

Rao, M.V.R. (1961) Some observations on fermented foods, in *Meeting the Protein Needs of Infants and Children.*, National Academy of Sciences. National Research Council, Publication, **843**, pp. 291–3.

Roelofsen, P.A. & Talens, A. (1964) Changes in some B vitamins during molding of soybeans by *Rhizopus oryzae* in the production of tempeh kedelee. *Journal of Food Science*, **29**, 224–6.

Sanchez-Marroquin, A. (1953) The biochemical activity of some microorganisms of pulque. Mem. Congr. Cientif. Mex., 4th Centenario Univ. Mex., pp. 471–84.

Sanchez-Marroquin, A. (1977) Mexican pulque – a fermented drink from Agave juice. Symposium on Indigenous Fermented Foods, Bangkok, Thailand.

Saono, S., Brotonegoro, S., Basuki, T., Sastraatmadji, D.D., Jutono, Badjre, I.G.P. & Gandjar, I. (1977) Tempe. Symposium on Indigenous Fermented Foods, Bangkok, Thailand.

Schwartz, H.M. (1956) Kaffircorn malting and brewing studies. I. The kaffir beer brewing industry in South Africa. *Journal of the Science, Food and Agriculture*, **7**, 101–5.

Shurtleff, W. & Aoyagi, A. (1979) *The Book of Tempeh*, Harper and Row, New York.

Shurtleff, W. & Aoyagi, A. (1980) *Tempeh Production*, New Age Foods, Lafayette, California.

Steinkraus, K.H. (1983) Lactic acid fermentation in the production of foods from vegetables, cereals and legumes. *Antonie van Leeuwenhoek*, **49**, 337–48.

Steinkraus, K.H. (1986) Production of vitamin B12 in tempe, in *Proceedings of the Asian Symposium on Non-salted Soybean Fermentation*, July 15–17, 1985, Tsukuba Science City, Ibaraki, Japan, pp. 205–8.

Steinkraus, K.H., Yap, B.H., Van Buren, J.P., Providenti, M.I. & Hand, D.B. (1960) Studies on tempeh – An Indonesian fermented soybean food. *Food Research*, **25**, 777–88.

Steinkraus, K.H., Hand, D.B., Van Buren, J.P. & Hackler, L.R. (1961) Pilot plant studies on tempe, in Proceedings of Conference on Soybean Products for Protein in Human Foods. USDA, pp. 75–84.

Steinkraus, K.H., van Veen, A.G. & Thiebeau, D.B. (1967) Studies on idli – an Indian fermented black gram-rice food. *Food Technology*, **21**, 110–13.

van Pee, W. & Swings, J.G. (1971) Chemical and microbiological studies on Congolese palm wines (*Elaeis guineensis*). *East Afr. Agric. For. J.*, **36**(3), 311–14.

Wang, H.L. & Hesseltine, C.W. (1979) Mold-modified foods, In *Microbial Technology*, 2nd edn (eds H.J. Peppler and D. Perlman), Academic Press, New York, Vol. 2, pp. 95–129.

Williams, W.L., Stiffey, A.V. & Jukes, T.H. (1956) Microbiological and chick assay of vitamin B12 activity in feed supplements and other natural products. *Agricultural & Food Chemistry*, **4**, 364–7.

20 Production of industrial enzymes and some applications in fermented foods

A. GODFREY

20.1 Introduction

The involvement of enzymes in the processing of foodstuffs is almost as old as the history of mankind. Many traditional fermentations rely upon the hydrolytic actions of indigenous or deliberately added enzymes. These enzymes derive from the metabolic activities of acceptable microbes or extracts of plant materials or animal glands. In recent and scientifically informed times these activities have been identified and frequently isolated to provide enhanced potency and controlled action resulting in a substantial industrial production of commercial enzymes.

This chapter begins with a short account of the development of that knowledge and the evolution of the industrial enzyme industry. There follows a discussion of the extraction of animal and plant enzymes and an account of the principles of the production of enzymes by microbial fermentation.

The concluding section outlines the widening use of industrial enzymes for the efficient conversion of complex fermentation feedstocks and describes the main features of the use of enzymes in many food fermentations.

20.2 Brief history of industrial enzymes

The origins of the use of enzymes are hidden in the depths of ancient history and the development of truly industrial processes does not appear until the late 19th century. The production of fermentable materials and their conversion to foods and beverages has been an activity of civilized human society for a very long time. The curdling of milk for cheese making is referred to in the ninth song of the Homeric Odyssey and again in the fifth song of the Iliad, where several uses for an extract of the fig (*Ficus* spp.) are described. These almost certainly represent the application of the proteolytic enzyme ficin in a crude extract.

By the time humans had begun to adopt a settled and agricultural society the modification of cereals and fruits rich in carbohydrates to induce an alcoholic fermentation had been discovered and developed. In his remarkable treatise, *The Ascent of Man*, Dr. Jacob Bronowski (1973) indicates that the turning point for the development of an agricultural society occurred in the region of the ancient settlements at Jericho around 10 000 years ago. With the rapidly developing application of highly focused genetic engineering in modern bioscience, it is fascinating to find that a natural, chance hybridization produced a cross between wild goat grass and wild wheat to give a plump-headed cereal that no longer had the 'sail' on the seed which consequentially prevented the natural wind-induced dispersion of its grains. Man could harvest the grain for his food, and at the same time become responsible for the survival of the hybrid cereal by sowing a portion for the next season's harvest. Many starchy plants such as rice, manioc, and later wheat, barley, oats, rye, maize and potato were used to produce alcoholic beverages. Conversion of the starch to fermentable sugars was achieved by the action of mixed cultures of microorganisms deliberately introduced to moist mashes of the starchy materials. The improved fermentations achieved if the starchy substances were first chewed can now be seen to have been the result of salivary amylase action. The microbial contributions will have been extracellular carbohydrases and proteinases.

This empirical development in ancient food fermentations incorporated an element of enzymic hydrolysis of complex polymers, coupled with either deliberate or inevitable inoculation with yeasts and bacteria that subsequently produced a stable material of a nutritious and palatable nature.

The preparation of concentrated and gradually more specific sources of these enzymes forms the basis of fermented products for many centuries, leading to the cultures for koji and saké fermentations in the Orient (see Chapter 12); the preparation of tea, coffee and cocoa (see Chapter 5); the development of fish and soybean condiments (see Chapters 12 and 16); the spread and diversification of cheeses and baked products (see Chapters 7, 8 and 10); the expansion of bacterial acid fermentations of vegetable and meat proteins (see Chapters 2 and 15); and the development of the fermentation industries based on cereals, with particular expansion following the discovery of the malting process that gave rise to a self-digesting cereal product (see Chapters 1 and 9).

By the middle of the 19th century, detailed investigation of the activities of microorganisms was established, and in 1878 the German chemist, Willy Küne, had coined the term 'enzyme' to describe the chemical entities responsible for continued catalysis observed with cell-free extracts of yeast. About the same time the industrial commercialization of an enzyme, rennet, extracted from calf stomachs, had begun under the direction of the Dane, Christian Hansen.

By the turn of the present century, the Japanese scientist, Takamine, had isolated fungal carbohydrases and proteinases and demonstrated their use to improve the traditional rice koji for fish and soybean hydrolyses. The product 'Takadiastase' is still used as a digestive-aid, and surface-culture methods developed by Takamine still form the basis of many commercial enzyme productions. The most notable of these are the fungal pectinolytic enzymes used by wine producers and fruit processors (see Chapter 9). Pancreatic enzymes were isolated by Dr. Otto Röhm around 1913 and soon became effectively used in the leather industry as a popular replacement for animal dung.

More recent progress in industrial enzymes derives from both the rapid expansion in biochemical knowledge about individual enzymes and their roles in metabolism, and from the discovery and rapid expansion of antibiotic production. Large-scale and highly specific microbial fermentations were developed. From the work of Fleming in the 1930s until around 1955, the transfer of technology to the enzymes industry was gradual, but by 1965 huge quantities of bacterial amylases and proteinases were being produced by submerged (deep) fermentations. Progress since 1965 has been extremely rapid and diverse, so that there are now commercial enzymes for most hydrolytic processes of relevance to industrial needs. Continuous production processes and also continuous enzyme reactions have been developed for the biggest industrial sectors such as starch conversion for syrups and fermentation feedstocks, chemical biotransformations of important intermediates in pharmaceuticals, and speciality food ingredients production. Although much has been claimed for the benefits of using immobilized enzymes in continuous processing, either as stirred reactions or in fixed-bed columns, there are still only a small number of industrial-scale applications. The particular limitations that demand a reaction mixture that is free of particulates, and the maintenance of sterility for a long operating time, have limited their practical use. The largest application is for the continuous isomerization of glucose syrups to a mixture of fructose and glucose for food and beverage sweetening. Other important, but relatively small-volume applications are largely in the production of highly specialized chemicals such as certain amino acids and pharmaceuticals including semi-synthetic antibiotics. The application of genetic manipulation of microorganisms is now accelerating the practical development of enzyme production by the transfer of productive genes from difficult to grow, or pathogenic, microbes into well-established production strains. The ingenuity of molecular biologists in protein engineering for the post-production modification of enzyme characteristics to create catalytic performances that meet industrially preferred conditions is starting to generate another round of expansion of the range of industrial enzymes.

A summary of the history of industrial enzymes is given in Table 20.1.

It is currently estimated that industrial enzymes are produced by some 300 companies throughout the world (excluding China, for which no production data are available). Almost three-quarters of the production by weight comes from Europe. The production of high-value, smaller-volume

Table 20.1 A summarized history of industrial enzymes

Pre-history	Cheese production Carbohydrate fermentation for beverages
Pre-biological sciences	Oriental fermentations – beverages, sauces, pickles, wines, beers, malting of cereals, milk and cheese products
1874	Chr. Hansen – commercial rennet
1878	W. Kühne – term 'enzyme' proposed
1890	Takamine – takadiastase (fungal amylase)
1900–1930	Malt extract for textile desizing Start of animal enzyme extraction on commercial scale
1930	Pectinases for fruit processing
1954	Submerged fermentation – proteinases for detergents; amylases for textiles and starch
1960s	Diversification of fermented enzyme types – wider range for starch and detergent industries; development of fungal milk coagulants
1970s	Further expansion of fermentation enzymes – immobilized enzymes; thermostable and thermolabile varieties produced
1980s	New, more refined enzymes; multi-activity specifically formulated products; energy-saving systems, bio-fuels production; chemical bio-transformations and diagnostics; environmentally significant developments; practical genetic manipulations
1990s	Rapid development of large numbers of speciality enzymes for R&D, diagnostics and industrial processing; enzymes from GMO; enzymes modified by protein engineering; large expansion of bio-transformations by enzymes

speciality enzymes is distributed throughout the world with considerable focus in Japan.

A full discussion of industrial enzymes and detailed examples of applications will be found in the text *Industrial Enzymology*, 2nd edition (1996). Readers who wish to identify specific commercial examples of the enzymes mentioned in this chapter are recommended to consult the detailed Data Indexes of this text, where they will find many commercial enzymes described in detail, together with the addresses of the producers.

20.3 Modern approach to enzyme production

Since the first edition of this book the number of different enzymes available as commercial preparations in processing quantities has risen from 50 to more than 100. In addition, there are more than 500 different enzymes available in small commercial quantities for research, diagnostic and biomanipulation purposes. The total of identified enzymes now approaches 4000. The majority of industrial enzymes are of microbial origin and the most significant developments in the second half of this century have been with this group. Improvements in biochemical understanding, coupled with substantial improvements in fermentation and downstream separation methods, have brought many more enzymes into economic use, and the ways in which they can be applied have also expanded.

Microbial enzymes are either fermented in very large volumes to produce low cost, but relatively impure (mixed enzyme) commodities for primary processing such as starch conversion, brewing and distilling, textile and detergency, or in smaller scale with far greater attention to downstream purification to create speciality catalysts for highly targeted reactions including flavours and food functionality modifications.

Some consistently important enzymes are still extracted from animals and higher plants (Table 20.2). The quantities are generally far lower than for the microbials, and the unit prices are higher. However, some large-scale processing still holds to these enzymes for preference; in particular, the traditional use of calf rennet for milk coagulation represents more than 10% of all enzyme sales. The limitations imposed by the agricultural cycles, the occasional severe shortages caused by disease or unusual weather conditions, and the increasing intensity of international controls over the movement of plant and animal materials to contain diseases all combine to make it difficult to expand an existing process, or develop a major new one, that depends upon these sources.

20.4 Enzymes from animal materials

Almost exclusively, animal enzymes of commercial significance are proteinases extracted from intestinal tissues or glands such as the pancreas or parotid.

Table 20.2 Enzymes for food fermentation applications from animals and plants

Enzyme	E.C. no.	Source
Alpha-amylase	3.2.1.1	Bovine/porcine pancreatic glands
Proteinase	3.4.21.1	
Lipase	3.1.1.1	
Rennets (chymosins)	3.4.23.4	Fourth stomach of calf, kid and lamb
Esterases (lipase variants)	3.1.1.1	Parotid glands of calf, kid and lamb
Catalase	1.11.1.6	Bovine liver
Lysozyme (muramidase)	3.2.1.17	Hen egg albumen
Alpha- and beta-amylases	3.2.1.1	Malted cereals, soybeans, unmalted barley
	3.2.1.2	
Peroxidase	1.11.1.7	Horseradish root (*Armaracia rusticana*)
Proteases	3.4.22.4	Bromelain (*Ananas* spp.)
	3.4.22.3	Ficin (*Ficus* spp.)
	3.4.22.2	Papain (*Carica papaya*)
	3.4.23.1	Pepsin from porcine/bovine stomach
	3.4.21.4	Trypsin from porcine/bovine pancreas
	3.4.21.1	Chymotrypsin (as trypsin)
Lipoxygenase	1.13.11.12	Soybeans

Pepsin is the carboxyl-acid proteinase extracted from porcine gastric mucosa. It is noted for its very acid pH optimum between 1.8 and 3. It has a very broad specificity which relates to its main role as a primary protein-hydrolysing digestive enzyme. Rennet (chymosin) is also a carboxyl-acid proteinase but with extremely high specificity in the pH range 5–7. It is used almost exclusively for milk coagulation where it acts to break one particular bond pair of amino acids in kappa casein to release the bulk of the casein to calcium mediated, non-enzymic, coagulation. This specificity ensures that very little further proteolysis takes place in the cheese curd, which in turn ensures almost no development of bitter tastes that might be generated if the proteins were more vigorously degraded.

Rennet is extracted from the abomasum (fourth stomach) of unweaned ruminants, notably calf, kid and lamb. The majority is extracted from calves, but the extracts from kid and lamb may be produced to contain additional lipase activity which is relevant to the production of speciality cheese flavours characteristic of many of the Mediterranean countries (see Chapter 10).

Pancreatin is a mixture of several enzymes obtained by extraction of bovine, or more particularly porcine, pancreatic glands. It contains the proteinases trypsin and chymotrypsin, together with lipase and alpha-amylase. The proteases are also prepared without the lipase activity as both mixtures and individual enzymes of high purity. It should be noted that generally these enzymes do not obtain certification under various religious codes. By rigorous controls and under strict religious supervision, it is possible that some enzymes of animal origin may be certified kosher for example.

20.4.1 Source materials

Generally, the appropriate tissues are excised immediately post-slaughter from healthy animals, fit for human consumption and veterinarian certified, and rapidly deep frozen while a sufficient quantity is accumulated for efficient transport to the processing plant. During thawing, the frozen material is roughly broken in a coarse mincing or grinding machine and then, when completely thawed, it is further subjected to grinding or comminution while still very cold.

20.4.2 Extraction

Rennets must be extracted in a way that will not release non-specific proteinases that would interfere with cheese quality due to excessive protein degradation. This is achieved by less than total homogenization of the tissues, and extraction by washing with chilled aqueous solutions of common salt (food-grade) ranging from 1 to 5% and containing dilute hydrochloric acid to activate the enzyme. The saline extract is recovered by filtration. The fluid is combined with any subsequent wash liquors that have been applied to the filter cake, or by re-filtration of resuspended residues, and concentrated. Modern processing increases the yield of active enzyme by employing ultrafiltration membrane methods rather than the older vacuum evaporation process. The concentrate will be adjusted to commercial strengths by the addition of salt solutions such that the activity meets the required use rates for the intended markets, and the salt level is sufficient to prevent microbial growth. Typically the salt level will be in the range 15–18% by weight. The pH is raised to between 5 and 6 to further ensure enzyme stability. Where permitted, the addition of preservatives and caramel colouring is made at this stage.

In the case of the pancreatic enzymes, and where a pure pepsin or rennet is required, the extraction will use an appropriate dilute buffer to maintain the maximum enzyme activity and to limit solubilization of other tissue components. The first liquor separated from the tissues will be treated to precipitate the desired enzymes, or to precipitate unwanted materials, leaving the enzymes in solution. Various and quite specific concentrations of salts such as sodium or ammonium sulphate or sodium chloride are typically employed for this stage. A further purification of the enzyme can be obtained by repeated 'salting out' by this method, or by the careful addition of chilled solvents such as ethanol or acetone.

Finally, the precipitated enzymes can be carefully dried under vacuum and standardized with diluting carriers such as sodium chloride or lactose. For liquid preparations, the powder extract will be dissolved in a stabilizing solution, usually containing sodium chloride and permitted preservatives. Some solid preparations are obtained by spray- or freeze-drying from liquid concentrates, followed by standardization.

Throughout the extraction, great care is taken to keep the enzyme extracts cold and to avoid dramatic changes of pH or concentration of added chemicals. This is achieved by the use of chilled liquids and continuous stirring during any additions.

It will be noted from the many publications of technical data on industrial enzymes that the pH at which an enzyme is most stable is not necessarily the same as the pH for maximum activity. It is therefore important to ensure that preparations are at the best pH for stability during supply and storage, and that statements of activity are clearly identified with regard to the physical conditions of the assay employed so that best use may be made in practice.

For the maximum activity of extracted animal enzymes, it is generally necessary to activate the precursor zymogen by addition of dilute acid or alkali, or a specific proteinase to the extracting buffers.

20.5 Enzymes from plant materials

Traditionally there are three classical proteinases produced from plants, bromelain, ficin and papain. Malted cereals are also a source of some carbohydrases of which alpha- and beta-amylases are the most common. Malted wheat and barley are the main cereal sources of these enzymes, although a very useful beta-amylase is obtained from the soya bean. The soya bean also yields a lipoxygenase which has applications in baking, although extraction and purification of this enzyme for use in other industries has not proven economic. By using an appropriate activation it is also possible to obtain beta-amylase from unmalted barley.

Bromelain, ficin and papain are all thiol proteinases and should be used in reducing conditions since they are inactivated by oxidizing agents. They are extracted from the various plant tissues by processes that essentially parallel those for extracting animal enzymes.

Bromelain is obtained from the juices resulting from crushing either the stems or fruits of various varieties of pineapple (*Ananas* spp.). The enzymes from stem and fruit have slightly different properties regarding the specific amino acid bonds that they will hydrolyse. Both are active between pH 5 and 8, unusually for proteinases, the optimum varying for each protein substrate. The temperature maximum for practical applications is 55°C.

Ficin is extracted from the latex sap expressed from wounds in the stem of the tree (*Ficus* spp.). The pH optimum lies between 5 and 7 and the maximum performance temperature is 65°C.

Papain is obtained from the latex of the fruits of various varieties of papaya (*Carica* spp.). It has characteristics similar to those of ficin, although it has the broadest range of protein degradation of the three plant proteinases. Traditionally it is the most widely accepted and used of the three. Current thinking describes at least five isoenzymes making up the usual complex proteinase that is used commercially. Caricain is a well-characterised isoenzyme with a significantly lower maximum performance temperature than that found for the regular papain products. Papain is active between pH 5 and 7 on most proteins, but the considerable thermotolerance of this enzyme places some limitations on its widest application in modern food processing. It will be found active up to 70°C, and may need up to half an hour at above 90°C to completely inactivate it, particularly when in the presence of substantial amounts of substrate, a condition typical of food processing. The use of these plant proteinases for food processing requires positive approval of legislative bodies in many countries. Bromelain and papain are obtained from the normally edible part of the plant which increases their acceptance, while ficin is obtained from a non-edible part. Ficin is therefore frequently prohibited for use in a food application.

Alpha-amylase and beta-amylase from cereals are extracted by the application of the typical 'mashing systems' of the brewing industry, with careful attention to reduced temperature limits so that inactivation is minimized. The cereals are milled in the typical brewery manner and then mashed with warm water in the range 40–50°C. After a period at this temperature, the liquor is drained and concentrated by vacuum evaporation or ultrafiltration membrane.

Separation and purification of the two amylases can be achieved by raising the temperature to 60°C, when the beta-amylase is more rapidly inactivated, or lowering the pH to 4, which more rapidly inactivates the alpha-amylase. Concentration and standardization can produce a liquid product requiring stabilization with permitted preservatives and stabilizing salts or sugars, or powders by spray- or freeze-drying. A number of enzymes with more specialized application are also extracted from plant tissues, including pectin-degrading systems from citrus fruits, some lipases from oil-bearing seeds, an alpha-galactosidase from coffee bean mucilage, and lipoxygenase from soya bean. A notable plant enzyme is horseradish peroxidase, which is a key catalyst in many diagnostic and monitoring test systems used in the food and medical sectors.

20.6 Principal steps in extraction of animal and plant tissues

20.6.1 Collection or harvesting

Account must be taken of the need for accurate identification of plant species or the selection of the appropriate enzyme-bearing part of the animal. The animal must have been fit for human consumption.

20.6.2 Storage

Storage and transportation before extraction will relate to the potential of microbial spoilage and the loss of the target enzyme(s). General autolysis of most tissues will be accompanied by the proliferation of microorganisms that were naturally present or may have been introduced during stage one. Careful drying of plant tissues often enables storage to be practicable for many months. Freezing of animal materials is particularly appropriate.

20.6.3 Preparation for extraction

Initial preparation for extraction usually requires the material to be crushed, milled, ground or comminuted in an appropriate aqueous buffer and generally under chilled conditions around 0–4°C. The equipment for the method of choice is varied and can generally be developed by minor modifications of traditional food processing plant. The maintenance of low temperatures is the most important modification.

20.6.4 *Enzyme extraction* (Figure 20.1)

The main process generally consists of the solubilization of the target enzyme, together with many other substances that are also soluble in the liquid phase selected, and the subsequent concentration and purification of the enzyme. Primary purifications are achieved by 'salting out' or solvent precipitation. More intensive purification involves further selective precipitations with the possible addition of bulk chromatography to specifically adsorb and remove impurities, or adsorb and then desorb the target enzyme.

20.6.5 *Presentation for market*

Commercial products will require concentration and standardization of the enzyme. Standardization is usually by dilution of a concentrate with an

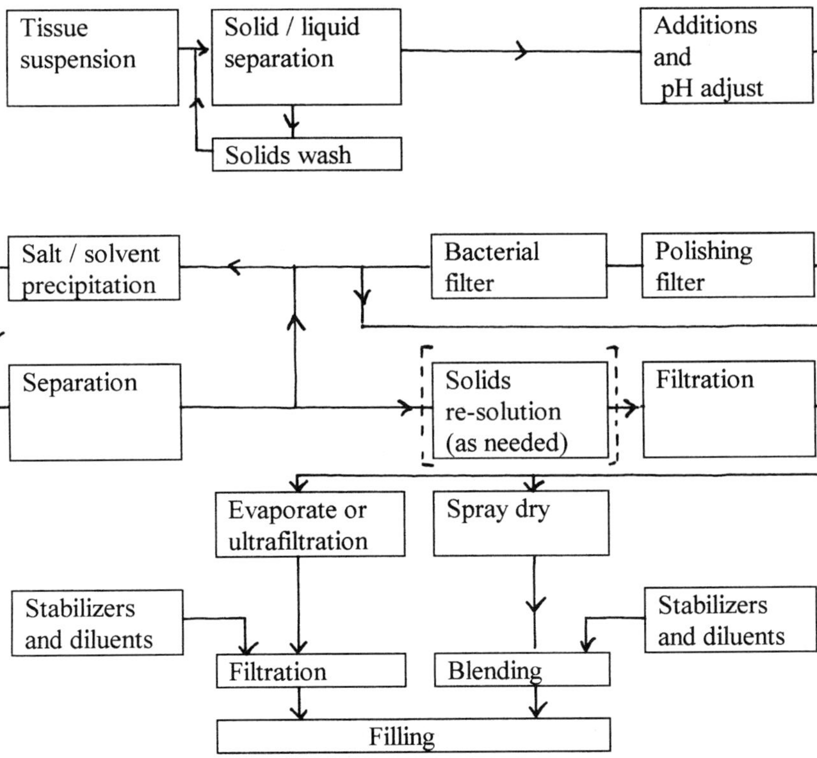

Figure 20.1 Schematic flow diagram for the recovery of animal and plant enzymes.

Table 20.3 Enzymes for food fermentation applications from microbial sources

Enzyme	Source genus	E.C.no.
α-Acetolactate decarboxylase	*Thermoanaerobium*	4.1.1.?
Amylases (diastases)	*Aspergillus*	3.2.1.1
	Bacillus	3.2.1.1/3.2.1.2
	Endomyces	3.2.1.1
	Rhizopus	3.2.1.1/3.2.1.2
Amyloglucosidase (glucoamylase)	*Aspergillus*	3.2.1.3
	Rhizopus	3.2.1.3
Anthocyanase (α-glucosidase)	*Aspergillus*	3.2.1.21
	Penicillium	3.2.1.21
Catalase	*Aspergillus*	1.11.1.6
	Micrococcus	1.11.1.6
	Penicillium	1.11.1.6
Cellulase	*Aspergillus*	3.2.1.4
	Humicola	3.2.1.4
	Penicillium	3.2.1.4
	Trichoderma	3.2.1.4
Galactomannanase	*Aspergillus*	3.2.1.78
	Bacillus	3.2.1.78
	Humicola	3.2.1.78
	Trametes	3.2.1.78
	Trichoderma	3.2.1.78
α-Galactosidase (melibiase)	*Aspergillus*	3.2.1.22
	Saccharomyces	3.2.1.22
β-Galactosidase (lactase)	*Aspergillus*	3.2.1.23
	Bacillus	3.2.1.23
	Escherichia	3.2.1.23
	Saccharomyces	3.2.1.23
	Kluyveromyces	3.2.1.23
β-Glucanase	*Aspergillus*	3.2.1.6
	Bacillus	3.2.1.6
	Humicola	3.2.1.6
	Penicillium	3.2.1.6
	Saccharomyces	3.2.1.6
	Trichoderma	3.2.1.6
Glucose isomerase	*Aspergillus*	5.3.1.5
	Bacillus	5.3.1.5
	Mycobacterium	5.3.1.5
	Streptomyces	5.3.1.5
Glucose oxidase (notatin)	*Aspergillus*	1.1.3.4
	Penicillium	1.1.3.4
Hemicellulase (Pentosanase)	*Aspergillus*	3.2.1.8/3.2.1.32
	Humicola	3.2.1.8/3.2.1.32
	Trichoderma	3.2.1.8/3.2.1.32

Table 20.3 *Continued*

Enzyme	Source genus	E.C.no.
Lipase (inc. esterase)	*Achromobacter*	3.1.1.3
	Alcaligenes	3.1.1.3
	Aspergillus	3.1.1.3
	Bacillus	3.1.1.3
	Candida	3.1.1.3
	Mucor	3.1.1.3
	Penicillium	3.1.1.3
	Rhizopus	3.1.1.3
Lipoxygenase	*Rhizopus*	1.13.11.12
Microbial Milk Coagulants	*Aspergillus* (GMO)	3.4.23.6
	Cryphonectria	3.4.23.6
	Endothia	3.4.23.6
	Escherichia (GMO)	3.4.23.6
	Kluyveromyces	3.4.23.6
	Mucor	3.4.23.6
	Rhizomucor	3.4.23.6
Pectinase	*Aspergillus*	3.2.1.15
	Humicola	3.2.1.15
	Penicillium	3.2.1.15
	Trichoderma	3.2.1.15
		most preps. contain other types.
Peptidase (amino)	*Aspergillus*	3.4.11.11
	Rhizopus	3.4.11.11
(carboxy)	*Aspergillus*	3.4.17.?
	Rhizopus	3.4.17.?
(leucine amino)	*Lactobacillus*	3.4.11.1
	Rhizopus	3.4.11.1
Peroxidase	*Aspergillus*	1.11.1.7
	Penicillium	1.11.1.7
Proteinase (serine)	*Bacillus*	3.4.21.4
(alkaline)	*Aspergillus*	None allocated
(carboxyl-acid)	*Aspergillus*	3.4.23.6
	Mucor	3.4.23.6
(metallo)	*Aspergillus*	3.4.24.4
	Bacillus	3.4.24.4
	Streptomyces	3.4.24.4
Pullulanase	*Bacillus*	3.2.1.41
	Klebsiella	3.2.1.41
Tannase	*Aspergillus*	3.1.1.20
Xylanase (hemicellulase/ pentosanase)	*Aspergillus*	3.2.1.32
	Humicola	3.2.1.32
	Trichoderma	3.2.1.32

appropriate carrier such as salt, lactose, starch, maltodextrin or sugars. Liquid products may be best concentrated by ultrafiltration followed by standardization with permitted preservatives and a substance to increase the solids concentration, so decreasing the water activity of the system and prevent microbial spoilage.

20.7 Enzymes from microbial sources

The participation of many, largely unspecified, enzymes in the processes of fermentation of foods is a powerful stimulus for the development and use of isolated, specified, food-grade enzymes for the development of food systems, with or without a fermentation stage in the food production itself.

In the general case, the fermenting microbe utilizes the available nutrients by producing extracellular (secreted) enzymes that hydrolyse polymers in the medium. The small fragments or monomeric components so produced are then suitable for transport into the cells to participate in its internal metabolism. This activity parallels the digestive action of higher animals and man. Industrial exploitation of this external digestive competence has been extensive since the mid-1960s, resulting in the collection of hydrolytic enzymes now available in commercial bulk quantities and of relevance to food and beverage fermentations (Table 20.3). In addition, there are a small number of industrial enzymes other than the hydrolases that have both food and non-food application. These include glucose isomerase, glucose oxidase, catalase and peroxidase.

Investigations of the metabolic activities of microorganisms in the past half century have brought a number of enzymes into commercial scale supply that derive from inside the cell, particularly some that are membrane-bound. Few of these find application in fermented food production, but extended shelf life has been achieved by the use of glucose oxidase/catalase systems to remove oxygen from many foods and beverages. The extraction of both extracellular and intracellular enzymes has been described by Atkinson & Mavituna (1991) and a detailed description of the commercial production of extracellular enzymes was given by Aunstrup (1979) and of intracellular enzymes by Lilly (1979).

20.8 General principles for industrial production of microbial enzymes

Little specific and detailed information has been passed to the public domain regarding the production of any one enzyme. This is due to the highly competitive nature of the enzyme production industry in which the factors affecting the production economics are considered highly confidential. However, in the development of positive regulatory control of materials added to or used in food processing it is known that full disclosures are made, in confidence, to regulating authorities in order to obtain approvals for particular enzyme preparations.

20.8.1 Production strains

The species used are considered to present the least risk of toxin production during fermentation as well as being non-pathogenic to humans. The genera used for producing enzymes of relevance to food and beverage fermentations are given in Table 20.3.

Bacillus species are comparatively easy to isolate and, to counter concerns associated with the potential of spore formation, many have been selected as non-sporing strains. The *Aspergilli* are similarly placed among the fungi, although the formation of conidial spores is desirable here for the ease of inoculation of large fermentors. In all cases, the selected production strain will have a highly enhanced enzyme-producing ability when compared with the 'wild' parent strain. These selected strains undergo stringent screening to ensure they do not produce toxins or antibiotics, even if the fermentation conditions are accidentally altered a long way from the selected production criteria, to meet the increasing standards for food applications for the enzyme(s) produced.

20.8.2 Fermentation

The choice of method lies between 'solid state' (sometimes referred to as 'semi-solid') and submerged or 'deep' fermentation. In rare cases, the organism will dictate the choice due to non-production or low yields by one route. Generally, however, the nature of the final enzyme product and its performance objective determine the method. Enzymes from solid-state cultures are frequently found to be complex mixtures, often including amylase, proteinase, lipase, and non-starch carbohydrases in proportions that are regulated by the whole cultivation operation. If a high level of a single enzyme activity is required, it is commonly produced by submerged fermentation.

The introduction of genetically manipulated production strains is altering the traditional ways to make a choice of fermentation method. Manipulated strains with very large numbers of genes coding for the required enzyme, or with genes for unwanted enzymes deleted, may be used in either fermentation system. The choice for these will be one of practice at the factory, or, if both methods are available, be based on overall economics, facility of isolation and purification, and productivity.

(a) Semi-solid fermentation. Semi-solid fermentations are a refinement of the traditional oriental 'koji' systems described in Chapter 21. Originally used for fungal enzyme production, the semi-solid systems are typically used for producing the non-starch carbohydrases (cellulases, lactases and pectinases), speciality proteinases, lipases, and some starch enzymes such as alpha- and gluco-amylases. The main characteristics are the great flexibility of scale and the simplicity of the basic production plant. The fungus is grown on a moist, particulate mash using wheat, soya or rice as the primary material. Cultivation is performed in shallow trays of up to 10 cm depth of substrate. These trays are stacked in suitable boxes or rooms that permit

moist air to be circulated over the trays. Alternatively, the substrate may be contained in slowly rotating cylinders with fine mesh screens for walls that permit free circulation of moist air through the gently tumbling medium. The tumble method has required considerable development to achieve commercially attractive productivity and the particular techniques are closely guarded secrets.

More significant developments have come by using forced aeration of relatively deep beds of medium up to 2 m deep. Control and maintenance of humidity and even distribution of air throughout the bed is essential.

Each specific production organism will require optimization of the medium composition and the precise conditions. The amount of air and the temperature at which it is supplied to the cultivation chambers is regulated to achieve the optimum moisture levels and remove excess heat from the fermentation. In cases where very fast fermentation occurs it may be necessary to employ cooling systems in the chamber design.

The culture medium (Table 20.4) is prepared so that it contains a large amount of the least expensive source of carbohydrate and protein. Wheat or rice bran and soybean meal plus supplements of trace elements, phosphates and in some cases additional nitrogen sources such as ammonia may be used. In many cases the yield of a particular enzyme can be raised by the addition of a specific stimulating substance. In general these are not inducing substrates for the enzyme since a production strain will be selected to have constitutive production of the required enzyme.

Sterilization of the medium before inoculation is usually by live steam injection in a rotating screw cooker. Inoculation is by introducing a spore suspension into the cooled tumbling medium which is then filled into trays or fermentation tumblers.

Table 20.4 Representative media for semi-solid fermentations for enzyme production[a]

Enzyme type	Medium component	Parts
Lactase	Wheat bran	10
	0.2 M HCl (containing traces of Zn, Fe and Cu)	6
Lipase	Wheat bran	3
	Soybean meal	1
	Water	3
Carbohydrases/ proteinases	Wheat bran	3
	Rice bran	2
	Soybean meal	1
	Water	4

[a]These are examples only – many productions will use other ingredients and ratios according to the found best productivity of the organism chosen. This is particularly true when cultivating microbes that have been subject to genetic manipulation.

(b) Submerged (deep) fermentation. Submerged or 'deep' fermentation has been claimed to be more economic for the production of large volumes of commodity enzymes. The media (Table 20.5) will be based on suspended insoluble nutrients and inexpensive additional sources of soluble nitrogen, phosphates and trace elements. The medium must be selected to support good growth while being as cheap as possible. Soybean meal, starches and starch hydrolysates supported by corn steep liquor and yeast hydrolysates dominate the list of ingredients. The specific performance may be stimulated by the addition of particular factors selected for each organism.

Industrial enzyme fermentation uses very large fermentors as large as $150\,m^3$.

Controls to monitor pH, temperature and in some cases dissolved oxygen are typical. When a suspended medium, and/or a high input air flow rate are employed, it is often important to operate at substantially below the design volume of the fermentor and to incorporate foam detection and automatic

Table 20.5 Representative media for submerged fermentation for enzyme production

Enzyme type	Medium component	$g\,l^{-1}$
Bacillus protease	Starch hydrolysate	50
	Soybean meal	20
	Casein	20
	Na_2HPO_4	3.3
	(alternatively)	
	Ground barley	100
	Soybean meal	30
	Na_2CO_3 (to adjust to pH 9–10)	
Fungal proteinase	Starch	30
	Corn steep liquor	5
	Soybean meal	10
	Casein	12
	Gelatine	5
	Distillers' dried solubles	5
	KH_2PO_4	2.4
	$NaNO_3$	1
	NH_4Cl	1
	$FeSO_4$	0.01
Bacillus amylase	Potato starch	100
	Ground barley	50
	Soybean meal	20
	Sodium caseinate	10
	$Na_2HPO_4{\cdot}12H_2O$	9
Fungal amylase	Corn starch	24
	Corn steep liquor	36
	NaH_2PO_4	47
	$CaCl_2$	1
	KCl	0.2
	$MgCl_2{\cdot}6H_2O$	0.2

antifoam treatment. Bulk medium is prepared separately, with pH adjustment followed by heating to sterilize. The sterile medium is pumped into fermentation vessels that have been pre-sterilized, usually with live steam.

Sterile filtered air is blown into the vessel through diffusers to obtain the correct solution rates and waste gas removal for aerobic fermentations.

The selection of medium components has importance for the downstream systems for enzyme recovery. Trial fermentations are always performed to help in the selection of the medium so that viscosity, total solids, cell mass, efficient medium conversion and the balance of gases and foaming are all considered. In many production fermentations the organism produces the desired enzyme at the end of the major growth stage. This has tended to limit the production to batch operation. When production of the target enzyme can be stimulated by the addition of specific compounds it may be convenient to run a two-stage fermentation by adding the trigger substance as the culture approaches maximum cell density in the late growth phase.

Intracellular enzymes have become of considerable interest for industrial processing due to the more selective and specific reactions that they catalyse. The use of enzymes for diagnostic systems, and more recently for highly specific biotransformations, has further increased the desire for commercial production of intracellular enzymes. Regio-, stereo- and chirally specific reactions are important to the fine chemicals and pharmaceuticals industries, and from initial developments in these sectors there is now an expanding application into speciality items for the food and beverage sectors. Natural flavourings and colourants can be developed and modified to particular properties by the use of enzymes catalysing oxido-reductive, and stereo or chiral selectivity. The fermentation systems are essentially the same as for batch fermentation of extracellular enzymes, but the process concentrates on obtaining maximum cell mass at economic cost. In a few cases, feeding the fermentation with one or more key substances can create a continuous system with separation of part of the cell mass for extraction. In some cases the desired enzyme is firmly bound to cell walls or internal cell membranes and may be isolated directly as an immobilized system.

Recovery of the enzyme from the culture system is necessary to ensure that the final preparation meets the requirements for purity and stability dictated by the end application. The number of stages and the loss of activity during processing are kept to the minimum. Throughout the recovery process attention is paid to the avoidance of contamination by harmful microorganisms or undesirable chemical substances.

Most of the equipment used is typical of traditional food processing design and this helps in establishing compatible standards of hygiene.

Extracellular enzymes produced in semi-solid fermentations are washed out with other solubles from the bran. Countercurrent and percolation systems are used, with the minimum amount of aqueous buffer of pH and ionic strength chosen to suite the enzyme being extracted. Usually the process is performed with chilled fluids in the range 5–10°C. In some situations the bran may be partially dried before extraction so that a more concentrated extract can be obtained. Once the bran is extracted, the processing of the aqueous fluid is essentially the same for enzymes from semi-solid and submerged fermentations.

20.8.3 Broth purification

As shown in Figure 20.2, the main unit processes are simple and well-known unit steps. The aqueous extracts contain the enzyme(s), residues of suspended and soluble medium components, and the cells of the fermented microorganism. The first step is to separate the solids by filtration or centrifugation aided by the use of flocculants such as calcium salts, polyelectrolytes and aluminium salts typically used in modern water treatments. It is also usual to incorporate some diatomaceous earth or other filter-aid into the stirred liquor when filtration by rotary vacuum filters is used. For centrifugation, high-speed continuous-disc machines are preferred.

Concentration of the aqueous enzyme is a compromise between energy efficiency and activity loss. Low temperature vacuum evaporation is often used for the more robust enzymes, while ultrafiltration membrane methods are helpful for the sensitive enzymes because a low temperature (c. 5°C) can be used.

Additional purification is usually needed to eliminate microorganisms and to reduce the preparation to the lowest practical contamination with other enzymes produced during the fermentation. The rapid introduction of production from genetically manipulated strains of microbes has tended to increase the purity and concentration within the primary broth. There is either a considerable multiplication of the gene for the target enzyme, or the introduction of the enzyme from one microbe into a very competent producing strain of another. Polishing filtrations remain important for ultimate hygienic quality, and a series of precipitations may be used to select the desired enzyme. The addition of inorganic salts such as ammonium sulphate to a specified concentration will precipitate a range of proteins which may include the desired enzyme or leave it in the soluble phase. Further solution and precipitations may be performed with different concentrations of precipitant to achieve desired purification. Organic solvents that lower the dielectric constant of the system and so reduce the solubility of proteins are also used to selectively precipitate enzymes. The most effective treatments are performed using chilled solvents and adding them to the aqueous broth whose pH has been adjusted to the iso-electric value of the enzyme being processed.

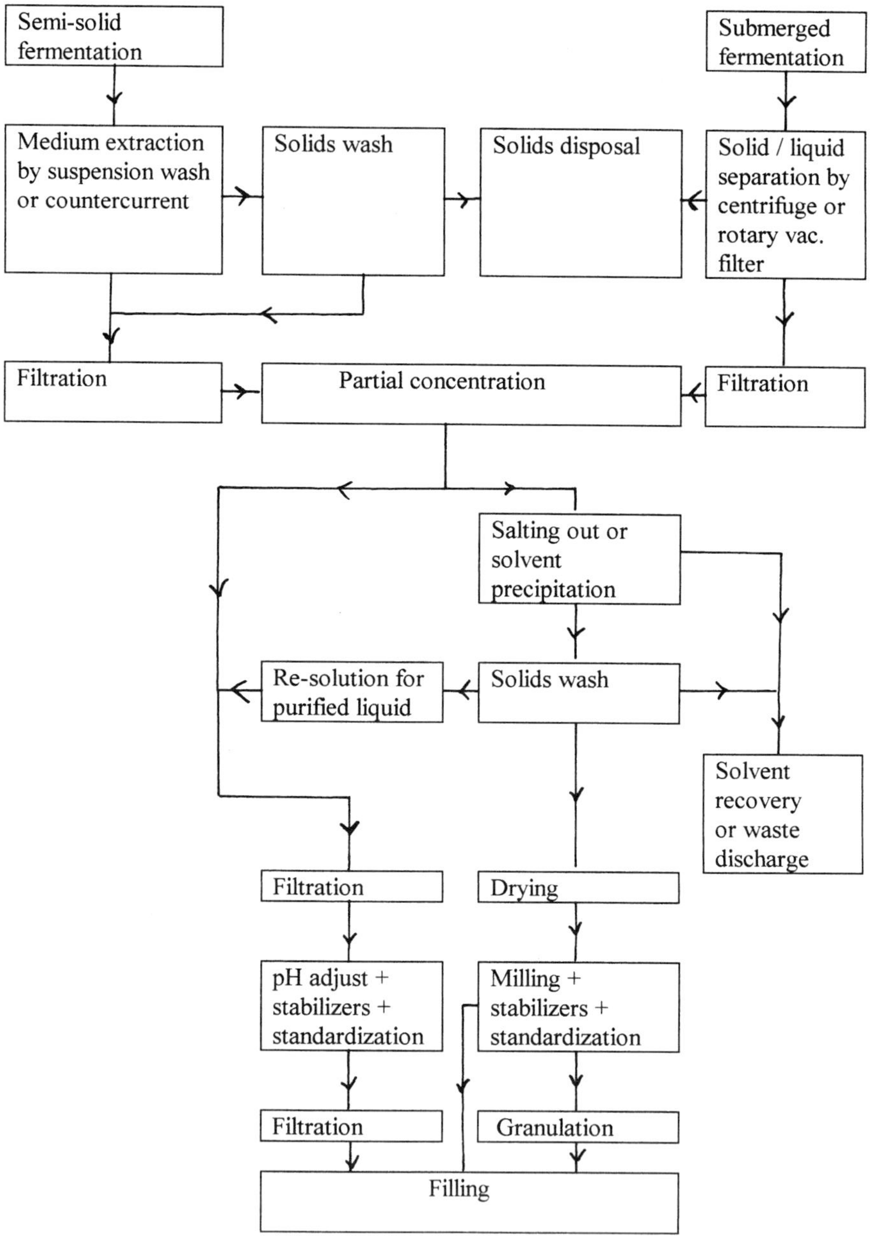

Figure 20.2 Schematic flow diagram for extracellular microbial enzyme recovery.

Purified liquid enzymes are standardized by dilution with solutions of stabilizing salts, polyalcohols or sugars and any permitted preservatives that are considered necessary. In the limited applications where a dry enzyme product is required, it is now recognized that specialized spray-drying techniques should be used to create dust-free granulates. The inhalation of protein dust is likely to increase the risk of allergic response on further exposure. This determines that maximum precautions and care should be taken when handling enzymes in powder form. Granulations to micro or larger granules will include incorporation of standardizing materials such as sugars, starch, starch derivatives or inorganic salts.

20.9 Industrial enzymes – applications in fermented food production

All fermentation processes use a wide range of conversions performed by the enzymes of the active organisms in the ferment. Some of the gross effects of these conversions can also be performed without the use of the whole organism, but rather by the introduction of industrial enzymes under appropriate conditions. In other circumstances, an exaggerated extension of the traditional fermentation can be created by adding enzymes to the whole ferment. Flavour, aroma and texture changes are examples of this extension.

The major use of enzymes in the production of fermented foods lies in the preparation or additional modification of the substrates for those fermentations. A variety of economically important fermentations are based on starch as primary raw material (mostly cereals) for yeast fermentation for beers, spirits and vinegars. Fermented breads also rely on yeast. A more diverse range of fermented foods can also be supported by the addition of exogenous enzymes and these include cider, wine, cheese, yoghurt, fermented meats and protein sauces from fish and soybean. Additionally, some more recent biotechnical activities have been applied to particular requirements in producing fermented bean and pulse products, speciality tea and coffee products and variations of cocoa extractions. These will all be discussed in the remainder of this chapter.

20.9.1 Enzyme use rates

The information provided here regarding the use rates for commercial enzymes is a very approximate indication. The potency of enzymes from different producers as well as from different strains and species of microorganism will vary and, in doing so, will require a different dose. The duration of the reaction that can be permitted by other considerations will affect the dose of enzyme required. The longer the reaction, the lower the enzyme dose. Enzymes perform at their best when used under conditions of temperature and pH that are optimal and peculiar to each enzyme. The closer to the optima, the lower the enzyme dose required for a given degree of conversion.

20.9.2 Substrates for alcoholic fermentation

The traditional sources for enzymes to hydrolyse starches to fermentable sugars have been koji and malted cereals (see also Chapter 21). The complex mixtures of enzymes produced by these ferments are used to hydrolyse further quantities of less costly starches deriving from cereals and roots such as potato and cassava.

For many countries these methods are expensive, especially if the raw materials for the enzyme production have to be imported. This is particularly important where malt is required.

(a) Starch hydrolysis. In order to hydrolyse starch economically at a useful aqueous concentration it is processed in two or three steps starting with heating to ensure that gelatinization takes place. This combination of physical and chemical changes renders the starch susceptible to enzyme depolymerization. It is now very typical to perform this gelatinization step by including a bacterial alpha-amylase in the suspended starch so that initial hydrolysis can take place as the gelatinization occurs. This minimizes the development of viscosity in the starch slurry and allows a higher initial starch-in-water concentration to be used than could be otherwise considered. After gelatinization, the slurry is cooled and a second quantity of amylase added to bring the depolymerization to the soluble dextrin stage. The maximum efficiency is obtained when a thermostable bacterial amylase is used and the starch processed rapidly through a direct steam-injection 'jet cooker'. This last method reduces the time for this initial step from between 2 and 4 hours to only some 40 to 60 minutes, without the need for a second amylase addition. To obtain fermentable sugars, a further depolymerization is required. This is performed by beta-amylases or amyloglucosidases and pullulanases to saccharify the dextrins. In many examples of processing from whole milled grains or tubers, there is an economically attractive amount of starch tightly bound to, or firmly trapped by, structural polymers of the plant material. These are cellulose, hemicellulose and lignified materials. The introduction of fungal cellulases and xylanases (specific hemicellulases), at the earliest opportunity after preparing a slurry of the milled materials, is used to release as much of this starch as possible. In many practical situations the enzymes (1 to 2 $kg\,tonne^{-1}$) are simply added to the slurry and stirred for some time before the gelatinization step is started. In the case of maize, the practice of steeping the corn allows addition at a stage where there is already a holding time. A further smaller amount of enzyme may be added to the milled corn before it is cooked. The

general scheme for the industrial conversion of starch for fermentation feedstock is given in Figure 20.3.

(*b*) *Brewing.* This process is accomplished in the traditional brewery by a simple infusion mashing of milled malt with warm water at 63–65°C for between 2 and 3 hours (Figure 20.4). For low-quality malt, the introduction of several industrial enzymes can compensate for the lack of endogenous catalysis and produce a satisfactory wort that will ferment adequately (Figure 20.5). Alpha-amylases together with beta-glucanases and neutral proteinases from *Bacillus* species, and some fungal glucanases from *Aspergillus* species will be used in the mash at combined doses ranging from 0.2 to 1.5 $kg\,t^{-1}$ malt.

Proteinase is required to ensure the release of sufficient soluble nitrogen, in the form of amino acids, to support the required yeast activity and produce colour and flavour. In all-malt brewing the need for additional saccharifying enzymes is unlikely. If the product is to be particularly highly fermented, or is to be the basis for a distillation, then amyloglucosidase from *Aspergillus niger* may be beneficial when used between 0.5 and 1.5 litres t^{-1} starch in the system.

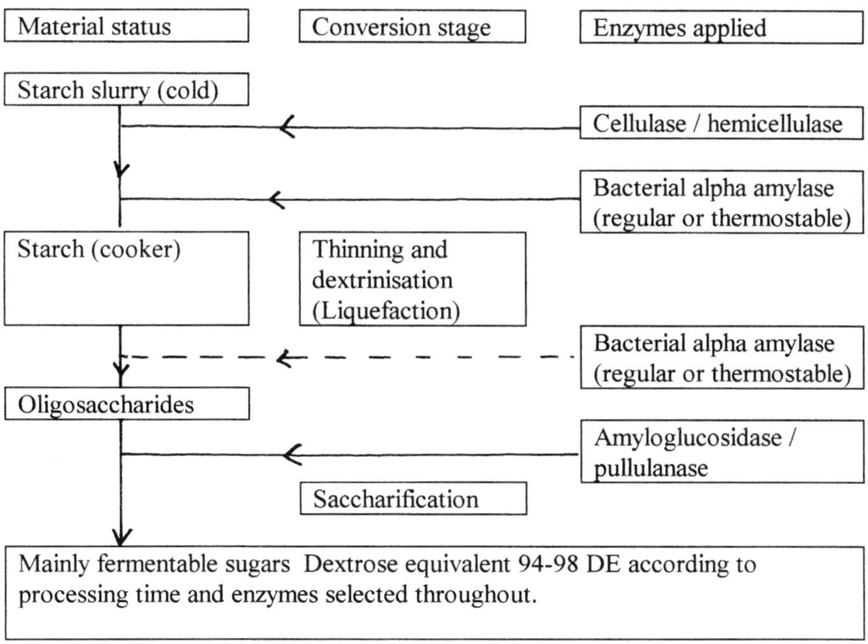

Figure 20.3 The generalized conversion of starch to fermentation feedstock showing points of application of industrial enzymes.

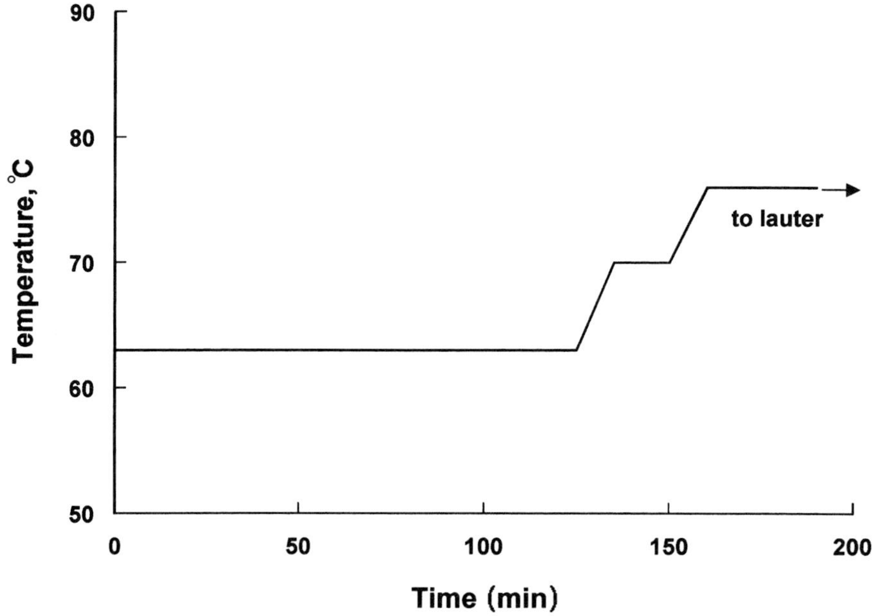

Figure 20.4 Mash diagram for an all-malt mash.

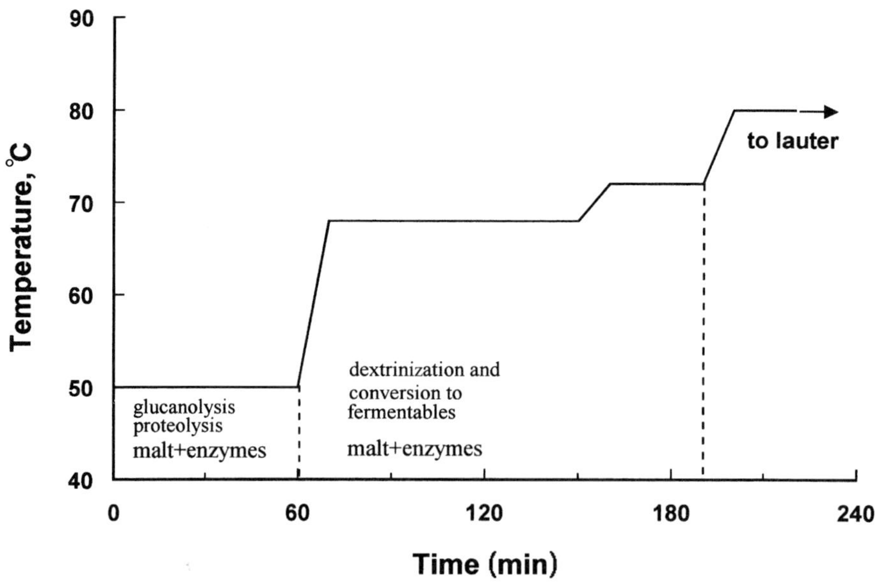

Figure 20.5 Mash diagram for enzyme-aided malt or malt barley mashing.

(c) Malt plus adjunct brewing and distilling. The greatest economic improvements will be achieved when a substantial quantity of the malt is replaced by other starch sources such as maize, wheat, barley, cassava, or sugar syrups derived from them. For these replacement mashes, it is usual to process the adjunct (non-malt starch source) separately by a 'decoction' mash programme. The finely ground adjunct is mashed and heated in the presence of a thermostable bacterial amylase (Figure 20.6). This may be done in a batch cooker, or most economically, by direct steam injection into a slurry of the adjunct in a jet cooker. In many grain distilleries the batch cooking will be performed at high pressure and then cooled to around 80°C before adding a less thermostable bacterial amylase for the dextrinization stage. The typical thermostable amylases come from *Bacillus licheniformis* and the regular types from *Bacillus subtilis.*

When producing beer using the combined malt and adjunct materials, the liquefied and dextrinized adjunct mash will be combined into a separate malt mash to obtain saccharification before the wort is drawn off. In distillery operations, the adjunct mash is cooled to around 60°C and dosed with amyloglucosidase to give a partial saccharification before it is cooled and fermented. The mash may also be cooled directly to fermentation temperature and the amyloglucosidase and yeast added together. This latter method will give a slightly slower overall fermentation because the yeast does not have the initial benefit of a prepared glucose supply coming from the presaccharification by amyloglucosidase enzyme.

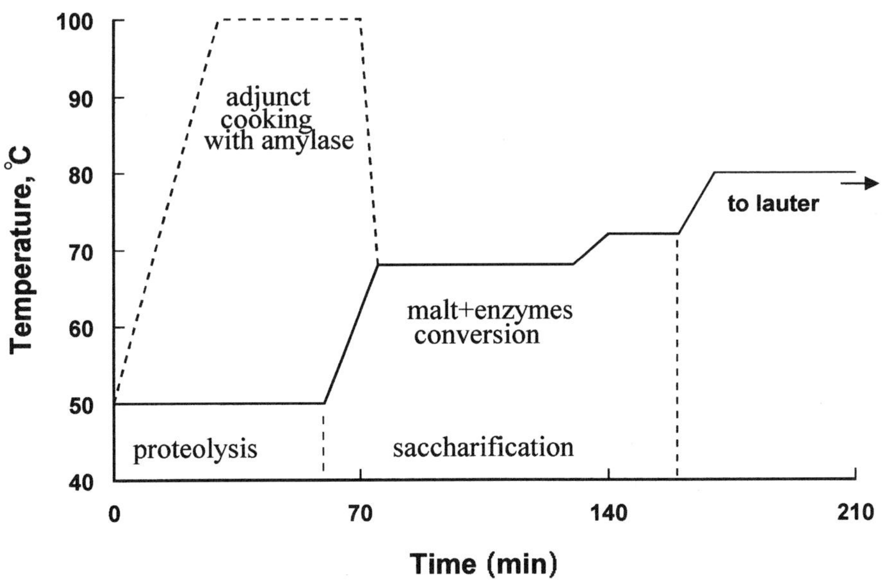

Figure 20.6 Mash diagram for adjunct decoction mash.

In all uses of non-malt adjuncts it is important to monitor the amount of soluble nitrogen in the fermentation and adjust this upward by adding neutral bacterial proteinase to hydrolyse proteins and release free amino acids.

The brewing process, particularly for long-matured products such as quality lager has a tendency to a taste defect characterized by the presence of higher than usual amounts of diacetyl, which can be removed by application of the commercially available enzyme alpha-acetolactate decarboxylase.

Some examples of enzymes for alcohol production are given in Table 20.6.

(d) Genetic manipulation of organisms (GMOs). Recent developments in the brewing and distilling industries have recognized the potential of using genetic manipulation to introduce saccharifying enzymes into the fermentation yeasts. A number of such modified yeast strains have been produced and reports of trials indicate that they perform very well. However, the viability and stability over very many generations of such yeasts has yet to be established. There are also studies to introduce the whole sequence of starch conversion enzymes into suitable yeasts so that a

Table 20.6 Microbial enzymes applied to distilling and brewing practice[a]

	Enzyme type	Enzyme ($kg\,t^{-1}$ starch)
Distilling		
Liquefaction	Bacterial alpha-amylase (regular)	1.5–2.0 (batch)
	Bacterial alpha-amylase (thermostable)	0.2–0.6 (batch)
	Bacterial alpha-amylase (thermostable)	0.2–0.3 (continuous)
Saccharification	Fungal alpha-amylase (maltogenic)	0.1–0.2
	Fungal amyloglucosidase	1.5–3.0
	Bacterial pullulanase	0.5–1.0
Brewing		
Malt improvement	Bacterial alpha-amylase	0.5–1.0
	Bacterial beta-glucanase	0.5–1.0
	Bacterial neutral proteinase	0.3–1.5
	Fungal beta-glucanase	0.2–0.5
High adjunct content mashing	Bacterial alpha-amylase	2.0–3.0
	Bacterial beta-glucanase	2.0–3.0
	Bacterial neutral proteinase	1.2–1.5
	Fungal beta-glucanase	1.5–2.5
	Fungal amylase	1.0–3.0
	Fungal glucoamylase	2.5–6.0
	Bacterial pullulanase (for high attenuations)	0.2–0.5
Adjunct cooking	Bacterial amylase (thermostable)	0.2–0.6

[a]Note: quantities are typical – because activities of different preparations may vary, always consult suppliers' data.

fermentation of economic speed and yield may one day be possible starting with a raw starch suspension. The saving in energy, the environmental impact of plant construction, and reduced numbers of pieces of plant for cleaning are seen as desirable. The quality and acceptability of the products may be a different matter, as will the acceptance of GMOs in the wider and less-regulated areas of industry. The use of such yeast materials for by-products such as yeast extracts and flavourings may also require considerable rethinking (see also Chapter 20).

(e) Vinegar fermentation. This subject is dealt with in greater detail in Chapter 1. The production of an alcoholic feedstock for the second, 'acetic', fermentation from starch sources follows the processes outlined in the preceding section. In some countries other materials such as bananas or pineapple canning wastes are utilized for vinegar production. The starch and partially degraded polymers of under-ripened bananas can be effectively released as fermentables for alcohol production by treatment of the pulped material with amyloglucosidases and pullulanases. These enzymes typically perform well at the naturally low pH values and temperatures of yeast fermentations which can be initiated directly on the pulp. The added enzymes release sugars steadily throughout the fermentation.

Additional release of fermentables and starches for conversion can be achieved by treating the pulp with pectinases. Like the amyloglucosidases, these are typically produced by *Aspergillus* species.

(f) Wines and ciders. There is an increasing use of enzymes in the preparation of extracts of fruits, predominantly grapes, for wines. The maximization of juice yields is achieved by the application of various 'cocktails' of pectinolytic enzymes. Depolymerization of pectins and structural components of plant cell walls such as cellulose and hemicelluloses can increase juice yields by as much as 20%. The doses of enzyme are very small, in the range 0.2 to 1.0 $kg\,t^{-1}$ fruit. Selected glucosidase enzymes can be used to increase the release of aroma and flavour components. The use of powerful yield-increasing enzymes with fruits having coloured skins or flesh can cause considerable colour extraction. If this colour is undesirable, anthocyanases can be used to remove colour for the production of white wines.

It is frequently found that apples release some starch into the juice. This can be treated with amylases and glucoamylases in small doses according to the amount of starch present. As with wine production, for ciders, the enzymes can be introduced at the same time as the yeast and will release sugars to the fermentation. These enzymes are typically produced from *Aspergillus* species.

Additional discussion of the production of alcoholic beverages will be found in Chapter 9.

20.9.3 *Baking*

Yeast-raised doughs constitute a food fermentation system, and there are several applications of enzymes that contribute to the quality of the final product in terms of texture, colour and shelf life. Most of these have little or no effect on the actual fermentation process, except for those that modify starch to release fermentable sugars. The typical enzyme used world-wide to improve dough fermentations is fungal alpha-amylase. This is generally obtained from either *Aspergillus niger* or *Aspergillus oryzae.*

It is claimed that the action of proteinases and lipases, when added to dough for benefits other than the fermentation, also contribute modestly to the yeast fermentation by providing improved nutrition. This effect is marginal. However, there is some evidence that flavour components produced by the yeast metabolism can be stimulated by the easily metabolized molecules resulting from the action of these enzymes.

A detailed discussion of yeasts and breads will be found in Chapters 7 and 8.

20.9.4 *Cheese*

The main application of industrial enzymes in the manufacture of cheese is the use of various types of rennet for the primary coagulation of milk. The production of cheese is thoroughly discussed in Chapter 10.

(a) Rennets. Chymosin (calf rennet) is the traditional coagulant. It is sometimes mixed with or replaced by similar extracts of lamb or kid stomachs to provide the appropriate coagulum for the special and highly flavoured cheeses typical of Mediterranean countries. Chicken pepsin also has a place in cheese production, particularly for kosher certification. Additionally, the dilution of calf rennet, due to its high cost, with adult bovine pepsin or with porcine pepsin is a regular practice. There are now some extremely successful and economic replacements for animal rennets prepared from a number of microorganisms.

Many microbes produce proteinases that can be demonstrated to coagulate milk, but most of them are broad spectrum and non-specific in their action. This means that such types of proteinase will continue to hydrolyse the casein of the curd to yield unacceptable flavours and textures. In some cases the result may be a rapid and complete solution of the curd. A small number of fungal enzymes have, however, proved to perform in a manner that imitates the action of calf rennet on the kappa casein forming the outer layer of the casein particles and to proceed no further. This action is required so that a minimum of damage is done to the casein protein while exposing the casein to calcium-mediated non-enzymic coagulation. These are classed separately from the rennets (chymosins) as Microbial Milk

Coagulants (MMCs). Notable examples of these fungal enzymes are obtained from *Mucor miehei*, *Mucor pusillus Lindt*, *Rhizomucor miehei*, *Endothia parasitica* and *Cryphonectria parasitica*. A useful presentation of the influence each different type has on the main characteristics of flavour, texture, bitterness and yield is given by Cheeseman (1981).

The use of *Bacillus* species for milk coagulants has often been investigated, but there is usually too little specificity to be attractive for the industry. It has long been known that many plants also produce proteinases and some of these have been used as milk coagulants. Historically plants such as the insectivorous species, the Sundews (*Drosera* spp.), Ladies bedstraw (*Galium verum*), pineapple (*Ananas comosus*), papaya (*Carica papaya*), fig (*Ficus glabra*) and cardoon (*Cynara cardunculus*) have also been used. From the modern cheesemaker's point of view, the preferred microbial enzymes are available in bulk, at equivalent strength to calf rennet (or multiples of that strength), in powder or liquid form with excellent storage stability and at substantially lower cost. Recent developments have used 'protein engineering' techniques to improve the performance of some of these microbial enzymes by making them more heat-labile. This improves their equivalence to calf rennet and ensures that usual heat treatments of the resulting cheese whey completely inactivate them. This is important if the whey is to go forward for the preparation of infant formulations where a residual coagulation activity would be unacceptable.

(b) Genetic manipulation in production of MMCs. Genetic modification by the introduction of some of the coagulant enzymes from microbial strains into more suitable organisms for bulk production has resulted in several products being produced in such hosts as *Aspergillus niger*, *Aspergillus oryzae*, *Escherichia coli* and *Kluveromyces* species. The economics of production and sales price of the enzyme from genetically manipulated organisms, together with a high yield in high purity make these developments extremely attractive to the large industrial cheesemaker (see also Chapter 20).

(c) Cheese flavour enhancement. Interest to produce cheeses rapidly but with well-balanced flavour has grown in recent years. The cost of long maturation times in cooled storage have become a significant part of cheese production costs. In addition, the requirement for cheese products with significantly higher flavour intensity (from 3 to 400 times the flavour strength of mature basic cheeses), has stimulated the use of additional enzymes specifically to enhance the flavour. A further discussion of aspects of cheese flavour is given by Law & Goodenough (1995).

Modification of fats for enhanced cheese flavour. The release of free fatty acids in regular cheeses arises by the action of the microbial 'starter' organisms added to the cheese during their prolonged fermentation. If the milk is pasteurized, which destroys almost all the milk lipase activity, this aspect of flavour development can be very slow. The addition of microbial lipases is now a common practice where a rapid and intense flavour profile is required. The most traditional sources of additional lipase enzymes have been those obtained from the pre-gastric (parotid) glands of lamb, kid and sometimes calf. The use of pancreatic (porcine) lipases is considerable, but the enzyme is usually a mixture containing broad-spectrum proteinases which can adversely affect the flavour, and also the use of porcine material precludes kosher acceptance.

There is now a considerable variety of microbial lipases available on a commercial scale. Each of these will have a particular characteristic degradation of the triglycerides of the milk fat. The resulting spectrum of free fatty acids will create a characteristic flavour profile. The higher the amount of short-chain fatty acids that are released, the more 'piccante' or 'blue' will be the flavour. Certain regional speciality cheeses have very particular flavours; these can also be copied and enhanced by an appropriate selection of one or more microbial lipases. The most accepted lipases for these applications will be produced from selected strains of *Mucor*, *Aspergillus*, *Rhizopus* and *Pseudomonas* microbes.

The use rates of lipases will be a combination of the potency of the enzyme preparation and the available time for the reaction. The longer the reaction time, the lower the enzyme needed for a particular result.

Modification of protein for enhanced cheese flavour. It has been established that the cultivation and extraction of microbes used as 'cheese starters' can yield active proteinase and peptidase enzymes that can contribute to accelerated development of flavour. The use of these 'normally present' enzymes in a higher than normal concentration allows for a considerable acceleration of the ripening process, but usually does not lead to a truly remarkable increase in the flavour intensity. The application of carefully selected fungal proteinases and peptidase with much higher potency, and in combination with appropriate lipases can increase the flavour of a cheese to at least 10 times that of a regularly fully matured example. Used in this way the flavour is usually fully acceptable and does not have a tendency to develop bitterness associated with over-hydrolysis of the protein.

To obtain the very high intensity flavours required for flavouring other goods with cheese, e.g. snack foods, where only a small weight of the flavouring can be applied per unit product, it is necessary to achieve an aggressive modification. This is usually the production of cheese powders or Enzyme-Modified Cheeses (EMC) whereby regular cheese material is emulsified and slurried to allow the action of both proteinases and lipases over relatively short time periods, e.g. 4–8 hours. Care is needed to select the correct balance of enzymes and to include a peptidase capable of degrading any bitter peptides produced by the main flavour-generating proteinases. In all applications, the amount of enzyme will have to be determined in relation to the enzyme(s) potency, the required flavour development, and the available time for reaction. A summary of the main changes that occur and contribute to flavour development in cheese is given in Figure 20.7. The actual components and their proportions will vary in different cheese types, and according to the maturity of the cheese. It is strongly emphasized by many trials and practices that the most suitable method of introducing enzymes that will enhance the flavour of cheeses is to incorporate them into the milk before coagulation. To minimize losses into the whey that would be incurred when using soluble enzymes, it is most effective to have the enzymes in encapsulated form so that they remain with the coagulum and are released soon after whey separation. The use of small capsules ensures an excellent distribution of the enzymes throughout the curd.

(d) Lysozyme – control of spore-forming bacteria. In cheese where a secondary fermentation to produce gas is a fundamental part of the cheese character, e.g. Edam, Gouda, Emmenthaler, it is essential to rigorously discourage the development of bacteria that might lead to a butyric fermentation, but not interfere with the desired propionic fermentation. Typically these butyric contaminants would be *Clostridia* spp. The natural antiseptic/antibiotic lysozyme is produce by many animals. Lysozyme (muramidase) is an enzyme that attacks and damages the cell walls of many bacteria, causing the osmotic pressure within them to rise with resultant bursting of the cells.

Lysozyme from hen egg white is the most usual commercial preparation and it is used in the production of many cheeses to prevent 'late blowing' in high pH, brined cheese types. The dose depends on the potency of the

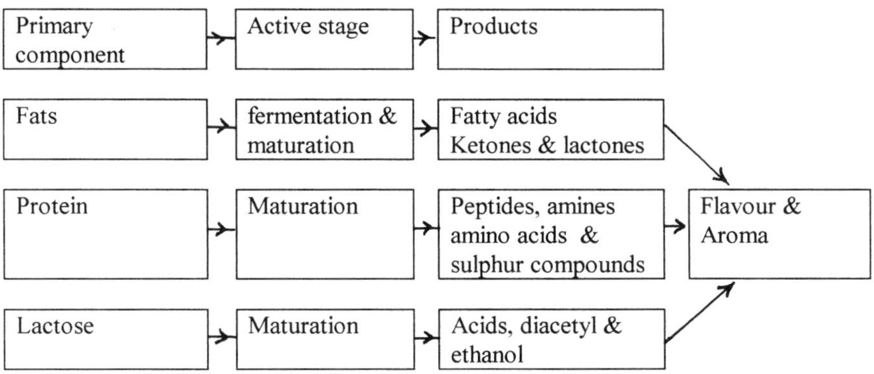

Figure 20.7 The contributing conversions to cheese flavour.

lysozyme preparation and also on the severity of contamination of the milk. Typically a regular lysozyme preparation would be used at 25 to 35 mg l^{-1} milk (Wigley, 1996).

(e) Lactase in the cheese industry. The predominant application of lactase (beta-galactosidase) is in the hydrolysis of the lactose in whole milk to provide reduced lactose levels and so make milk and milk-based drinks and other products more widely acceptable among the large proportion of humans who lack the enzyme and have partial or complete lactose intolerance. In the 1960s and 1970s there were confident expectations that the conversion of very large amounts of cheese whey lactose into sugar syrups and fermentation feedstocks would be a very major use of this enzyme. There has never been an economically acceptable lactase process cost that generated a product able to compete with other fermentables or sweet syrups. Some conversion of whey by the action of lactase enzyme, or the use of a yeast that produces the enzyme directly, is used to produce alcohol. As the lactose concentration is only around 4% in whey, it is usual to use the low alcohol concentration that is produced for vinegar production.

It is reported that cheese ripening may be accelerated a little by lactase application to hydrolyse a little of the lactose in the milk before coagulation. It is presumed that the sugars it releases stimulate the cheese cultures. It is noted to be more effective for blue cheeses, while other cheeses may have altered textures which may be commercially beneficial. The curd for cottage cheese may be firmer and there can be less losses through 'fines'. It is not widely adopted. This principle is also applied to the use of lactase to treat milk that is intended for the fermentation of 'starter' cultures. This is somewhat more widely used. There are some similar benefits when lactase is used to treat buttermilk, sour cream and in yoghurt production. The main commercial lactases are produced from *Kluyveromyces* and *Aspergillus* species.

(f) Catalase in the cheese industry. The catalase enzyme decomposes hydrogen peroxide to water and molecular oxygen. Where it is permitted and selected, the successful use of peroxide to sterilize milk requires that a small but definite excess of peroxide be added. Catalase is added after the sterilization is complete and it acts to remove the excess peroxide.

A summary of the cheese process with the points of addition of industrial enzymes is given in Figure 20.8.

20.9.5 Protein-based fermentations

(a) Condiments. The production of soya bean and fish protein fermented foods requires the modification of the proteins to provide nutrition

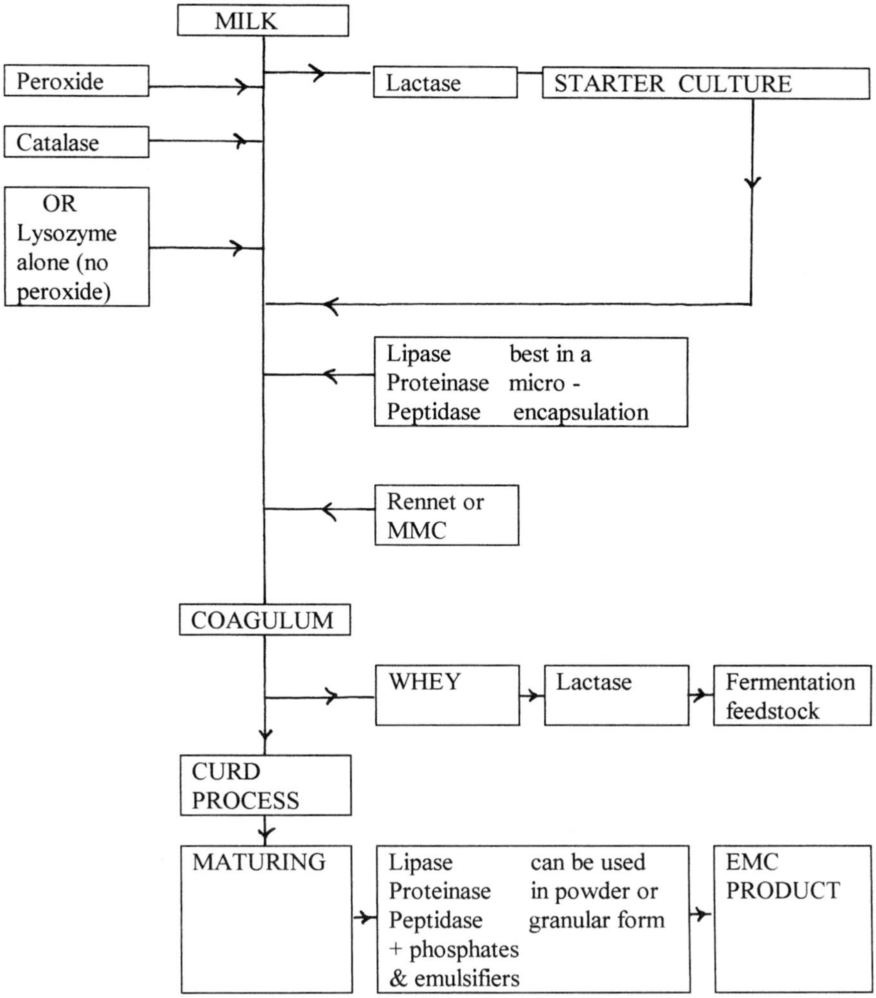

Figure 20.8 The points of addition of industrial enzymes in cheese production.

of the microorganisms of the fermentation which in turn provide the texture and flavour characteristic of the end-product. In each case, the ingredients of choice are finely ground and mixed with an appropriate amount of water. Inoculation with cultures and a usually lengthy fermentation follows (see also Chapters 2 and 14).

To increase the protein conversion, to accelerate the fermentation, and in some cases to assist in compliance with regulations controlling fungal fermentations in some countries, proteinases and peptidases may be added to the mash before inoculation. Traditionally the enzymes used would be the plant extract proteinases bromelain or papain. More recently, the use of microbial proteinases and peptidases has been found to give a lower risk of contamination with unwanted microbes that are often associated with low-cost examples of the plant proteinases. By varying the amount of enzyme, the degree of solubilization before and during the fermentation can be regulated to produce pastes and sauces. Microbial proteinases from *Bacillus* species are usually very powerful, but can tend to create a bitterness in the final product. Proteinases from some fungal sources usually have associated peptidases which effectively act as debittering agents by hydrolysing the peptide fragments that would otherwise contribute a bitter taste. Some of these fungal enzymes also contain glutaminase which acts to release free glutamic acid. This glutamic acid will largely be converted to sodium glutamate in the system and so provide a further positive flavour enhancement. These fungal proteinases are produced from various strains of *Aspergillus* and *Rhizopus*. Most recently, interest has focused on the wider acceptability of products from fermented beans and pulses by looking at the possibility of reducing the 'flatulence factors'. These are a small number of small oligosaccharides and disaccharides that are not fermented by the food product organisms, but may be fermented in the human digestive tract, producing excessive gassing and discomfort. Stachyose, raffinose and verbascose are examples of these oligosaccharides. All of these contain galactose bonded in the alpha configuration. Alpha-galactosidase is now commercially available from fungal sources and it has been demonstrated to have beneficial effects when introduced into the processing of raw materials from beans and pulses.

(b) Fermented meats. There is a growing use of selected fungal proteinases and peptidases in the preparation of fermented meat products. The function is to increase the available soluble nutrition for the active inoculum in the preparation. Selection of enzymes with less than full broad-spectrum attack on meat proteins ensures that the texture of the final product is not damaged by solubilization of the mass. There is some evidence that the use of these enzymes can also improve the emulsification of fat within the system. Typically these enzymes are obtained from *Aspergillus* and *Rhizopus* species. The use of bacterial enzymes is contra-indicated by their aggression and tendency to generate bitter peptides.

Where whey or skim milk powders are incorporated in the formulation of the meat product, there is indicative use of lactase (beta-galactosidase) to break the lactose component into glucose and galactose. These sugars stimulate the inoculum and so cause the pH to fall faster than would be the case without the use of the enzyme. This also avoids the need to make a separate addition of glucose for the same purpose (see p. 652). Fermented meat products are described in Chapter 14.

20.9.6 Silage fermentations

For several years there has been a positive application of several enzymes in the preparation of silage. The main objective has been to accelerate the initial fermentation so that the pH is reduced rapidly to avoid undesirable *Clostridial* activity which is rapidly curtailed when the pH falls below 4.5. The use of amylases to break down starches and release sugars acts as a stimulant to most fermentations within the harvested crop. The addition of glucose oxidase can then utilize some of this glucose with consumption of oxygen in the silage mass. The removal of oxygen from the silage is a benefit in gaining control of the fermentation by selecting positively for anaerobes and against aerobes. The product will be gluconic acid which further encourages a reduction in pH.

The inclusion of various enzyme 'cocktails' containing cellulase, xylanase, hemicellulase, pectinase and amylase, in different proportions according to the harvested forage crop, can produce rapid breakdown of the plant cells and the release of their contents as nutrients for the fermentation. This potential acceleration is valuable, but must be balanced with complete control of the liquid in the silage so that there is no loss of valuable materials by run-off. This run-off is also a serious potential environmental pollutant and must be avoided strenuously. An excellent and wide-ranging discussion of many aspects of biotechnology in the feed industry, including probiotics, silage and compound feed modification was presented by Pearse-Lyons (1987). This author has yet to find a better, broad presentation of this subject. In recent years, the farming community has recognized that there is a greater economic benefit in harvesting the crop and making the silage at the correct time and under good weather conditions than the reliance on additional materials such as chemicals and enzymes. With the declining numbers of animals to be fed there is a far lower pressure on the production of silage, and the emphasis is now more on making it correctly without additives. The production of silage is discussed in Chapter 3.

20.9.7 Tea, coffee and cocoa

These are discussed in detail in Chapter 5. The potential benefit of the addition of exogenous enzymes in the processing of these materials is extremely limited. While the processing is not strictly a fermentation, there is evidence of the contribution of microbial activity during the processing of the crops.

(a) Tea. If tea is to be extracted to prepare 'instant teas and iced teas' then there is a desire to eliminate the development of some insoluble

materials that tend to float on the surface of the beverage. These materials are known as 'tea cream'. Composed of a collection of phenolic carbohydrates, they can be rendered more soluble, and into a stable solution, by processing the extracted tea with the enzyme tannase. This is a polyphenol attacking enzyme which is prepared from *Aspergillus* fungi. The details of the methodology remain obscure, probably as part of the 'trade secrets' of tea processors.

(b) Coffee. The natural mucilage and gums that surround the coffee bean in the pod are rapidly broken down when the pods are harvested. This is due to the presence of a considerable amount of the enzyme alpha-galactosidase.

In the extraction of coffee for its solubles there is a tendency to release polysaccharides related to the plant gums. When the liquor is evaporated before spray- or freeze-drying these can cause a rapid rise in viscosity and decrease the efficiency of the whole process. The application of very small amounts of a fungal enzyme, galactomannanase, as an addition to the dilute extract before it is evaporated can dramatically reduce this problem. The economics of the production of coffee solubles can be improved if a selected group of pectinases is applied to the crushed beans suspended in warm water for up to an hour before the hot extraction takes place. This can also be applied to the residual mass after a first hot extraction has been made. A further extract, having good aroma and flavour, but low colour, can then be obtained.

(c) Cocoa. Unlike the coffee fruit, the cocoa bean pod does not have an effective alpha-galactosidase to degrade the gums. The release of the cocoa bean from its pod is facilitated by allowing a natural infection of microbes to ferment the gums and mucilage surrounding the beans. This allows the beans to be obtained more easily and in a cleaner, drier condition. This is important for their storage and transport without deterioration and spoilage. The use of complex 'carbohydrase cocktails' as a spray onto the harvested pods can accelerate this process of degradation of the internal mucilage. The economics of such a treatment have not been exposed to date. There is little evidence of significant use of enzymes for this treatment.

20.10 Endogenous enzymes in fermented food materials

Throughout this chapter the discussion has been about the application of exogenous, industrially produced enzymes to modify components of the food materials or to improve the performance of the fermentation. At harvest, natural materials have many enzymes that remain capable of activity for longer and shorter times. If the structure of the cells is damaged, contents will leak out and their enzymes will be able to become active in a more general way. Many of the steps taken at the harvesting and early processing stages include a component of control over these endogenous enzymes. Drying, heating, blanching, chilling and freezing are examples of such actions. The nature of the endogenous enzymes to be found in food materials, and many aspects of their actions and ways to control or eliminate the undesirable ones is an important part of food technology. An excellent modern presentation of these will be found in the chapters of *Enzymes in Food Processing* (Tucker & Woods, 1995).

References

Atkinson, B. & Mavituna, F. (1991) *Biochemical Engineering and Biotechnology Handbook*, 2nd edn, Macmillan. ISBN 0 333 42403 4.

Aunstrup, K. (1979) *Applied Biochemistry and Bioengineering. 2. Enzyme Technology*, Academic Press, New York.

Bronowski, J. (1973) *The Ascent of Man*. British Broadcasting Corporation. ISBN 0 563 10498 8.

Cheeseman, G.C. (1981) Rennet and Cheesemaking, in *Enzymes and Food Processing*, Applied Science Publishers. ISBN 0 85334 935 5.

Godfrey, T. & West, S.I. (eds) (1996) *Industrial Enzymology*, 2nd edn, Macmillan Press. ISBN 0 333 59464 9.

Law, B.A. & Goodenough, P.W. (1995) Enzymes in Milk and Cheese Processing, in *Enzymes in Food Processing* (eds G.A. Tucker & L.F.J. Woods), Blackie Academic and Professional. ISBN 007514 0249 4.

Lilly, M.D. (1979) *Applied Biochemistry and Bioengineering. 2. Enzyme Technology*, Academic Press, New York.

Pearse-Lyons, T. (1987) *Biotechnology in the Feed Industry*, Alltech Technical Publications. Nicholasville, Kentucky, USA.

Tucker, G.A. & Woods, L.F.J. (eds) (1995) *Enzymes in Food Processing*, Blackie Academic and Professional. ISBN 007514 0249 4.

Wigley, R.C. (1996) Cheese and Whey, in *Industrial Enzymology*, 2nd edn, Macmillan Press. ISBN 0 333 59464 9.

21 Koji

N. LOTONG

21.1 Definition and scope

The term koji derives from the Chinese character meaning mouldy grains, which is pronounced differently in the various languages deriving from Chinese, for instance 'shui' in Chinese, 'koji' in Japanese and 'ku' in Korean (Mheen, 1972). Koji preparation is considered as an essential step in fermentation of a variety of fermented foods of the Orient. Basically, the process is the solid-substrate cultivation of moulds to produce hydrolytic enzymes on seeds, such as soybean, or on other cereals. Koji therefore serves as a source of a variety of enzymes which catalyse the degradation of solid raw materials to soluble products providing fermentable substrates for yeast and bacteria in the subsequent fermentation stage. The use of koji is considered analogous to that of malt in the Western World (Yong & Wood, 1974). Apart from key enzymes, various chemicals essential for the second fermentation stage or those imparting flavour to the products are also produced during koji preparation.

Koji can be called after the cereals used, for instance kome (rice) koji and muji (barley) koji. They also are named according to the colours of spores or pigments produced by the cultivated moulds such as ki-koji (yellow) for koji cultured with *Aspergillus oryzae*, kuro-koji (black) for *A. awamori* culture and beni-koji (red) for koji produced by *Monascus purpureus*. There are also names indicating the end use such as miso koji, shoyu (soy sauce) koji or saké koji.

According to the written records, the art of koji making originated in China, where soy sauce and chaing, the forerunner of Japanese miso, were produced before 1000 BC (Yong & Wood, 1974; Abiose *et al.*, 1982). The method was introduced to Japan during the 6th century by Buddhist priests (Hesseltine, 1965). Although it originated in China, researches and development directed toward improvement of the process have been extensively carried out in Japan. As a result, at present, the practice of koji preparation in the latter country is remarkably different from that first introduced and the process has become known world-wide by the Japanese term 'koji'.

As far as cultivation of moulds on cereals for sources of enzymes is concerned, rice wine and sweetened rice are other groups of fermented products produced for centuries. Although it is believed to have originated in China as well, other countries have their own traditional processes, which have been handed down through generations. Among the fermentations of these products, the Japanese 'saké' brewing is very well documented, in both historical and technological aspects, as was intensively reviewed by Kodama & Yoshizawa (1977). Rice wine and sweetened rice fermentations in China and other Asian countries are similar in methods of rice koji preparation, by which a common type of dry inoculum, called differently in local names, is used as starter.

Within the meaning of the term koji, ang-kak (red rice) should also be included as a kind of koji (Nokano, 1972), which provides a source of pigments and flavour substances for a number of fermented products (Hesseltine, 1965).

Apart from being a step of the two-stage fermentation, the term 'koji process' has also been used for some of the single-stage solid-substrate fermentation of fermented foods. Examples of these are tempeh, a traditional Indonesian food prepared by inoculating *Rhizopus oligosporus* (*Rhyzopus microsporus* bar. oligosporus) on soybean (tempeh kedala) or coconut meal (tempeh bonkrek); ontjom, prepared by fermentation of *Neurospora sitophila* on peanut press cake; sufu, the soybean cheese fermented by *Mucor* spp. (Cannel & Moo-Young, 1980; Aidoo *et al.*, 1982).

The ease and other advantages of the koji process have generated numerous researches leading to prospective application and technological improvement. The process has now been developed to modern and advanced solid-substrate fermentation technology and has been used to produce a variety of non-food products such as commercial enzymes, toxins and antibiotics (Hesseltine, 1972, 1977) and bioinsecticide (Degranges *et al.*, 1991). The technology concerning solid-substrate fermentation has been reviewed by Aidoo *et al.* (1982), Cannel & Moo-Young (1980), Pandy (1992) and Dorta *et al.* (1994).

Although the term 'koji' process has been applied to almost all solid-substrate fermentations as earlier mentioned, this review aims to confine itself only to the koji process, from which it serves as a step to provide enzymes and other essential chemicals for the subsequent fermentation stages, as it originally meant.

21.2 Soybean koji

21.2.1 Raw materials

In preparation of soybean koji, either soybeans alone or mixtures of soybeans and wheat or some other carbohydrate source are used.

(a) Soybean. In general, soybean (*Glycine max* (L.) Merr.) is composed approximately of 38% protein, 31% carbohydrate, 18% fat and 8% water.

It also contains high levels of some minerals and vitamins (Aykroyd & Doughty, 1964). Soybean protein contains mainly glycinin, phaseolin, legumelin and soy legumelin (Windholz *et al.*, 1976).

Originally, whole soybean was used as starting raw material in preparation of soy sauce koji. Since 50 years ago, most of the soy sauce fermentation in Japan has used defatted soybean meal to replace the whole beans. Its use has been shown to reduce the cost of production, as well as to increase the nitrogen solubilization (Yokotsuka, 1985). With much success, defatted soybean meal also has been used in Korea, Taiwan (Tang, 1978) and China (Yue, 1990). In the fermentation of tau-si, the condiment composed of black and solf soybean pieces, the black-coated variety of soybean is required (Tang, 1978).

There have been reports on the use of other raw materials substituted for soybeans. An example is in the Philippines, where sauce has been prepared from a mixture of soybeans and coconut paring meal (Soriano *et al.*, 1967), soybeans and copra (Baens-Arcega, 1966) and mecan pea (Crisostomo *et al.*, 1974). Work on substitution by winged bean seeds has been reported in Thailand (Bovornsombat, 1981). In their review of soy sauce fermentation, Yong & Wood (1974) reported on the use of soybean hull and acorn in Japan.

(b) Wheat. Either whole wheat or wheat flour has been commonly used for production of koji of high-quality soy sauce. Yokotsuka (1964) has noted that the roles of wheat in soy sauce brewing are as follows:

(i) To decrease the moisture of cooked beans from 60% to about 45% which is unfavourable for growth of undesirable microorganisms but just adequate for mould growth.
(ii) To assist in obtaining the maximal growth of the koji moulds.
(iii) To serve as precursors of sugars, alcohols and organic acids.
(iv) To provide lignin and glycosides, the precursors of vanillic flavour of the products.
(v) To serve as a rich source of glutamic acid.

While whole wheat is essential in preparation of the Japanese soy sauce koji, wheat flour is more common in those prepared in the South East Asian countries. Beside wheat flour, rice flour or mixtures of both also have been used commercially in this region (Abdon, 1978; Lan *et al.*, 1978; Lotong, 1978; Tang, 1978). Substitution of wheat flour by tapioca or glutinous rice flour also has been reported in Indonesia (Hartadi & Jutono, 1979). In the Philippines, the use of banana flour has been tried on the experimental scale. Lee *et al.* (1972) reported that good-quality soy sauce has been obtained from koji prepared with mixtures of corn and wheat or barley and wheat. Recently wheat bran has been used successfully in China (Yue, 1990).

(c) Soybean to wheat ratio. It is generally believed that the 50:50 by weight ratio of soybean to wheat produces best quality Japanese soy sauce. This ratio is normally used by the manufacturers in Japan (Yong & Wood, 1974). In preparation of the Chinese type soy sauce koji, ratios of soybean to wheat higher than 50:50 are preferable. The ratio of 2:1 has been reported as optimal in production of soybean koji in the Philippines (Soriano & Pardo, 1978). In Thailand, the amount of wheat flour or rice flour used in preparation of soy sauce or soybean paste koji is variable from 15 to 50% of soybean dry weight (Bhumiratana *et al.*, 1980), whereas 40–50% is normally used in Indonesian soybean koji.

21.2.2 Koji preparation

(a) Japanese soy sauce koji. Fermentation of Japanese soy sauce has gone through steps of development, starting from the traditional method to the present-day automated process, which results in considerable improvement of the products. There has been an extensive number of publications on different aspects of Japanese soy sauce fermentation which were previously reviewed in detail by Yong & Wood (1974). This chapter aims to compare the traditional method of koji preparation to that of the recently developed process commercially practised currently.

Traditionally, koji is prepared by soaking the beans for 10–12 hours (Yokotsuka, 1960). It is the customary practice in Japan to soak the beans with changes of water at intervals of several hours or using running water. The beans are then cooked under pressure at $15\,\text{lb/in}^2$ for 1 hour. At the same time as the soybeans are being treated, the whole wheat is roasted and cracked into 4–5 pieces. Today, for large-scale koji production, a continuous cooker has been developed, by which high temperature, short time cooking ($7.0\,\text{kg/cm}^2$ for 15 s) is carried out. Using this method, protein digestion of soybean increases to 92–93% compared with 68% in the conventional method. Roasting of wheat is also done in a continuous roaster at 170–180°C for a few minutes and it is then cracked by a roll mill.

After cooling to less than 40°C, the cooked beans and roasted wheat are mixed and inoculated with 0.1–0.2% tane koji (seed koji) composed of the selected strains of *A. oryzae* or *A. sojae* (Hesseltine, 1965; Yong & Wood, 1974). In general practice, mixed cultures consisting of 80% of *A. orzyae* and 20% *A. sojae* are used (Yokotsuka, 1985). The mixture is then spread onto shallow wooden trays and incubated in the culturing room, where temperature is usually kept at 25–35°C. As a result of mould growth, the temperature of the material rises. Therefore stirring of the mixture during koji incubation is required. Stirring is usually done twice at about 20–40 hours (Yokotsuka, 1960). After 3 days, koji moulds grow throughout the materials. At this time the material becomes compacted and yellowish-green in colour as a result of sporulation of the moulds. This material is called 'koji'. The conventional method of koji preparation as described is still being used in small-scale industry, whereas in a medium- to large-scale factory automatic koji making process is applied for soy souce brewing.

The modern process includes a continuous cooker, automatic inoculator, automatic mixer and large perforated shallow vats in closed chambers equipped with forced-air devices, temperature and humidity controls and mechanical devices for turning the substrate during incubation. Another type of automatic koji-making equipment on the market is a two-floor system cylindrical koji incubator with a perforated rotary disc in each floor, and it is also equipped with all necessary automatic control devices (Hesseltine, 1972; Fukushima, 1978).

There have been reports on advantages of using automatick koji-making equipment over that of traditional methods, claiming increases in soy sauce yield from 65% to almost 95%. It also has been reported that the sauce obtained from koji produced by the automatic system is superior in aroma, since the system prevents koji from being contaminated with undesirable microorganisms (Fukushima, 1978).

(b) Miso-dama. Miso-dama is another soybean koji used for fermentation of a special type of miso called mame miso. Whole soybean is the only raw material used. In preparing miso-dama, the steamed beans are formed into balls of about 1.3–1.4 cm diameter. After inoculation with tane koji, the balls are incubated at 30°C for 3–4 days (Shibasaki & Hesseltine, 1962; Kikuchi *et al.*, 1975).

(c) Koji produced by natural inoculation. In South East Asia, soybean koji for the fermentation of soy sauce and soybean paste is mostly produced by the traditional method which is believed to have been introduced by the Chinese in ancient times. In general commercial practice, the koji fermentation depends upon chance inocula. It is only recently that intensive research directed toward the improvement of soy sauce fermentation, especially on microbiological aspects, have been carried out. As a result, improvements in koji making have been at the early stage of industrial implementation by which inocula of the selected mould prepared in various forms have been introduced.

Preparation procedures. In general, the koji processes of each country in the region are somewhat similar with only slight modification from factory to factory. The general procedure follows these steps (Abdon, 1978; Lan *et al.*, 1978; Lotong, 1978; Tang, 1978). Whole soybeans are washed and soaked for various periods (3 hours to overnight). The beans are then cooked in boiling water or in a steam cooker until soft. After draining, they are left to cool to less than 40°C and mixed with wheat flour, rice flour or other materials as previously described. In common industrial practice, the flour is used without roasting. However, in the Philippines roasted flour is preferred (Soriano & Pardo, 1978). Traditionally the bean and flour mixture is then spread onto wooden bamboo trays about 3–5 cm thick. The trays are stacked on wooden racks and incubated in a room without either temperature or moisture controls. During various periods of incubation, the mixture in each tray is stirred for better aeration as well as to decrease the koji temperature. This is usually done on the second day of incubation, but some factories do so at other periods. Recently, locally designed koji rooms with inferior automatic devies compared with those used in Japan have been operated commercially.

Koji moulds naturally inoculated from the air and from the bamboo trays, previously used and not washed or sterilized between batches, overgrow the mixture. The mycelial growth penetrates throughout the mixture and covers the whole surface of grains, forming a compact koji cake after 2 days. Usually, the koji is further incubated for 3–5 more days, during which period the koji cake becomes covered with abundant yellowish green or greyish spores or a mixture of both (Figure 21.1) depending on incubation temperature and humidity.

Microorganisms involved. Characterization of the microorganisms involved in the chance inoculation koji fermentation in South East Asia, reveals that in any of the factory-produced koji, moulds of the *Aspergillus flavus–A. oryzae* group and the Zygomycetous group (e.g. *Rhozopus* spp., *Mucor* spp. and *Syncephalastrum* spp.) are predominant (Lotong, 1978; Bhumiratana *et al.*, 1980; Chen, 1980). *Penicillium* spp., *Scopulariopsis brevicaulis* and *Endomycopsis* spp. have been found as a minor population

Figure 21.1 Soy sauce koji produced by conventional method in Thailand.

in koji produced from some factories (Devahudee, 1978; Chen, 1980). It has been reported that most of the *Aspergillus* spp. isolated are *A. oryzae*. However, strains of *A. flavus* var. *columnaris* have been isolated from Thai soy sauce koji (Bhumiratana *et al.*, 1980). Ten & Ho (1980) reported that 9.5% of *Aspergillus* spp. isolated from koji prepared in West Malaysia belong to the *A. wentii* group.

Ratios of the *Aspergillus* spp. to those of Zygomycetous organisms are known to be variable among batches depending on incubation temperature and humidity. However, green koji common in cool weather are preferable over grey ones common during warm weather. Characterization of microorganisms in green koji has shown that it is dominated by *Aspergillus* spp., whereas those in grey koji belong to the *Phycomyces* (Suwanmena *et al.*, 1978; Bhumiratana *et al.*, 1980). This fact, as well as knowledge acquired concerning the microbiology of the Japanese soy sauce (Yong & Wood, 1974) where koji fermentation is achieved by inoculation with *A. oryzae* or *A. sojae*, indicate that *Aspergillus* spp. alone are the microorganisms responsible for fermentation of soy sauce koji in this region and that the presence of *Rhizopus* spp. or other genera of this class is due to the uncontrolled environments.

Several cultures of *Aspergills* capable of producing high protease and amylase activities have been selected. While most of the selected strains have been found to belong to *A. oryzae*, one strain isolated from Thai soy sauce koji has been reported to be a non-toxigenic strain of *A. flavus* var. *columnaris* and has subsequently been shown to produce improved quality soy sauce when used as inoculum alone (Bhumirataha *et al.*, 1980).

Selection of mutants of *A. oryzae* with better enzymatic activities and producing light-coloured conidia has also been reported (Nga *et al.*, 1978; Kalayanamitr *et al.*, 1987). As some of the selected mutants of originally non-toxic parents were found acutely toxic to weanling rats, although all were negative for aflatoxin production, it is advised that during mutagenic or other genetic manipulated strain development for the food industry, one can not assume that mutants of a safe parent mould are also safe (Flegel, 1988). Improving the soy sauce koji mould has also been successful using the technique of protoplast fusion between an alpha-amylase hyper-producer of *A. oryzae* and a protease hyper-producer of *A. sojae* (Ushijima *et al.*, 1991).

Appropriate technologies for producing large quantities of spore mass have been developed. Recently, most manufacturers in this region, particularly in Thailand, have changed their koji preparation from that depending on natural inoculation to the use of seed inocula.

21.2.3 Koji enzymes

Various classes of enzyme have been detected in soybean koji, including amylase, cellulase, invertase, lipase and protease. Among these, those with amylolytic activities play an important role in the degradation of starch and oligosaccharides to provide substantial fermentable sugars for the subsequent fermentation stage, through which taste and aroma are enhanced. While amylase hydrolyses starch of wheat or other kinds of flour to produce disaccharides and monosaccharides, invertase has been found to hydrolyse sucrose present in the bean into glucose and fructose (Yong & Wood, 1975). It has been shown that reducing sugar of the koji increases remarkably after 18 hours incubation, whereas amylase activity is not detectable until after 25–30 hours. The appearance of amylase is accompanied by a second increase in reducing sugar. Therefore, the early increase in reducing sugar is believed to have resulted from inversion of sucrose present in soybean by the enzyme invertase (Yong & Wood, 1975).

Soybean cell wall also has been reported as another source of carbohydrate which is hydrolysed to monosaccharides by various polysaccharidases. Using koji extract as a source of enzymes in soybean hydrolysis, Kikuchi (1976) and Kikuchi & Yokotsuka (1973), reported that they found considerable increase in D-galacturonic acid, L-rhamnose, D-xylose, L-arabinose, D-galactose and D-glucose.

While the saccharolytic enzymes take their major role in providing the fermentable sugars, the proteolytic enzymes play a very important part in solubilization of soybean protein into peptides and amino acids, through which the product yield in terms of soluble nitrogen is increased (Nakadai, 1985).

Studies on kinetics of total protease production in soy sauce koji have shown that the largest increase in protease activity appeared in the late phase of fungal growth. Normally the activity reaches its maximum at about 48 hours (Yamamoto *et al.*, 1972; Impoolsub *et al.*, 1981). There have been a number of studies on purification and identification of components of the enzyme mixtures (Misaki *et al.*, 1970; Nakadai *et al.*, 1973a,b,c; Kundu & Manna, 1975; Impoolsub *et al.*, 1981). Three kinds of protease have been reported to exist in most of the crude enzymes produced in koji, namely acid, neutral and alkaline protease. The existence of the semi-acid protease in the crude enzyme has been shown by Matsushima (1958). Semi-alkaline protease also has been isolated from an enzyme preparation of *A. oryzae* (Misaki *et al.*, 1970).

Since the pH of the koji and the early mash stages is nearly neutral, the practical role played by acid protease is believed to be minimal. Most studies concerning proteases produced by the koji moulds are therefore focused on neutral and alkaline proteases. The activity profiles of the two proteases isolated from most of the koji are similar in their optimal pH. The neutral proteases have been reported to have optimal pHs of 5.5–7.0, while those of the alkaline proteases are 8.0–10.5 (Nakadai *et al.*, 1973a,b,c; Kundu & Manna, 1975; Impoolsub *et al.*, 1981). In China, where a certain strain of *A. oryzae* is used in 98% of soy sauce factories, the main protease produced is neutral protease which plays a significant role during early mash fermentation. In enhancing protein digestibility at late fermentation which the pH drops to 5.5 or less, *A. oryzae* koji is combined with koji prepared using *A. niger* at a ratio of 8:2 (Yue, 1990).

Inhibitory effects of sodium chloride on the activity of proteases (Nakadai *et al.*, 1973a) and (Impoolsub *et al.*, 1981) and glutaminase (Maramatsu, 1994) have been reported. At the salinity normally used in mash fermentation (18–20%), the respective activities of the alkaline and neutral protease of *A. flavus* var. *columnaris* are only 10 and 15% of the original (Impoolsub *et al.*, 1981). Similar results have been reported by Nakadai *et al.* (1973a,b) on the enzyme produced by *A. oryzae*. Moreover, Impoolsub *et al.* (1981) have shown that the residual proteolytic activities of the enzymes produced by *A. flavus* var. *columnaris* in 18% salt brine are rather stable during 45 days' incubation and that the activity of the neutral protease is comparatively more stable. From these facts, they noted the substantial protein hydrolysis by the koji enzymes should be accomplished before the mash incubation. In order to attain the maximum proteolysis of koji, a suggestion has been made for incubating the koji for 1–2 hours at a temperature optimal for protease activity (50–55°C) after the regular 2 days' incubation.

Besides a number of proteases, koji moulds have been reported to produce both aminopeptidases and acid carboxypeptidases (Nakadai *et al.*, 1972a,b, 1973d, e.f.g; Nakadai & Nasuno, 1977) and it has been shown that hydrolysis of soybean protein to free amino acids in soy sauce brewing involves two-step enzyme reactions. Proteases, the endo-type enzymes, hydrolyse protein to large molecular peptides. The breakdown of the large peptides into lower molecular peptides and free amino acids is carried out by the action of peptidases (Nakadai *et al.*, 1972a,b).

In their studies on the use of enzyme preparations from various sources such as *Streptomyces* spp., *Bacillus* spp., *Rhizopus* spp. and koji moulds for the enzymatic hydrolysis of soybean protein, Nakadai & Nasuno (1976) have shown that enzymes of koji mould origin were best for liberation of amino acids. They pointed out that the superiority of the enzymes produced by koji moulds is considered to be mainly due to their high content of composite peptidases. A number of *A. oryzae* isolates with higher production of peptidases in addition to other key enzymes have been reported as the superior strains in producing enzymes for soy sauce manufacturing (Nakadai & Nasuno, 1977).

Glutaminase is another enzyme that has been reported to play an important role in soy sauce quality, since it attacks glutamine, yielding additional glutamic acid, the amino acid essential for good taste of the sauce. Therefore, beside being efficient in proteolytic activities, it is desirable that the koji moulds should at the same time be capable of producing high glutaminase activity in order to give high-quality soy sauce (Yamamoto & Hirooka, 1974a; Tochikura *et al.*, 1983). Glutaminase-hyperproductive mutants of *A. sojae* have been reported and it has been suggested that they be used in brewing of soy sauce (Yamamoto & Hirooka, 1974b). Recently, koji prepared using mixed cultures between high protease activity *A. oryzae* and a strain of the same species strong in glutaminase enzyme has been introduced. Koji obtained had twice the glutaminase activity of that using single strain (Ukei *et al.*, 1994).

21.2.4 Application of soybean koji

(a) Soy sauce fermentation. Beside serving as a source of enzymes, soybean koji is used as sole substrate in the mash stage (moromi) of soy sauce fermentation. In this process the mature koji mass is mixed with brine (20–22% salt) to form the mash. During koji preparation, soybean protein and wheat starch are partially hydrolysed by various enzymes produced in it. The remaining undigested raw material is further hydrolysed in the mash to provide more fermentable substrates. The mash concurrently undergoes lactic acid and yeast fermentation which results in production of compounds characteristic of the flavour and aroma of the finished products (Yong & Wood, 1974; Yokotsuka, 1985).

In traditional Japanese soy sauce brewing, the mash is held for over 12 months under uncontrolled conditions (Yong & Wood, 1974). However, in the large-scale commercial practice of today, mash fermentation can be reduced to 6–8 months by inoculation of *Tetragenococcus halophilus* (*Pediococcus halophilus*), *Candida versatilis* and *Saccharomyces rouxii* and the mash is incubated under controlled temperature (Ziemba, 1974; Fukushima, 1978; Yokotsuka, 1985).

Realising, as mentioned earlier, that sodium chloride has a strong inhibitory effect on protease activitiy, a new method of short-term soy sauce making on the industrial experimental scale has been reported in Japan, where the time of the moromi stage was divided into two functionally independent processes, namely autolysis of koji and brine fermentation. For this process soy bean koji was autodigested at various temperatures in the absence of sodium chloride. A large amount of total nitrogen was obtained in koji autolysate which was done at 55–60°C for 4 days, whereas autolysis at 45–50°C produced a higher amount of glutamic acid. As koji autolysed at 60°C resulted in darker moromi, the optimal temperature for autodigestion was decided as 55°C. The autodigested koji was then subjected to two-step brine fermentation. The first step started by inoculation of *T. halophilus* and *C. versatilis*. After 16 days, an equal amount of freshly prepared koji autolysate was added and *S. rouxii* was inoculated to induce major alcohol fermentation. Alcohol fermentation was completed in around 2 months. The moromi was then aged and the overall fermentation process completed in 3 months, yielding good-quality soy sauce as compared with that obtained using the conventional method (Maramatsu, 1994). As practised in a few factories in Thailand, protein digestibilty is increased by preincubation of koji with small amount of water for 2 hours prior to addition of brine to make up the desired salt concentration. In this country, an industrial-scale experiment showed that the preincubation period could be extended to 5–6 hours and that the process resulted in shortening of fermentation time from 36 days to 21 days (Aeaungkoon, 1995).

(b) Soybean paste fermentation. In addition to its use as starting materials in soy sauce fermentation, the Chinese type soybean koji is also used in production of soybean paste. In general, the mash fermentation is carried out under direct sunlight for 1–3 months. The original brine is drained off as first-grade soy sauce. More brine may be added, and the mash is further incubated for 1–2 months to yield lower grades of sauce. In Thailand, there has been another practice, by which the solid soybean sediment separated from the first draining is sold as second-grade tauchieo (soybean paste). For fermentation of first-grade tauchieo, and taoco, a similar condiment produced in Indonesia and Malaysia, the mature koji is fermented in brine as with soy sauce production, except that less brine is added (Hartadi & Jutono, 1979; Flegel *et al.*, 1981a).

(c) Sufu fermentation. Sufu (soybean cheese) is a highly flavoured, soft, cream cheese-type fermented product made from soybean curd. Among several kinds of sufu produced, only the yellow sufu is fermented using soybean koji as source of enzymes (Yuan, 1994). Cubes of soybean curd are placed into dressing mixture containing koji, fermented soybean or soybean paste and steamed rice and allowed to ferment for 2–3 months. Other processes for sufu production have been reviewed by Su (1966).

(d) Fish sauce fermentation. Besides the use of koji enzymes in soybean substrate degradation, soybean koji also has proved applicable as a source of supplement proteases to shorten the time required for fish sauce fermentation (Togano *et al.*, 1978; Park *et al.*, 1979). Following this idea, fish meal (Hayakawa *et al.*, 1993) and wheat bran (Klomklang, 1995) koji have been produced using soy sauce koji mould. By application of fish meal koji for fish sauce fermentation, a product with a pleasant aroma was obtained. It is possible that the advantage of using koji for digestion of fish protein is due to the capability of *A. oryzae*, the koji mould in utilizing trimethylamine and trimethylamine oxide, the main components of fishy odour as reported by Kunimoto *et al.* (1992). An experiment using wheat bran koji for fish sauce fermentation showed that maximum soluble nitrogen was obtained in 42 days (Klomklang, 1995).

21.3 Rice koji and similar products

21.3.1 Mould enzymes used in the fermentation of alcoholic foods and beverages

Among the traditional fermentations existing for centuries in the Orient, the use of fungal enzymes produced in cereals for saccharification of starchy materials is one of their unique characteristics. In obtaining the enzymes, two main different methods are known. One is the preparation of rice koji used in saké and shochu brewing in Japan, in which spores of *A. oryzae*, *A. awamori* or *A. kawachii* in the form of tane koji, are used as inocula. In another process common among other Asian countries, inocula in the form of starter cakes cosisting of mixed cultures of moulds of the Mucorales and yeasts are used. Details concerning starter cakes and tane koji are described under 'Seed inocula' (section 21.4).

Two related fermented products are obtained through the enzyme application. One is alcoholic beverages prepared by saccharification of a variety of starchy materials such as rice, glutinous rice, corn, cassava or potatoes. The saccharified products are simultaneously fermented to alcohol. Among these beverages, however, those prepared from rice or glutinous rice are the most popular. The products are generally called rice wine or rice beer and are known by different local names (Table 21.1). Another related product presented in the same table is sweetened rice (sweet–sour mild alcoholic rice dessert) prepared by partial liquefaction of glutinous rice.

A similar process for obtaining fungal enzymes, using wheat bran as substrate, was introduced to the Western World by Takamine (1914) and known as the Taka-koji or mould bran process. The mould bran preparation was used to saccharify grains in distilleries in place of barley malt.

Table 21.1 Rice wine and sweetened rice of Asian countries

Country	Rice wine	Sweetened rice	References
China	Shaohsing	Lao-chao	Inui *et al.* (1965); Hesseltine (1965)
Japan	Saké	Amazake	Kodama (1970)
Korea	Cheongju		Mheen (1972)
Thailand	Sato, nam-khao, krachae, Ou	Kaomag	Kamnuanta (1981)
Indonesia	Arak, brem	Tape ketan	Saono *et al.* (1974)
Philippines	Tapoi	Binuburan	Soriano *et al.* (1980)
Malaysia	Arak beras	Tapai	Ismail & Karim (1980)
Tibet	Chang		Batra & Millner (1976)
Burma	Congee		Batra & Millner (1976)
India	Murcha, bakhar, pachwai		Batra & Millner (1976)
Nepal	Ranu		Batra & Millner (1976)
Laos	Shoum		Batra & Millner (1976)

Vietnam Ruou can Ruou nep Vo *et al.* (1993).

Although yields of alcohol obtained by using the mould bran as saccharifying agent were reported to be higher than those produced by using malt, the process was not adopted (Prescott & Dunn, 1959).

(a) Saké rice koji. Saké is a popular Japanese traditional beverage. It is prepared by saccharification of rice by enzymes produced in rice koji, followed by mash fermentation with saké yeast and lactic acid bacteria.

Preparation of rice koji. Preparation of rice koji is an essential step in saké brewing. The subject has been intensively discussed by Kodama & Yoshizawa (1977), on which the following account is mainly based. In saké brewing, the proper treatment of rice used in any fermentation steps including koji making is considered significant. These include polishing of the rice to remove protein, fat and minerals which are present in excess in the peripheral layer of brown rice. These are unfavourable materials for best quality saké. Generally, only 60–70% of the original weight is retained in the polished rice. In preparation of rice koji, the rice is washed and steeped for about 12 hours at low temperature (Kiku-Masamune Saké Brewing Co. Ltd). Usually 25–30% water is absorbed during washing and steeping. The rice is then steamed for 30–60 minutes. As a result of adequate steaming, rice with about 35% total water gain is obtained. The steamed rice is cooled to near 40°C before tane koji of *A. oryzae* is inoculated. Recently an experiment using a transformant of *A. oryzae* with glucoamylase gene of *A. shirousamii* has been reported (Shibuya *et al.*, 1992). Usually, the proportion of the inoculum used is 60–100 g/1000 kg rice. In general, about 20% of the total steamed rice is required for koji making (Akita, 1994). According to the conventional method, the mixture is transferred to and heaped in the koji room where temperature is controlled at 26–28°C. In order to expel the heat and carbon dioxide resulting from mould respiration, the heap is mixed at 10–12 hours (Yoshida & Taguchi, 1979). At about 20 hours, when white spot of mycelium is distinctly seen and the temperature has risen to 32–34°C, the mouldy rice is dispensed into the perforated-bottom shallow wooden boxes and covered with sheet cloth. At this stage, the mould grows very rapidly resulting in a steady increase of temperature, reaching 38°C after 36 hours, and 40–42°C by 45 hours. In order to overcome the heating, the mixture is occasionally stirred. At this late phase of koji culturing, the material becomes compacted, with white mycelium covering the surface and penetrating into the rice kernel. The characteristic koji aroma is evident. In general, the process is completed by 40 hours.

At present, although the conventional process as described is still operating in many breweries, in large commercial applications the koji is prepared in various types of koji-making machines, by which the uniformity of growth, temperature, aeration and humidity is controlled by automatic devices. By using this modern technology, culturing time can be shortened by 6–8 hours. Studies leading to a better process control of sake koji making have been made continuously; for instance, determination of physical properties of steamed rice (Wakai *et al.*, 1995), mycelial growth monitoring (Honda *et al.*, 1995) and studies on the relationship of mycelial propagation and enzyme production (Chou & Rwan, 1995).

Roles of rice koji in saké brewing. During koji culturing, various hydrolytic enzymes are produced and accumulated. Rice is partially hydrolysed, yielding sugars, amino acids and peptides. Vitamin B_{12}, along with some precursors of the characteristic saké aroma, are detected (Kodama, 1970; Suzuki *et al.*, 1980, 1982).

Rice koji plays significant roles both in preparation of yeast starter (moto) and mash fermentation. The substrate used for culturing of the starter is obtained by mashing rice koji with steamed rice and water. Usually a ratio by weight of 3:6:10 is used. In starting the fermentation, the starter is inoculated into the substrate consisting of steamed rice, rice koji and water in a large vessel (moromi mash). In commercial operations the substrate is periodically added on the third and fourth day after inoculation (Kiku-Masamune Saké Brewing Co. Ltd). The fermentation is carried out at a temperature below 15°C for 20–25 days.

Among various enzymes detected in koji, α-amylase, the liquefying enzyme, and α-glucosidase and glucoamylase, the saccharifying enzymes, play major roles in the degradation of rice, providing substantial sugars for simultaneous fermentation by yeast and lactic acid bacteria (Morimota *et al.*, 1995).

Although both acid and alkaline protease are found in koji, liberation of peptides and amino acids in the mash, which is of low pH range, is believed to be achieved through the reaction of acid protease (Kageyama & Sugita, 1955).

In the traditional saké brewing, lactic acid is produced in the mash by lactic acid bacteria derived from koji. However in the large-scale commercial practice of today, lactic acid is added. The acid is considered essential to inhibit growth of nondesirable microorganisms and lower the pH of mash.

The characteristic aroma of saké results from combined flavour components of a number of compounds produced during fermentation (Kodama, 1970). It has been reported that precursors of saké-aroma substances exist in rice koji extract. In order to identify these precursors, Suzuki *et al.* (1980) fractionated the koji extract into six fractions. Each fraction was added to a saccharified mash of steamed rice, followed by fermentation with saké yeast. The fermented mash was subjected to a panel organoleptic test for strength of saké aroma. It was found that the strongest saké aroma was developed by adding a particular fraction and that the saké prepared by adding this fraction is much higher in ethyl leucinate. In the subsequent paper, the same authors (Suzuki *et al.*, 1982) have reported that leucinic acid was found in rice koji but not in steamed rice and that it was produced only by koji mould, *A. oryzae*, not by the saké yeast. They also have reported that saké yeast converted leucinic acid to ethyl leucinate, while the koji mould lacked the ability to do so. From these findings, it has been proposed that leucine is converted to leucinic acid, the precursor of saké aroma, by the koji mould and accumulated in the koji. The compound is subsequently converted to ethyl leucinate by saké yeast in the mash fermentation. Moreover, they have noted that abilities to produce high yields of these compounds vary among strains of koji moulds and saké yeasts.

Another distinguishing characteristic of saké brewing is the accumulation of high alcohol concentration, reaching 20% or more. Several reasons, based on a number of research papers explaining this phenomenon, have been summarized by Kodama (1970). The influence of a special factor extracted from the mycelia of the koji mould is one of the proposed reasons. The factor has later been called the 'high alcohol concentration-producing factor' and found to be a phospholipid-containing unsaturated fatty acid bound to a macromolecule complex such as albumin or methylcellulose (Hayashida, 1974).

Analysis of the proteolipid showed phosphatidylcholine to be the major phospholipid. The respective supplementation of the phosphotidylcholine–albumin complex and the phosphotidylcholine–methylcellulose complex to the basal synthetic medium resulted in the promotion of yeast growth and accumulation of a high concentration (20%) of alcohol. Examination by electron microscopy revealed the deposition of lipid in the cytoplasm of saké yeast grown anaerobically in the presence of these proteolipids. Furthermore, it was found that spheroplasts prepared from this yeast were stable in 20% alcohol, whereas those obtained from cells grown without the alcohol concentration factor were ruptured. These findings suggested that the accumulation of high alcohol concentration is related to lipid metabolism of yeast in the presence of proteolipid provided by the koji moulds (Hayashida *et al.*, 1976; Hayashida & Ohta, 1978).

(b) Saccharolytic enzymes obtained by using starter cakes as inocula, and their applications. In the traditional fermentation of varieties of rice wine in which starter cake, the dry inocula of various local names (see Table 21.4), is used as inoculum, similar processes of enzyme preparation and saccharification are followed. Generally, polished glutinous rice is washed, steeped for 6–12 hours and steamed. After being cooled to room temperature, the rice is then inoculated with starter cake powder. The material is loosely packed in shallow trays and incubated at room temperature for 24–48 hours. During incubation, mycelial growth occurs and the saccharolytic enzymes are produced, followed by partial saccharification of the rice. Since

the starter cake consists of a mixed flora of moulds and yeasts, alcohol is therefore simultaneously produced. Moreover, the moulds, *Amylomyces rouxii* and/or *Rhizopus* spp., which predominate in most of the starter cakes, can also by themselves convert the liberated sugars to a certain amount of alcohol (Cronk *et al.*, 1977). In comparison therefore, this stage should be called koji preparation; the koji so produced is certainly different from saké koji.

Mash processes by which the described activities continuously proceed to yield greater liquefaction and higher alcohol concentration are slightly different in various countries. According to Hesseltine (1965) and Cronk *et al.* (1977), the Indonesian arak or brem is obtained by allowing the inoculated rice to incubate longer. In fermentation of the Indian bakhar and of sato or nam-khao, the Thai rice wine, after 24–48 hours incubation in shallow trays, the material is transferred to earthenware jars and water is added; the mash is then further fermented (Hesseltine, 1965; Batra & Millner, 1976; Kamnuanta, 1981). Mash fermentation of sato varies from 7 to 14 days depending on the alcohol concentration required, which normally ranges from 7 to 14% (Chaownsungket, 1978). The solid residue is then removed by filtration.

Ou, the special Thai rice wine, is similarly fermented, except that the steeped rice is mixed with rice husk before steaming. Traditionally, this beverage is directly served in the jar through a bamboo straw without removal of solid residue (Kamnuanta, 1981).

In making sweetened rice, procedures similar to those used in early stages of rice wine fermentation are followed, but 24–48 hours incubation is enough for partial liquefaction (Djien, 1972; Cronk *et al.*, 1977; Soriano *et al.*, 1980). In the case of kaomag (Thai sweetened rice) preparation, steamed rice should be rinsed with water to remove the gluten before it is inoculated with starter powder. Moreover, starter cake obtained from different recipes from that of rice wine is used.

Apart from serving as a dessert, kaomag is used for imparting flavour and providing fermentable sugars in the fermentation of a variety of fermented fish such as plachao and plapangdang (Deangsubha, 1979).

Recently, starter cake has been used to substitute diastic malt extract as a saccharifying agent in producing rice-based weaning food. It was reported that besides reducing dietary bulk, the product obtained using starter cake alone had some antimicrobial effect on tested enteric pathogens, but this was much improved when the inocula was used in conjunction with L-lactate-producing *Lactococcus lactis* or *Lactobacillus bavaricus* (Yusof *et al.*, 1995).

(c) Saccharolytic enzymes used in distilling industries in Thailand. Unlike rice wine fermentation which is practised on a home or cottage scale, white spirit and whisky distilleries are operated in larger scale

industries. A similar process in obtaining saccharolytic enzymes to that used in fermentation of rice wine was once followed. Owing to the difficulties in controlling the activities of the unknown mixed cultures contained in starter cakes, pure cultures of the selected efficient strains of *Rhizopus* spp. (mainly *R. oryzae*) have been used for enzyme production by most of the manufacturers.

In commercial runs, enzyme preparation procedures similar to those of the mould bran process are used. Koji produced for this purpose is called pang-cheu. For its preparation, a mixture of broken rice, broken glutinous rice and rice bran is used as substrate. In some distilleries, glutinous rice is not included. It has been reported that less glucoamlyase is produced in substrate containing glutinous rice (Tawatpun, 1983). Rice is ground into a coarse powder, mixed with rice bran and steamed. After cooling, the material is thoroughly mixed with spore suspension and water(400–500 ml/kg dry material). The inoculated material is spread in wooden bamboo trays and incubated for 36–40 hours at ambient temperature. By the end of incubation, loose texture koji cakes are formed. As a usual practice, the koji is broken into pieces and dried in the sun for 6–8 hours, then ground into a powder.

21.3.2 Ang-kak

(a) Application and microorganisms involved. Ang-kak, the Chinese term for red rice, is a product made by culturing certain strains of *Monascus* spp. on rice grains. It has also been called by other different names: anka, ang-quac, anka koji and beni koji (Palo *et al.*, 1961). The product has been widely used in Asian countries for imparting flavour and colouring to a variety of fermented products such as red soybean cheese and several kinds of fermented fish. It is also used in brewing of red rice wine and red Chinese liquor (kaoliang). Use as a medical therapeutic agent is also reported (Lee, 1979).

It has been pointed out that although several strains of the fungus have been isolated from various sources, most of them are not suitable for ang-kak production because they do not produce the desired deep purplish-red pigment and only certain strains of *M. purpureus* have this property (Church, 1920; Hesseltine, 1979). However, Lin (1973) isolated from koji of kaoliang liquor a strain of *Monascus* sp. suitable for pigment production by submerged culture. It is identified as a new species, *M. kaoliang* (Lin & Iizuka, 1975). In a Japanese patent, *M. anka* is used as a pigment producer (Ishihashi, 1978). Through a series of strain manipulations, a number of mutants which produced large amounts of pigment were selected (Lin & Iizuka, 1982; Chansiripotha, 1988; Boonprab, 1996). Moreover, a mutant producing only yellow pigment was obtained. Production of the pigment could be made using both solid-substrate fermentation on rice and submerged culture in cassava medium (Yongsmith *et al.*, 1993).

The main components of *Monascus* colour are the red pigments monascorubrin and rubropunctatin and the yellow pigments monascin, ankaflavin and monascoflavin (Nishikawa, 1932; Lin, 1973: Manchand *et al.*, 1973) and purple pigments rubropuntamine and monascorubramine (Hiroi & Shima, 1975). This fact may explain that most strains give orange red colours except for certain ones whose pigments contain more of the red components, so giving a deep red colour. While most of the *Monascus* colours are reported as intracellular pigment, one produced by the hyperpigment-producing mutant is excreted from the mould (Lin & Iizuka, 1982).

Besides its pigments, *Monascus* spp. also produces various hydrolytic enzymes such as glucoamylase (Iizuka & Mineki, 1977, 1978; Lotong & Suwamarit, 1990) and protease (Tsai *et al.*, 1978).

(b) Preparation. One of the most important parameters for the success of ang-kak fermentation is the low initial moisture content. As the fermentation starts at low moisture, the culture dries out to such a level that the mould grows poorly and pigmentation does not occur before fermentation is completed. To overcome this problem, an appropriate amount of water is occasionally added throughout the fermentation period (Su, 1980; Steinkraus *et al.*, 1983).

In the industrial practice in Taiwan, as described by Su (1980), the ang-kak fermentation process follows these steps: 42 kg of polished rice is washed, steeped in water for 3 hours, then steamed for 1 hour. After cooling, the steamed rice is inoculated with 24 kg of seed culture, mixed with 75 litres of water in a tank and incubated. To minimize heat accumulation, the rice is occasionally stirred during incubation. After incubating for 4 days, the mash begins to develop a red colour and is used as seed culture for 1500 kg rice. After washing, the rice is steamed under pressure at 0.2 kg/cm^2 for 1 hour, cooled to 40°C and sprayed with 20% water, then steamed for another 30 minutes and cooled to 36°C. A 35-litre portion of seed mash is then added to it and the mixture is placed in a large tank. During incubation, there is a gradual rise of temperature up to 42°C. At this stage the koji is spread into koji bamboo trays and incubated. To prevent the culture from drying the koji tray is steeped in water and drained. This is done three times during the 8-day incubation. The koji is then dried until it contains 7–10% moisture. The yield of dried angkak produced is about 48% of the rice used.

According to Palo *et al.* (1961), corn as well as rice of all varieties except glutinous rice may be used to produce the red colour. Ishihashi (1978) patented a process of red pigment production by *M. anka* using bread meal as substrate. By comparison, pigment productivity of the same strains (*M.*

kaoliang and its hyperpigment-producing mutant) using various cereal meals as substrates, Lin & Iizuka (1982) demonstrated that mantou (steamed bread) meal gives a higher amount of pigment than bread meal and much higher than wheat meal, corn meal or rice meal.

It has been reported that production of ang-kak on an experimental scale also needs low initial moisture content of c.30% (Hesseltine, 1965; Sooksan & Gongsatidi, 1977; Su & Wang, 1977). According to the method used in these studies, steamed rice is cultured with *M. purpureus* in Erlenmeyer flasks and allowed to ferment for 3 weeks at 25–32°C with occaional stirring and water addition.

Recently, a method of producing ang-kak in plastic bags with paper board closures (detail of preparation of the bags is described in section 21.4.2) has been developed (Lotong & Suwanarit, 1990). With this method, fermentation carried out with rice containing 26–32% initial moisture gave a good-quality product. The advantage of the use of plastic bags as culture vessels is that addition of water during the course of fermentation, as required in both the traditional process and that conducted in the laboratory was not necessary. Since the culture bags were laid on their sides, most of the water evolved from respiration accumulated as condensate over the inside surface of the bags. Therefore the substrate moisture could be easily maintained by occasionally shaking the bags so that the condensate was reabsorbed by the grains. Because of the flexibility of the plastic, packing of rice could be overcome by external manipulation. Thus, contamination was completely eliminated. The ang-kak obtained contained pigments of the same absorption spectra as those of the market products. It was also found that the experimental products were superior to the commercial products in respect of the intensity of both red (500 nm) and yellow (370 nm) pigments. In this study, it was also reported that the fermentation period was shortened to 10–12 days.

This study also elucidated the mechanism by which the initial moisture content of rice played a role in regulation of pigment formation during the fermentation of ang-kak; high initial moisture enhanced glucoamylase production. As a result, glucose was liberated in an amount that inhibited pigmentation. Moreover, attempts have been made to select for glucose derepression mutants which produced higher pigment levels with a faster rate, and at high moisture content. It was reported that, in addition to resistance to glucose repression, the mutants were able to regulate the liberated glucose at a concentration above which the mould could not tolerate. By subse quently exhibiting a sharp drop of glucoamylase production, maximal activity was achieved (Boonprab *et al.*, 1996).

21.3.3 Miso rice koji

Among several kinds of miso produced in Japan, rice-base (kome) miso is the most common type, representing about 80% of the total production

(Abiose *et al.*, 1982). It is a product prepared by a two-step fermentation. The first step involves rice koji preparation to produce enzymes, followed by fermentation of the koji along with steamed soybean, salt and seed miso (Shibasaki & Hesseltine, 1962; Hesseltine, 1965). The subject of miso has been reviewed in various aspects by Abiose *et al.* (1982) and Ebine (1986).

For a production run, rice koji is made by steeping and steaming polished rice and then inoculating it with appropriate tane koji (usually *A. oryzae*). In general, 1 g inoculum is used per kg of rice. The material is then placed in wooden trays and incubated at 30°C for 40–50 hours. During incubation the mass is stirred twice in order to prevent overheating and to maintain uniform temperature, moisture and aeration. To quote Shibasaki & Hesseltine (1962) 'the miso koji at its best looks like a white felt of mycelium or a solid cake of mouldy rice'.

21.4 Seed inocula

Production of spore masses for use as inocula is considered an important aspect in producing koji for food fermentation. Various methods for use in fungal spore production have been discussed by Hesseltine *et al.* (1976). This review aims to describe some particular forms of inocula used as seed cultures in koji preparations referred to in this chapter.

21.4.1 Tane koji

Tane koji, which means seed koji, is a spore mass of koji mould produced on rice grains. It has long been used as inocula in koji preparation for a variety for Japanese fermented foods. Although the art of tane koji culturing and its application has been known by the Japanese since the 14th century, it was not until about 100 years ago (1876) that the koji fungus was first characterized by two German scientists, O. Korchelf and H. Ahrburg. The fungus was, however, firstly reported as *Eurotium oryzae* Ahrburg (Korchelf, 1876). Korchelf's work was published in Japanese by his assistant, S. Matsubara two years later (Hasegawa, 1979). Since then, numerous publications concerning morphological comparison of the koji moulds have appeared (Kibi, 1926; Young & Wood, 1974). Mycological studies reveal that they are members of the large group of *A. flavus* – *A. oryzae* (Thom & Church, 1926).

Tane koji has long been produced commercially in Japan. Most of the miso, saké, shochu and amazake (sweetened rice) and small soy sauce industries buy tane koji, while the large soy sauce manufacturers produce their own (Hesseltine *et al.*, 1976). By the mid-1970s, only a small number of tane koji producers were still operating (Hasegawa, 1979). The procedures used in the preparation of tane koji vary slightly from company to company. The following details of the steps in preparation are based on papers reviewed by Nakano (1972) and Hesseltine *et al.* (1976) as well as information given by manufacturers in Japan.

In general, brown rice is used as culture material. The rice is lightly pounded by a wooden pestle to scratch the surface or outer epidermis of the kernels. After soaking with running water for about 10 hours, it is cooked in a steamer for 1 hour, then cooled to about 40°C. The ash of twigs and leaves of certain kinds of trees, for example camellia, oak and chestnut, is added at an amount equal to 2% of rice dry weight, so that the rice grains appear to be covered with ash dust. The selected strains of *A. oryzae* or *A. sojae* are then inoculated. In most cases, mixed cultures of two or more strains are used. Selection of culture combination is based on type of fermentation for which the tane koji is to be applied. The inoculated rice is incubated in a deep wooden box kept at 28–30°C overnight. The mass is then transferred to several small shallow wooden trays and further incubated (Figure 21.2). Normally, the surface of the culture is covered with spores by the third day.

a

b

Figure 21.2 a Production of tane koji in wooden trays. b Preparing trays of koji for drying.

At this stage, a wooden rake is pressed into the culture and pulled across to provide better aeration. After incubation for another day, the trays are stacked in pairs, so that their surfaces are facing one another. Two or three layers of cloth sheets are sandwiched between each pair, so that the moisture is absorbed. Two more days incubation are adequate for maximum sporulation. The spore mass is then dried at 45–60°C for one day. The final rice–spore mixture of 5–6% moisture content is then packaged in paper bags of about 500 g, normal retail size or in cans (Figure 21.3).

21.4.2 Production of seed inocula for soybean koji in South East Asia

(a) Production of seed koji in plastic bags in Thailand. An innovation in the method for producing koji mould spore inoculum by introducing the use of plastic bags as culture vessels, has been developed in Thailand (Lotong & Suwanarit, 1983). The method has been continuously improved during the past decade and the technology has been successfully transferred to the soy sauce industry. As a result, at present approximately 70% of soy sauce manufactures in this country have changed their koji preparation process from that using the traditional method in which fermentation depends upon natural inocula to the use of seed inoculum of selected strains of soy sauce koji mould. Moreover, by following the introduced technology, commercial seed koji has been recently produced and has become one among several biotechnology-based industries in Thailand (Lotong, 1992, 1994).

Figure 21.3 Various kinds of tane koji, packed for sale.

Preparation of flask-like plastic culture bags. With this method, it is necessary that bags are made of autoclavable plastic (polypropylene) of sufficient thickness that, when laid on the side with closures attached they do not settle down onto the koji, be used. In each bag an appropriate amount of broken rice blended into coarse meal, or wheat bran and water, are added. In forming a self-contained flask-like plastic bag for growing mould, the open end of the bag is inserted through a tubular closure. The protruding bag end is then folded back over the closure and is tightened in place with a rubber band or with adhesive tape. Because of its rigidity, the closure serves like the neck of a flask and can be plugged with cotton. The substrate-filled, cotton-plugged bags are autoclaved and then cooled to room temperature. Each bag is inoculated with the local selected strains of soy sauce koji mould. Maximum viable spore count of about 10^9 per g of material can be obtained after a 5-day incubation at temperatures ranging from 28 to 35°C, which is the normal range of ambient temperature in this region. After removing the cotton plugs and closures, the spores can be dried by keeping at 50°C or in the direct sun for 6 hours, without significant decrease in viability. The spores can be packed in their original bags and stored at room temperatures or in a refrigerator for at least 3 months. Stages of sporulation in plastic bags are shown in Figure 21.4. The spores can be directly used as seed inoucla of soy sauce koji.

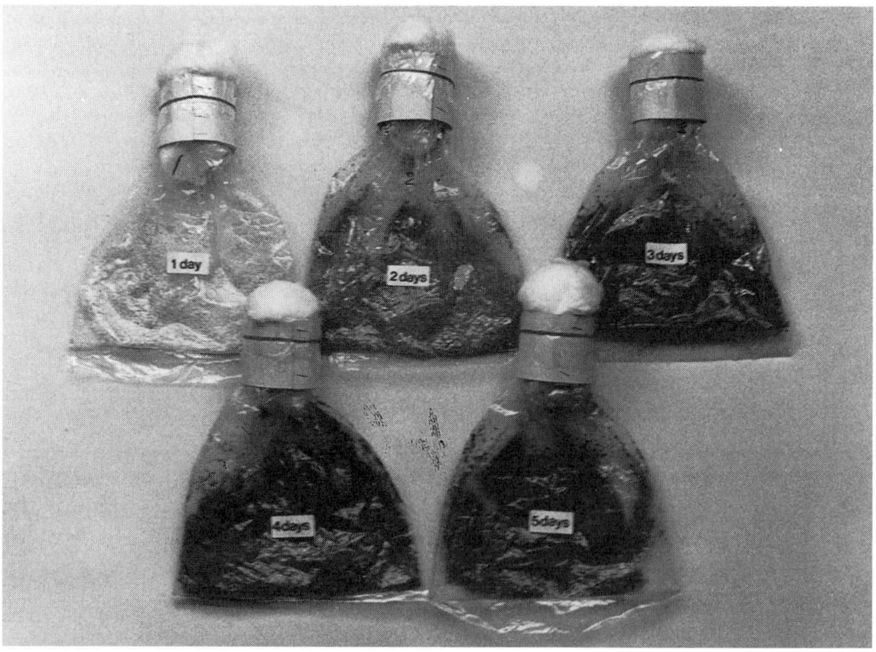

Figure 21.4 Stage of sporulation of soy sauce koji mould grown in plastic bags from 1–5 days.

As scales of soy sauce production in Thailand range from those of cottage level to large factories, they require different quantities of seed inocula to start koji fermentation. This can be managed by using different sizes of plastic bags with closures of appropriate diameter and the correct amount of substrate as examples suggested in Table 21.2. However, the 3.5-cm diameter closures made of hard plastic (high-density polypropylene) are commonly available for mushroom cultivation. For larger closures to fit the sizes of the bags, paper or stainless steel closures have been used successfully. The type of paper board used for making cake boxes works well. The paper is cut into strips of 5 × 22 cm. By overlapping and clipping the two ends of each strip, a 6-cm diameter tubular closure is formed (Figure 21.5). This paper closure can be used about 5–6 times (Bhumiratana *et al.* 1988; Lotong, 1992).

Types of seed koji. These plastic bag incubators are used to make two types of seed koji. One type, the primary seed koji, produces a spore powder stock which can in turn be used to produce the other type, the secondary seed koji, which is used to mix directly with the factory koji,

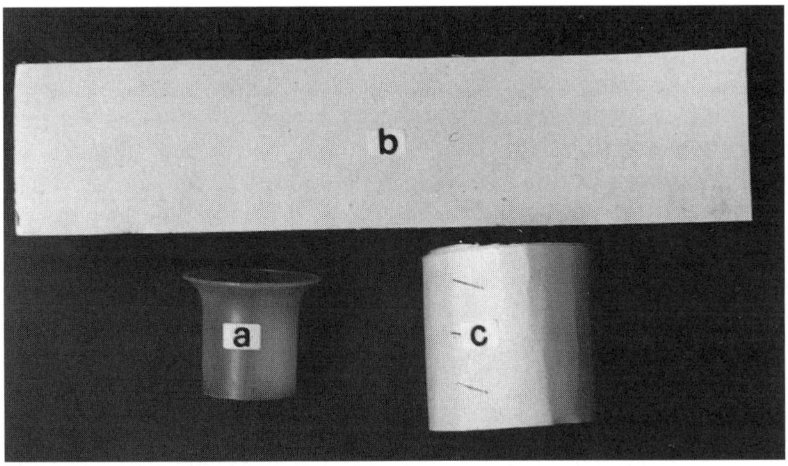

Figure 21.5 Closures for spore cultivation bags: (a) autoclavable plastic closure; (b) strip of paper or cardboard; (c) paper cardboard closure.

Table 21.2 Appropriate quantities of substrate and closure diameters for various bag sizes

Coarse rice meal (g)	Closure diameter (cm)	Bag size (cm)
30	3.5	10 × 20
100	6.0	22 × 36
400	10.0	49 × 60

These seed kojis differ in the source of their inoculum and moisture contents. The primary seed koji is grown directly from a pure culture, or from a factory stock culture. The secondary seed koji is grown as a factory subculture of the primary seed koji. The latter is prepared using substrate of lower moisture content than the former as shown in Table 21.3. The primary seed koji should also be incubated for a longer time (7–10 days). A loose dry spore mass so obtained is ready for use as the primary spore powder stock. Such stock bags can be kept for several months in the refrigerator, and used one at a time to inoculate 50 bags of secondary seed koji. Due to the high moisture content in secondary koji bag, the mould will grow much faster than it did in the primary koji bag. Maximum spore mass will be obtained within 4–5 days. The secondary seed koji is used to add directly to the factory koji in the process of soy sauce fermentation. In general practice, 100 g of seed inoculum is used to produce 100–500 kg. of factory koji.

Factory stock cultures. By following the scheme for preparation of primary seed koji from a pure stock vial, one may run out of the stock culture within a few months. At that time, a new batch of pure spores would then have to be requested from the original source. To avoid this continued necessity, a technique for producing factory stock cultures has been introduced following the procedure used for obtaining spore mass for distilling industries (Section 21.4.4), in which mould is grown on inside surface of bottles, only that smaller bottles are used. To each of 70-ml volume bottle, 7 g rice and 3 ml water is added and left for about 30 minutes. This allows the rice kernels to absorb the added water and swell. Each container is then swirled so that the rice kernels will spread to form a single layer over its inside surface and the bottles are sterilized. After cooling, the spore suspension is inoculated and the bottles are incubated at room temperature. Actually, spore production will reach its maximum within only one week. However, at that time, the moisture is still too high for storage. During further incubation, the culture eventually sufficiently dries. The stock culture so prepared (Figure 21.6) can be kept for at least one year in the refrigerator. This stock culture is used to produce primary seed koji as earlier described by inoculating a few grains to one bag.

Table 21.3 Proportion of substrate to water used for cultivation of seed koji

Type of seed koji	Incubation time (days)	Rice meal/water (g/ml)	Wheat bran/water (g/ml)
Primary seed	7–10	10:1.5	10/3
Secondary seed	4–5	10:2–3	10/10

Figure 21.6 Factory stock culture of soy sauce koji mould.

(b) Production of seed koji in some other South East Asian countries. In order to achieve the objective of introducing selected strains as seed inocula for soy sauce koji fermentation, a variety of methods for producing large quantities of spore mass has been developed. In Singapore, Tang (1978) prepared spore inoculum by growing the selected koji moulds or mutants thereof in sterilized koji materials, which consist of wheat bran, soy bean flour, wheat flour and water. The cultures are incubated for 4 days. Such grown cultures are supplied to local manufacturers, who had accepted that the use of pure inoculum would be an improvement over the ancient practice of relying upon organisms present in the environment. At the factories, the cultures are mixed with soybean and wheat flour and placed on shallow flat trays and incubated for 3–4 days in a cabinet. This culture is then used as a starter to inoculate a very much larger quantity of soybean–wheat flour mixture to produce koji.

Powder inoculum has been developed on an experimental scale in Indonesia. The preparation steps start by boiling 250 g of rice in 250 ml water. After draining off the excess water, the cooked rice is transferred to a 1-litre glass beaker, covered with a piece of paper and sterilized. The sterilized material is then transferred to a perforated aluminium tray. Spores of locally selected strains of *A. oryzae* are inoculated and incubated at 30°C for 6 days. The spore–rice mixture is then dried under the sun for one day after replacing the aluminium cover with paper. The semi-dry spore–rice mixture is subjected to further drying in an oven at 40°C for 6 days. The dried material is then blended and the powder sieved through a 48-mesh particle size sieve. The resulting powder contains approximately 7×10^8 spores per gram, with a germination rate of 43%. For use as a seed inoculum, an amount of 0.3% of the spore powder is recommended (Poesponegoro *et al.*, 1978).

Fresh inoculum obtained by a building-up process has also been produced in pilot scale studies in Indonesia. Starting from a pure culture in the test tube, spores are successively inoculated to 50 g of sterilized beans in an Erlenmeyer flask, and from the flask to 1 kg of beans in a perforated aluminium tray. The so-cultured beans are then used as seed inoculum for 200 kg of cooked beans for koji fermentation (Hartadi & Kabirun, 1980).

21.4.3 Dry starter cakes

In Asian countries several similar starter cakes composed of mixed cultures of filamentous fungi, comprising mostly the Mucorales and yeasts, have long been used as inocula in fermentation of starchy raw materials into sweet alcoholic foods (sweetened rice) or alcoholic beverages (rice wine). The technology concerning preparation of these inocula is believed to have originated in China (Yamazaki, 1932). The product is called by various names in Chinese such as chiu-yueh, peh-yueh (Wang & Hesseltine, 1970), chi-chu (Liu *et al.*, 1959) and known in the Western World as chinese yeast cake. Similar starter cakes produced in other Asian countries are called by various local names (Table 21.4).

(b) Raw materials and method of preparation. In the preparation of the starter cake, rice or wheat is ground and thoroughly mixed with certain spices. Water is added to make a dough-like material. The dough is then shaped into small balls (2–6 cm diameter) and placed on a wooden bamboo tray. Powder from an old starter is then sprinkled over the balls. After covering the tray with a cloth, it is incubated at room temperature for 2–5 days. During incubation, the dough is slightly raised, with fungal mycelium covering the whole surface. As selective inhibitors, the spices play their major role in preventing growth of undesirable microorganisms. After drying in the sun, these cakes can be kept for many months (Yamazaki, 1932; Stitnimagarn, 1950; Saono *et al.*, 1974). A photograph of loog-pang (Thai starter cake) is shown in Figure 21.7. Chinese yeast is normally produced in a large cake (25 cm diameter) and ground into coarse powder (Seng, 1983).

There has been a variation in the spices used in starter cake recipes. Most of the producers regard their recipes as trade secrets passed from generation to generation. However, in general, a mixture of garlic, pepper, rhizomes of *Alpinia galanga* and *Capsicum frutescens* is used in preparation of the Indonesian ragi (Saono *et al.*, 1974). Several combinations of spices

Table 21.4 Starter cakes produced in some Asian countries and their products

Local name	Country	Product	References
Chiu-yueh, peh-yueh	China	Sweetened rice,	Wang & Hesseltine (1970)
Chi-chu	China	Rice wine, spirit	Liu *et al.* (1959)
Pekka	Taiwan	Spirit, rice wine	Inui *et al.* (1965)
Pab (phap)	Tibet	Spirit, rice wine	Batra & Millner (1976)
Levain	Khasia, Sikkim	Spirit, rice wine	Batra & Millner (1976)
Bukhar, murcha	India	Rice wine	Batra & Millner (1976)
Ramur	India	Spirit	Pederson (1979)
Nuruk	Korea	Spirit	Mheen (1972)
Chi-chu	Korea	Spirit	Kozaki (1976)
Ragi tapeh	Indonesia	Sweetened rice	Ochse (1931)
Ragi beras	Indonesia	Rice wine	Djien (1979)
Ragi puyam	Indonesia	Sweetened cassava, Sweetened rice	Saono *et al.* (1974)
Ragi tempeh	Indonesia	Tempeh	Saono *et al.* (1974)
Ragi tapai	Malaysia	Rice wine	Djien (1979)
Jupiang	Malaysia	Rice wine	Djien (1979)
Bubod	Philippines	Fermented Toddy	Soriano *et al.* (1980)
Loog-pang kao-mag	Thailand	Sweetened rice	Krisanamaravisit (1951)
Loog-pang lao	Thailand	Spirit, rice wine	Stitnimagarn (1950)
Loog-pang namsom	Thailand	Vinegar	Mahintratep *et al.* (1987)
Banh men	Vietnam	Rice wine, Sweetened rice	Lan Dung (1969)

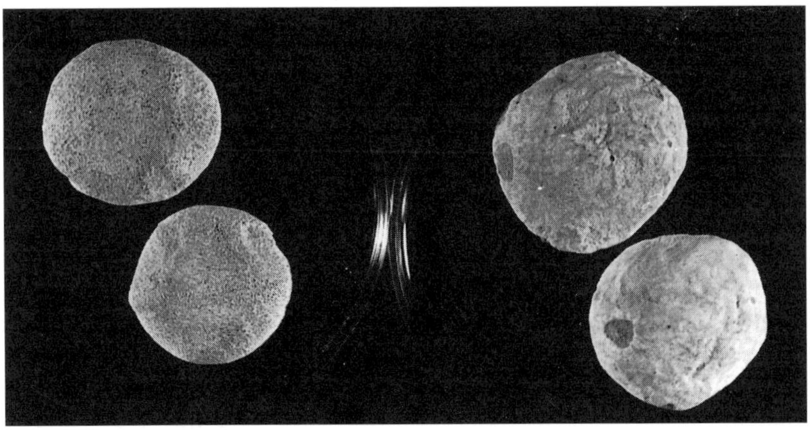

Figure 21.7 Loog-pang (Thai starter cakes): left: loog-pang kao mag; right: loog-pang lao.

have been formulated in loog-pang recipes. Examples of those commonly used are shown in Tables 21.5 and 21.6.

(b) Microflora. Numerous filamentous fungi and yeasts have been isolated from starter cakes from various sources. Among the moulds, *Amylomyces rouxii*, formerly known as *Chlamydomucor oryzae* (Ellis *et al.*,

Table 21.5 Ingredients used for loog-pang kaomag[a] preparation

Allium sativum (garlic)	420 g
Zingiber officinale (ginger)	120 g
Glycyrrhiza glabra (root)	180 g
Piper nigrum (pepper)	60 g
Piper chaba (flower)	120 g
Alpinia galanga (rhizome)	60 g
Rice	12 kg

[a] Starter cake used in fermentation of kaomag (sweetened rice).
From Krisanamaravisit (1951).

Table 21.6 Ingredients used for loog-pang lao[a] preparation

Allium sativum (garlic)	40 g
Zingiber officinale (ginger)	40 g
Alpinia sianensis (rhizome)	20 g
Myriopteron extensum (root)	40 g
Piper nigrum (pepper)	6 g
Piper chaba (flower)	6 g
Allium ascalonicum (red onion)	20 g
Rice	2.5 kg

[a] Starter cake used in fermentation of Thai rice wine.
From Stitnimagarn (1950).

1976), *Rhizopus* spp. *Mucor* spp., are those isolated from most of the samples. *Aspergillus oryzae*, *A. niger* and *Zygorrhynchus moelleri*, however, have been reported in some samples. A list of the moulds isolated from various starter cakes is shown in Table 21.7.

Hansenula malanga and *Endomycopsis fibuligera* are the yeasts common in many starter cakes. In addition, *Saccharomyces cerevisiae* has been found in samples of Thai loog-pang, whereas *Candida* spp. are those normally isolated from the Indonesian ragi (Table 21.8).

Lactobacilli and the coccoid lactic acid bacteria have been isolated from some samples of starter cakes. *Pediococcus pentosaceus* is the acid producer commonly found in ragi, loog-pang and bubod (Kozaki, 1976; Hadisepoetro *et al.*, 1979; Pakdisupapol & Kootin, 1981).

Among microorganisms isolated from various sources, only certain genera are found to play leading roles in the fermentation of sweetened rice and rice wine. These include *Amylomyces rouxii*, *Rhizopus* spp., *Endomycopsis* spp., *Hansenula* spp. and *Saccharomyces* spp. Based on abilities to produce high activities of α-amylase and/or amyloglucosidase, a number of strains of *Amy. rouxii*, *Rhizopus* spp. and *Endomycopsis* spp. were selected (Djien, 1972; Sukhumavasi *et al.*, 1975; Chatisatien, 1977; Cronk *et al.*, 1977; Chaownsungket, 1978; Kulapreecha, 1978; Pakdisupapol & Kootin, 1981). Furthermore, a number of investigations have been carried out to search for proper combinations of the selected moulds and

Table 21.7 Moulds isolated from various starter cakes

Starter cake	Moulds	References
Chinese yeast	*Rhizopus javanicus*	Liu *et al.* (1959)
	R. chinensis, *Amylomyces rouxii*	Wang & Hesseltine (1970)
Ragi	*Amy. rouxii*, *Mucor dubius*, *M. javanicus*, *R. oryzae*, *Aspergillus niger*	Djien (1965); Dwidjoseputro & Wolf (1970)
	R. stolonifer, *M. rouxii*	Saono *et al.* (1974)
	R. cohnii, *Zygorrhynchus moelleri*	Hadisepoetro *et al.* (1979)
	A. oryzae, *A. flavus*, *R. oligosporus*, *R. arrhizus*	Dwidjoseputro & Wolf (1970)
	Fusarium spp.	Saono *et al.* (1974)
Loog-pang	*Amy. rouxii*, *Mucor* spp., *Rhizopus* spp., *A. oryzae*	Chatisatien (1977); Chaownsungket (1978); Kulapreecha (1978); Pakdisupapol & Kootin (1981)
Murcha	*M. fragilis*, *M. rouxii*, *R. arrhizus*	Batra & Millner (1976)
Bubod	*Rhizopus* spp., *Mucor* spp.	Kozaki (1976)

Table 21.8 Yeasts isolated from various starter cakes

Starter cake	Yeasts	References
Chinese yeast	*Endomycopsis* spp.	Wang & Hesseltine (1970)
Ragi	*Torula indica*, *Hansenula anomala*, *Saccharomyces cerevisiae*	Djien (1965)
	Endomycopsis chodati, *E. fibuligera*	Djien (1972)
	H. subpelliculosa, *H. malanga*, *Candida guilliermondii*, *C. humicola*, *C. intermedia*, *C. japonica*, *C. pelliculosa*	Dwidjoseputro & Wolf (1970)
Loog-pang	*Endomycopsis fibuligera*	Sukhumavasi *et al.* (1975); Chatisatien (1977); Chaownsungket (1978); Pakdisupapol & Kootin (1981)
	Saccharomyces cerevisiae	Chaownsungket (1978)
Murcha	*H. anomala*	Batra & Millner (1976)
Bubod	*Endomycopsis* spp. *Saccharomyces* spp.	Kozaki (1976)

yeasts by which high-quality products are produced. Djien (1972) demonstrated that fermentation of glutinous rice by mixed cultures of *Amy. rouxii* and *E. fibuligera* gives good characteristic tape (Indonesian sweetened rice). The product prepared by using combinations of *Amy rouxii* and one of these yeasts: *E. fibuligera*, *C. lactosa*, *C. melinii*, *C. parapsilosis*, *H. anomala*, *H. malanga* or *H. subpelliculosa*, has been shown to be palatable, but lacked the desirable flavour associated with tape prepared by using

Amy. rouxii and *E. burtonii* (Cronk *et al.*, 1977). In fermentation of kaomag (Thai sweetened rice) by pure cultures, Pakdisupapol & Kootin (1981) found that kaomag produced by *Amy. rouxii* and either *E. fibuligera* or *H. anomala* is comparable in taste and aroma with that fermented by starter cake. Chaownsungket (1978) obtained quality sato (Thai rice wine) by using a combination of *Saccharomyces* spp. and either *Amy. rouxii* or the selected strains of *Rhizopus* spp. These findings would lead to an improvement in activities as well as to a more standard quality of the starter cakes, if applied on the commercial scale.

It is noteworthy that, although tane koji and starter cakes are similar in usage, they are distinctly different both in preparation procedures and microorganisms involved. Tane koji consists of *A. oryzae*, *A. soji* or *A. awamori* grown on cooked rice grains, whereas starter cakes consist of mixed cultures of moulds of the *Mucorales* and yeasts grown on raw rice meal. Moreover, it has been demonstrated that the latter contains in it a certain amount of saccharolytic enzymes (Pichayangura, 1983).

21.4.4 Spore mass produced on rice adhering to inside surfaces of flasks

This method of spore production is used in a variety of fermentation industries in Taiwan, Thailand and probably other Asian countries. The following describes the procedure in obtaining a spore mass of *Rhizopus* spp. used in distilling industries.

Ordinary rice or glutinous rice can be used as substrate. The rice is steeped for 3–6 hours and placed in flasks or bottles. The container is swirled so that rice kernels are spread over its inside surfaces. It is plugged with cotton and steamed. After cooling, spore suspension is inoculated and incubated for 4–5 days at either ambient temperature or 35°C. During incubation aerial mycelial growth occupies almost all the space in each flask and spores are produced abundantly.

21.5 Conclusion

The procedures described in this revew are typically of uncertain origin, probably ancient, and highly traditional in their methodology. Their importance in the Oriental production of foods remains considerable, despite the impact of Western practices and ideas. Improvements in methods, hygiene and microbiological control – notably the use of pure cultures and sterile procedures – are now taking place, and the pace of change may be expected to be maintained or even to increase, to the benefit of both producers and consumers. To this end, it is very desirable that technical information should be disseminated as widely as possible to manufacturers operating on all scales from the smallest to the largest.

References

Abdon, I.C. (1978) Soy sauce industries in the Philippines. *Report of the ASEAN Workshop on Soy Sauce Manufacturing Techniques*, 22–24 January, Singapore.

Abiose, S.H., Allan, M.C. & Wood, B.J.B. (1982) Microbiology and biochemistry of miso (soy paste) fermentation. *Advances in Applied Microbiology*, **28**, 239–55.

Aeaungkoon, S. (1995) Improvement of soy sauce brine fermentation. *MSC Thesis, Kasetsart University*, Bangkok, Thailand.

Aidoo, K.E., Hendry, R. & Wood, B.J.B. (1982) Solid substrate fermentations. *Advances in Applied Microbiology*, **28**, 201–37.

Akita, O. (1994) Control of yeast in sake making. *Proceedings of the International Workshop on Application and Control of Microorganisms in Asia*, 14–18 March, Tokyo, Japan.

Aykroyd, W.R. & Doughty, J. (1964) Legumes in human nutrition, *FAO Nutritional Studies No.* 19, Rome, Italy.

Baens-Arcega, L. (1966) Feasibility of mould process production of sauce from copra and soybean mixture with authentic Philippine strain of *Aspergillus oryzae* (Cehburg) Coh. *Araneta Journal of Agriculture*, **12**, 80–103.

Batra, L.R. & Millner, P.D. (1976) Asian fermented foods and beverages. *Developments in Industrial Microbiology*, **17**, 117–28.

Bhumiratana, A., Flegel, T.W., Glinsukon, T. & Somporan, W. (1980) Isolation and analysis of moulds from soy sauce koji in Thailand. *Applied and Environmental Microbiology*, **39**, 430–5.

Bhumiratana, A., Flegel, T., Lotong, N. & Suwanarit, P. (1988) *A Manual of Soy Sauce Production for Small Scale Manufacturers in the ASEAN Countries*. Text and Journal Corp. Co. Ltd., Bangkok, Thailand.

Boonprab, K., Suwanarit, P. & Lotong, N. (1996) Selection of glucose derepression mutants for the improvement of ang-kak production and their regulation of pigmentation. *Proceeding of Kasetsart University Annual Conference*, 30 January–1 February, Bangkok Thailand.

Bovornsombat, S. (1981) Feasibility study on soy sauce production by using winged bean seeds. *MSc thesis, Kasetsart University*, Bangkok, Thailand.

Cannel, E. & Moo-Young, M. (1980) Solid state fermentation systems. *Process Biochemistry*, **15**, 2–7.

Chansiripotha, S. (1988) Improvement of Monascus pigment by mutation. *MSc Thesis, Kasetsart University*, Bangkok, Thailand.

Chaownsungket, M. (1978) Selection of yeast and mould strains for rice wine production. *MSc Thesis, Kasetsart University*, Bangkok, Thailand.

Chatisatien, C. (1977) Selection of mould and yeast strains in loog-pang for kaomag fermentation. *MSc Thesis, Kasetsart University*, Bangkok, Thailand.

Chen, Y.F. (1980) Solid substrate fermentation in Malaysia. *Report of the Second ASEAN Workshop on Solid Substrate Fermentation*, 27–29 November, Kuala Lumpur, Malaysia.

Chou, C.C. & Rwan, J.H. (1995) Mycelial propagation and enzyme production in koji prepared with *Aspergillus oryzae* on various rich extruded and steamed rice. *Journal of Fermentation and Bioengineering*, **79**(5), 509–12.

Church, M.B. (1920) Laboratory experiment on the manufacture of chinese angkak in the United States. *Journal of Industrial Chemistry*, **12**, 45–6.

Crisostomo, L.C., Corpuz, L. & Cells, F.F. (1974) A comparative study on the utilization of mecan pea for soy sauce preparation. Handout copy.

Cronk, T.C., Steinkraus, K.H., Hackler, L.R. & Mattick, L.R. (1977) Indonesian tape ketan fermentation. *Applied and Environmental Microbiology*, **33**, 1067–73.

Deangsubha, W. (1979) Solid substrate fermentation in Thailand. I. Fish fermentation. *Report of the First ASEAN Workshop on Solid Substrate Fermentation*, 8–13 May, Bandung, Indonesia.

Desgranges, C., Vergoignan, C., Georges, M. & Durand, A. (1991) Biomass estimation in solid state fermentation. *Applied Microbiology and Biotechnology*, **35**, 200–5.

Devahudee, P. (1978) Study on the yeast in soy sauce and soybean paste. *MSc thesis, Kasetsart University*, Bangkok, Thailand.

Djien, K.S. (1965) *Research di Indonesia* (ed. R.M. Soemantri), pp. 1945–65. Cited in Saono *et al.* (1974).

Djien, K.S. (1972) Tape fermentation. *Applied Microbiology*, **23**, 976–8.

Djien, K.S. (1979) Some aspects concerning the microbiological safety of traditional fermented foods in Tropical Asia. *Proceedings of International Symposium on Microbiological Aspects of Food Storage, Processing and Fermentation in Tropical Asia*, 10–13 December, Bogor, Indonesia.

Dorta, B., Bosch, A., Arcas, J. & Ertola, R. (1994) Water balance in solid-state fermentation without forced aeration. *Enzyme* and *Microbial Technology*, **16**(7), 562–5.

Dwidjoseputro, D. & Wolf, E.T. (1970) Microbiological studies of Indonesian fermented foodstuffs. *Mycopath. Mycology Applied*, **41**, 211–22.

Ebine, H. (1986) Miso, in *Legume-Based Fermented Foods* (eds N.R. Reddy, M.D. Pierson & D.K. Salunkhe), CRC Press, Inc., Florida, pp. 47–68.

Ellis, J.J., Rhodes, L.J. & Hesseltine, C.W. (1976) The genus *Amylomyces Mycologia*, **68**, 131–43.

Flegel, T.W., Bhumiratana, A. & Srisutipruti, A. (1981a) Problematic occurrence of tyrosine crystal in the Thai soybean paste tao chieo. *Applied and Environmental Microbiology*, **41**, 746–51.

Flegel, T.W., Bhumiratana, A., Impoolsub, A., Takada, N., Oshima, Y. & Okada, H. (1981b) Studies on *Aspergillus flavus* var. *columnaris* in soy sauce koji, *Proceedings of the JSPS-NRCT Seminar on Microbial Utilization of Renewable Resources*, 5–8 January Songkhla, Thailand.

Flegel, T.W. (1988) Yellow-green *Aspergillus* strains used in Asian soybean fermentation. *ASEAN Food Journal*, **4**(1), 14–30.

Fukushima, D. (1978) Soy sauce manufacturing techniques. *Report of the ASEAN Workshop on Soy Manufacturing Techniques*, 22–24 January, Singapore.

Hadisepoetro, E.S., Takada, N. & Oshima, Y. (1979) Microflora in ragi and usar. *Journal of Fermentation Technology*, **57**, 241–59.

Hartadi, S. & Jutono (1979) Case study: taoco. *Report of the First ASEAN Workshop on Solid Substrate Fermentation*, 8–13 May, Bandung, Indonesia.

Hartadi, S. & Kabirun, S. (1980) Soy sauce inoculum preparation. *Progress Report of ASEAN Project.*

Hasegawa, T. (1979) An historical view of the development of Japanese industrial culture collections. *Proceedings of the International Conference on Global Impacts of Applied Microbiology*, 21–26 November, Bangkok, Thailand.

Hayakawa, K., Ueno, Y., Nakanishi, S., Honda, Y., Komuro, H., Kikushima, S. & Shou, S. (1993) Production of fish sauce from fish meal treated with koji mold. *Seibutsu Kogaku Kaishi*; *Journal of Fermentation and Bioengineering*, **72**(4), 245–51.

Hayashida, S. (1974) Function of the high concentration alcohol producing factor. *Agricultural* and *Biological Chemistry*, **38**, 200–6.

Hayashida, S., Feng, D.D. & Ohta, K. (1976) Composition and role of *Aspergillus oryzae* proteolipid as a high concentration alcohol producing factor. *Agricultural and Biological Chemistry*, **40**, 73–8.

Hayashida, S. & Ohta, K. (1978) Cell structure of yeast grown anaerobically in *Aspergillus oryzae* proteolipid – supplemented media. *Agricultural and Biological Chemistry*, **42**, 1139–45.

Hesseltine, C.W. (1965) A millenium of fungi, food and fermentation. *Mycologia*, **57**, 1–148.

Hesseltine, C.W. (1972) Biotechnology report: solid-substrate fermentation. *Biotechnology and Bioengineering*, **14**, 517–32.

Hesseltine, C.W. (1977) Solid substrate fermentation. *Process Biochemistry*, **12**, 24–7.

Hesseltine, C.W. (1979) Microorganisms involved in food fermentation in Tropical Asia. *Proceedings of the International Symposium on Microbiological Aspects of Food Storage Processing and Fermentation in Tropical Asia*, 10–13 December, Bogor, Indonesia.

Hesseltine, C.W., Swain, E.W. & Wang, H.L. (1976) Production of fungal spores as inocula for oriental fermented foods. *Developmants in Industrial Microbiology*, **17**, 101–15.

Hiroi, T. & Shima, A. (1975) Studies on structures of two pigments obtained from *Monascus* sp. *Journal of Japanese Society of Food and Nutrition*, **28**, 497.

Honda, H., Okkusu, E., Hanai, T., Nishida, Y. & Fukay, I. (1995) Estimation of cell concentration on rice koji by image-analysis. *Seibutsu Kogaku Kaishi Journal of Fermentation and Bioengineering*, **73**(5), 409–12.

Iizuka, H. & Mineki, S. (1977) Studies on the genus *Monascus*. I. Purification and properties of two forms of Glucoamylase from *Monascus kaoliang* nov. sp. F-1, *Journal of General and Applied Microbiology*, **23**, 217–30.
Iizuka, H. & Meneki, S. (1978) Studies on the genus *Monascus*. II. Substrate specificity of two glucoamylases obtained from *Monascus kaoliang* F-1. *Journal of General and Applied Microbiology*, **24**, 185–92.
Impoolsub, A., Bhumiratana, A. & Flegel, T.W. (1981) Isolation of alkaline and neutral protease from *Aspergillus flavus* var. *columnaris*, a soy sauce koji mould. *Applied and Environmental Microbiology*, **42**, 619–28.
Inui, T., Takeda, Y. & Iizuka, H. (1965) Taxonomical studies on genus *Rhizopus*. *Journal of General and Applied Microbiology*, **2**, 1–12.
Ishihasi (1978) Japanese patent, 53–17678.
Ismail, M.M. & Karim, A. (1980) The role of microorganisms in food fermentation with special reference to Malaysian fermented foods. *Report of the Second ASEAN Workshop on Solid Substrate Fermentation*, 27–29 November, Kuala Lumpur, Malaysia.
Kageyama, K. & Sugita, O. (1955) *Journal of Fermentation Technology*, **33**, 109–13. Cited in Kodama (1970).
Kalayanamitr, A., Bhumiratana, A., Flegel, T.W., Glinsukon, T. & Shinmyo, A. (1987) Occurrence of toxicity among protease, amylase, and color mutants of a nontoxic soy sauce koji mold. *Applied and Environmental Microbiology*, **53**(8), 1980–2.
Kamnuanta, J. (1981) Sato and Ou. *Report of Workshop on Development in Spirits and Ethanol Production*, 29–30 August, Kasetsart University, Kampangsan, Thailand.
Kibi, M. (1926) *J. agric. Chem. Soc. Jap.*, **12**, 885. Cited in Yong & Wood (1974).
Kikuchi, T. (1976) Changes in soybean cell wall polysaccharide during soy sauce fermentation (Food chemical studies on soybean polysaccharide, part IV). *Journal of the Agricultural Chemistry Society of Japan*, **50**, 29–33.
Kikuchi, T. & Yokotsuka, T. (1973) Acid and enzymatic hydrolysis of an acidic polysaccharide from soy sauce. *Agricultural and Biological Chemistry*, **37**, 973–9.
Kikuchi, K., Ohta, T. & Ebine, H. (1975) Changes in lipid components of misodama koji. *Journal of Fermentation Technology*, **53**, 869–75.
Kiku-Masamune Saké brewing Co Ltd. Brochure.
Klomklang, W. (1995) *Study on fermentation of fish sauce by using halophilic bacteria with koji*, MSc Thesis Kasetsart University, Bangkok, Thailand.
Kodama, K. (1970) Saké yeast, in *The Yeasts*, Vol. 3 (eds A.H. Rose & J.S. Harison), Academic Press, London, pp. 225–82.
Kodama, K. & Yoshizawa, K. (1977) Saké, in *Economic Microbiology*, Vol. 1, *Alcoholic Beverages* (ed. A.H. Rose), Academic Press, London, pp. 423–75.
Korchelf, O. (1876) *Mitt Dtsch. Ges. Natur.-u. Völkerkunde Ostasiens*, **16**, 240–58. Cited in Hasegawa (1979).
Kozaki, M. (1976) Fermented foods and related microorganisms in South East Asia. *Proceedings of the Japanese Association of Mycotoxicology* **2**, 1–9.
Krisanamaravisit (1951) Kaomag. *Samitsan*, **7**, 75–9.
Kulapreecha, S. (1978) *Effects of culture media on sporulation of Rhizopus* spp. *isolated from loog-pang*, MSc Thesis, Chulalongkorn University, Bangkok, Thailand.
Kundu, A.K. & Manna, S. (1975) Purification and characterization of extracellular proteases of *Aspergillus oryzae. Applied Microbiology*, **30**, 507–13.
Kunimoto, M., Terao, K. and Kaneniwa, M. (1992) Consumption of trimethylamine and trimethylamine N-Oxide by *Aspergillus oryzae. Nippon Suisan Gakkaishi; Bul. Japanese Soc, Sci. Fisheries*, **58**(8), 1471–6.
Lan Dung, N. (1969) Isolation and application of yeasts for livestock-raising in Vietnam. *Agricultural Science and Technology*, **10**, 663–8.
Lan, Y.Q., Merican, S., Aziz, A. & Mohamed, R. (1978) Survey of some soy sauce factories in Peninsular Malaysia. *Report of ASEAN Workshop on Soy Sauce Manufacturing Techniques*, 22–24 January, Singapore.
Lee, S.T. (1979) The Anka, in *Pen cha kan mu* (Chinese herbal medicine). Yeh Book Co., Taipei, pp. 1518–93.
Lee, J.M., Kim, Y.S., Hong, Y.M. & Yu, J. (1972) Studies on substitution of raw material for soy sauce, Part III. Use of cornand barley. *Korean Journal of Food Science and Technology*, **4**, 182–6.

Lin, C.F. (1973) Isolation and culture conditions of *Monascus* sp. for the production of pigment in submerged culture. *Journal of Fermentation Technology*, **51**, 407–14.
Lin, C.F. & Iizuka, H. (1975) Abstract of the Annual Meeting of the Agricultural Chemistry Society of Japan. Cited in Lin & Iizuka (1982).
Lin, C.F. & Iizuka, H. (1982) Production of extracellular pigment by a mutant of *Monascus kaoliang* sp. nov. *Applied and Environmental Microbiology*, **43**, 671–6.
Liu, P.W., Chen, S.H., Chin, T.H. & Chen, C.C. (1959) Taxonomical studies of saccharifying mould from chiu-chu of shao-shing wine, via *Chemical Abstracts*, **53**, 22722.
Lotong, N. (1978) Soy sauce production in Thailand. *Report of the ASEAN Workshop on Soy Sauce Manufacturing Techniques*, 22–24 January, Singapore.
Lotong, N. & Suwanarit, P. (1983) Production of spore inoculum of soy sauce koji moulds in plastic bags. *Applied and Environmental Microbiology*, **46**, 1224–6.
Lotong, N. & Suwanarit, P. (1990) Fermentation of ang-kak in plastic bags and regulation of pigmentation by initial moisture content. *Journal of Applied Bacteriology*, **68**, 565–70.
Lotong, N. (1992) *Seed Inocula and Their Production Technology.* (in Thai), Funny Publishing Co., Bangkok.
Lotong, N. (1994) Food fermentation and related industries in Thailand. *Proceedings of the International Workshop on Application and Control of Microorganisms in Asia*, 14–18 March, Tokyo, Japan.
Mahintratep, B., Laopaiboon, R. & Lotong, N. (1987) Production of dry inoculum for vinegar fermentation. *Kasetsart Journal*, **21**, 278–88.
Manchand, P.S., Whalley, W.B. & Chem, F.C. (1973) Isolation and structure of anka-flavin, a new pigment from *Monascus anka. Phytochemistry*, **12**, 2513–32.
Maramatsu, S. (1994) The new method of short-term shoyu (Soy sauce) making. *Proceedings of the International Workshop on Application and Control of Microorganisms in Asia*, 14–18 March, Tokyo, Japan.
Matsushima, K. (1958) *Journal of Agricultural and Chemical Society of Japan*, **32**, 215. Cited in Impoolsub *et al.* (1981).
Mheen, T.I. (1972) Korean fermented foods. *Selected paper from the UNESCO Work Study on Waste Recovery by Microorganisms*, 1–8 May, University of Malaya, Kuala Lumpur, Malaysia.
Misaki, T., Yamada, M., Okazaki, T. & Sawada, J. (1970) Studies on the protease constitution of *Aspergillus oryzae*: I. Systemic separation and purification of protease. *Agricultural and Biological Chemistry*, **34**, 1383–92.
Morimota, Y., Kitamoto, K., Fujita, Y., Gomi, K. & Kumagai, C. (1995) Activity of alpha-glucosidase in sake koji and its role during sake brewing. *Seibutsu-Kogaku Kaishi*; *Journal of Fermentation and Bioengineering*, **73**(2), 97–104.
Nakadai T., Nasuno, S. & Iguchi, N. (1972a) The action of peptidase from *Aspergillus oryzae* digestion of soybean protein. *Agricultural and Biological Chemistry*, **36**, 261–8.
Nakadai, T., Nasuno, S. & Iguchi, N. (1972b) The action of peptidase from *Aspergillus sojae* on soybean protein. *Agricultural and Biological Chemistry*, **36**, 1239–43.
Nakadai, T., Nasuno, S. & Iguchi, N. (1973a) Purification and properties of alkaline protease from *Aspergillus oryzae. Agricultural and Biological Chemistry*, **36**, 2685–94.
Nakadai, T., Nasuno, S. & Iguchi, N. (1973b) Purification and properties of neutral protease I from *Aspergillus oryzae. Agricultural and Biological Chemistry*, **37**, 2695–701.
Nakadai, T., Nasuno, S. & Iguchi, N. (1973c) Purification and properties of neutral protease II from *Aspergillus oryzae. Agricultural and Biological Chemistry*, **37**, 2703–8.
Nakadai, T., Nasuno, S. & Iguchi, N. (1973d) Purification and properties of leucine amino peptidase I from *Aspergillus oryzae. Agricultural and Biological Chemistry*, **37**, 757–65.
Nakadai, T., Nasuno, S. & Iguchi, N. (1973e) Purification and properties of leucine amino peptidase II from *Aspergillus oryzae. Agricultural and Biological Chemistry*, **37**, 767–74.
Nakadai, T., Nasuno, S. & Iguchi, N. (1973f) Purification and properties of leucine amino peptidase III from *Aspergillus oryzae. Agricultural and Biological Chemistry*, **37**, 775–82.
Nakadai, T., Nasuno, S. & Iguchi, N. (1973g) Purification and properties of acid carboxypeptidase IV from *Aspergillus oryzae. Agricultural and Biological Chemistry*, **37**, 1237–51.

Nakadai, T. & Nasuno, S. (1976) Enzymatic hydrolysis of protein by various enzyme preparations. *Journal of Fermentation Technology*, **54**, 872–84.
Nakadai, T. & Nasuno, S. (1977) Hyperproduction of peptidase by mutant of *Aspergillus oryzae. Journal of Fermentation Technology*, **55**, 273–6.
Nakadai, T. (1985) The roles of enzymes produced by shoyu koji molds. *Journal of Japan Soy Sauce Research Institute*, **11**(2), 37–69.
Nakano, M. (1972) Synopsis on the Japanese traditional fermented foodstuffs. *Selected papers from UNESCO/ICRO Work Study on Waste Recovery by Microorganisms*, 1–8 May, University of Malaysia, Kuala Lumpur, Malaysia.
Neuville, H. (1902) *Les Ferments Industriels d'Extreme-Orient*, Masson, Paris, 192pp. Cited in Batra & Millner (1976).
Nga, B.H., Tang, T.C. & Lim, C. (1978) Genetics of microorganisms in soy sauce fermentation. *Report of the ASEAN Workshop on Soy Sauce Manufacturing Techniques*, 22–24 January, Singapore.
Nishikawa, E. (1932) Studies on the biochemistry of mould. I: The pigments of *Monascus purpureus Went. Journal of Agricultural and Chemical Society of Japan*, **8**, 1007–14.
Ochse, J.J. (1931) Vegetables of the Dutch East Indies. *Archipel Drukkerij, Buitenzorg.* Cited in Saono *et al.* (1974).
Pakdisupapol, S. & Kootin, S. (1981) Fermentation of sweetened rice by pure culture, in *Microbial Utilization of Renewable Resources, Vol. 2, JSPS-NRCT Seminar on Agro-Industry Including Microbial Technology*, 5–8 January, Songkla, Thailand.
Palo, M.A., Vidal-Adeva, L. & Maceda, L. (1961) A study on ang-kak and its production. *Philippines Journal of Science*, **89**, 1–22.
Pandy, A. (1992) Recent proces developments in solid-state fermentation. *Process Biochemistry*, **27**, 109–17.
Park, J.H., Kung, S.N., Yun, U.S. & Shim, W.M. (1979) Studies on the fermentation of fish sauce by koji. *Chugang Uihak*, **37**(2), 93–101.
Pederson, P.S. (1979) *Microbiology of Food Fermentation*, 2nd edn, AVI Publishing Company, Westport Conn., USA.
Pichayangura, S. (1983) Existence of saccharolytic enzymes in loog-pang. Personal communication.
Poesponegoro, M., Tanuwidjja, L. & Rosstamsjah (1978) Studies on the fermentation process of soy sauce. *Report of the ASEAN Workshop on Soy Sauce Manufacturing Techniques*, 22–24 January, Singapore.
Prescott, S.C. & Dunn, C.G. (1959) *Industrial Microbiology*, 3rd end, McGraw-Hill, New York, p. 836.
Saono, S., Ganjar, I., Basuk, T. & Karsono, H. (1974) Mycoflora of ragi and some other traditional fermented foods of Indonesia. *Annales Bogorienses*, **5**, 187–204.
Seng, C.C. (1983) Brewing of spirit from cereal. *Alcohol Fermentation Society of Thailand*, **2**, 51–8.
Shibasaki, K. & Hesseltine, C.W. (1962) Miso fermentation. *Econ. Bot*, **16**, 180–95.
Shibuya, I., Tamura, G., Goto, E., Ishikawa, T. & Hara, S. (1992) Characteristic of koji prepared from the transformant of *Aspergillus oryzae* with glucoamylase gene *of A. shirousamii*, and its utilization for sake brewing. *Journal Fermentation and Bioengineering*, **73**(5), 415–17.
Sooksan, R. & Gongsakdi, S. (1977) Thai red rice anka. *Symposium on Indigenous Fermented Foods*, Bangkok, Thailand.
Sooksan, R. & Yoshida, T. (1987) Solid state culture of *Monascus purpureus* for the pigment production. *Annual Report of ICBiotech*, **10**, 357–60.
Soriano, M.R., Gonzalez, O. & Avelino, E. (1967) Studies on the preparation of soy sauce from coconut paring meal. *Philippines Journal of Science*, **96**, 129–37.
Soriano, M.R., Navarro, N.S. & Parel, S.O. (1980) Solid substrate fermentation technology in the Philippines. *Report of the Second ASEAN Workshop on Solid Substrate Fermentation*, 27–29 November, Kuala Lumpur, Malaysia.
Soriano, M.R. & Pardo, D.P. (1978) Studies on the improvement of soy sauce manufacture in the Philippines. *Report of the ASEAN Workshop on Soy Sauce Manufacturing Techniques*, 22–24 January, Singapore.

Steinkraus, K.H., Cullen, R.R., Pederson, C.S., Nellis, F.L. & Gavitt, B.K. (1983) *Handbook of Indigenous Fermented Foods*, Marcel Dekker, New York, pp. 547–71.

Stitnimagarn, W. (1950) Rice wine starter cakes. *Science. N.Y.*, 383–7.

Su, Y.C. (1966) Sufu, in *Legume-Based Fermented Foods* (eds N.R. Reddy, M.D. Pierson & D.K. Salunkhe), CRC Press, Inc., Florida, pp. 69–84.

Su, Y.C. & Wang, W.H. (1977) Chinese red rice-anka. *Symposium on Indigenous Fermented Foods*, Bangkok, Thailand.

Su, Y.C. (1980) Traditional fermented foods in Taiwan. *Proceedings of the Conference on Oriental Fermented Foods.* 10–14 December, Taipei, Taiwan, Republic of China.

Sukhumavasi, J., Kato, K. & Harada, T. (1975) Glucoamylase of a strain of *Endomycopsis fibuligera* isolated from Mould Bran (loog-pang) of Thailand. *Journal of Fermentation Technology*, **53**, 559–65.

Suwanmena, K., Suwanarit, P. & Lotong, N. (1978) A study of koji mould in soy sauce fermentation. *6th Annual Conference of Agricultural and Biological Science*, 2–5 February, Kasetsart University, Bangkok, Thailand.

Suzuki, M., Yoneyama, H. & Koizumi, T. (1980) The fractionation of some precursors of saké-aroma substances from rice koji. *Hakkokogaku*, **58**, 377–83.

Suzuki, K., Yoneyama, H. & Koizumi, T. (1982) The mechanism of formation of ethyl leucinate, an important aroma component in saké. *Hakkokogaku*, **60**, 19–25.

Takamine, J. (1914) Enzymes of *Aspergillus oryzae* and the application of its amyloclastic enzyme to the fermentation industry. *Ind. Engng Chem.*, **6**, 824–8.

Tang, T.C. (1978) Development in soy sauce fermentation. *Report of the ASEAN Workshop on Soy Sauce Manufacturing Techniques*, 22–24 January, Singapore.

Tawatpun, K. (1983) *Solid state cultivation of Rhizopus oryzae for glucoamylase production.* MSc Thesis, Chulalongkorn University, Bangkok, Thailand.

Ten, S.K. & Ho, C.C. (1980) Characterization of *Aspergillus* from the koji phase of soy sauce fermentation in the ASEAN region. *Report of the Second ASEAN Workshop on Solid Substrate Fermentation*, 27–29 November, Kuala Lumpur, Malaysia.

Thom, C. & Church, M. (1926) *The Aspergilli*, Williams and Wilkins, Baltimore.

Tochikura, T., Ito, M., Yano, T., Yamamoto, K., Tachiki, T. & Kumagai, H. (1983) Enzymes of soysauce koji and their application to food processes. *Microbial Utilization of Renewable Resources, Vol. 3, JSPS-NRCT Seminar on Agro-Industry Including Microbial Technology*, 22–26 March, Osaka, Japan.

Togano, T., Nakamura, M. & Sanchez, P.C. (1978) Fish sauce in S.E. Asia. *5th International Congress in Food Science and Technology*, Kyoto, Japan.

Tsai, M.S., Hseu, T.H. & Shen, Y.S. (1978) Purification and characterization of an acid protease from *Monascus kaoliang. Int. J. Pept. Protein Res.*, **12**, 293–302.

Ukei, T., Noda, Y., Teramoto, Y., Ohba, R. & Ueda, S. (1994) Practical soy-sauce production using a mixed koji-making system. *Journal of Fermentation and Bioengineering*, **78**(3), 262–4.

Ushijima, S., Nakadai, T. & land Uchida, K. (1991) Interspecific electrofusion of protoplasts between *Aspergillus oryzae* and *Aspergillus sojae*. *Agricultural and Biological Chemistry*, **55**(1), 129–36.

Vo, H.N., Ngo, K.S., Trinh, T.H. (1993) Filamentous fungi in rice alcohol production. *Journal of Science and Technology*, **2**, 36–40.

Wakai, Y., Mizuma, T., Miyazaki, N., Kiyokawa, Y. & Yanagiuchi, T. (1995) Dynamic viscoelasticity of steamed rice for sake brewing. *Seibutsu-Kokagu Kaishi*; *Journal of Society of Fermentation and Bioengineering*, **73**(3), 191–7.

Wang, H.L. & Hesseltine, C.W. (1970) Sufu and lao-chao. *Journal of Agriculture and Food Chemistry*, **18**, 572–5.

Windholz, M., Budavari, S., Stroumtsos, L.Y. & Fertig, M.N. (1976) *The Merck Index. An Encyclopedia of Chemicals and Drugs*, 9th end. Merck Co. Inc.

Yamamoto, Y., Yanagida, F. & Suminoe, K. (1972) Changes of enzyme activities during preparation of Haku-Shoun-koji. *Journal of Food Science and Technology* (*Nihon Shokuhin Kogyo Gakkai-shi*), **20**, 109–12.

Yamamoto, S. & Hirooka, H. (1974a) Production of glutaminase by *Aspergillus sojae*. *Journal of Fermentation Technology*, **52**, 564–9.

Yamamoto, S. & Hirooka, H. (1974b) Partial purification and properties of glutaminase from *Aspergillus sojae*. *Journal of Fermentation Technology*, **52**, 570–6.

Yamazaki, M. (1932) Studies on the Chinese yeast cakes 'Ch'u'. Some characteristics of Chinese yeast cakes. *Bull. Utsunomiya Agr. College*, **1**, 1–24.
Yokotsuka, T. (1960) Aroma and flavor of Japanese soy sauce. *Advances in Food Research*, **10**, 75–134.
Yokotsuka, T. (1964) *International Symposium on Oil Seed Protein Foods*, pp. 31–48. Cited in Yong & Wood (1974).
Yokotsuka, T. (1985) Fermented protein foods in the Orient, with emphasis on shoyu and miso in Japan, in *Microbiology of Fermented Foods. Vol. 1* (ed. B.J.B. Wood), Elsevier Applied Science Publishers, London, pp. 197–247.
Yong, F.M. & Wood, B.J.B. (1974) Microbiology and biochemistry of soy sauce fermentation. *Advances in Applied Microbiology*, **17**, 157–94.
Yong, F.M. & Wood, B.J.B. (1975) Sucrase from soy sauce moulds. *Transactions of the British Mycological Society*, **64**, 143–73.
Yongsmith, B., Tabloka, W., Yongmanitchai, W. & Bavavoda, R. (1993) Culture conditions for yellow pigment formation by *Monascus* sp. KB 10 grown on cassava medium. *World Journal of Microbiology*, **9**, 85–90.
Yoshida, T. & Taguchi, H. (1979) Fermented foods in Japan. *Proceedings of the International Symposium on Microbiological Aspects of Food Storage, Processing and Fermentation in Tropical Asia*, 10–13 December, Bogor, Indonesia.
Yuan, G.F. (1994) Isolation and identification of microorganisms using for fermented soy cheese (sufu) in Taiwan. *Proceedings of the International Workshop* on *Application* and *Control* of Microorganisms in Asia, 14–18 March, Tokyo, Japan.
Yue, X. (1990) Advances in the soy sauce industry in China. *Journal of Fermentation and Bioengineering*, **70**(6), 434–9.
Yusof, R.M., Baker, T.A., Morgan, J.B. & Adams, M.R. (1995) Effect of ragi and L-lactate-producing cultures on enteric pathogens in a rice-based weaning food. *World Journal of Microbiology and Biotechnology*, **11**, 654–7.
Ziemba, J. (1974) Authentic soy sauce brewed in Wisconsin. *Food Engnieering*, **46**, 58–60.

22 Food fermentation in the tropics

W.R. STANTON

22.1 The decade of the mycotoxins

In the history of the study of the microbe in relation to man's requirement for a regular supply of food, the 1960s may be described as the decade of the mycotoxins. The threat that the microbe presented to the developers, in terms of their evolving new arts in food fermentation, imposed a severe constraint on this aspect of food development, from the stultifying effects of which on research the industry has only recently begun to recover. The name *Aspergillus* is no longer unmentionable in relation to inclusion in the repertoire of food fungi. Indeed, in view of the steady increase of use of the genus for enzyme preparation, crude preparations as well as high-purity ones, its exclusion would now be hard to justify.

In the decade since the publication of the first edition of this book there have been many new developments which are mentioned in the appropriate chapters. From the point of view of this particular essay, *Food fermentation in the tropics*, the changes in the subject have not been so much in the use of new microbes, or substrates, but in the influence of the ideas presented in the first edition on the adoption of food fermentation practices in other parts of the world. The type of process refinement, quality control of the inoculum and standardization of product, all of which have occurred in the wine and cheese industries, for examples, have yet to have much impact on the process of manufacture of the traditional foods discussed in this chapter.

The period of scare-mongering, engendered by the identification of and study of the mycotoxins, seems to have been largely dissipated, and scares which beset the food industry have been generally applied to other aspects of the food industry such as additives and the rise of new microbial contaminants such as *Campylobacter*.

The dangers associated with *Listeria* in cheese manufacture have been loudly and often unjustifiably voiced, as was predicted many years ago, (Stanton & Da Silva, 1977). The food preparations which accepted that humans had to live with the microbial world have been largely unaffected by the furore over *Salmonella*, for example.

However, this braking effect on research had a particularly strong impact on the development of tropical fermented foods, though, had the same type of sensitivity to the danger to humans (of using microbes in food preparation) emerged a century earlier, the braking effect might have had influence on the development of cheese and other temperate zone food fermentations, so familiar to the community as to go unnoticed that they were also food fermentations in every sense of the word.

In the same way the 'millenia of food fermentation' (to quote Hesseltine's classical (1965) remark) have been part of the tropical way of life, just as much even if not more than they have in the temperate zone. The sophisticated techniques have been developed and practised by civilizations, the populations of which could at their peaks be measured in hundreds of millions. It is this vast archive of local foodcraft which is being currently unpacked and its content deciphered, not only to exploit the wealth of techniques for the food industry but also to derive inspiration from the basic concepts, elegant and superficially simple, which humanity has evolved to safeguard its food supply and extend the range of materials which might be exploited as sources of food.

22.2 Antiquity of food processes involving fermentation

To gain an impression of the antiquity of these processes and a feeling for how humans set about making the best use of resources, one should contemplate the Mayan artefact or pace the terraces of Borobodur, in Java (Figure 22.1). studying the bas-reliefs. Such contemplation helps dispel the textbook attitude to food in antiquity and the pre-eminence of Mohenjo-Daro and its successors, the Mediterranean cultures.

Some of these tropical civilizations had also achieved city-state political stability, with its concomitant need for a food reserve, under climatic regimes where chilling never occurred, drying was difficult, and precious food rotted overnight. The written history of civilization, as dispensed by the Western world, has grossly underrated the organizations present in some of the new world and far eastern civilizations. The physical structures of these worlds are still being unearthed and interpreted, but a start is yet to be made on the social organization which must have underlain these impressive artefacts that are being studied.

Sadly it has been a very troubled decade for much of mainland South-east Asia, from which that region is only just emerging from the time of writing. The author can only draw the readers' attention to the need for further study of this area from the point of view of its past food economies. It is this overnight rotting which is the essence of tropical fermentation and may be regarded biotechnologically as very modern in relation to current haste (high reaction rates) demanded in food engineering. Reaction rates in the tropics are optimal. Using 'tropical' conditions, biochemical transformations can be achieved at rates hitherto regarded as economically

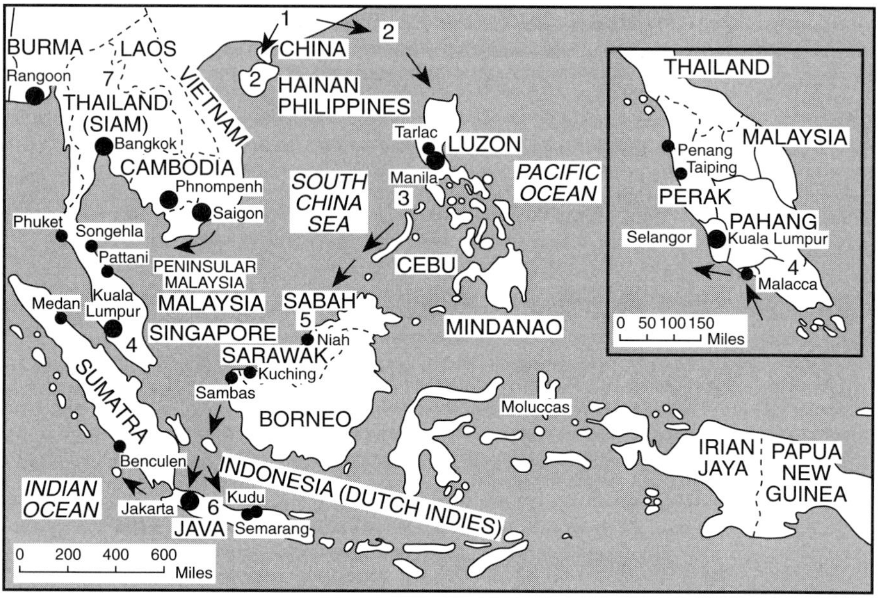

Figure 22.1 Spread of techniques of food fermentation to Southeast Asia from China.

1. Postulated centre of origin 3000–4000 years ago. See, for example, references to the production of a range of fermented foods and related products – liquors, pickles, sauces, dried salted and unsalted fish and meat, syrups, leaven for fermentation, bean relish with salt – the first reference to soya sauce in *Han Shu* as translated by Swann (1974).
2. Land routes within the empire to the 'vassal' kingdoms of Indochina (Needham & Bray, 1984).
3. Sea routes/migratory routes hugging the coasts of the South China Sea or traversing open sea by island hopping, a practice paralleled by the early voyagers in the Mediterranean.
4. An ancient porterage route across South Malaya.
5. Small victualling and trading ports along the north-west coast of Borneo, artefacts from which area show strong connections with Chinese trade.
6. Java: a territory rich in traditions of fermented foods, the raw materials, techniques and nomenclature providing evidence for the 'Chinese origin' hypothesis.
7. In addition to the coastal trade there is a strong tradition of overland migration into Thailand as well as *in situ* development of richly endowed indigenous cultures such as the Khmer.

unattractive. Space requirement, size of unit, amount of labour, become overheads in the product cost which are no longer unacceptable in terms of Atkinson's dictum of fermentation economics.

Apart from harnessing the weed microbe's aggressiveness – food microbes obey the natural laws of weediness which are characteristic of the higher plants used as food – the various fermentations have positive effects, one or more of which may be identified in analysis of the antecedents of individual food products. Perhaps one should mention here, as an advance in our understanding of the microbe, the extent to which we have increased our understanding of mutability of these organisms. This understanding has mainly been applied to those of direct adverse affect on humans. Nevertheless, the phenomenon occurs across the spectrum of organisms and is a feature of those possessing the necessary weediness of food organisms. As the author predicted in the 1960s, there are enzymes which will degrade, for example aflatoxin *in vivo*, but the microbial ecology question as to why aflotoxin occurs has still not been fully answered. The extent of these positive effects, and indeed their nature, has hitherto been largely unappreciated. They are listed in Table 22.1 and further discussion will be conducted here with reference to this set, these rules being the instructions for the assembly of the building blocks of the process engineering flow scheme.

At the village level (the 'small is difficult' level. correcting an intermediate technology impression when that philosophy is applied to fermentation processing), – 'small is beautiful' is a concept developed by Dr Schumacher in the 1960s as being the appropriate way to advance village technology – processes often occur concomitantly. The use of the phrase 'small is difficult' is a parody of the 'small is beautiful' movement of the 1960s, originated by Dr F. Schumacher. It is only with long study of some of these indigenous fermentations, the preparation of the raw material, the conservation of the inoculum, and the general conduct of the fermentation, that one realized that the skills and experience needed to conduct these fermentations properly without error is substantial. Hence the use of the term 'difficult' is not unwarranted.

In an industrially-defined process with stringent product quality control, each unit process has to be identified and understood in terms of its contribution to the final article and the true cost of its incorporation into the flow scheme. For example, the burden of the expenses of the developers of

Table 22.1 Virtues conferred on a potential food material by fermentative processing

Contribution to the economy of food-winning
- Preserves perishable raw material at low cost, aids dehydration.
- Salvages waste otherwise not usable as food.
- Reduces cooking time, hence reducing demand on the food-winning time by reducing fuel demand.
- Enhances nutritional value by improving digestibility, protein value, vitamin content.

Protective value
- Destroys toxic, undesirable or anti-digestive components of the raw material; this action increases the range of raw materials available as food.
- Adds positive antibiotic components, destroys harmful biota and protects against infection or infestation.

Psychological and social value
- Improves appearance and flavour, often imparting a meat-like flavour.
- Functions as appetite stimulant and condiment. A recommended introduction to these two subjects is the classical work on *Ecological Biochemistry* by Harborne (1982).
- Imparts texture, fibre and bite/chewiness, makes the product enjoyable.

single cell protein (SCP) (or MBP, microbial biomass protein) processes are largely attributable to the costly peripheral processes succeeding, and even preceding, the core fermentation.

This application of the analysis of the unit processes, necessary for producing a particular fermented food, is relatively advanced in, for example, the brewing industry, where the substrate components are pre-treated, the inherent activity of the malt is augmented, or even replaced by specific enzymes. There are still opportunities for improvement in for example cheese manufacture, with its heterogeneous substrate, varying through its mass in degree of aerobiosis and A_w (water activity) and having discrete fat globules offering special surface properties and a markedly different environment from the non-fat matrix. Indeed, as the author experienced, one may look in vain for guidance in tropical fermentations to the apparently analogous cheese fermentation. The problems are similar: the answers are not always forthcoming. The converse may hold in that some current studies of tropical fermentations may benefit traditional temperate zone solid-substrate fermentation.

22.3 Classification and uniqueness

One's first encounter with tropical fermentation is a mass of names from language groups alien to the Western world, obscure process descriptions and unfamiliar raw materials. The author has therefore attempted to construct an all-embracing classification (Figure 22.2) and relate the products to this framework. It will be seen that the complexity can be reduced to a few simple rules and environmental boxes, outside which are the 'no go' areas as far as the process engineer is concerned.

Rice may be replaced by sorghum, or wheat. Milk solids may be replaced by soybean curd, white cabbage by another brassica, the grape by a variety of fruits, but the rules are the same. Thus, *in general*, there is no uniqueness about tropical fermentations with the possible exception of thermophilicity, the need for exploitation of which grows as fermentations increase in scale in temperate zone processes.

22.4 Technology transfer; geographical restriction

The qualification 'in general' is important to the argument, because there are unique substrates, organisms and enzyme systems – used for example, to extract safe foods from starchy roots and tubers – which do not have temperate zone counterparts. The geographical division is more longitudinal than latitudinal, being between the western Old World and eastern Old World on the one hand and between the Old World generally and the New World.

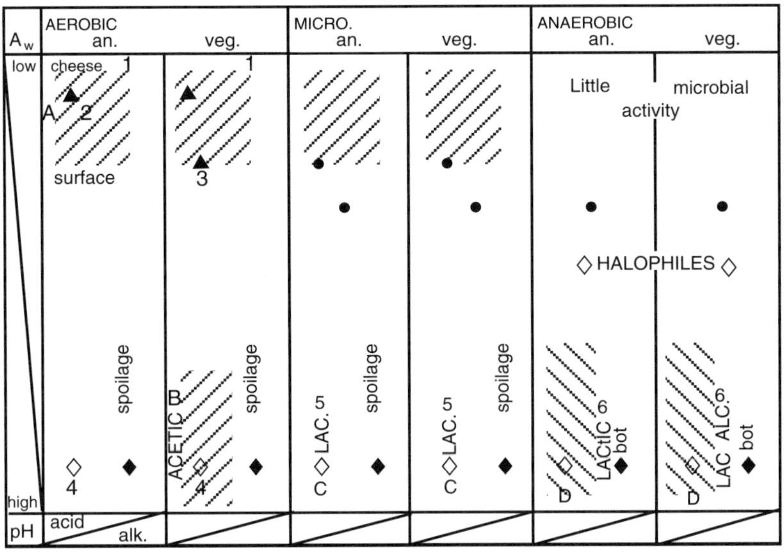

Figure 22.2 Types of fermentation.

1. Vegetable cheeses tend to be more porous than the animal milk cheeses; hence, the aerobic state may penetrate the full depth of a tempe, but may be confined to the surface of a pressed-curd cheese.
2. Just as in trockenläse grapes, fungal infection is used to accelerate drying in some meat and fish products, for example the moulded tuna fish of Japan, katsuobishi.
3. In contrast to bacterial and yeast cultures, fungal cultures can exist biphasically, that is, in a floating culture the aerial portion can transfer oxygen to the submerged hyphae, or into strands or pellets of solid material, a property which is unconsciously exploited in some tropical fermented foods.
4. Advantage is taken of the slime forming capacity of *Leuconostoc* in both idli (Indian rice/bean flour cakes) and in the slime-coated whole soya bean product of Japan, natto. In general, however, these environments cause spoilage.
5. To describe adequately the ecological niches for the lactic acid forming bacteria, temperature, ionic strength and several other factors are required, in addition to the pH and A_w depicted here. These requirements are well reviewed in the second edition of Pederson (1979).
6. The area to the alkaline side of the hatched boxes is the *Clostridium* niche, when competing organisms are present. The range of environments under which these and indeed many other virulently toxic organisms can be persuaded to grow under sterile conditions is much larger.

The north–south division is however, accentuated by the botanical differences noticeable at the family level. For example, Sterculiaceae, Palmae and Cycadaceae are typically tropical. Dioscoreaceae and Araceae are also tropical, but Oxalidaceae, Solanaceae and Euphorbiaceae span a wider range of environments, to mention some of the families contributing to the food source after fermentation.

The evidence for this hypothesis of transfer derives not only from biotechnological analysis of processes and products, but also linguistic analysis of process and product terms, for example those of the Indo-malayan language group. Linguistic analysis has been used extensively in the study of human migration, particularly across Europe, but has yet to be exploited fully in the author's special study area, Southeast Asia.

However, because some tropical societies have been subjected to a less frequent and complicated influence of migrating cultures than others, it is sometimes possible to see how particular food practices arose. This poses a biological difficulty for von Daniken-type hypotheses, who has been a prolific writer of science fiction. The question being, 'If man acquired food fermentation techniques from space, then the extra-terrestrials must also have taught many animal species the virtue of fermenting the harvest', since there is a continuum of food-gathering fermentation techniques from other animals to man. Man has often benefited from watching how other animals cope with potential sources of food.

At present, human development is progressing fast towards exterminating (euphemistically called culturally assimilating) indigenous human societies so their lands may be converted to cattle ranches or plantations. These indigenous societies are the repositories of knowledge of how to make best use of the materials of their environment. It is a mistaken assumption that indigenous diets were monotonous, inadequate in quality, insufficient in quantity, or the food *as eaten* was riddled with poisonous substances. Both in terms of numbers of starving people, and the severity of mal- and undernutrition to which they are being subjected, there are more starving now than there have ever been in the course of human evolution hitherto. Admittedly there is archaeological evidence of malnutrition, (Cohen, 1977), and for the fall of civilization which had overrun their environment, but never on today's scale. Fermentation, control of rotting by controlled rotting, was one response to the threat to the food supply.

Man is an opportunistic eater, catholic in taste for food. Le Gros Clark, for example, summarizes his review on food habits by stating 'It is doubtful then, whether palaeolithic or mesolithic hunters would have adopted any food practices or taboos that might be deleterious to the health of some members of their communities'. Perhaps one should qualify this by saying 'to those members of benefit to the selfish gene'.

22.5 Distribution of existing fermented foods

Using the general rules elicited in the first part of this chapter, it is possible to characterize the range of fermented foods more discretely in terms of four determinants. These are: people; environment; substrate and microorganism.

22.5.1 People

As we have already discussed, there may be a misunderstanding of the relative cultural attainment of different people, attainment having been measured more on durable artefacts than the, more difficult to measure, complexity of social institutions. In the author's experience these institutions are less well documented in Southeast Asia than further north, for example. The cultures of the humid tropics have left relatively few artefacts and this, as Harris (1980) has stressed, has overemphasized the cultural advance of the Near East in comparison with Southeast Asia, lowland central America and even sub-Saharan Africa. Food fermentation has, as we have hypothesized, pre-human origins. One may assume the skills were there and thus their absence did not provide a great restriction to the distribution of the fermentation practice. Indeed, the author has found deep in the jungles of Southeast Asia fermentation practices adhering to all the rules, including addition of fermentation-controlling agents available in the form of secondary components of local jungle plants, examples being the excellent sherry-like liquor produced by the Semelai of Malaya from the tree fruit *Baccaurea* (Stanton, 1972) and the use of the inner bark of a *Shorea* species as a hop-substitute in a fermented sugar cane liquor by the same people. Only by close observation of the technique can these practices be detected, and sadly, acculturation is proceeding at an alarming rate among the people of the dwindling tropical rainforest areas of both the New and Old Worlds.

22.5.2 Environment

While ethnographers may extol the ability of man to master the environment, students of food fermentation are more cautious. The environment is all important. For example, tempe manufacture at two rainforest towns in Indonesia, only 100 km apart, is quite different. Jakarta, at near sea level, is hotter and drier and undergoes a wider diurnal fluctuation in temperature than Bogor at 500 m elevation, with the result that a better quality, slower fermenting, but more ripened (without undue sporulation) product can be produced at Bogor. These subtle differences have had unforeseen effects, in the author's experience, in the conduct of the Unesco training courses (1972–78) in food fermentation in South East Asia, in that results were not always as expected, and attempts to simulate the requisite environments in Northern Europe or the USA for research and demonstration have, until understanding and experience have been increased, proved difficult and a test of microbiological skill.

Understanding the environment is obligatory to producing successful fermentations, differentiating the conditions for the water fungi (Mucorales) from the 'dry' fungi (e.g. Aspergillales), the spore formers from the true lactobacilli and *Saccharomyces cerevisiae* from the other yeasts and yeast-like forms. It is only from the fresh emphasis given to the study of the microenvironment of fermented foods, and much work still needs to be done, that this element of successful fermentation is being mastered. In a modern laboratory the small-scale fermentation can be

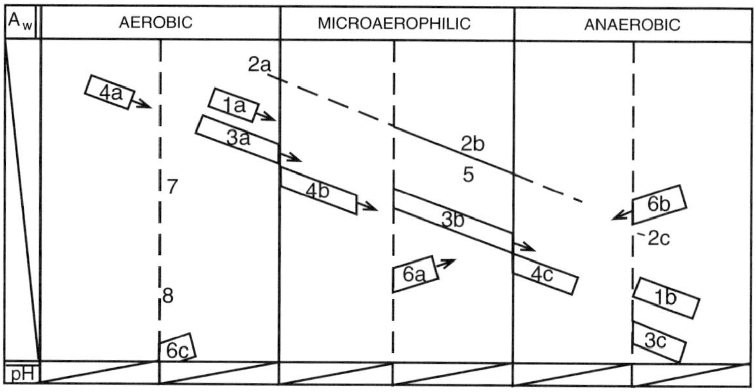

Figure 22.3 Fermentation environments of named processes and products.
1a. The koji stage of sake manufacture. Characteristically aerobic, neutral and moderate (fungal) water activity.
1b. The moromi stage. Acid, anaerobic, often operated as a fed batch process from 1a.
2a. Ragi, the multi-organism starter cakes found throughout Southeast Asia, Produced by a koji-type fermentation but varying according to the fermentation desired, the different types being identifiable according to shape, size and colour coding, i.e. several different kueh-mueh ragi (fermentation starter cakes) may be found on a single market stall.
2b. Whilst ragi has a postulated common origin with sake (the koji stage), the semi-solid leaf-wrapped paste, or lump-substrate fermentations, acquire their characteristic flavour through microaerophilic lactic fermentations.
2c. Tape/tapai is the final stage of the paste fermentation, using rice, cassava, sweet-potato, taro (*Colocasia*, etc.), starch, banana. The end product varies in acid and alcohol but the drained liquor of the alcoholic rice tape is used as a drink (brem), often starting as a purple rice (a similar application of anthocyanins is found in the *Sorghum* brews of the West African Cameroons).
3. The soya sauce fermentation differs from sake in that in the Chinese process a complex fermentation is encouraged with a surface fermentation of halophiles (3b) and an anaerobic fermentation in the bulk of the liquor operating simultaneously (3c).
4. Fish sauces vary as to the importance of autolytic processes from the fish enzymes and the action of halophilic anaerobic/microaerophilic fermentations. In the author's view the microbial population, though not large, is significant in imparting the characteristic aromas (e.g. nampla of Thailand, patis of the Philippines and shoisuru of Japan).
The general trend of all the above fermentations is from aerobic/low water activity/neutral to mildly acid to anaerobic/increased water activity/acid.
5. Poi of the Pacific, idli of India, tuwo and kenkey of West Africa are all short fermentation bacterial sourdoughs which are cooked, often by steaming, before being eaten. They do not have the multistage complexity of the previous processes.
6. Vinegar manufacture is in marked contrast to all the processes 1–5 in that the second, high water activity stage is strongly aerobic.
7. Tempe is a solid substrate fermentation consumed at the end of the first stage. It contrasts with 1–3 in occurring at a higher water activity.
8. If the water activity is permitted to be very high then the solid substrate soya bean fermentation assumes a typical *Leuconostoc* spoilage niche, though the process has affinity with tempe (and possibly a common origin) in that it is a solid consumed at the end of the first stage of fermentation. This is typical of natto from Japan. It is placed close to 6c as it has affinity with the Philippine jelly-like product also from a slime-forming organism, known as nata (di coco).
To summarise, this diagram shows that disparate named processes fall into a very limited number of classes and that similar concepts have evolved, possibly separately, in geographically remote localities. Some, apparently separate, may however have a common origin.

adequately isolated from the vicissitudes of the environment, but difficulties re-emerge on scale-up, not only technical but economical.

22.5.3 Substrate

It is the substrate which differentiates the foods most clearly, but their diverse origins obscure the fact that from a process engineering viewpoint there are foods from different geographical regions with many features in common. This concept of common denominators may be seen by superimposing named fermentations on Figure 22.2 as has been done in Figure 22.3.

Old World tropical fermentations employ rice, sorghum and millet, a variety of peas and beans, yams and aroid yams, tree legumes, palms and cycads, while the New World fermentations employ maize, cassava, some palms, cocoa, agave and various fruits not found in the Old World.

The situation is obscured by the coconut, which is quaternary pantropical, and the fact that the sweet potato and cassava are now ubiquitous. Rice and bananas are ancient transferred substrates and some bizarre situations occur in which an Old World crop is subjected to a New World fermentation and, more commonly, the converse, typically the *ogis* of West Africa, derived from maize hut owing their origins to the *tuwos* or sorghum-fermented porridge. This replacement is a fermented food analogy of the groundnut (*Arachis*) bambarra groundnut (*Voandzeia*) substitution.

The geographical barriers are however quite rigid, as evinced by the complete lack of technology transfer of soybean and its fermentation products to Africa, in contrast to the spread of the banana for example.

22.5.4 Microorganisms

As an act of devil's advocacy, one might describe the microorganisms as the least important determinant in defining the distribution of fermented foods, in spite of the emphasis placed on them in the textbooks. Food fermentation arose by acquiring weed microorganisms from the surroundings to complement the higher plant weeds, growth of which was encouraged by humans for food. The microorganisms colonized the substrate when the latter were appropriately prepared (Stanton, 1969). Not that early people did not have and conserve culture collections of microorganisms, the properties of which were quite distinct from those of their wild counterparts. They did, but were unable to prevent those organisms changing as a consequence of their migration from one environment to another, as has recently been demonstrated in comparative studies of the soya sauce organisms and their Southeast Asian counterpart products the *kacups* of Indonesia.

Discrimination of techniques in favour of 'good' foods, selecting new substrates and acquiring new organisms, was probably a rapid event even in human evolution, with minimal hazard attached to it. Many may not have, or have lost, the basic capacity to be 'as sick as a dog' – a useful food discrimination device – but it is not beyond the imagination to envisage food-discrimination capacity in our genetic make-up. Hence, as Japanese workers in particular have reiterated, the low incidence of aflatoxin-producing activity in their range of *Aspergillus*-based foods is the result of an effective subconscious selection process.

22.6 How they evolved

Stress has been placed on the nature of the philosophical frame within which the practitioners of tropical fermented food techniques operate, on the understanding of the framework and the forces which cause a particular process to take place, or a particular organism to acquire dominance on the substrate.

As so aptly pointed out by Schothorst (1980–81) fermentation is, more than any other food preservation process, dependent on the 'normal' (*sic*) behaviour of microbial populations. Since bacteria, yeasts and fungi are living organisms, they are liable to genetic change. Conversely, they have a great ability to adapt to new environments.

However, this means that in theory, fermented foods can never be considered completely safe when they are not produced under carefully controlled conditions, the trend being to divorce the organism from the substrate by substitution by enzyme preparations. This in fact may be the way to overcome resistance to development of new village-scale fermentations.

Nevertheless, more or less natural processes have evolved with relatively few cases of food-induced disease. Indeed, it is the so-called modern frozen and pre-packed foods which have given rise to serious outbreaks of poisoning, a phenomenon well described by Silliker *et al.* (1980).

Fermented foods have had safety factors built into them consonant with village-scale operation. Careful study of the systems will, in the author's view, allow application of the techniques to other parts of the world, spanning the old geographical barriers. South–south and south–north transfer are as important as north–south transfer.

This philosophical approach is not solely pie-in-the-sky. Understanding the rules governing these fermentations is a prerequisite to developing new food fermentations, whether at the industrial, or the 'more difficult' household level.

22.7 Differentiation by environment and substrate

Figure 22.2 classifies the various tropical food fermentations according to: substrate (animal or vegetable); degree of aeration of the system (aerobic, micro-aerophilic or anaerobic); water-activity permitted by the fermenting mass, whether at the surface, interstitially, or in its interior; hydrogen-ion concentration.

Temperature is omitted, as it is assumed that the conditions are ambient, that is within the range 20–35°C.

Figure 22.2 illustrates, on the basis of the above parameters, the broad areas in which particular types of fermentation occur, showing how the fungal environment is the aerophilic environment with low water-activity, the yeasts occupy all environments at a higher water-activity, but nevertheless occupy a range of water-activities, since yeasts are a 'form taxon', embodying the physiological properties of a wide range of fungi. Indeed, fungi will occupy micro-aerophilic and high A_w environments, but normally compete ineffectively with the other microorganisms under these conditions.

Bacteria occupy the whole spectrum of environments bounded by high water-activity. Relatively harmless dominants are depicted by a hollow symbol.

It must be borne in mind that, in this classification, consideration is being given to the organisms in competition one with another, as this is how they are found in fermented foods. In axenic conditions, that is in the absence of competition, some organisms can occupy and grow in a wide range of environments. Even fungi (e.g. *Rhizopus*) may grow under almost anaerobic conditions, producing organic acids, alcohols and esters which may have valuable effects on a final product, but this type of situation is exceptional.

Setting the fermentalions in a frame, bounded with 'ecological' coordinates, permits graphical display of differentiation of the different types of tropical fermentation. It becomes apparent that complex and sequential fermentations are relatively common and demonstrates clearly how processes have evolved in which the enzymatic capacity of an aerobic organism is exploited in an anaerobic situation. Notice that, whereas the aerobic phase (top left of Figure 22.3) operates in a neutral environment, the anaerobic phase is always more or less acid.

The *katsuobishi* – a Japanese food (top left of Figure 22.3) – dried tuna fillet allowed to mould with a *Penicillium* species, is of interest in that it uses the same technique in drying a fresh-food as is used in the mould-ripening of grapes.

The *koji* family of proccesses (*koji*, Japan; *ragi* and *tempe*, Indonesia) spring, as is hypothesized earlier in this chapter, from a common origin. They are of several types:

(i) Those that are eaten at the end of the mould stage, e.g. tempe.
(ii) Those that are used as the raw material for a second stage, e.g. rice koji for saké and soya bean koji for soya sauce. (The real starters for these are tané kojis or seed kojis, but they are like the principal fermentation

in that they are also solid-substrate fermentations. The main distinction is that tané kojis are often developed under strictly aseptic conditions, using broken rice.)

(iii) Those that furnish only a small part of the final fermenting mass, yet are not solely an inoculum but a source of enzymes, the Far Eastern equivalent of active malts in a grain (as distinct from pure malt) whisky mash. Typical of this class is ragi, a material made by dry fermenting broken rice (c 60% moisture) then milling, forming into biscuits or pellets and then drying thoroughly to <12% moisture. It is a product found throughout Southeast Asia, the equivalent of the dried yeast of the Western world, but in any one market several types may be found. Each may be identified by the shape, size, colour and sometimes marking of the biscuit, being square, flat-round, spherical or lozenge-shaped and often bearing a distinctive colour code. This distinguishes the types according to whether they are to be used for rice, cassava, plantain (banana) or fish-rice fermentations for such end-products as tapai, tapé, tapé ketan, *or* budu.

From the point of view of industrialization, ragi production is focal for development, in that control of quality at this stage can provide effective control and hence ensure satisfactory quality and safety of a whole range of tropical village-scale fermentations. It provides a combined enzyme–active material and inoculum relatively cheaply.

Existing texts on food microbiology cover adequately the discussion on what microorganisms are likely to develop when a substrate is held under particular conditions of pH, A_w and temperature. The separation of microorganisms according to the condition of the substrate is clearly depicted

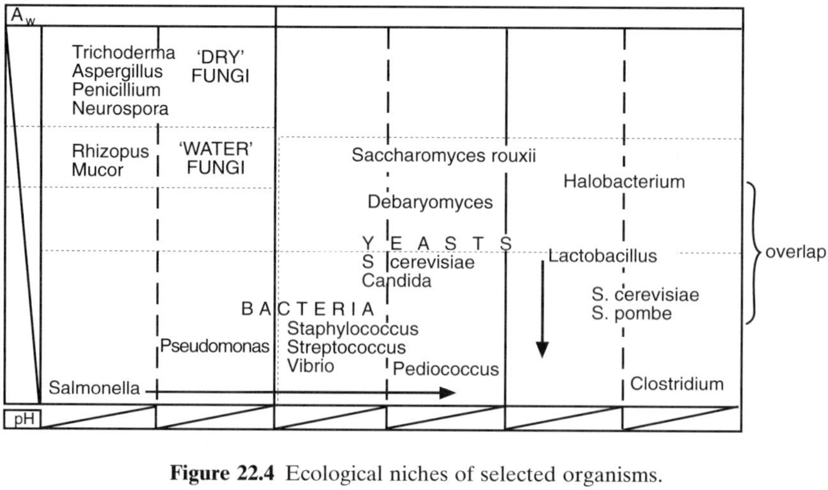

Figure 22.4 Ecological niches of selected organisms.

in Figure 22.4. Thus, in general the fungi occupy the top of the chart, the yeasts the middle and the bacteria the lower portion (high water activity environment). However, practice has shown that to achieve the desired fermentation, *viz.*, to encourage the growth of the correct element of the ambient (endemic) microflora, the substrate must be presented in a specific manner, the parameters for which have been achieved by trial and error over a very long time, but much work needs to be done on challenging substrates with known mixed populations of yeast or lactobacilli, for example, and then studying the relative growth rates of the elements of these mixed populations.

Again, in practice, the situation is confounded by the fact that past practitioners have found that additions of starch, spices, leaf or bark extracts, various salts, or even waters of a particular stream (peaty, sulphurous), may be the required feature of the recipe to achieve a satisfactory result.

The industrial microbiologist circumvents the problem by using high concentrations of pure cultures, but such an approach is usually not possible, or (if possible) dangerous, in the context of the village-scale fermentation. The political solution to an inherently dangerous situation, in terms of likelihood of serious food poisoning. has been to ban the making of a particular fermented food. This was done, for example, in Indonesia recently for the tempe-like product made from coconut waste called locally tempe bongkrek. However, Ko (1979) has stated how the poisoning, arising due to a misfermentation, may be easily – and has been traditionally – avoided. The danger arose not from the practice but from the misuse of the practice and is paralleled in Western societies by the misuse of frozen meat portions, resulting in massive outbreaks of salmonellosis, for example.

Environmental microbiology is still a young science. One may conclude that the reasons why particular organisms occupy particular niches are often unknown. Perhaps the current study of extreme thermophiles (Baross & Deming, 1983) will throw light on the question of the interaction between higher-order biochemical structure and environment.

22.8 Discussion

22.8.1 Dispensing with living organism

In a book devoted to microorganisms, the idea of developing processes to dispense with them may seem heresy, but it is a natural consequence of our increasing understanding of microbial processes and the ever widening range of their applications. Further, the idea is a way out of the impasse posed by Schothorst (1980–81) that fermented foods can in theory never be considered to be completely safe.

In rejecting the microbe, one is only separating the living, reproducing organism from the substrate and replacing it in the process being operated by an enzyme mixture performing the desired reaction, but excluding the hazard. This development allows high standardization of process and hence product. It may save time and space and thus confer economic advantage. Further, the substitution may reduce complexity of the requisite machinery and hence may even be of advantage at the village scale.

The development predicted above has already happened in other sectors, e.g. brewing, baking, fruit juice extraction and cheese manufacture. Will the change take place in the craft of tropical fermented food manufacture? The use of enzymes in soy sauce and their experimental use in fish sauce manufacture shows this possibility, and indeed it is difficult to identify a single fermented foodstuff to which it might not be applied. It is a logical extension to the perforated polythene and tempe starters already used. The kit marketing technique is already operated in the developed world, not only for brewing and wine making, but yoghurt and various soya-based foods, even to the extent of having its own journal, *Soy Foods* (100 Heath Road, COLRAIN, MA 01340, USA). A good introduction lies in the new (4th) edition of Prescott and Dunn's classical work (Reed, 1982).

22.8.2 Contribution to microbiology

The question arises 'Are there foods mentioned in this essay of importance in their own right, or as processes for application to other purposes, or as unexplored territory from which to garner industrial microorganisms?' In the author's view these tropical fermented foods are significant on all three counts and, although the foods are ancient, they are thoroughly modern in their elegance as 'exercises' in the manipulation of microbial ecosystems.

Simulation of the environment and the materials available for a village in rural Java, for example (deliberately chosen because it is in such a community, with its inheritance of the arts of the ancient high civilizations of China and India, that the richness of the world of tropical fermented foods is fully displayed), is no easy matter. On reflection, this difficulty of simulation has also been experienced by those involved in industrializing cheese and fermented sausage manufacture. One should therefore not be surprised at the technical skill demanded; nor, on the other hand, should one dismiss facilely new innovations as 'too sensitive for the village to operate'. The skill of the innovator is taxed to the full to master the traditional techniques to a pitch where they can be exploited industrially and he then has to endeavour to introduce these to communities which would benefit by the addition of fermented food to their economy. Really elegant techniques are deceptively simple, as Hesseltine demonstrated with his perforated polythene tempe bags.

22.8.3 Implications of fermented food philosophy for its contribution to the future world food economy

The application of the 'green revolution' was seen in the 1970s as the panacea to the world's food problem, inheriting the accolade from the petroprotein of the 1960s. However, doubt has recently been cast on this latest panacea, because of its sensitivity to disaster and its dependence on costly and high-energy inputs.

Early man – famine-threatened man (Read, 1946) – did not discard potential food, nor did he waste food, if wastage could be avoided. His ability was limited in part by technology and in part by resources. It was not, as Read (loc. cit.) and others have shown, through want of trying.

Traditionally, in seasonal environments, summer climates have been recognized as the time when food goes bad easily. Milk sours overnight, meat goes off, shellfish are dangerous (no 'R' in the month). Some parts of the world have had to contend with this problem throughout the year. The 'diarrhoea dialogue' is protracted (to borrow from the title of the journal devoted specifically to diarrhoea in tropical countries). Fermented food discussion is an alternative dialogue concerned with living with the microbial world, the statements of which have been selected by long-term practice.

22.8.4 Caution in transfer

As the research on fermented foods has shown, the concept of 'simplicity' is deceptive, and attempts to use new substrates, or by-products, have proved far more difficult than originally envisaged. Not only are the strains of the microorganisms cultivars or land races, in the truest sense, in that they are closely attuned to the substrate – hence they may or may not grow on alternative substrates – but each new experimental interaction of substrate–microorganism–environment reopens the question of whether the fermentation is at its termination food-safe. Thus, it is not sufficient, from either a technical or a legal viewpoint, to employ a fungal strain reputed not to produce a mycotoxin, since the conditions under which mycotoxins are produced are still ill understood (Ciegler, 1983). A strain of *Aspergillus flavus* may be a potential producer *in vitro* of aflatoxin and yet not exhibit this property *in vivo*, as shown for example by Alvarez-Barrea *et al.* (1962), using standard NRRL strains.

Other technical and economic hazards may also arise in applying these fermentations, especially when the substrates are waste materials. These hazards are not confined to previous contamination with microbial toxins, but may also be due to biocide residues, or heavy metals. To the technician, these constraints may be obvious, but others such as the social constraints on acceptability and the economic constraint on the true cost of collecting the feedstock and preparing it for fermentation may be overlooked. This last-named hazard has beset the work on use of lignocellulosic wastes as feedstocks.

Notwithstanding these difficulties the tropical food fermentations, isolated as they have been in the past by geographical barriers, offer a viable contribution to future world food supply.

Acknowledgement

The author wishes to acknowledge the kindness of Professor J.R. Norris, of the Lord Zuckerman Research Centre (Cadbury-Schweppes), University of Reading, for providing study facilities in 1983.

References

Alvarez-Barrea, V., Pearson, A.M., Price, J.F., Gray, J.I. & Aust, S.D. (1962) Some factors influencing aflatoxin production in fermented sausage. *Journal of Food Science*, **47**, 1773–5.

Baross, J.A. & Deming, J.W. (1983) Growth of 'Black Smoker' bacteria at temperatures of at least 250°C. *Nature*, **303**, 423–6.

Ciegler, A. (1983) Do mycotoxins function in ecological processes? *Journal of Food Safety*, **5**, 23–30.

Cohen, M. (1977) *The Food Crisis In Prehistory: Overpopulation and the Origins of Agriculture*, Yale University Press, New Haven, p. 341.

Harborne, J.B. (1982) *Introduction to Ecological Biochemistry*, Academic Press, London, New York, Toronto, Sydney, San Francisco.

Harris, D.R. (ed.) (1980) Tropical savanna environments, in *Human Ecology in Savanna Environments*, Academic Press, New York, pp. 3–37.

Hesseltine, C.W. (1965) A millenium of fungi, food and fermentation. *Mycologia*, **57**, 149–97.

Ko, S.D. (1979) Indigenous fermented foods, in *Fermented Foods*; *Economic Microbiology*, Vol. 7, (ed. A.H. Rose), Academic Press, New York, pp. 15–36.

Needham, J. & Bray, F. (1984) in *Science and Civilization in China, Vol. 6: Biology and Technology, Part 2: Agriculture* (ed. J. Needham), Cambridge University press, Cambridge.

Pederson, C.S. (1979) *Microbiology of Food Fermentation*, 2nd edn, AVI, Westport, Connecticut.

Read, B.E. (1946) Famine foods, in *Chiu Huang Pen Ts'ao*, British Council/Henry Lester Institute of Medical Research, Shanghai/London, p. 90.

Reed, R.G. (1982) *Revision of Industrial Microbiology*, by S.C. Prescott and P.G. Dunn, Macmillan, New York.

Schothorst, M. van (1980/81) Safety aspects of food fermentation. *Nestlé Research News*, pp. 53–62.

Silliker, J.H. *et al.* (eds) (1980) *Microbial Ecology of Foods*, Academic Press, New York, p. 997.

Stanton, W.R. (1969) Some domesticated lower plants in south-east Asian food technology, in *Domestication and Exploitation of Plants and Animals* (eds P.J. Ucko & G.W. Dimbleby), Duckworth, London, pp. 463–9.

Stanton, W.R. (1972) Microbially produced foods in the tropics, in *Proceedings; Fourth International Symposium on Conversion and Manufacture of Foodstuffs by Microorganisms, Kyoto, 6–9 December*, 1971. Saikon Publ. Co., Kyoto, pp. 133–9.

Stanton, W.R. & Da Silva, E.J. (eds) (1977) *GIAM (Global Impacts of Applied Microbiology): State of the Art 1976*, University of Malaya Press, Kuala Lumpur.

Swann, N.L. (1974) Food and money in ancient China, in *Han Shu 24* (the earliest economic history of China to AD25), translated and annotated by N.L. Swann. Octagon Books, New York.

23 African fermented foods

S.A. ODUNFA AND O.B. OYEWOLE

23.1 Introduction

Fermented foods constitute a significant component of African diets. Many fermented foods are known, some serve as main course meals, others as beverages while others are highly prized food condiments. Those which serve as main meals and beverages are usually products of carbohydrate-rich raw materials. Some of the most important ones in this group include 'gari' from cassava, 'ogi' and 'mahewu' from maize and 'kaffir' beer from sorghum. Those which serve as food condiments are usually made from the fermentation of protein rich seeds. These include 'iru' from African locust bean; 'ugba' from African oil bean and 'ogiri' from melon seeds among others. All are known to be good sources of proteins and vitamins.

There are various reasons why foods may be fermented. Perhaps the most important one is to produce a variety in flavours of existing foods. Others are to mask some undesirable flavours, to make some inedible foods edible, and to preserve some foods.

Within the past few years, there had been increased scientific interest in traditional fermented foods, but the age-old techniques used for their production have not improved as a result. The importance of fermented foods in the nutrition of Africans is now better appreciated, as a result of which efforts are now being made to industrialize some of the processing in some cases (Steinkraus, 1989). For convenience, African fermented foods will be treated under the following commodity groups:

- Fermented non-alcoholic starchy foods.
- Alcoholic beverages.
- Fermented vegetable proteins.
- Fermented animal proteins.

23.1.2 History

Fermented foods have a long history in Africa. However, the absence of writing culture in most of Africa makes their origin difficult to trace. By the medieval ages, when most of northern and western Africa was conquered by the Muslim Arabs, many records of the presence of fermented foods were made by the Arab travellers, mostly merchants and geographers. By this time, in the 8th to 16th centuries, the art of fermenting some foods had been perfected and had been part of the culture of the people. Unfortunately the Arabs' world of knowledge was not extended to the forested regions of West and Central Africa; hence there were hardly any records for these regions.

Perhaps the most documented of the fermented foods is sour milk. Fairly frequent references were made to it by the Arab authors. Sour milk was consumed all over the Guinea savannah of West Africa and the Northern Africa by the Negro and Berber people of West Africa. Ibn Batoutah, an Arab traveller describing his journey in 1352 from Walata to the town of Mali recorded that the travellers arriving at the villages on the road were met by women selling various local products including sour milk. During his stay in the town of Mali (1392–1393), he was sent a hospitality gift by the local King, Mansa Sulayman '. . . a magnificent meal, according to Sudanic notions which included a gourd filled with sour milk'. He also reported that people at Walata (present-day Sudan) consumed porridge made of millet and sour milk. Al-Bakri, an Arab geographer, in 1068 noted a town of Zafqu in Senegal where the people were snake worshippers and jars of sour milk were brought to the snakes as offerings. A very important drink, dakno, containing sour milk mixed with honey was often served to important travellers by the governor of Walata. Souring of milk was popular presumably to preserve the milk, a highly perishable product as much milk was produced by these people from cattle, goats and camels. Sour milk is still popular today in many parts of Africa where the products are called different local names (e.g. leban in Morocco) (Tantaoui-Elaraki *et al.*, 1983; Direr, 1991).

Next to sour milk in historical importance are alcoholic drinks. The earliest records were made by al-Idrisi in medieval times; he obtained his information from anonymous travellers and merchants and reported that the people of West Africa prepared an alcoholic drink from millet. This was by adding boiling water to millet flour (presumably to gelatinize the starch constituent), filtering it and subjecting it to natural fermentation. This kind of beer was known throughout West Africa as dolo, kimbi, merissa, etc. Another alcoholic drink was mentioned by Al-Muhallabi before AD 996 and described by Golberry in the account of his travels of the land of Bambuk (1785–1787) inhabited by the Mandigo group of Senegal. According to Golberry, the people made this alcoholic drink as follows. They put millet into an earthen pot filled with water and kept it there until it turned sour. They then added honey and exposed it to the sun for 10 days. After that time, they filtered the contents through a sieve made of leaves, obtaining a strong mead with a very pleasant flavour.

Various alcoholic drinks played an important part on various solemn occasions and were often offered as gifts. Al-Bakri reported in 1068 that the people of the ancient Ghana empire brought offerings, including alcoholic drinks, to their dead. As early as this time, brewing was already a specialized art in Ghana. Oral history indicates that these brewers visited various places on festive occasions and retailed beer in small calabashes. So specialized was brewing in Ghana that the brewers who were in the king's service were buried with the king on his death in the belief that they would continue to give him such specialist service in the next world.

Early reports indicate that a more elaborate preparation in the medieval age which is in line with the present-day brewing was practiced by these traditional brewers. Millet or sorghum grain was malted by soaking in water and placing in a pit for a short time to allow sprouting. The grains were then pounded to a thick meal and cooked in an earthen pot; on filtering, a light sweet liquid was obtained which was then slowly fermented in calabashes. After 1–2 days, a mildly intoxicating drink was ready.

Sour porridges are quite common throughout Africa, south of the Sahara. The practice of soaking cereal grains to let them sour was long established. An Arab author, Al-Omari, between 1342 and 1349 reported that grains were wetted and pounded in mortars and soaked in water to make them sour. This was to make the flour more savoury, and so to add more flavour to the porridge prepared from it. A German traveller, Nachtigal, recorded that thirst was quenched in Bornu very largely by a drink made by allowing millet or sorghum to ferment in water for a short time. The supernatant water tasted sharp and sweet.

Dirar (1993) did extensive work on the origin of food fermentations in Sudan. It was evident that the practice had been developed over centuries. According to one of his sources, the writings of Strabo (7 BC), the Greek philosopher showed that the Ethiopians has been brewing merissa-like beer since 7 BC. Special utensils for making the staple kisra bread were discovered in archaelogical sites which were traced back to 550 BC. Although wine from grapes largely displaced date wine in a certain era of Sudanese history, evidence from ancient inscriptions showed that ship loads of date wine, beer and other precious items were offered as gifts to the ruler of a powerful Sudanese kingdom of Yam in recognition of his numerical military strength. Date wine brewing developed after AD 560. The Sudanese intricate pottery designs are linked to food and drink fermentation history. The long-necked, narrow-mouthed pottery styles reminiscent of the ancient practice of wine brewing were collected from excavation sites of Meroitic civilization of 690 BC to AD 560; this persisted till the Christian era about AD 560. Ancient drawings made in 1860 showed wine-drinkers, horned and tailed men dancing and drawing wine from a typical wine jar.

23.2 Fermented non-alcoholic starchy foods

23.2.1 Cassava products

Cassava (*Manihot esculenta* Crantz), a root crop, is the most abundant starch food in the tropics. The world production of cassava in 1990 was estimated at about 153 million tons, 46% of which was produced in Africa (FAO, 1991). About 90% of the cassava produced in Africa is estimated to be used for human food (Cock, 1985), and most of this is fermented before consumption.

Cassava products are common in West, East and Central Africa. In West Africa, the common fermented cassava products include 'gari', a sour farinaceous meal; 'fufu', a fine paste starchy food; 'lafun' or 'kokonte' a cassava flour product in Nigeria and Ghana respectively; 'agbelima' a Ghanaian cassava dough; and 'attieke', a fermented and granulated, steam-cooked cassava food. Other products of cassava fermentation in West Africa include 'kumkum', 'myiodo' and 'atangana' in Cameroons and 'placali' in Côte d'lvoire. Products of cassava in East and Central Africa are largely in the form of flours, although fermented doughs called 'Makaka' and 'Maphumu' (Malawi), and others like 'Chikawangue' and 'Cingwade' are available. The village production of some of these cassava products is illustrated in Figure 23.1.

There are wide variations in the cassava fermentation process from one region to another. The various fermentation processes had been broadly categorized into solid-state fermentation and submerged fermentation (Oyewole, 1992). The solid state, which is typified by the West African 'gari' production, involves an initial grating of peeled cassava roots. The grated pulp is filled into cloth bags and tied securely. Heavy stones or hydraulic pressure are placed or exerted on the sack to press out the cassava juice. The grated cassava root is left in this position for 3–5 days, depending on the degree of fermentation desired. The caked fermented mash is sifted through hand-made raffia sieves to remove coarse fibres and ungrated lumps. The pulp is afterwards roasted in a shallow metal pot over an open fire and is constantly stirred to keep the particles separate, a process now commonly referred to as 'garification'. The final product is dry, farinaceous and cream-coloured and could be further sieved to produce 'gari' of different particle sizes.

In another variation of the solid state fermentation, peeled cassava tabers are spread in open-air to dry. As they dry out, they are colonized by moulds (Essers & Nout, 1989). The dried products are milled into flour, before being cooked into a stiff dough. The product is called 'kokontey' in Ghana. In Tanzania the cassava is heaped, allowed to go mouldy and then sundried. The product is 'makopo'.

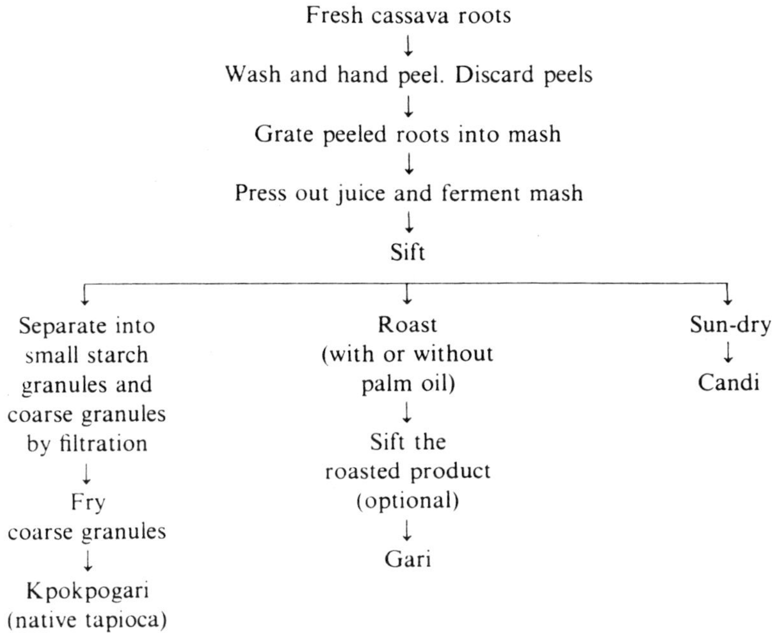

Figure 23.1 Flow diagram of village processing of cassava for gari, kpokpogari and candi production. (From Odunfa, 1985.)

In the submerged fermentation process typified by the 'fufu' production method (Oyewole & Odunfa, 1990a), the cassava roots are peeled and cut into convenient pieces. The pieces are then submerged in water for up to 3–5 days, during which period fermentation occurs. The fermented cassava is then washed and mashed. The mash is sieved through small baskets to remove the lignified central strands. The sievate is pressed to drain off the water and the resultant stiff wet mash could be boiled in water and pounded to a soft dough, or roasted over simmering fire or sundried before being milled to flour. Among the fermented products of cassava, more studies have been focused on 'gari' and 'fufu' production.

(a) Gari. **Gari** is the most important of the fermented cassava products. It is the staple food of about 100 million people in West Africa. In Southern Nigeria, it contributes about 60% of the calorie intake (Akinrele *et al.*, 1962). The original technology, often referred to as 'traditional' or 'indigenous' is believed to have been introduced over a century ago into West Africa by the immigrant freed slaves from Brazil who were used to 'farinha de Mandioca', an analogue of gari in South America. The traditional method of production, done mainly by village women is illustrated in Figure

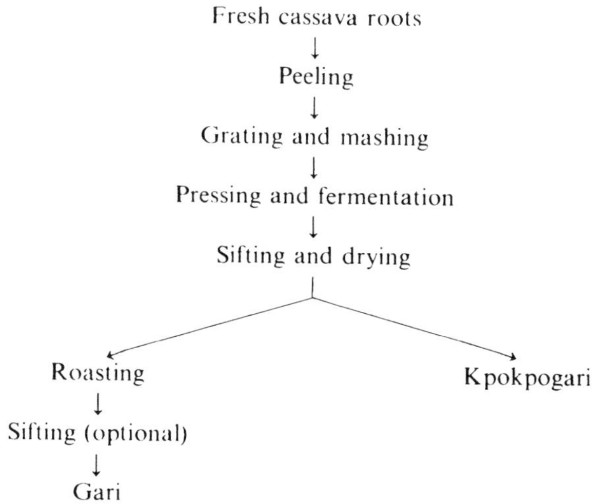

Figure 23.2 Village processing of gari. (From Odunfa, 1985.)

23.2. Efforts have been made to mechanize the gari production technology (Onyekwere *et al.*, 1989; Idowu, 1990). The development has been mainly in grating, granulating and frying operations. The fermentation process however, is still the traditional spontaneous inoculation despite the various studies carried out on the microbiology of the process.

Earlier reports showed that the fermentation of cassava during 'gari' production is anaerobic and exothermic and is a two-stage process (Collard & Levi, 1959; Akinrele, 1964). In the first phase *Corynebacterium manihot* is said to attack the starch of the root leading to the production of various organic acids (such as lactic and formic) and lowering the pH. In the second stage, the acid condition is said to stimulate the growth of a mould, *Geotrichum candida* which begins to proliferate rapidly, bringing about further acidification and the production of a variety of aldehydes and esters that are responsible for the taste and aroma of 'gari'. A number of other organisms have also been found to increase in population during cassava fermentation for 'gari'. These include various genera of the lactics (*Leuconostoc*, *Lactobacillus*) and some yeasts of the *Candida* species (Table 23.1) (Okafor, 1977). These are in fact more important in the fermentation. The media used by earlier workers was selective against the lactic acid bacteria.

Reports from an American laboratory contend that *Streptococcus faecalis* is the primary fermentative organism in acidic cassava fermentation. They claim that *Corynebacterium manihot*, though it was also isolated, grew slowly and lacked significant acid-producing capabilities (Abe & Lindsay, 1978). This view was also supported by Ngaba & Lee (1979). The

Table 23.1 Succession of the major organisms encountered in cassava pulp fermentation for gari production

	Days						
Organism	0	1	2	3	4	5	6
Leuconostoc	7.2×10^6	4.0×10^8	6.4×10^8	9.2×10^8	1.4×10^9	1.2×10^9	3.6×10^6
Alcaligenes	2.4×10^5	4.8×10^5	5.4×10^3	0	0	0	0
Yeasts	1.2×10^5	2.0×10^5	3.5×10^5	4.8×10^5	4.6×10^5	5.2×10^5	5.6×10^5

From Okafor (1977).

fermentation process is now recognized as a lactic process involving the activities of other microorganisms, including the yeasts, all of which have different roles to play (Ofuya & Nnajiofor, 1989). However, while the microorganisms are known to be involved in the acidification and flavour development processes, their role in the detoxification of cassava has recently been confirmed. Many of the lactic acid bacteria isolated from cassava have been confirmed to be capable of producing detoxifying linamarase enzyme (Okafor & Ejiofor, 1990; Oyewole & Odunfa, 1991; Giraud *et al.*, 1992). However, many experts maintain that the linamarase present in the cassava tuber is the principal agent for hydrolysing the cyanogenic glycoside, and that the lactic acid bacteria are chiefly important in producing acid which aids in driving off the weakly ionized hydrogen cyanide and in providing a satisfactory pH for the enzyme (Vasconcelos *et al.*, 1990).

During gari fermentation, the pH, linamarase activity and total cyanide decrease (Ikediobi & Onyike, 1982). The major change during the fermentation is the increase in acids. Detailed studies by thin-layer and gas-liquid chromatography detected acetic, propionic, butanoic, lactic, oxalic and pyruvic acid together with higher fatty acids in the gari flavour. However, the most predominant acid produced is lactic acid. It is chiefly produced by *Streptococcus* spp. and *L. plantarum*; the latter was found dominant after 48 hours fermentation (Ngaba & Lee, 1979). The lactic acid, along with some pyrazines which are formed from sugars and amino acids during the roasting process, constitute the main flavouring constituents (Dougan *et al.*, 1983). Oxalic and succinic acids are products of fungal metabolism. The higher fatty acids are similar to those found in raw cassava; being weak acids and insoluble in water, they have no sour taste and hence do not contribute to the flavour.

The optimum temperature for the fermentation of gari is 35°C, and it may proceed satisfactorily up to 45°C. Frequent mixing of the mash and exposure to light accelerate fermentation, particularly the second stage (Akinrele, 1964). The main significance of cassava fermentation is that it yields a product that can be stored for a prolonged period. The shelf life of gari is about 6 months, depending on the moisture content. For long shelf life a moisture content of below 12% is recommended (Adeniji, 1976).

Although most of the toxic cyanide in cassava is broken down during fermentation, gari still has some cyanide (20–36 ppm). High cyanide level in gari was reportedly implicated in certain nervous diseases (Osuntokun, 1969, 1972, 1973) and indirectly in goitre cases in West Africa (Ekpechi *et al.*, 1966).

(b) Fufu. Fufu (foofoo) production is widespread from the Ivory Coast to Congo Brazzaville (Figure 23.3). It is produced through the submerged fermentation of cassava roots. The fermentation is known to involve mainly lactic acid bacteria (Abioye, 1981; Okafor *et al.*, 1984; Oyewole & Odunfa, 1990b). The spectrum of the lactic acid bacteria has been characterized and the role of the microorganisms in the retting and detoxification of cassava in 'fufu' production had been confirmed (Oyewole & Odunfa, 1991; Giraud *et al.*, 1993; Raimbault, 1995; Westby & Choo, 1994). In fufu production, cutting of cassava into smaller pieces has been found to accelerate the rate of fermentation and the quality of the product (Oyewole & Odunfa, 1992b). A temperature range of 30–50°C and a soaking period of less than 60 hours was found best for the natural fermentation process. The retting process leading to the softening of the cassava roots during fufu production is known to be due to some pectinolytic microorganisms (Oyewole & Odunfa, 1992a; Ampe *et al.*, 1995).

23.2.2 Cereal grain products

Cereal grains constitute an important group of substrates for fermented foods in tropical Africa. The common cereals which are used include maize

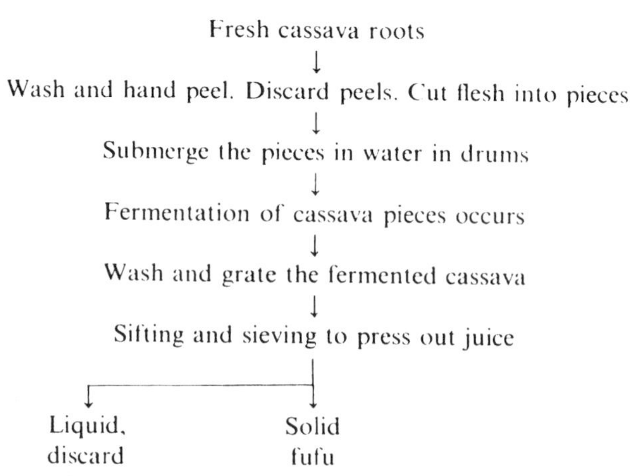

Figure 23.3 Flow diagram for fufu production. (From Odunfa, 1985.)

(corn), sorghum (syn, Guinea corn, Kaffir corn) and millet (*Eleusine* and *Pennisetum* species). Other less-known cereals with good fermentation potentials include Acha (*Digitaria exilis*) a West African crop also known as fonio and (*Ensete ventricosum*) known as finds (Gambia), Fnndi (Senegal), founde (Mali) and Podgi (Benin) (BOSTID, 1992, 1996). The average chemical compositions of the cereals are indicated in Table 23.2. In general, profound biochemical and nutritional changes occur during the fermentation of cereals. A wide variety of products are obtained as a result of these processes and these can be categorized into the following groups.

1. Gruels (thin porridges) and non-alcoholic beverage
2. Dough/thick porridges
3. Baked products
4. Alcoholic beverages

(a) Gruels and non-alcoholic beverages. These are fermented cereal products which on processing yield acidic, non-alcoholic and of high water content fluidy gruels (porridge). The common cereal gruels include 'Ogi' in West Africa, 'Akasa' or 'Koko' in Ghana, 'Uji' in Kenya, 'Mahewu' or 'Magou' is South Africa and 'Abreh' in Sudan. The best documented of these gruels is ogi.

Ogi. This is a sour gruel obtained as a result of the submerged fermentation of some cereals. The common cereal used in West Africa is maize in the southern part, while sorghum and millet are used in the north where it is drier. Ogi is normally prepared as a water suspension and cooked before consumption. The cooked product is usually a gel of variable degree of stiffness. The fluid or semi-solid cooked Ogi is called by different names in different localities such as 'eko'; 'akamu', 'kafa', while the stiff gel is called 'agidi' in Nigeria. Agidi is cooked when wrapped in leaves, then allowed to set to a stiff jelly.

Ogi is an important indigenous, traditional weaning food common in the whole of West Africa. It is consumed as a breakfast meal by many and it

Table 23.2 Average chemical composition of feeding stuff

Food	Starch (%)	Protein (%)	Fat (%)	Vitamins (mg/ml) Ascorbic acid	Thiamine	Riboflavin	Niacin
Cassava root	94.1	2.6	0.5	3.5	0.04	0.02	0.6
Maize	82.6	10.2	4.1	11.4	0.5	0.08	2.0
Guinea corn	76.2	14.2	3.2	29.1	–	–	–
Millet	83.2	8.0	5.0	0	0.3	0.04	1.8
Acha	78.3	7.5	1.2		0.47	0.10	1.9

From Oyenuga (1968), Oyewole & Akingbala (1993) and BOSTID (1996).

serves as the food of choice for the sick in many cases. The traditional preparation of Ogi is presented in Figure 23.4. White maize, sorghum or millet grains are washed and steeped in clean water for 1–2 days. The softened grain is then wet-milled into a fine slurry which is subsequently sieved through a fine wire sieve. The pomace is retained on the sieve and later discarded for animal feed, while the starch which has been separated with the water is allowed to sediment in a pot. The sievate is allowed to ferment for 2–3 days during which it becomes sour. The fermented starch paste is ogi. For consumption, the paste is added to some quantity of water and boiled with continuous stirring to make a gruel. Aspects of the microbiological and biochemical changes during ogi preparation from maize were studied by Akinrele (1970). He found that the grain fungal flora predominantly of *Aspergillus, Penicillium and Fusarium* spp. were eliminated during the early steeping period. After 24 hours of steeping, the predominant microorganisms included *Lactobacillus plantarum, Aerobacter cloacae* and *Corynebacterium*. The pH of the steep liquid decreases to about 4. This gives rise to conditions favourable for subsequent souring of the maize mash by *L. plantarum* and yeasts, *Saccharomyces* spp., *Candida mycoderma* and *Rhodotorula.* Christian (1970) found only lactic acid bacteria in the fermenting mash. Fields *et al.* (1981) contend that only lactic acid bacteria were necessary for the fermentation. The predominant acids produced were lactic acid, 0.6%, acetic acid, 0.1% and butyric acid, less than 0.02% Altogether, 10 volatile acids were identified by gas chromatography (Banigo & Muller, 1972). The predominant microorganism in the

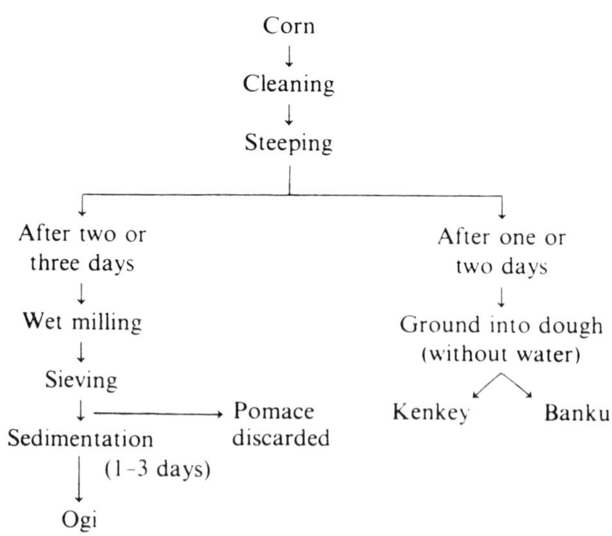

Figure 23.4 Flow diagram for ogi, kenkey and banku production. (From Odunfa, 1985.)

fermentation responsible for the production of lactic acid is *L. plantarium*; *Corynebacterium* hydrolyses the corn starch to form organic acids while *S. cerevisiae* and *Candida mycoderma* contribute to the flavour development.

A microbiological study of the fermentation of sorghum for ogi production showed that the main microorganisms involved were *Lactobacillus* spp. particularly *L. plantarum* and two yeasts, *Candida krusei* and *Debaryomyces hansenii* (Odunfa & Adeyele, 1984).

The methods of processing of various grains for ogi production encourage significant nutrient losses. Losses may occur during steeping, milling and sieving (Lagunna & Carpenter, 1951). Much of the protein in cereal grains is located in the testa and germ which are sifted off during processing. These losses have been evaluated by several workers. Losses of fibre, protein, ash as well as thiamine, riboflavin, niacin, folic acid, pantothenic acid and iron have been reported. On the positive side, Hamad & Fields (1979) reported increases in the protein quality and relative nutritive value (determined using *Tetrahymena pyriformis*) of millet by 14% and maize by 13%. Available lysine, presumably produced by lactic acid bacteria, also increased significantly; phytic acid was completely removed (Oke, 1967). Some workers found increases in riboflavin and niacin.

Attempts have been made to modify the processing of ogi to enhance its nutritional qualities. Banigo *et al.* (1974) dehulled the maize grain in a Palyi mill, milled and then fermented it using cultures of *L. plantarum*, *Streptococcus lactis* and *Saccharomyces rouxii*. The only nutrient loss described occurred on dehulling. Akinrele (1970) showed that ogi produced from a mixed culture of *Lactobacillus* and *Aerobacter* had better nutrients than the unfermented grain and the ogi produced traditionally.

The Federal Institute of Industrial Research, Lagos, developed a protein-enriched ogi for infant food by supplementing it with 10% soya flour (Akinrele, 1970; Akinrele *et al.*, 1970). Apart from the nutritional improvement of ogi, soy flour increased and accelerated the production of organic acids. The acids formed considerably reduced the heavy bean flavour of soya.

Koko and kenkey. Koko and kenkey are two important fermented cereal products in Ghana. The Koko is prepared in much the same way as ogi except that it is fermented as a dough (referred to as 'mbar') and not as a suspension. Figure 23.5 shows the flow diagram of its production. Less material is discarded during the preparation, hence the nutrient loss is not as high as in ogi. Christian (1970) examined the koko dough at an advanced stage of fermentation and found yeasts and lactic acid bacteria. There were homofermentative *Pediococcus cerevisiae* and heterofermentative *Leuconostoc mesenteroides* and *Leuconostoc fermenti*. The latter organisms

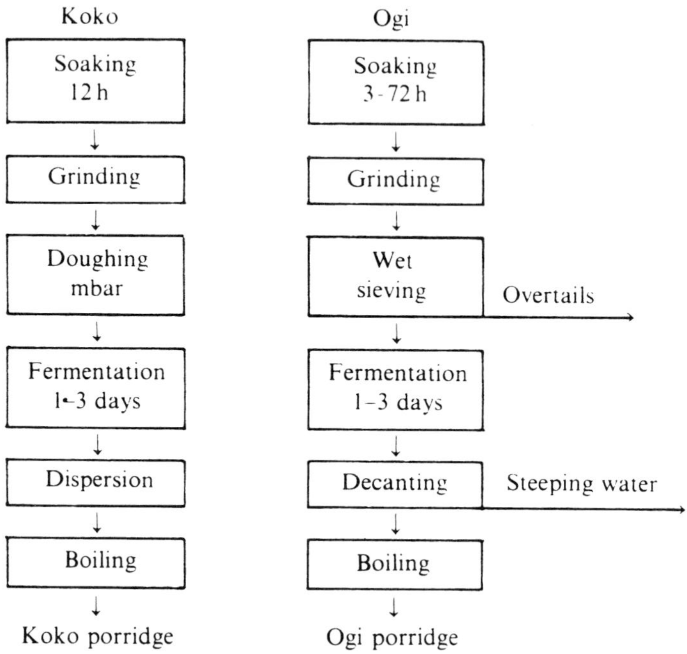

Figure 23.5 Preparation of ogi and koko porridges. (From Muller, 1980.)

probably accounted for the 'dough rise', a leavening action similar to that of the Indian fermented food idli, a mixture of rice and black gram (Mukhergee *et al.*, 1965).

The use of various starter cultures was also investigated. Fermented dough as starter was most satisfactory. Use of starter improved the production of the flavour constituent. Fermented dough used as starter was better than either freeze-dried dough or a pure culture of *Lactobacillus acidophilus* (Andah & Muller, 1972).

Studies have been carried out to determine the most effective methods for preservation (Andah & Muller, 1972). Dehydration by roller-drying, spray-drying, freeze-drying and canning were tried. The volatile flavour constituents were lost by the first two methods. The third method is satisfactory, but it is too expensive for commercial exploitation. The heat treatment involved in the last method leads to gelatinization of the starch and hence is unsatisfactory. Satisfactorily dehydrated ogi was obtained by spray-draying (Adeyinka, 1968).

Kenkey is also fermented then ground into a dough (Figure 23.4) and divided into two parts. One part is cooked into a thick porridge. The other uncooked part is then mixed with the cooked portion to have a mixture. The mixture is moulded into balls and wrapped in dried corn husk or plantain leaves. They are then boiled. There are many varieties of 'kenkey', depending on the localities. In Northern Ghana, sorghum is used for the dough instead of maize.

Uji. This is a sour cereal gruel of East Africa. The basic cereal used for uji production is maize, but mixtures of maize and sorghum or millet in the proportion of 4:1 are also used (Mbugua, 1981, 1984, 1987, 1985). The raw cereal is finely ground and slurried with water in a concentration of 30% (w/v) and allowed to ferment for 2–5 days at room (25°C) temperature. The product is diluted to about 8–10% solids and boiled. It can be further diluted and sweetened with sugar before consumption. Uji production is an acidic fermentation process during which the pH of the slurry reduced to 3.5–4.0 in 40 hours (Mbugua, 1981). Spontaneous uji fermentation is characterized by the sequential growth of the dominant microorganism, where coliforms grow rapidly initially and are later succeeded by the lactic acid bacteria. Early acid production at high rates by the lactic acid bacteria rapidly restricts coliform activity, thereby eliminating the problems of off-flavours and flavour instability. *Lactobacillus plantarum* is mainly responsible for souring of uji, although early activity of heterofermentative strains of *L. fermentum*, *L. cellolbiosus* and *L. buchneri* during the fermentation is evident (Mbugua, 1981). Sucrose is the major fermentable sugar in the uji flours.

Mahewu (Magou). This is a sour maize meal similar to ogi, taken by the Bantus of South Africa. It is prepared by boiling water and adding one part of mash of mealie maize to 9 parts of boiled water. The suspension is cooked for 10 minutes and allowed to cool to 50°C and then transferred to a fermentation vessel. Some wheat flour (about 5% of the maize meal used) is added; this is the source of inoculum. Fermentation takes place in a warm sunny place within 24 hours. The main organism in native-made mahewu is *Streptococcus lactis* (Hesseltine, 1979).

Mahewu is now manufactured on a modern industrial scale as a dry product. The industrial process is outlined in Figure 23.6. The details of the process were given by Hesseltine (1979). Industrial concerns produce liquid mahewu brewed at room temperature, using starters which are pitched daily. Now, pre-cooked, ready-mix powders are manufactured to satisfy the growing Bantu urban population.

The advantages of mahewu over ogi are that: (i) the initial wild fermentation by fungi, etc. is eliminated by boiling the maize meal and the water for steeping; (ii) the mahewu is pre-cooked and requires only mixing before consumption; and (iii) the mahewu has coarse maize particles while ogi contains very fine pasty maize particles.

The average preferable degree of acidity of mahewu is pH 3.5 and almost all the acid is lactic (about 0.4–0.5% titrable acidity).

Maize meal
↓
Mixed in warm water to give 8% solids
↓
Cooked at 15 psi for 15 min
↓
Cooled
↓
Inoculated
(5% wheat flour and adapted pure culture of *Lactobacillus delbrueckii*)
↓
Incubated 30–35 °C with wheat
(If *L. delbrueckii*, incubate at 45 °C)
↓
Fermentation 36 h with mixing only at the beginning of fermentation
↓
Heated for 10–15 min at 7 lb/in^2
↓
Sprayed or drum dried

Figure 23.6 Mahewu fermentation. (From Odunfa, 1985.)

The development of a method of producing mahewu on a large-scale industrial basis created the need for the development of starter cultures. Schweigart & Fellingham (1963) evaluated various lactic acid bacteria. The elevated temperature inhibited undesirable bacteria. However, this method has, as far as could be ascertained, not found industrial application. Van Noort & Spence (1976) both of Jabula Foods Ltd. S.A., produced a more satisfactory mahewu at room temperature by a combination of an acid-producing bacterium, a yeast and a non-acid-producing bacterium; the names of these organisms were however, not disclosed. Dehydrated starter cultures could be made by freeze- or spray-drying (Schweigart, 1971). Addition of buffering salts such as calcium carbonate before drying increased the capacity of the cultures to produce lactic acid. It was also established that dried cultures were less active than wet cultures.

Ogi differs from uji, mahewu and other fermented cereal gruels of eastern and southern Africa in that it is a fine paste product. The latter are coarse and gritty.

Abreh. This is a 'see-through' flaked product made from fermented sorghum in Sudan. Abreh is considered as a 'thirst-quenching product' (Dirar, 1991). In the preparation of Abreh, whole sorghum grains or grain meals are soaked in water overnight, followed by the wet-milling of the grains in about four stages to last for about 4 days, during which time fermentation continues. After this, about two-thirds of the dough are cooked into a thin porridge. The porridge is allowed to rest for 24 hours before mixing back with the remaining mother dough which in its turn acts as a starter for a second round of fermentation which lasts for about 4 days. The batter may be spiced and sieved before being baked on a hot-plate into its thin, dry, transparent flakes called abreh. Abreh flakes are added to water, into which they dissolve, before consumption. The batter may be coloured before baking to improve the look of the flakes.

(b) Baked products. Some acid-leavened breads and pancakes are produced from fermented cereals. In Sudan, the most popular is kisra a fermented sorghum 'bread' with a very sour taste (Perten, 1976; Dirar, 1992). Kisra can be made from either sorghum or millet flour which is ground 'in small country stone mill'. In Ethiopia, 'Enjera' is a popular fermented sour leavened pancake-like bread (Gashe, 1985). It is produced from the flour of the cereal called 'Tef enjera' alone or in combination with other cercals. In the processing, Tef flour with or without flours of sorghum or millet is soaked in water containing old fermented stock water as inoculum for 1–3 days. The batter that settles is collected and part of it is boiled in three parts of water before being mixed back into the original uncooked stock for a short secondary fermentation of 2 hours. The resultant batter is then steam-baked to get Enjera. A wide range of lactic acid bacteria are known to be involved in the fermentation process (Gashe, 1985).

23.3 Fermented alcoholic beverages

African fermented alcoholic beverages differ from most European beers in that they contain a mixture of acids and alcohols. They may be liquid, or thin gruels containing over 90% water, and have a sour taste. The fermentation often involves yeast and lactic acid bacteria. Consequently they are more nutritious in that they contain more vitamins and other essential growth factors.

The practice of acidic alcoholic fermentation is ancient in Africa. Virtually every African locality has its peculiar alcoholic beverage made from the cereal or sugary plant sap predominant in the area. Because the fermentation is often spontaneous and uncontrolled, acetic acid bacteria usually succeed the alcoholic fermentation, thereby giving unacceptable vinegar flavours. The beverages are often consumed in an actively fermenting stage and they have short shelf life.

African alcoholic beverages can be broadly listed under the following headings:

1. From sugary substrates; generally these are from sugary palm saps; they include palm wine and tej.

2. From cereal substrates.
3. From other non-cereal substrates.

A list of some of the alcoholic beverages is given in Table 23.3.

23.3.1 Palm 'wine'

Palm 'wine' is an important alcoholic beverage in West Africa. It is consumed by an approximately 10 million rural population in Southern Nigeria. Its popularity is due to the fact that it can be consumed in a variety of flavours varying from sweet unfermented sugary drinks with little alcohol to sour, well-fermented, vinegary alcoholic drinks. It is produced from sugary palm saps. The most frequently tapped palms are raphia palms (*Raphia hookeri* or *R. vinifera*) and the oil palm (*Elaeis guineense*). The sap is collected by cutting the unexpanded flower spathe and tying a gourd to it. The juice trickles into the gourd and up to 3 litres of sap per day can be collected (Steinkraus, 1979). The fresh palm juice is a sweet, clear, colourless juice containing 10–12% sugar and is neutral in reaction (Bassir, 1962).

Fermentation starts soon after the juice collects in the gourd. The organisms responsible are yeasts, mainly *S. cerevisiae* and *Schizosaccharomyces pombe*, and the bacteria *Lactobacillus plantarum* and *L. mesenteroides* (Faparusi *et al.*, 1973; Okafor, 1975a,b, 1978); other bacterial species vary. There are reports that the yeasts and bacteria originate from the gourd, palm tree, and tapping implements. However, the high sugar content of the juice would seem selectively to favour the growth of yeasts which might originate from the air. The fact that spontaneous fermentation also takes place in plastic containers lends further credence to this view. By 24 hours of fermentation the initial pH is reduced from 7.4–6.8 to 5.5 and the alcohol content ranges from 1.5 to 2.1%. In a 72-hour wine, the ethanol content is from 4.5 to 5.2%, and pH is 4.0. The organic acids present are lactic, acetic and tartaric acids. The changes that occur in the fermenting palm wine are shown in Table 23.4.

Bassir (1967) states that the ethanol content never rises above 7.1%. Thus, palm wine is not really a wine in a strict oenological sense. Like most African alcoholic beverages, palm wine has a very short shelf life. Accumulation of an excessive amount of acetic acid makes it unacceptable to consumers. The bark of a tree (*Saccoglottis gabonensis*) may be added as a preservative. The alkaloid and phenolic compounds which are extracted into the wine have antimicrobial effect (Okafor, 1975a; Ayanru & Bossey, 1976) and can halt the onset of the acetic acid fermentation.

Improved nutritional values have been reported in palm wine. Increases in thiamine, riboflavin and pyridoxine have been reported (van Pee & Swings, 1971). Vitamin B_{12} was also found to increase in fermented wine.

Table 23.3 Fermented alcoholic beverages in Africa

Product name	Country	Substrate	Starter
From sugary substrates			
Palm 'wine'	Nigeria (south) Ghana	Palm sap (oil palm, Raphia palm)	Yeasts (various)
Tej	Ethiopia	Honey	Yeasts
From starchy substrates			
Burukutu	Nigeria (north) Northern Ghana	Guinea corn and cassava	Yeasts and lactic acid bacteria
Pito	Nigeria (Bendel state) Ghana	Guinea corn and maize	Moulds, yeasts and *Lactobacillus* spp.
Kaffir beer	South Africa }	Kaffir corn (or maize)	*Lactobacillus* spp. and yeasts
Busaa (maize beer)	East Africa }	Maize	Yeasts and *Lactobacillus* spp.
Malawa beer	Uganda		*Candida krusei*
Zambian opaque maize beer	Zambia	Maize	Yeasts
Merissa	Sudan	Sorghum	Lactic acid bacteria, acetic acid bacteria
Sheketeh	Nigeria (south)	Maize	Unknown
Bouza	Egypt	Wheat (or maize)	Unknown
Talla	Ethiopia	Sorghum	Unknown
Kishk	Egypt	Wheat and milk	*Lactobacillus* spp., yeasts and *Bacillus* spp.
From other substrates			
Agadagidi (plantain wine)	South-western Nigeria	Plantain	Yeasts and lactic acid bacteria
Cacao wine	Nigeria	Cacao	Yeasts
Alcoholic spirits			
Chang'aa (Nubian gin)	Kenya	Molasses or cane sugar	Yeasts
Ogogoro (Akpeteshi)	Nigeria, Ghana	Palm wine	Yeasts

From Odunfa (1985).

Table 23.4 Progressive changes in the taste, alcoholic content, sugar and acid content of fermenting palm wine

Hours after collection at 28–30°C:	0	3	6–12	24
Taste	Fresh and sweet	Sweet	Slightly sour	Very sour
Alcohol (%)	3.78	4.84	6.32	6.70
Sucrose (%)	6.80	4.18	1.48	0.35
Acetic acid (%)	0.49	0.54	0.57	0.69

Data compiled from Chinnarasa (1968).

23.3.2 Tej

Diluted honey wine has been consumed in various parts of Africa for centuries. Tej is an example of this produced in Ethiopia; it is a yellow, sweet alcoholic wine. One part of honey is mixed with four parts of water. This is poured into a fermentation pot which has been previously smoked over an olive wood, hop stem fire. The pot is covered and incubated in a warm place for 2–3 days after which wax and scum are removed. Boiled hops are added and the mixture is further fermented for another 8 days. The resultant coarse wine is filtered through cloth to get rid of the sediments; the final product is cloudy due to the actively fermenting yeasts and other organisms (Steinkraus, 1979). Tej honey is relatively expensive hence its consumption is restricted to special occasions and holidays.

23.3.3 Burukutu

Burukutu is a popular alcoholic beverage in the Northern Guinea Savanna region of Nigeria, Benin Republic and Ghana. It is brewed from Guinea corn (*Sorghum bicolor*), the prevalent grain in this area.

The following account of the preparation is based on the report by Faparusi *et al.* (1973). Sorghum grains are steeped in water overnight. The grains are then put into a basket and the water drained off. The grains are spread in mats in a bed about 6 cm thick and covered with banana leaves. The grains germinate. They are watered on alternate days and turned over at intervals. Germination continues until the plumule attains a certain length, and this is usually within 4–5 days. The malted grains are spread out in the sun to dry for 1–2 days. The dried malt is ground into a powder.

In the next stage, gari (a farinaceous cassava powder described above) is mixed with a mixture of the ground malt and water. The resulting mixture, gari–malt–water roughly in the ratio 1:2:6, is stirred vigorously and left to ferment for 2 days.

At the end of the fermentation the mixture is boiled for about 4 hours and then left to mature for another 2 days. The resulting drink is a cloudy liquid.

The sorghum malt contains mainly yeasts and moulds. In the fermenting mixture the organisms isolated are yeasts, mainly *Saccharomyces cerevisiae* and *S. chavelieri* and the bacteria *Leuconostoc mesenteroides* and *Lactobacillus* spp.

The acidity of the fermenting mixture falls from about 6.4 to about 4.2 within 24 hours and becomes 3.7 after 48 hours. At the end of the 2-day maturing stage, the dominant microorganisms are *Acetobacter* spp. and *Candida* spp. (Faparusi *et al.*, 1973). The earlier yeasts and lactics are killed by boiling. Unlike some of the other alcoholic beverages, the premium burukutu is that with vinegary taste and odour. Up to 0.4–0.6% of acetic acid occurs in fully matured burukutu beer.

23.3.4 Pito

Nigerian consumption of pito is restricted to the Binis in mid-western part of Nigeria. It is also widely consumed in Ghana. The traditional method for preparing pito is described by Ekundayo (1969).

Cereal grains (maize, sorghum or a combination of both) are soaked in water for 2 days, after which they are malted by leaving for 5 days in baskets lined with moistened banana leaves. The malted grains are ground, mixed with water and boiled. The mash is allowed to cool and filtered through a fine mesh basket. The filtrate is then left overnight until it tastes slightly sour. It is boiled to concentrate it. A starter from the previous brew is added to the cooled concentrate and is left overnight again to ferment. The product is pito, a dark brown liquid with taste varying from sweet to bitter. It contains lactic acid, sugars, amino acids and 3% alcohol. The organisms responsible for souring are *Geotrichum candidum* and *Lactobacillus* spp., while *Candida* spp. is responsible for the alcoholic fermentation.

Lack of information on the predominant organisms and changes in pH makes the report on pito fermentation fairly incomplete. Sefa-Dedeh (1992) describes some field studies, and points out the great variability of this product.

23.3.5 Kaffir beer

Kaffir beer is the traditional beverage drink of the Bantu people of Southern Africa. It is the African fermented food on which most research work has been done, and in fact is the only one being produced by modern industrial processes today.

Kaffir beer is an opaque, effervescent beer with yeasty odour and fruity flavour. It is pink due to sorghum tannins. It contains an average alcohol content of 3.2% and lactic acid content of 0.3%. The beer must not contain acetic acid. If a beer contains little alcohol but is well soured, it is still very acceptable.

Kaffir beer is made from malted kaffircorn (sorghum) and unmalted maize and kaffircorn. Millets may be used in place of these. The method of producing kaffir beer is shown in Figure 23.7. The details of the process were described by Novellie (1968, 1976).

Cleaned and washed kaffircorn grains are steeped for 8–24 hours, then drained and spread out in layers to germinate (at about 30°C) for 5–7 days. The grains are watered for the first 3 days and they are turned and cool air used to reduce heat. Germination continues until the plumule is (2–5 cm) (1–2 in) in length. They are then sun-dried or dried in an oven at 50–60°C and ground. The finished malt contains β-amylase (30–40%), α-amylase (60–70%) and diastatic power of 11.6–17.2 (Lintner). The main steps in kaffir beer brewing are mashing, souring, boiling, conversion, straining and alcoholic fermentation. Warm water (50°C) is used for mashing. Malted sorghum is added to unmalted grains in the ratio of 1:4. Water and inoculum from the previous batch are added. Souring (lactic acid fermentation) takes place at 48–50°C for 8–16 hours. *Lactobacillus* is chiefly responsible. The finished sour mash has a pH of about 3–3.3. The mash is diluted with 2 volumes of water and boiled for 2 hours. It is then cooled to 40–60°C and additional malt added. Conversion proceeds for 2 hours and then the mash is cooled to 30°C.

After straining to remove coarse malt husks, alcoholic fermentation begins using top-fermenting yeast (*Saccharomyces cerevisiae*) as inoculum. Kaffir beer is ready in 4–8 hours at which time. It is still actively fermenting. It has a shelf-life of up to 40 hours.

Acetic acid production by *Acetobacter* is the principal cause of spoilage.

23.3.6 Busaa

Busaa is a Kenyan opaque maize beer. It is similar to the Malawa beer in Uganda and the kaffir beer in South Africa. In fact, there are numerous opaque beers in several African countries, each with a local name. The nutritional value of opaque maize beer is considered superior to that of clear lager beer due to higher contents of crude protein, thiamine and riboflavin.

Busaa is commonly prepared from maize endosperm grits and finger-millet malt, *Eleusine coracana*. Its preparation is smililar to that of kaffir beer. The major steps involved in the preparation are shown in Figure 23.8.

On average, Busaa when ready for consumption contains 0.5–1% lactic acid and 2–4 vol. % ethanol (Nout, 1980a,b).

At the maize souring, the microorganisms involved are lactic acid bacteria and a few yeasts; they are *L. helveticus*, *L. salivarius*, *Pediococcus*

Sorghum (Kaffir corn, red varieties)

↓

Cleaned and washed

↓

Steeped 8–24 h

↓

Germinated 5–7 days at 25–30 °C in depth of 3–4 ft and watered for first 3 days.

↓ Grain must be turned and cool air used to reduce heat

Malt (60–65 % moisture: roots and shoots 1–2 in. in length)

↓

Dried at 50–60 °C

↓

Ground

↓

Finished sorghum malt

↓ Water (50 °C) + unmalted cereal 10 % inoculum from previous batch

Fermentation, 48–50 °C, 8–16 h
(fermentation wash equals one-third volume of beer)

↓

Finished sour mash, pH 3–3·3

↓

Pumped to cooker and diluted with 2 vol of water

↓ Corn grits (occasionally sorghum grits may be added)

Cooked 2 h at atmospheric pressure

↓

Cooled to 40–60 °C

↓

Sorghum malt added

↓

Held for 2 h

↓

Cooled to 30 °C

↓ Top fermenting yeast (*Saccharomyces cerevisiae* NRRL Y-7184) added as pure culture

Strained to remove coarse particles of malt husks

↓

Fermentors, 30 °C, 8–24 h if bulk beer; 0–6 h if sold in cartons

↓

Finished kaffir beer

Figure 23.7 Kaffir beer fermentation on industrial scale. (From Hesseltine, 1979.)

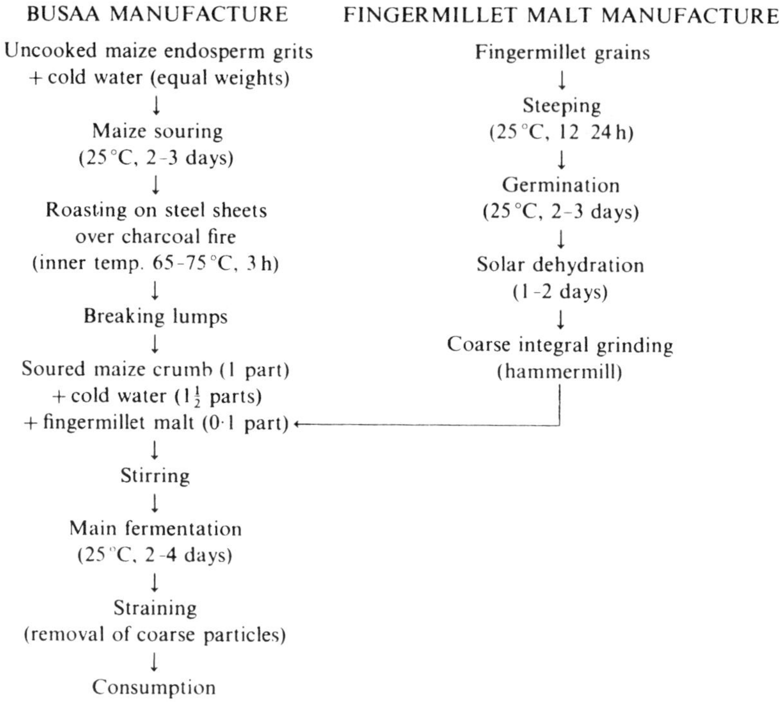

Figure 23.8 A flowsheet for traditional busaa manufacture. (From Nout, 1980a,b.)

damnosus, *P. partulus*, *Candida krusei* and *S. cerevisiae.* The main fermentation is also by a mixture of yeasts and lactics, *C. krusei* and *L. casei* var. *rhamnosus*.

Although the production techniques of kaffir beer and busaa are similar, there are some fundamental differences. The corn souring in Busaa is done before the malt is added. Secondly, souring in kaffir beer is done at about 50°C, thereby providing a selectively favourable temperature for the lactobacilli – perhaps that is why yeasts are excluded. In kaffir beer, the wort resulting from saccharification of maize by sorghum malt is boiled for 2 hours before inoculation with a quantity of previously manufactured beer. The boiling eliminates the lactic acid bacteria from the souring stage. Hence, it is possible to have pure alcoholic fermentation. In Busaa, the main fermentation is by both lactic acid bacteria and yeasts, many of which are carried over from the souring stage. Busaa beer fermentation is thus spontaneous and uncontrolled, having much in common with pito production.

Studies in process optimization of Busaa indicate that mashing for 1 hour at 50°C produced the highest degree of saccharification (Nout, 1980b). A mixed inoculum of *Saccharomyces cerevisiae* and *L. brevis* yielded an acceptable and stable product containing 3.25% ethanol and 0.8% lactic acid. To increase the shelf life, preservation studies showed that pasteurization is sealed bottles had good consumer acceptability. These studies might increase the popularity of Busaa which is at present restricted to the low-income groups.

In Zambia, the opaque maize beer is also the usual alcoholic beverage of farmers and urban workers of low income (Lovelace & Nyathi, 1977). Its production is more similar to kaffir beer. The village brewing process usually involves 3-day germination of maize millet or red sorghum. Dried, pounded malt is added to maize meal porridge and left for 2 days, boiling twice. Fresh malt is added for fermentation lasting 1–2 days. The process thus lasts a week and it is given the local name of 'seven-days'.

23.3.7 Merissa

Merissa is an alcoholic drink of the Sudanese. Though another opaque beer, merissa deserves special mention as a beer prepared by a relatively complex process. The brewing process consists of three distinct phases, 'ajeen' fermentation, a lactic souring of sorghum; 'deboba' fermentation, a starter activating phase, and the merissa fermentation, an alcoholic fermentation phase (see Figure 23.9).

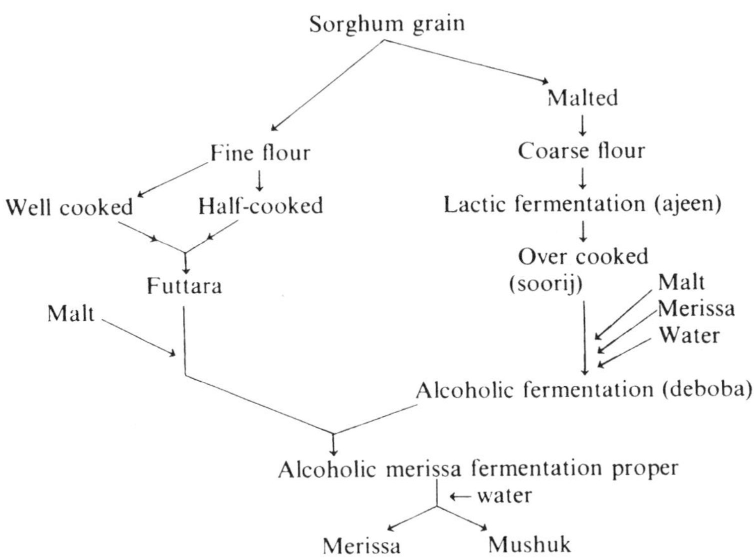

Figure 23.9 Diagrammatic representation of the 'merissa' fermentation process. (From Dirar, 1978.)

The following account of the brewing process is based on the report by Dirar (1978). Some sorghum grains are malted and milled into coarse flour. Ungerminated sorghum is milled into fine flour. One part of the malt flour is mixed with a small amount of water and fermented for 36 hours at room temperature. The fermented sourdough (called 'ajeen') is then cooked without further addition of water until the product is dark brown. This intermediate moisture product (called 'soorij') is extremely sour and has a pleasant, caramelized flavour.

'Soorij' is spread out on a container to cool. Malt (5%), water and an inoculum of previous merissa are added. The mixture is left to ferment for 4–5 hours. The resulting product (called 'deboba') is a vigorously fermenting, thick, dark suspension which is too sour to drink.

Meanwhile, the unmalted sorghum flour is cooked in two equal lots: one lot is half-cooked to a greyish brown paste, the other is well-cooked to a brown paste. The two are mixed together and spread out to cool. The product (called 'futtara') is a gelatinized solid material. After cooking, portions of it are mixed with about 5% malt flour and successively added to the 'deboba'. Fermentation continues for 8–10 hours before the product – merissa – is strained off.

The character of merissa fermentation is similar to that of other African alcoholic beverages. The 'ajeen' fermentation is by lactic acid and acetic acid bacteria and yeasts. The pH is 4, the ethanol content 1% and lactic acid 2.5%. In the final merissa stage of fermentation the significant change is the alcohol content which reaches about 6%. As with all other fermented drinks, there are many types of Merissa; Dirar (1993) provides an extensive discussion of the subject.

23.3.8 Sheketeh

This is another opaque maize beer taken in some restricted areas in Nigeria. The production is similar to that of busaa. Although there are no written accounts about this, the microbiology will most probably be essentially similar to other opaque maize beers.

23.3.9 Bouza

Bouza, a fermented alcoholic wheat beverage in Egypt, has been known since the Pharaohs (Morcos *et al.*, 1973b). The ancient Egyptian bouza was the common drink, but today it is consumed only by the lower-income urban population. Modern bouza is a thick, pale yellow alcoholic beverage with an agreeable taste and producing a sensation of heat. It is prepared from wheat (*Triticum vulgaris*). Coarsely ground wheat grains are placed in wooden basins and kneaded with water into a dough. The dough is cut into thick loaves which are baked very lightly. About one-quarter of the total wheat grains are moistened with water and germinated for 3–5 days, then sun-dried and ground. They are mixed with the bread loaves and soaked in water in a wooden barrel. Some bouza from a previous brew is added as an inoculum. The mixture is left to ferment for 24 hours at room temperature. The product is sieved to remove large particles, diluted with water, and is then ready to consume. The process is rather like that for kvass production in Russia and East Europe.

Like other opaque beers, the shelf life of bouza is very short; it is expected to be consumed within 24 hours. By 24 hours the pH is 3.9–4.0, and alcoholic content is 3.8–4.2%.

23.3.10 Ethiopian talla

Talla is a tan to dark brown home-processed beer with a smokey flavour. There are different varieties of talla with hops or a variety of spice. The smokey flavour is derived from inversion of the fermentation vessel over smouldering wood (Steinkraus, 1979).

Powdered hop leaves and water are placed in the fermentation vessel and allowed to stand for 3 days. Then ground barley or wheat malt is added along with pieces of flat bread which has been previously baked to coal black on the outside. On the fifth day, hop stems are added along with cereal flour made by boiling, toasting and milling sorghum or other grains. On day 5 water is added and fermentation continues for 5–7 days. It is then strained and served.

23.3.11 Sudanese hulu mur drink

Half a batch of sorghum grains are malted, milled and added to the other unmalted half. Both flours are kneaded into a dough (Agab, 1985). Nine ground spices are added and an inoculum of fermented kisra dough is added to the dough. After the dough is fermented it is baked into flat sheets. These sheets are used to flavour cold water and this serves as a refreshing soft drink. Hulu mur drink is popular especially during the Muslims' Ramadan fast festival. Dirar (1993) gives an interesting discussion of this product and its uses.

23.3.12 Kishk

Kishk is a fermented wheat milk mixture common in Egypt and most of the Arab world (Morcos *et al.*, 1973a). The milk used is the surplus which accumulates during the winter and is stored in earthenware containers for use in the summer. The process of preparation of kishk is shown is Figure 23.10. *Lactobacillus* spp. form the main flora of the sour milk. During other stages of kishk making, *Lactobacillus*, yeasts and *Bacillus subtilis* are the

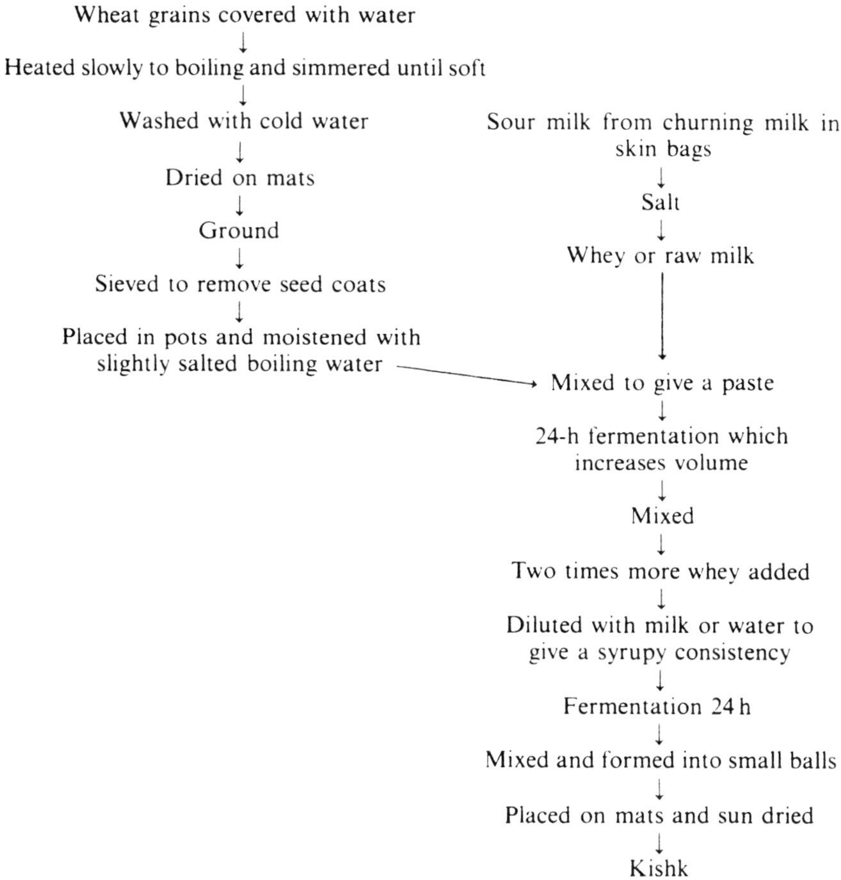

Figure 23.10 Kishk fermentation. (From Odunfa, 1985.)

dominant types. The kishk is usually oven-heated to improve its keeping quality. The balls are brownish and hard. When moistened in water, they rapidly break up.

Kishk is a highly nutritious food, its protein content is 23.5% and it has high digestibility, high biological value and good net protein utilization. Dirar (1993) gives an extended discussion.

23.3.13 Miscellaneous alcoholic beverages

Apart from cereals there are a few other cases when other starchy materials are fermented. One such is the spontaneous fermentation of bananas and plantains to produce a plantain beer called Agadagidi. The hot, humid climatic conditions in Southern Nigeria and Cameroons predispose harvested plantains and bananas to quick over-ripening and spoilage. These overripe bananas are peeled, sliced and soaked in containers to ferment. Yeasts and lactic acid bacteria occur in the fermentation; the microbes presumably utilize the abundant sugars in the ripened bananas. The alcoholic content in the beverage is very low, about 1.5% (Sanni, 1982). The nutrient content of the beverage produced is very high (Ketiku & Scott-Emuakpor, 1975; Sanni & Oso, 1988).

23.3.14 Alcoholic spirits

In many African countries, traditional spirits are obtained by the distillation of fermented local sugary substrates. However, information on these is hard to obtain since the production is illegal in many countries. Since colonial times the view is that the spirits may contain toxic alcoholic components due to lack of scientific quality control; hence, their production was prohibited. Despite this the spirits were produced secretly. However, since the independence of many countries, their production is being legalised and information is just beginning to appear on them.

An example of a local alcoholic spirit is Kenyan Chang'aa (Nubian gin) which is obtained by distillation of fermented sugar or molasses to which components of busaa beer may be added. It is reported that the art of distilling was introduced into Kenya by immigrating Nubians during the First World War. Alcohol content varies from 24.2% to 41.4%, the average alcoholic content of 33 commercial samples is 31.4%. Comparison with Swiss standards showed that the majority of the spirits were of low quality due to a high content of fusel alcohols.

Another local alcoholic spirit is ogogoro (Nig.) akpeteshi (Gh.), a native gin common in Southern Nigeria, Ghana and Cameroons. It is obtained by a crude distillation of palm wine. The alcoholic content varies from 26.8% to 39.9%. In both chang'aa and ogogoro fermentation, yeasts were identified as the fermenting organisms.

23.4 Fermented vegetable proteins

Protein-rich seeds are often fermented to make food condiments. The use of hydrolysed vegetable proteins in seasoning has long been recognized. In Sough East Asia, soybeans are fermented to a variety of products notably shoyu (soy sauce), natto, miso (all in Japan) and tempeh (in Indonesia).

In Africa, many other proteinaceous seeds are fermented to make food condiments. Fermentation markedly improves the digestibility, nutritive value and flavour of the raw seeds. Some of these seeds are in fact not consumed in their unfermented state.

In many countries in Africa protein/calorie malnutrition is a major problem; fermented vegetable proteins therefore have great potential as key protein sources and as basic ingredients for food supplementation. In Nigeria, a business paper, *Business Concord* (of May 6, 1983) reported that about N115 million (100m pounds sterling) was estimated to be spent in 1983 on sauce flavours imported into Nigeria. Sauces are essential features of Nigerian diet; these sauces with slimy vegetables need to be served with the sticky, starchy foods.

The following account of fermented vegetable proteins is limited to those produced in West Africa, particularly Nigeria. Table 23.5 shows a list of the fermented vegetable proteins on which some information is available.

23.4.1 Iru (Dawadawa)

Iru is by far the most important food condiment in Nigeria and many countries of West and Central Africa (Campbell-Platt, 1980; Odunfa, 1986). It is prepared from the seeds of African locust bean. *Parkia biglobosa* Welw. ex Oliv. (syn. *P. clappertoniana*). The African locust bean is a perennial tropical legume tree widespread in the savanna regions of West and

Table 23.5 Fermented vegetable proteins in West Africa

Product	Area of production	Substrate	Microorganisms involved	Reference
Iru	West Africa, especially the northern parts	Locust bean (*Parkia biglobosa*)	*Bacillus subtilis*, *B. licheniformis*, *Staphylococcus saprophyticus*	Odunfa (1981a)
Ogiri	South-western Nigeria	Melon seeds[a] (*Citrullus vulgaris*)	*Bacillus* spp. (predominant), *Escherichia*, *Proteus*, *Pedicoccus*	Odunfa (1981b)
Ogiri-nwan	South-eastern Nigeria	Fluted pumpkin bean (*Telfaria occidentale*)	*Bacillus* spp. (proteolytic)	Odunfa (unpublished)
Ogiri-igbo (same as ogiri-agbor)	Southern Nigeria	Castor oil seed (*Ricinus communis*)	Various *Bacillus* spp., *B. subtilis*, *B. megaterium* and *B. firmus*	Odunfa (unpublished)
Ogiri-saro	Sierra Leone	Sesame seed (*Sesamum indicum*)	*Bacillus* spp.	Odunfa (unpublished)
Ugba	Eastern Nigeria, Igbos	Oil bean (*Pentaclethra macrophylla*)	*Bacillus subtilis*, *Staphylococcus* and *Micrococcus* spp.	Oyeyiola (1981) Obeta (1983)

[a] Cotton seeds are used in some localities.
From Odunfa (1985).

Central Africa. It has been erroneously referred to as *P. filicoidea* in many literature references (Anon., 1970; Odunfa, 1981a). The confusion in the taxonomy of the *Parkia* spp. was resolved by Hugos (1962) and Lowe (1983).

Locust beans are not normally used for food in their natural state. The fermentation makes them edible by increasing their digestibility and removing some objectionable flavours.

Normally iru is used as a food condiment. However, in poor families it is added generously to sauces and soups and this serves as a low-cost meat substitute. As a rising population causes more demand for animal protein sources such as meat and eggs, more people will depend on iru or similar vegetable proteins to supplement their protein intake.

Iru preparation is, like other fermented foods, a traditional family art done in homes. Dried bean seeds are boiled in pots over firewood fires for about 12–24 hours to soften the tough, hard testa and the cotyledons. The seeds are then put in a mortar and pressed with feet to remove the softened testa. Some sand or other abrasive agent may be added to aid in the removal of the testas. The cotyledons are then washed thoroughly and boiled again for 1–2 hours. A softening agent called kuru (containing sunflower seed and kaun), may be added during the second boiling to aid softening of the cotyledons.

After boiling, the seeds are drained and spread in calabash trays in layers of about 10 cm depth. About two or three trays are stacked together and wrapped in many layers of jute sacks and coarse cotton cloth to provide warmth and a humid atmosphere. The seeds ferment there for about 36 hours or longer. Fermented beans have a greyish sticky mucilage covering and a strong ammoniacal smell, thus resembling Japanese natto.

The long boiling period can be shortened by pressure-cooking of the locust beans (Oyewole & Odunfa, 1990c) and a threshing machine has been developed to reduce the drudgery involved in the post-harvest processing of locust bean (Ajayi, 1991).

Iru fermentation is a solid-substrate fermentation. This is unlike most other food fermentations in Africa which are usually in liquid menstrum. The fermentation of iru is an exothermic process; the temperature of the fermenting mash increases gradually from the ambient temperature of 25–30°C to 45°C. The microorganisms isolated from samples prepared in the traditional way include different subspecies of *Bacillus subtilis*, *B. licheniformis* and a few *Staphylococcus* spp. (Odunfa, 1981a; Odunfa & Komolafe, 1989). Table 23.6 lists some changes found during the process.

The *Bacillus* strains from fermenting locust bean had been found to produce glutamic acid (Ogbadu *et al.*, 1990) and extracellular proteinases (Aderibigbe *et al.*, 1990).

Studies were conducted on the biochemical changes and the extent of nutritional improvement during iru fermentation. Locust beans, like most

Table 23.6 Physical and microbiological changes during fermentation of *Parkia* seeds to form iru

Time (h)	Moisture content (%)	Temp. (°C)	pH	Aerobic plate count on APT	Anaerobic plate count on APT
0	43	25	7.0	5.0×10^4	0
12	48	30	7.2	2.4×10^5	0
24	54	42	7.5	1.2×10^6	1.8×10^4
36	56	45	8.1	2.5×10^6	3.0×10^5

From Odunfa (1985).

legumes, contain some indigestible oligosaccharides, notably stachyose and raffinose. Anaerobic bacteria in the intestine metabolize them to different products including gases, thereby causing flatus. Odunfa (1982a) found that the quantities of these flatus-forming oligosaccharides decreased significantly during the first 24 hours of fermentation. This decrease was attributed to the activities of α- and β-galactosidase which hydrolysed the oligosaccharides to simple reducing sugars. The heat generated in the fermenting beans possibly provided the ideal temperatures for the optimal activity of α-galactosidase. Other nutritional improvements during fermentation include the extensive hydrolysis of the seed proteins into amino acids, hence making the beans more digestible. Higher protein and B vitamins (thiamine and riboflavin) were reported in the fermented samples than the unfermented samples, but the level of vitamin C was lower in the fermented product (Eka, 1980). Also toxic components like oxalate and phytic acid were reduced significantly in the fermented samples.

Iru does not keep very well if the fermentation is allowed to continue for long Periods, uninhibited. The methods of preservation vary from place to place. In south-western Nigeria, salt is added; in northern Nigeria and other parts of West Africa, the iru is dried in the sun or smoked.

Iru fermentation is very similar to that of natto prepared from soybean in Japan. In both foods the organism reponsible in *B. subtilis*, the physical changes accompanying the fermentation being quite similar. In fact in the middle belt of Nigeria, local soybean varieties are used in place of locust bean to make the iru. Iru (dawadawa) is now being manufactured industrially by Cadbury Nigeria, although utilizing the traditional feremention process without cultures. This product however, contains, various additives other than locust bean.

23.4.2 Ogiri

Ogiri generally refers to oily pastes made from oil seeds in West Africa; they are used as soup condiments and are generally strong-smelling. The fact that these pastes are called the same name by different ethnic groups, implies that they have a common origin. The different ogiris are shown in Table 23.5. The art of making ogiri in Sierra Leone was most probably introduced by the freed slaves, mostly of Yoruba origin, who settled there.

Different types of ogiri are named after each ethnic group; hence, ogiri-ijebu is consumed among the Ijebus of south-western Nigeria and ogiri-igbo among the Igbos of eastern Nigeria.

Ogiri is a product of the fermentation of seeds of melon (*Citrullus vulgaris* Schrad.). It is consumed by the Ijebu and Ondo tribes in the forest zone of sourth-western Nigeria. Melon seeds are boiled until they are very soft. They are wrapped tightly in leaves (forming packets about 30 cm in diameter) and left to ferment for about 5 days. Thereafter, the fermenting mash is smoked for about 2 hours and the mash is then ground. The resulting oily paste is the ogiri (Table 23.7).

The organisms found consistently in the fermenting mash were *Bacillus* spp. and some genera of Enterobacteriaceae (Odunfa, 1981b). Although Ogundana (1981) claimed to isolate some fungi (*Absidia corymbifera* and *Geotricum candidum*), this looks doubtful since the low oxygen tension and the alkaline pH of the fermenting mash would discourage fungal growth and acitivity.

A remarkable feature of the changes during fermentation is the increased activity of the proteinases as well as the increases in the amount of amino acids (Odunfa, 1982b). As in iru, the later stages in the fermentation are often accompanied by ammonia production.

23.4.3 Miscellaneous fermented vegetable proteins

Other fermented vegetable proteins are ugba from oil bean (*Pentaclethra macrophylla* Benth.), Ogiri-igbo from castor oil seed (*Ricinus communis* Schrad); and ogiri-nwan from fluted pumpkin beans (*Telfairia accidentale*

Table 23.7 Physical and microbiological changes during fermentation of ogiri

Time (days)	Temp. (°C)	pH	Aerobic plate count	Anaerobic plate count
1	28	7.2	1.2×10^4	0.6×10^4
2	29	7.5	2.8×10^4	0.8×10^4
3	30	8.4	5.4×10^5	2.4×10^5
4	30	9.1	7.1×10^6	4.8×10^5
5	30	9.1	1.7×10^7	5.6×10^6

Figures are means of three samples each analysed in triplicate.
From Odunfa (1985).

Hook.), Okpehe (Okpiye) from the seed of *Prosopis africana* and Owoh from cotton seed (*Gossypium*). These are popular in the rain forest of the mid-west and eastern part of Nigeria where *Parkia* trees do not occur.

The fermentations of these seeds follow a similar pattern. They are boiled until they are very soft, the length of time depending on the strength of the testa. They are then left to ferment in a pot for about 3–5 days. Various species of proteolytic *Bacillus* spp. are responsible for the fermentation. Salt may be added before the fermentation. This possibly explains the presence of the salt-tolerant *Staphylococcus* spp. Fermentation of castor oil seed increased the levels of free amino acids, fatty acids and iodine (Anosike & Equatu, 1980). Of special interest in the fermentation of castor oil seed is the toxicological aspect. This seed contains a toxic protein, ricin, and other toxic constituents, ricinine and ricinoleic acid (Merck Index, 9th edn). The toxicity of castor oils is eliminated during the fermentation (Odunfa, 1985).

The fermentation of cocoa (*Theobroma cacao*) is different from all the others outlined above for two reasons. First, the fermentation is neither an indigenous nor a traditional technology; it arises from the acquired agricultural culture of planting cocoa for export. Secondly, the fermentation does not result directly in food production but in giving aroma, flavour and colour to the beans which are used in production of other foods such as chocolates, cacao drinks, etc.

The common method of fermentation involves putting the wet beans in a basket lined with plantain leaves which are also used to cover the beans. The beans are left to ferment for 6 days, during which time considerable heat is generated.

The fermentation of cocao beans is both aerobic and anaerobic and occurs in many stages. During the first 24 hours, the abundant sugars in the acidic pulp are broken down into ethanol by yeasts, notably *Candida krusei*; lactic acid bacteria also ferment the sugars and citiric acid in the pulp to lactic acid. Following these reactions is the oxidation of the ethanol to acetic acid by acetic acid bacteria. The intense fermentation by these microorganisms generates enough heat in the beans to kill the bean embryo. In the latter stages of fermentation there is an increase in the proportion of aerobic spore-forming bacilli, notably *Bacillus subtilis*, *B. cereus* var. *mycoides* and *B. megatherium*. These are capable of generating unacceptable levels of volatile fatty acids if the fermentation continues for too long.

The death of the embryo and the continual contact with an acid liquor at a high temperature develop the flavour precursors which are responsible for the flavour of the roasted cacao beans (Hansen & Welty, 1968). During the fermentation the pH rises from about 3 to 7 and the temperature may reach 50°C. This process is further discussed in Chapter 5.

23.5 Fermented animal proteins

23.5.1 Fermented milk

A report by Nout (1981) indicates that 53% of the milk produced in the cooler Kenyan highlands is kept within the household and consumed as fermented soured milk ('maziwa lala'). The obvious aim of this fermentation is thus to preserve the milk in the absence of refrigeration facilities. Various additives such as wood ash, animal blood, urine or sometimes leafy vegetables may be added depending on the custom in the area. These additives have effects on the dominant microbes and the shelf life of the soured milk. Nout (1981) suggested that the addition of small amounts of urine, containing traces of H_2O_2 and thiocyanate, is analogous to the activated lactoperoxidase system (Bjorck, 1978) which inhibits catalase positive, Gram-negative bacteria in milk.

The flora of the fermented milk is characterized by *Streptococcus lactis. S. thermophilus and Lactobacillus casei.*

Another fermented milk product is Nono (Hausa), a native curdled milk commonly taken in the Guinea Savanna Zone in West Africa. The milk is fermented into a curd. Eka & Ohaba (1971) carried out a microbiological study of the nono and found a wide variety of bacteria among which non-haemolytic *Streptococcus* and *Lactabacillus* were abundant.

Other studies confirmed the involvement of a wide array of microorganisms, but with *Lactobacillus acidophilus* and *Lactococcus cremoris* predominating in milk fermentation for 'nono' production (Akinyanju, 1989). 'Nono' is a highly soured milk product (pH 4.1–4.4, and titratable acid 0.54–0.93%) (Bankole & Okagbue, 1992). The use of a mixed culture made of *Lactococcus diacetilactis*, *Lactococccus cremonis* and *Lactobacillus brevis* has been reported to produce a good-quality 'nono' (Faresi & Muzondo, 1989). Similar fermented milks are 'Iben' and 'raib' produced in Morocco (Hamama, 1992) and Zimbabwe (Feresu, 1992).

23.5.2 Fermented fish

In South East Asia, fermented fish products are very popular (Sundhaghul *et al.*, 1975; Bhumiratana, 1980; Del Rosario, 1980). However, there are few records of fermented fish products in West Africa. Guedj is reported as a fermented dried fish product popular in Senegal (Toury *et al.*, 1970). Unsold fresh fish is piled together for about 24 hours in the open air. During this period the fish undergoes putrefactive fermentation by its own enzymes and endogenous bacteria. It is then eviscerated (sometimes the larger species are filleted to shorten the period of drying), and soaked in salty sea water in wooden buckets. This water is changed once a week when it has become too dirty. Finally the fishes are spread out on straw mats to dry in the sun for 2–4 days.

Another product is 'Bonome' (stink fish), a Ghanaian fermented fish product similar to guedj (Nerguaye-Tetteh *et al.*, 1978).

23.6 Summary

The organisms involved in the fermentation of African foods are restricted to a few groups of yeasts and bacteria. The cassava is fermented by lactic acid bacteria and some yeasts. The cereals for gruel and alcoholic beverages are equally fermented mainly by yeasts and lactic acid bacteria; the proportion of the yeasts is more in the alcoholic beverages. The vegetable proteins are fermented mainly by proteolytic spore-forming bacilli. Information on the fermentation of proteins of animal origin is scanty and few generalizations can be made at this stage.

An important difference between the African fermented foods and those of South East Asia is the fact that moulds are not common in the African fermentations. Mould fermentation of soybean, rice and other substrates is quite common in South East Asia. Shoyu (soy sauce), miso, tempeh and sufu are all products of mould fermentations. Moreover, whereas koji or starter cultures are features of Japanese and many Asian food fermentations, starter cultures are rarely used in Africa.

With the notable exception of mahewu and kaffir beer among the Bantus of southern Africa, the production of African fermented foods is largely a family art characterized by chance inoculation, poor hygiene, variable quality and short shelf life. Interest in studies on a scientific basis of these fermentations has begun. There are now in fact a few reports on the attempts to mechanise some aspects of the production. For example, the mechanization of the production of gari in Nigeria is successful and is gradually growing. These developments are desirable and will consolidate or resuscitate the popularity of these foods among the indigenes.

References

Abe, M.O. & Lindsay, R. (1978) Evidence for the lactic streptococcal role in Nigerian acidic cassava, *M. esculenta* Crantz. *Journal of Food Protection*, **41**, 781–4.

Abioye, J.O.K. (1981) *A microbiological study on the fermentation of cassava for the production of 'fufu' a staple food in Southern Nigeria*, MSc Thesis, Department of Botany, University of Ibadan, Nigeria.

Achi, O.K. (1992) Microorganisms associated with natural fermentation of *Prosopis africana* seeds for the production of okpiye. *Plant Foods for Human Nutrition*, **42**, 297–304.

Adeniji, M.O. (1976) Fungi associated with the deterioration of gari. *Nigerian Journal of Plant Protection*, **2**, 74–7.

Aderibigbe, E.Y., Schink, B. & Odunfa, S.A. (1990) Extracellular proteinases of *Bacillus* spp. isolated from fermented African locust bean, Iru. *Food Microbiology*, **7**, 281–93.

Adeyinka, O. (1968) *Feasibility of mechanized processing of 'Infant Ogi' from Nigerian cereals.* Federal Ministry of Industries, Lagos, Nigeria.

Agab, K.A. (1985) Femented food products; 'hulu-mar'' drink made from *Sorghum bicolor. Food Microbiology*, **2**, 147–55.

Ajayi, O.A. (1991) Design of a thresher for locust bean. *Agricultural Mechanization in Asia, Africa, and Latin America*, **22**, 21–4.

Akinrele, I.A. (1964) Fermentation of cassava. *Journal of the Science of Food and Agriculture*, **15**, 589–94.

Akinrele, I.A. (1970) Fermentation studies on maize during the preparation of a Traditional African starch cake food. *Journal of the Science of Food and Agriculture*, **21**, 269.

Akinrele, I.A., Cook, A.S. & Holgate, R.A. (1962) The manufacture of gari from cassava in Nigeria. *1st Int. Congress of Food Science and Technology, Proceedings*, Vol. 4, pp. 633–44.

Akinrele, I.A., Adeyinka, O., Edwards, C.C.A., Olatunji, F.O., Dina, J.A. & Koleoso, O.A. (1970) *The development and production of soy-ogi, a corn based complete protein food FIIRO*, Research report no. 42.

Akinyanju, J.A. (1989) Characterization and production process of 'Nono' – A Nigerian fermented milk food. *Chem. Mikrobiol. Technol. Lebensm.*, **12**, 14–9.

Alozie, T.C., Rotimi, C.N. & Oyibo, B.B. (1980) Production of aflatoxin by *Aspergillus flavas* (UBMI) in some Nigerian indigenous beverages and food stuffs. *Mycopathol. Mycol. Appl.*, **70**, 125–8.

Ampe, F., Keleke, S., Robert, H. & Brauman, A. (1995) The role and origin of pectin-degrading enzymes during cassava retting, In *Cassava Food Processing* (eds T.A. Egbe, A. Brauman, D. Griffon & S. Treche), Orstom, France, pp. 331–44.

Andah, A. & Müller, H.G. (1972) Studies on *koko*, a Ghananian fermented maize porridge. *Gh. Jnl. Agric. Sci.*, **6**, 103–8.

Anon (1970) *List of Foods Used in Africa*, 2nd edn, FAO, Rome.

Anosike, E.O. & Equatu, C.K. (1980) Biochemical changes during fermentation of castor oil (*Ricinus communis*) seeds for use as a seasoning agent. *Qual. Plant Pl. Foods Hum Nutr.*, **30**, 181–5.

Ayanru, D.K.G. & Bossey, S. (1976) Effects of *Sacoglottis gabonensis* Bark in palm wine yeasts. *Niger. J. Pl. Prot.*, **2**, 64–9.

Banigo, E.O.I. & Müller, H.O. (1972) Carboxylic acid patterns in ogi fermentation. *Journal of the Science of Food and Agriculture*, **23**, 101–11.

Banigo, E.O.I., de Man, J.M. & Duitchaever, C.I. (1974) Utilisation of high lysine corn for the manufacture of ogi using new improved system. *Cereal Chemistry*, **52**, 559–72.

Bankole, M.O. & Okagbue, R.N. (1992) Properties of 'Nono', a Nigerian fermented milk food. *Ecology of Food and Nutrition*, **27**, 145–9.

Barber, L., Ibiama, E.A. & Achinewhu, S.C. (1989) Microorganisms associated with fermented fluted pumpkin seeds. *(Telferia occidentalis) International Journal of Food Science and Technology*, **24**, 189–93.

Bassir, O. (1962) Observations on the fermentation of palm wine. *West African Journal of Biology and Applied Chemistry*, **6**, 20–5.

Bassir, O. (1967) Some Nigerian wines. *West African Journal of Biology and Applied Chemistry*, **10**, 42–5.

Bhumiratana, A. (1980) Traditional fermented foods in Thailand, in *Proceedings of the Conference on Oriental Fermented Foods*, FIRDI, Taiwan, pp. 58–70.

Bjorck, L. (1978) Antibacterial effect of the lactoperoxidase system on psychrotrophic bacteria in milk. *J. Dairy Res.*, 109–18.

BOSTID (1992) Application of biotechnology to traditional fermented foods. Report of an Ad-Hoc Panel of the Board on Science and Technology for International Development, (BOSTID). National Academy Press, USA, pp. 3–7.

BOSTID (1996) *Lost Crops of Africa, Vol. 1, Gramp.* National Research Council, USA. National Academy Press, pp. 59–76.

Campbell-Platt, G. (1980) African locust bean (*Parkia species*) and its west African fermented food product, *dawadawa. Ecol. Food Nutrit.*, **9**, 123–32.

Chinnarasa, E. (1968) *The preservation and bottling of palm wine*. Federal Institute of Industrial Research, Nigeria. Research Report No. 38.
Christian, W.F. (1970) Lactic acid bacteria in fermenting maize dough. *Ghana. J. Sci.*, **10**, 22–8.
Cock, J.H. (1985) *Cassava – a new potential for a neglected crop*, West View Press, Boulder and London, pp. 1–12.
Collard, P. & Levi, S. (1959) A two-stage fermentation of cassava. *Nature*, **183**, 620–1.
del Rosario, R.R. (1980) Traditional Filipino fermented foods, In *Proceedings of the Conference on Oriental Fermented Foods*, FIRDI, Taiwan, pp. 71–87.
Dirar, H.A. (1978) A microbiological study of Sudanese Merissa brewing. *Journal of Food Science*, **43**, 1683–6.
Dirar, H.A. (1991) The indigenous fermented foods and beverages of Sudan, in *Applications of Biotechnology to Food Processing in Africa*, United Nations Industrial Development Organisation Paper 92-55931, pp. 23–40.
Dirar, H.A. (1992) Sudan's fermented food heritage, in *Application of biotechnology to Traditional Fermented Foods*. Report of an Ad-Hoc Panel of the Board on Science and Technology for Nutritional development. National Academy Press, USA, pp. 27–34.
Dirar, H.A. (1993) *The indigenous fermented foods of the Sudan. A study in African food and nutrition*, CAB International Wallingford, Oxon, UK.
Dougan, J., Robinson, J.M., Sumar, S., Howard, G.E. & Coursey, D.G. (1983) Some flavouring constituents of cassava and of processed cassava products. *Journal of the Science of Food and Agriculture*, **34**, 874–84.
Eka, O.V. (1980) Effect of fermentation on the nutrient status of locust beans. *Food Chemistry*, **5**, 303–8.
Eka, O.V. & Ohaba, J.A. (1971) Microbiological examination of Fulani milk (*Nono*) and butter (*man shanu*). *Nigerian Journal of Science*, **11**, 113–22.
Ekpechi, O.L., Dimiriadou, A. & Fraser, R. (1966) Goiterogenic acitivity of cassava, a staple Nigerian Food. *Nature*, **210**, 1137–8.
Ekundayo, J.A. (1969) The production of pito, a Nigerian fermented beverage. *Journal of Food Technology*, **4**, 217–25.
Ekundayo, J.A. (1980) An appraisal of advances in biotechnology in Central Africa, In *Fungal Biotechnology*, Vol. 4 (eds J.E. Smith, D.R. Berry & B. Kristiansen), London, Academic Press, pp. 224–70.
Essers, A.J.A. & Nout, M.J.R. (1989) The safety of dark, moulded cassava flour compared with white. A comparison of traditionally dried cassava pieces in North East Mozambique. *Tropical Science*, **29**, 261–8.
FAO (1991) *Food Outlook*, May, 1991.
Faparusi, S.I., Olofinboba, M.O. & Ekundayo, J.A. (1973) The microbiology of burukutu beer. *Z. Allg. Mikrobiologie*, **13**, 563–8.
Feresu, S. (1992) Fermented milk products in Zimbabwe, In *Application of Biotechnology to Traditional Fermented Foods*. Report of an Ad-Hoc Panel of the Board on Science and Technology for International Developments. Natioal Academy Press, USA, pp. 80–5.
Feresu, S. & Muzondo, M.I. (1989) Factors affecting the development of two fermented milk products in Zimbabwe. *MIRCEN Journal*, **5**, 349–55.
Fields, M. (1980) Improvement in nutritive quality of cereals and legumes by fermentation and germination. *Bulletin Technico Labal.*, **1**, 14–6.
Fields, M.L., Hamad, A.M. & Smith, D.K. (1981) Natural lactic acid fermentation of corn meal. *Journal of Food Science*, **46**, 900–2.
Gashe, B.A. (1985) Involvement of lactic acid bacteria in the fermentation of Tef (*Eragrotis tef*), an Ethiopian fermented food. *Journal of Food Science*, **80**, 800–1.
Giraud, E., Gosselin, L. & Raimbault, M. (1992) Degradation of cassava limamarin by lactic acid bacteria. *Biotechnology Letters*, **14**, 593–8.
Giraud, E., Gosselin, L. & Raimbault, M. (1993) Production of a *Lactobaillus plantarum* starter with linamarinase and amylase activities for cassava fermentation. *Journal of the Science of Food and Agriculture*, **62**, 77–82.
Hamad, A.M. & Fields, M.L. (1979) Evaluation of the protein quality and available lysine of germinated and fermented cereals. *Journal of Food Science*, **44**, 456–9.

Hamama, A. (1992) Moroccan traditional fermented dairy products, In *Application of Biotechnology to Traditional Fermented Foods.* Report of an Ad-Hoc Panel of the Board on Science and Technology for International development. (BOSTID). National Academy Press, USA, pp. 73–9.

Hansen, A.P. & Welty, R.A. (1968) Microflora of cacao beans. *Mycopathol. Mycol. Appl.*, **44**, 309–16.

Hesseltine, C.W. (1979) Some important fermented foods in mid-Asia, the Middle East and Africa. *Journal of the American Oil Chemists' Society*, **56**, 367–74.

Hugos, T.H. (1962) A revision of the genus *Parkia* R. Br. (Mim) in Africa. *Acta Botanica Neerlandica*, **11**, 231–65.

Idowu, I. (1990) Cassava: Gari technology makes progress in Nigeria's agro-industrial development. *Der Tropenlandwirt Zeitsdorift fur die landwirtschaft in den Tropen and subtropen*, **91**, 51–64.

Igbeka, J.C. (1995) Recent development in cassava frying operation and equipment used for gari production in Nigeria, In *Cassavta Food Processing* (eds T.A. Egbe, A. Branman, D. Griffon & S. Treche), ORSTOM, France, pp. 581–93.

Ikediobi, C.O. & Onyike, E. (1982) Linamarase activity and detoxification of Cassava (*Manihot esculenta*) during fermentation. *Agriculture and Biological Chemistry*, **46**, 1667–9.

Ikenebomeh, M.J., Kok, R. & Ingram, J.M. (1986) Pocessing and fermentation of the African locust been (*Parkia filicoidea* Welw.) to produce Dawadawa. *Journal of the Science of Food and Agriculture*, **37**, 273–82.

Ketiku, O.A. & Scott-Emuakpor, M.M. (1975) The nutrient content of plantain and banana beverages. *Food and Nutrition of Africa*, **14**, 59–60.

Lagunna, J. & Carpenter, N.J. (1951) Raw versus processed corn in niacin-deficient diets. *Journal of Nutrition*, **45**, 21–8.

Lovelace, C.E.A. & Nyati, C.B. (1977) Estimation of fungal toxins, zearalonone, and aflatoxin contaminating opaque maize beer in Zambia. *Journal of the Science of Food and Agriculture*, **28**, 288–92.

Lowe, J. (1983) Personal communication.

Mbugua, S.K. (1981) *Microbiological and biochemical aspects of* uji *(an East African sour cereal porridge) fermentation and its enhancement through applications of lactic acid bacteria*, PhD Thesis, Cornell University, Ithaca, USA.

Mbugua, S.K. (1987) The nutritional and fermentation characterisics of Uji produced from dry-milled maize flour and whole wet-milled maize flour and whole wet-milled maize. *Chem. Microbiol. Technol. Lebensm.*, **10**, 154–61.

Miracle, P.M. (1966) *Maize in Tropical Africa*, Wisconsin Press, London.

Morcos, S.R., Hegazi, S.M. & El-Damhoughy, S.T. (1973a) Fermented foods in common use in Egypt. I. Nutritive value of kishk. *Journal of the Science of Food and Agriculture*, **24**, 1153–6.

Morcos, S.R., Hegazi, S.M. & El-Damhoughy, S.T. (1973b) Fermented Foods in common use in Egypt. II: The Chemical composition of bouza and its ingredients. *Journal of the Science of Food and Agriculture*, **24**, 1157–61.

Mukhergee, S.K., Albury, M.N., Pederson, C.S., Van Veen, A.G. & Steinkraus, K.H. (1965) The role of *Leuconostoc mesenteroides* in leavening the batter of Idli, fernented food of India. *Applied Microbiology*, **13**, 227–31.

Müller, H.G. (1970) Traditional cereal processing in Nigeria and Ghana. *Ghana Journal of Agricultural Science*, **3**, 187–95.

Müller, H.G. (1980) Fermented cereal products of tropical Africa, in *Advances in Biotechnology* II (eds M. Moo-Young & C.W. Robinson), Pergamon Press, New York, pp. 541–5.

Nerguaye-Tetteh, G., Eyeson, K.K. & Tetteh-Marmon, J. (1978) Studies on 'Bonome', a Ghanaian fermented fish product. *Ghana Journal of Agricultural Science*, 21–6.

Ngaba, R.R. & Lee, J.S. (1979) Fermentation of cassava. *Journal of Food Science*, **44**, 1570–1.

Njoku, H.O., Ogbulie, J.N. & Nnubia, C. (1990) Microbiological study of the traditional processing of African oil bean (*Pentaclethra macrophylla* Bebntham) for ugba production. *Food Microbiology*, **7**, 13–26.

Nout, M.J.R. (1979) The manufacture and composition of Chang'aa (Nubian gin). *Lebensm. Wiss. Technol.*, **12**, 212–16.

Nout, M.J.R. (1980a) Microbiological aspects of the traditional manufacture of Busaa, a Kenyan opaque maize beer. *Chem. Mikrobiol. Technol. Lebensm*, **6**, 137–42.
Nout, M.J.R. (1980b) Process development and presrvation of Busaa, a Kenyan traditional opaque maize beer. *Chem. Mikrobiol. Technol Lebensm.*, **6**, 175–82.
Nout, M.J.R. (1981) The role of lactic acid fermentation in the context of food manufacture in Kenya, In *GIAM VI. Global Impacts of Applied Microbiology* (eds S.O. Emejuaiwe, O. Ogunbi & S.O. Sanni), Academic Press Inc., London, pp. 169–75.
Novellie, L. (1966) Bantu beer – popular drink in South Africa. *International Brewer and Distiller*, **1**, 27–31.
Novellie, L. (1968) Kaffir beer brewing, ancient art and modern industry. *Wallerstein Lab. Comm.*, **31**, 17–29.
Novellie, L. (1976) Beverages from sorghum and millets, in *Proc. International Association for Cereal Chemists*, Symposium, Vienna, Austria, pp. 73–7.
Obeta, J.A.N. (1983) A note on the micro-organisms associated with the fermentation of seeds of African oil bean seed (*Pentaclethra macrophylla*). *Journal of Applied Bacteriology*, **54**, 433–5.
Odunfa, S.A. (1981a) Microorganisms associated with fermentation of African locust bean (*Parkia filicoidea*) during *iru* preparation. *Journal of Plant Foods*, **3**, 245–50.
Odunfa, S.A. (1981b) Microbiology and amino acid composition of Ogiri – a food condiment from fermented melon seeds. *Die Nahrung*, **25**, 811–6.
Odunfa, S.A. (1982a) Carbohydrate changes in fermenting locust bean during *iru* preparation. *Quality of Plant Foods for Human Nutrition*, **32**, 3–10. (See also *Journal of Plant Foods*, 1982, **4**, 105–10.)
Odunfa, S.A. (1982b) Biochenical changes during melon fermentation for ogiri production. *Quality of Plant Foods for Human Nutrition*, **32**, 11–8. (See also *Journal of Plant Foods*, 1982, **4**, 111–6.)
Odunfa, S.A. (1985) African fermented foods, In *Microbiology of Fermented Foods, Volume 1* (ed. B.J.B. Wood), Elsevier Applied Science Publishers, London, pp. 155–91.
Odunfa, S.A. (1985) Microbiological and toxicological arpects of fermentation of castor oil seeds for 'Ogiri' production. *Journal of Food Science*, **50**(6), 1758–9.
Odunfa, S.A. (1986) Dawadawa, In *Legume-Based Fermented Foods* (eds N.R. Reddy, M.D. Pierson & D.K. Salunkhe), CRC Press, Boca Raton, Florida, pp. 173–89.
Odunfa, S.A. & Adeyele, S. (1985) Microbiological changes during the traditional production of *ogi-baba* a West African fermented sorghum gruel. *J. Cereal Science*, 3, 173–80.
Odunfa, S.A. & Komolafe, O.B. (1989) Nutritional characteristics of *Staphylococcus* species from fermenting Africa locust bean (*Parkia biglobosa*). *Die Nahrung*, 33, 607–15.
Ofuya, C.O. & Nnajiofor, C. (1989) Development and evaluation of a starter culture for the industrial production of gari. *Journal of Applied Bacteriology*, **66**, 37–42.
Ogbadu, L.J., Okagbue, R.N. & Ahmad, A.A. (1990) Glutamic acid production by *Bacillus* isolates from Nigeria fermented vegetable proteins. *World Journal of Microbiology and Biotechnology*, **6**, 377–82.
Ogundana, S.K. (1981) The production of ogiri, a Nigerian fermented soup condiment. *Lebensm-Wiss. Technologie*, **13**, 334–6.
Okafor, N. (1975a) Preliminary microbiological studies on the preservation of palm wine. *Journal of Applied Bacteriology*, **38**, 1–7.
Okafor, N. (1975b) Microbiology of Nigerian palm wine with particular reference to bacteria. *Journal of Applied Bacteriology*, **38**, 81–8.
Okafor, N. (1977) Microorganisms associated with cassava fermentation for gari production. *Journal of Applied Bacteriology*, **42**, 279–84.
Okafor, N. (1978) Microbiology and biochemistry of oil-palm wine. *Advances in Applied Microbiology*, **24**, 237–56.
Okafor, N. (1990) Traditional alcoholic beverages of tropical Africa: strategies for scale-up. *Process Biochemistry International*, **25**, 213–20.
Okafor, N. & Ejiofor, A.O. (1990) Rapid detoxification of cassava mash fermenting for garri production following inoculation with a yeast simultaneously producing linamarase and amylase. *Process Biochemistry International*, **25**, 82–6.
Okafor, N., Ijioma, B. & Oyolu, C. (1984) Studies on the microbiology of cassava retting for foo-foo production. *Journal of Applied Bacteriology*, **56**, 1–13.

Okagbue, R.N. & Bankole, M.O. (1992) Use of starter cultures containing *Streptococcus diacetilactis*, *Lactobacillus brevis* and *Saccharomyces cerevisiae* of Nigerian nono. *World Journal of Microbiology and Biotechnology*, **8**, 251–3.

Oke, O.L. (1967) Chemical studies on the Nigerian foodstuff – ogi. *Food Technology*, **21**, 98–100.

Onyekwere, O.O., Akinrele, I.A. & Koleose, O.A. (1989) Industrialization of Ogi frmentation, In *Industrialization of indigenous fermented foods* (ed. K.H. Steinkraus), Marcel Dekker, pp. 329–61.

Osuntokun, B.O. (1969) Cassava diet and a chronic degenerative neuropathy; epidemological study. *Nigerian Journal of Science*, **3**, 3–15.

Osuntokun, B.O. (1972) Chronic cyanide neutrotoxicity and neuropathy in Nigerians. *Pl. Fds. Ham. Nutr.*, **2**, 215–66.

Osuntokun, B.O. (1973) Ataxic neuropathy associated with high cassava diets in West Africa, In *Chronic Cassava Toxicity*, IDRC, Ottawa, pp. 127–38.

Oyenuga, V.A. (1968) *Nigeria's Foods and Feedingstaffs: Their Chemistry and Nutritive values*, 3rd edn, Ibadan University Press, Ibadan, Nigeria.

Oyewole, O.B. (1992) Cassava processing in Africa, in *Application of Biotechnology to Traditional Fermented Foods*. Report of an ad-hoc panel of the Board of Science and Technology for international development. National Academy Press, USA, pp. 89–92.

Oyewole, O.B. & Akingbala, J.O. (1993) *Acha (Digitaria exilis)*, a little known cereal with high potential, In *Lost Crops of Nigeria* (eds J.A. Okojie & D.U.U. Ukali), UNAAB, pp. 323–6.

Oyewole, O.B. & Odunfa, S.A. (1990a) Optimization of cassava fermentation for 'fufu' production. Effects of single starter cultures. *Journal of Applied Bacteriology*, **68**, 49–54.

Oyewole, O.B. & Odunfa, S.A. (1990b) Characterization land distribution of lactic acid bacteria in cassava fermentation during 'fufu' production. *Journal of Applied Bacteriology*, **68**, 145–52.

Oyewole, O.B. & Odunfa, S.A. (1990c) Effect of cooking on water absorption and ease of dehulling in preparation of African locust bean for 'iru'. *International Journal of Food Science and Technology*, **25**, 461–3.

Oyewole, O.B. & Odunfa, S.A. (1991) Characterization of lactobacilli in fermenting cassava and their evaluation as cassava starter cultures, In '*Traditional African Foods-Quality and Nutrition*' (eds A. Westby & P.J.A. Reilly), IFS, Sweden, pp. 145–50.

Oyewole, O.B. & Odunfa, S.A. (1992a) Extracellular enzyme activities during cassava fermentation for fufu production. *World Journal of Microbiology and Biotechnology*, **8**, 71–2.

Oyewole, O.B. & Odunfa, S.A. (1992b) Effect of processing variables on cassava fermentation for 'fufu' production. *Tropical Science*, **32**, 231–40.

Oyeyiola, G.P. (1981) *Studies on the microorganisms isolated from ugba, a Nigerian indigenous fermented food* MSc dissertation, Department of Botany, University of Ibadan, Nigeria.

Perten, H. (1976) UNDP/FAO sorghum processing project in the Sudan, In *Tropical Product Institute Conference Papers*, Vienna, pp. 53–5.

Raimbault, M. (1995) Importance of lactic acid bacteria in cassava fermentation, In *Cassava Food Processing* (eds T.A. Egbe, A. Branmar, D. Griffon & S. Treche). ORSTEM, France, pp. 259–75.

Roellofsen, P.A. (1958) Fermentation, drying and storage of cacao beans. *Advances in Food Research*, **8**, 225–96.

Sanni, A. (1982) *Studies on the plantain wine fermentation*. MSc Thesis, Department of Botany, University of Ibadan, Nigeria.

Sanni, A.I. & Oso, B.A. (1988) Nutritional studies on agadagidi. *Die Nahrung*, **2**, 169–77.

Schweigart, F. (1971) The drying of lactic acid bacteria cultures for *mahewu* production. *Lebensm. Wiss. Technol.*, **4**, 20–3.

Schweigart, F. & Fellingham, J.A. (1963) A study of fermentation in the production of *Mahewu*, an indigenous sour maize beverage of Southern Africa. *Milchwissenschaft*, **18**, 241–6.

Sefa-Dedeh, S. (1992) Cereal based traditional beverages from Ghana: process and product characteristics of Pito, In *Proceedings of Regional Workshop on Traditional African Foods – Quality and Nutrition* (eds A. Westaby & P.J.A. Reilly), International Foundation for Science, Stockholm, Sweden, pp. 1–7.

Steinkraus, K.H. (1979) Nutritionally significant indigenous foods involving an alcoholic fermentation, In *Fermented Food Beverages in Nutrition* (eds C.F. Gastineau, W.J. Darby & T.B. Turner), Academic Press, London, pp. 35–59.

Steinkraus, K.H. (1989) *Industrialisation of Indigenous Fermented Foods*, Marcel Dekker Inc., New York.

Sundhagul, M., Daengsubha, W. & Suyanandana, P. (1975) Thailand's traditional fermented food products: a brief description. *Thai Journal of Agriculture Science*, **8**, 205–19.

Tanaka, T. (1976) *Tanaka's Cyclopaedia of Edible Plants of the World* (ed. Sasuke Nakao), Keigaku Publishing Co., Tokyo, Japan.

Tantaoni Elaraki, A., Berrada, M., El marrakch & Berramon, A. (1983) Etude du iben marocam. *Le Lait*, **63**, 230–45.

Toury, J., Wane, A., Giorgi, R. & Cros, J. (1970) Fish in the diet in Senegal: quantitative and qualitative aspects. Methods of presentation. (Joint FAO/WHO/OAU-STRC Bulletin) *Food and Nutrition in Africa*, **8**, 6–13.

van Noort, G. & Spence, C. (1976) The Mahewu Industry, *S.A. Food Review*, October, pp. 129–33.

van Pee, W. & Swings, J.G. (1971) Chemical and microbiological studies of Congolese palm wines (*Elaeis guineensis*) *East African Agriculture and Forestry Journal*, **36**, 311–14.

Vasconcelor, A.T., Twiddy, D.R., Westby, A. & Reilly, P.J.A. (1990) Detoxification of cassava during gari preparation. *International Journal of Food Science and Technology*, **25**, 198–203.

Westby, A. & Choo, B.H. (1994) Cyanogen reduction during lactic fermentation of cassava, In *International Workshop on Cassava Safety* (eds M. Bokenga, A.H.A. Essers, M. Ponlters, H. Rosling & O. Tewe), *Acta Horticulture*, **375**, 209–55.

Wood, B.J.B. (ed.) (1985) *Microbiology of Femented Foods*, Vol. 1, Elsevier Applied Science Publishers, London.

24 Fermented foods of the Indian subcontinent

R. SANKARAN

24.1 Introduction

Fermented foods form an important part of the diet in South East Asia, the middle East and Africa. Because of their high nutritive value, organoleptic characteristics and easy digestibility, they are of late becoming increasingly popular in the developed world also. They are the results of the activity of a few species of microorganisms among the thousands of species of bacteria, yeasts and moulds. These are the lactic acid bacteria, acetic acid-producing bacteria and certain alcohol-producing yeasts and moulds. It is a relatively low energy preservation process which increases product shelf life and reduces the need for refrigeration or other energy intensive operations in food preservation (Wood, 1985).

India, being a vast country, has been the home of innumerable sects of human population and the diversity in the climatic conditions has resulted in a large number of traditional fermented foods. Many of these remain highly secretive in their preparation being passed on from generation to generation and tend to be regionalized, with many of them being made only on home scale. In some regions of China and India, the hereditary nature of the profession is not only encouraged, but also protected by tradition (Hesseltine, 1979). In many places they are still being manufactured according to the traditional method using simple equipment and are produced at cottage industry scale by means of natural microflora from the staple ingredients and surroundings. The practice of preparing and consuming in a particular way depended mostly upon the traditional method inherited from generation to generation, in many cases as a closely guarded secret with the result that the technique of some of them are even lost to the world. However, the interest and inquisitiveness of many researchers has brought out the much-needed information on these foods and created an awareness of their production, consumption and beneficial effects. This has paved the way for further research on these foods, in some cases leading, to modernization and industrialization of the process.

There is a wide variety of fermented foods comprising products of cereal, pulses, vegetables, meat and milk. Of these, milk-based foods tend to dominate followed by cereals, pulses and meat-based foods. In all these products, the process of fermentation is initiated by lactic acid bacteria, acetic acid bacteria and certain alcohol-producing yeasts. In general, fermented foods are characterized by the accumulation of metabolites of microbial activities such as lactic acid, acetic acid, trace amounts of ethanol and other compounds collectively contributing to the characteristic flavours. These metabolites contribute to the flavour, bouquet and texture and thereby to the overall acceptability of the final fermented foods, as well as providing preservative action. Asians and Africans are pioneers in utilizing fermentation as a means to preserve a wide variety of foods. Of late, much interest has been generated in fermented foods of plant origin on health reasons. In the context of the growing consumer awareness towards obtaining products rated for their quality criteria as well as 'natural foods' subjected to minimal processing and devoid of chemical preservatives, fermented foods become very important.

24.1.1 Historical

Fermentation is one of the oldest methods for food preservation, and one which originated several centuries ago. Exact reports about the origin of fermented foods are lacking, but the process dates back to antiquity. The transition from hunting and gathering food to organized food cultivation and production is believed to have taken place some 10000 to 15000 years ago in the Middle East. It is believed that it was purely by chance that fermentation became a method of preservation. People of Asia, more precisely the Indonesians, were the pioneers in fermenting vegetable proteins to make meat-like flavours and textures. Koreans are credited for the production of acid-fermented vegetables. People of Egypt developed wheat breads leavened with yeasts, while Indians discovered methods for souring and leavening cereal-legume batters (Campbell-Platt, 1987).

24.2 Classification

On the basis of the substrate used, the fermented foods can be divided into:

- fermented milk products
- fermented cereal-legume/legume foods
- fermented cereal foods
- fermented vegetable and fruit products
- fermented meat and fish products
- miscellaneous foods

Since there is not much difference in the way that alcoholic and non-alcoholic beverages are prepared in India, only those specific to India will be discussed, and then too very briefly. A list of products with the dominant microorganisms, substrates and references is given in Table 24.1.

24.3 Fermented milk products

Milk is the most important foodstuff for a mammal and has always been the food of the newborn. Considering the microbial quality of milk, it is not surprising that centuries ago, many communities acquired a taste for sour milk and with time techniques were developed to ensure that the process of souring followed a particular pattern. The constant use of the same vessels or the addition of fresh milk to an on-going fermentation would have led to the gradual evolution of locally popular products. The value of these products would have differed from region to region depending on the indigenous microflora which is turn was dependent upon the climatic conditions of the area (Tamime & Robinson, 1988).

Fermented milk products are known for their taste, nutritive value and therapeutic properties. Although the origin of the cultured dairy products is not very clear, they date back to the dawn of civilization. It is rather difficult to be precise about the period when they were first prepared. Being a tropical country, it must have been used from prehistoric times since sour milk is more stable and advantageous than fresh milk. It preserved the high-quality nutrients present in milk in a relatively more stable form (Mann, 1977). There are references to cultured dairy products in some of the earliest writings of civilization such as the Bible and sacred books of Hinduism (Oberman, 1985).

Milk from eight species of domesticated mammals, cow, buffalo, sheep, goat, horse, camel, yak and zebra have been used to make traditional fermented milk products (Kroger *et al.*, 1989). From a biological point, fermented milks are characterized by the accumulation of microbial metabolic products. It was realized very early that microbial metabolites such as lactic acid and ethyl alcohol and dozens of other chemicals collectively called flavour substances were not altogether unpleasant and even contributed to overall preservative action (Kroger *et al.*, 1989).

In the early part of the 20th century, Metchnikoff (1908) claimed that, owing to lactic acid and other products present in sour milk fermented by lactic acid bacteria, the growth and toxicity of anaerobic spore-forming bacteria in the large intestine are inhibited. Lactic acid is biologically active and capable of suppressing harmful microorganisms, especially the putrefactive ones and has a favourable effect on the vital activities of human beings. In fact, Metchnikoffs' theory of longevity considerably influenced the spread of fermented milk products to many countries. Further, he emphasized the need to carry out exclusive studies on the biochemical and physiological properties of fermented milk. Kroger *et al.* (1989) in their paper have put forth an imaginative situation as to what would have been the picture had fermented milk been considered spoiled and inedible, and had not entered the human diet in the past. The speculation was that mankind's development would have taken an entirely different course.

Table 24.1 Indian fermented food products with their substrate and microorganisms

Product name	Substrate	Dominant microorganisms	Nature of the product	Used as	Reference
Adai	Rice and legumes	Lactic acid bacteria	Pancake	Breakfast/snack	Campbell-Platt (1987)
Ambali ganji	Ragi and milk, sugar/salt	*Leuconostoc mesenteroides* *Lactobacillus fermentum*	Thick gruel/balls	Breakfast/main food	Ramakrishnan (1979)
Apong	Rice	Natural flora	Beverage	Beverage/refreshing drink	Taye (1983)
Bhallae	Blackgram	*L. fermentum* *L. mesenteroides* and yeasts	Deep-fried patties	Snack after soaking in curd/spiced tamarind water	Soni & Sandhu (1990a)
Bhatura	White wheat flour		Flat, deep-fried leavened bread spongy cake. Spongy pancake	Breakfast food/main food	Campbell-Platt (1987)
Chaang	Ragi, wheat, rice, barley	*S. ovarum* *Endomycopsis fibuligera* *Candida* spp.	Beverage	Refreshing drink	Bhatia *et al.* (1977); Rehana *et al.* (1991)
Dhokla	Rice and B.G. dhal	*Lactobacillus fermentum* *L. lactis*, *L. delbruckeii*	Spongy cake	Snack/breakfast	Joshi *et al.* (1989)
Dosa	Rice and B.G. dhal	*Leuconostoc* spp. *Lactobacillus* spp. *Saccharomyces* spp.	Spongy pancake	Breakfast food, staple food	Soni & Sandhu (1990a)
Idli	Rice & B.G. dhal	*L. mesenteroides* *S. faecalis*	Steamed spongy pancake	Breakfast food, staple food	Soni & Sandhu (1990a)
Jalebi	Wheat flour and dahi	*L. fermentum* *L. buchnerii* *Streptococcus lactis*	Crispy, deep-fried pretzels	Confectionery snack food	Batra (1981)
Kallappam	Rice and toddy	Yeasts	Steamed spongy pancake	Breakfast/snack	Lakshmi (1996); Saraswaty Vasudevan (1996)

Khaman	Bengal gram		Spongy cake	Breakfast food	
Kulcha	White wheat flour		Spongy cake, flat baked bread	Staple food with side dish	
Madhu	Rice	*Mucor* spp., *Rhizopus* spp.			Dahiya & Prabhu (1977)
Nan	White wheat flour	*S. cerevisiae* Lactic acid bacteria	Flat baked bread	Staple food with side dish	Campbell-Platt (1987)
Papadam	Blackgram	*L. mesenteroides* *L. fermentum*, *S. faecalis* *S. cervisiae*	Circular tortilla-like wafers	Condiment	Batra & Millner (1974); Reddy *et al.* (1982)
Rabadi	Maize flour and buttermilk	*Bacillus* spp. *Micrococcus* spp. *Pediococcus acidolactis*	Cooked spongy cake	Breakfast/main meal	Ramakrishnan (1979)
Rice beer	Rice	*A. oryzae* *Rhizopus* spp. *Mucor* spp.	Beverage	Refreshing drink	Ram Ayyar & Hutchinson (1925)
Ruhi	Rice	*Mucor* spp., *Rhizopus* spp.			Dahia & Prabhu (1977)
Uthappam	Rice and coconut	Yeasts	Steamed spongy pancake	Breakfast/snack	
Vadai	Blackgram	Yeasts and lactic acid bacteria	Deep-fried patties	Snack food	Soni & Sandhu (1990a)
Vella-yappam	Rice	Yeasts	Steamed spongy pancake	Breakfast/snack	Saraswaty Vasudevan (1996)
Wadiyan	Blackgram	*Candida*, *S. cerevisiae* *L. mesenteroides*	Ball-like hollow, brittle or irregular-shaped	Spicy condiment with vegetables, legumes or rice	Batra (1981); Soni & Sandhu (1990a)

The main advantages of fermentation are:

1. The ability to produce lactic acid and acetic acid, aroma compounds and polysaccharides which give the products a characteristic taste, texture as well as improved nutritional quality.
2. Increased digestibility through metabolic activity.
3. The growth of lactic acid bacteria over a wide range of temperatures results in the preparation of a large variety of products.
4. Enhanced keeping quality by suppression of the growth of spoilage and pathogenic bacteria.
5. Help in preserving foods between production and consumption.
6. Possible possession of anti tumour and anti-carcinogenic effects.

Considering the importance and added health benefits derived from fermented milks since 1966, International Dairy Federation (IDF) has been involved in recording and updating the consumption pattern of fermented milks. However, not all countries who have been members of IDF are providing the data. From the statistics of 1985, it is of interest to observe that the major producers of fermented milks are not necessarily those countries with the highest capital consumption figures, but those with the most inhabitants (Kroger *et al.*, 1989). India leads among the 26 countries with 69%.

24.3.1 Cheese

Steinkraus (1983) and Campbell-Platt (1987) have made extensive attempts in bringing into focus the large number of fermented foods of varying sensory attributes. Based on type of fermentation: two categories are recognized:

- Lactic fermentation
- Lactic yeast fermentation

Lactic fermentation is the most popular; starter cultures are mainly lactic acid bacteria, wherein most of these products are made using a mixed starter culture and in some instances, a single culture is used to obtain the desired product. On the other hand, in some products, the growth of yeasts occurs which results in production of ethanol and CO_2, giving an effervescent characteristic to those specific products.

Traditionally, fermented milk products have been consumed as beverages, as meal components, or as ingredients in cookery. However, changing social patterns as well as new food innovations by food technologists have created a multitude of new products for the shelves of modern supermarkets (Kroger *et al.*, 1989). By controlling the conditions of fermentation – primarily the growth temperature – the natural microflora can be altered to produce a wide variety of products. Besides, use of desired microorganisms as in the case of controlled fermentation, would greatly enhance the chances of obtaining products with desirable and acceptable sensory attributes (Steinkraus, 1983).

24.3.2 Dahi/dudhi/curd

This is one of the most popular Indian fermented milk products, liked for its milk acidic taste and pleasant flavor (Laxminarayana & Shankar, 1980). It is prepared both under household and commercial conditions. The product is obtained by lactic fermentation of pasteurized or boiled cow's or buffalo's milk through the action of single or mixed cultures of lactic acid bacteria (Bureau of Indian Standards, 1980; Prevention of Food Adulteration Act, 1995). A good quality Dahi is of firm and uniform consistency with a sweet aroma and a clear acid taste. The surface is smooth and glossy and usually a cut surface is trim and free from cracks and gas bubbles.

Dahi probably originated by treating the surplus milk for conservation. The Aryans made a virtue of it so much so that it is a must in the Indian daily menu. The Indian medical treatise, *Sushruta Samhita* describes Dahi as a food which promotes appetite and increases strength. Dahi is an important component of 'Panchamritham', referred to mythology by the sage 'Charaka'. Curd is said to create relish for foods, promote appetite, operate as emollient and increase strength. According to 'Sushruta Samhita', there are three kinds of curds: sweet, acid and extremely acid. Sweet curd increases slimy secretions of organs and quality of fat and *Kapham* in the body, acid curd deranges *Pitham* and *Kapham*, and extremely acid vitalises the blood.

The inoculum used is a mixture of lactic streptococci and lactobacilli. Generally, a very small portion of the curd from a previous fermentation is used as the inoculum. The commonly associated lactic flora in the inoculum are *Lactobacillus lactis* ssp. *lactis*, *L. lactis* ssp. *cremoris*, *S. thermophilus*, *L. delbrucki* ssp. *bulgaricus*, *L. acidophilus* and *L. helveticus*.

Extensive research has been carried out at NDRI (Karnal and Bangalore) covering important aspects such as determining the lactic microflora in home-made and market samples of *dahi*; isolation of lactic cultures from dahi and suitability of lactic cultures in obtaining a stable product of acceptable quality attributes; good keeping quality; and biochemical and microbiological changes occurring during fermentation (Srinivasan & Banerjee 1946; Laxminarayana *et al.*, 1952; Tiwari *et al.*, 1969; Attaie *et al.*, 1987).

Proper heat treatment is essential in preparation of good-quality Dahi. Boiling is the most common heat treatment in India, with even a few seconds boiling resulting in 99.8% destruction of the microflora of milk and thus providing favourable conditions for the growth of starter cultures.

Steaming milk for 10 minutes has resulted in *dahi* of firm texture and pleasant aroma. Usually, a proper heat treatment eliminates numerous defects in the product which might accrue as a result of improper processing.

(a) Standardization of dahi preparation. Extensive studies have been carried out on the selection of organisms for the preparation of dahi with desirable characteristics. According to Banerjee (1960), a mixed culture of *L. lactis* ssp. *lactis*, *S. thermophilus* and *Leuconostoc cremoris* is best for sweet Dahi, while a mixed culture of *S. thermophilus* and *L. delbrueckei* ssp. *bulgaricus* is best for Dahi of high acidity. Baisya & Bose (1975) found the culture of *L. lactis* ssp. *lactis biovar diacetilactis* to impart desirable flavour to dahi because of higher diacetyl and volatile acid production, as against the cultures of *S. thermophilus* and *L. lactis* ssp. *cremoris*. The various combinations of lactic acid bacteria used in dahi are not only to obtain a product of good quality in terms of texture, flavours and taste, but also to improve upon the nutritive value of dahi. The quantity of starter culture added generally varies from 1–5% (v/v) depending on the acidity desired and also the temperature of incubation.

Under household conditions, incubation of inoculated milk for 12–16 hours at atmospheric temperature is desirable, while under laboratory conditions the incubation temperature varied from 25–45°C and the time from 8–16 hours, depending upon the type of starter and the desired acidity. An actively growing culture produces 0.9–1.0% acidity (lactic acid) after 8–10 hours of incubation. A good-quality Dahi has a pH of 4.6–5.0 and titratable acidity 0.8–1.0. Milk stored too long before inoculation with starter often results in poor flavour and a broken curd.

Studies have been aimed at finding out the population and types of lactic acid bacteria occurring in samples of dahi available in different regions of India. Mohanan *et al.* (1984) reported the lactic acid bacteria count to be in the range 132–246 × 10^6/ml. The samples showed the presence of a wide range of lactic acid bacteria species; nearly 46 isolates were those of *S. thermophilus*, four each were of *L. lactis* ssp. *lactis*, *L. lactis* ssp. *cremoris* and *L. helveticus*, 62 isolates were of *L. delbruecki* ssp. *bulgaricus*, and eight isolates of *L. casei*.

Since dahi is prepared under both household and market conditions with the involvement of personnel handling the product, there are ample opportunities for foodborne pathogenic and spoilage bacteria to occur as contaminants. Studies carried out in several parts of India have indicated the presence of *E. coli*, *S. aureus*, *Str. pyogenes* and spore formers in market samples (Tiwari & Singh, 1964, 1966; Mohanan *et al.*, 1984; Lekshmi *et al.*, 1989). In some recent studies (Misra *et al.*, 1993) the presence has been reported of pathogenic bacteria of serious public health concern in samples of curd collected locally in and around Bhubaneswar city. The pathogenic bacteria encountered were *Str. pyogenes*, *Staphylococcus* spp., *E. coli*, *Proteus* spp., *Salmonella*, *Shigella* and *Bacillus* spp.

24.3.3 Lassi

This is a popular lactic fermented milk beverage prepared from dahi and is a highly refreshing drink during summer months. The set curd is broken with an agitator and required amounts of water, sugar or salt (according to salt or sweet required) and essence are added into the broken curd. The mix is then homogenized well so that it can be sipped through a straw from the package.

Here also, as a lot of handling is involved, undesirable bacterial contamination can often occur. Lassi sold in small shops or by vendors has been shown to contain a high incidence of aerobic spore formers. Commonly encountered bacteria are *Bacillus subtilis* and *B. pumilus*. In many instances these organisms occur as post-processing contamination (Pillai *et al.*, 1993).

24.3.4 Buttermilk

Buttermilk is a medium-acid, slightly viscous cow's or buffalo's milk beverage with a buttery flavour.

(a) Preparation. The milk is boiled, cooled and inoculated with 1% starter culture and incubated under ambient conditions overnight (16–20 hours), churned to remove butter and stored under refrigeration. Salt, chopped ginger and green chillies are added to give extra flavour. It is diluted further with water and forms a refreshing drink, particularly in summer months. However, buttermilk is an intermediary product in the manufacture of butter and has no market value.

(b) Microbiology and biochemistry. Starter cultures contain mixtures of homofermentative *Streptococcus cremoris* and *S. lactis* which produce lactic acid, and flavour-producing *S. lactis* ssp. *diacetylactis* and *Leuconostoc cremoris* which produce diacetyl and some acetoin in addition to lactate. Diacetyl gives the butter flavour. The final pH is 4.4–4.6 and acidity 0.7–0.9% as lactic acid.

(c) Nutritive value. Typical analytical value is as follows: moisture 50–93%. Per 100 g dry matter; protein 34–38 g; fat 1–10 g; carbohydrate (lactose) 48–52 g; fibre 0; ash 6–9 g; energy 360–400 Kcal or 1.5–1.7 MJ; Ca 1200 mg; P 1000 mg; Fe 0.6–1 mg; K 1500 mg; Na 1200 mg; Mg 120 mg; Zn 50 mg; thiamine 0.4 mg; riboflavin 1.7 mg; niacin 0.7–1.1 mg; pantothenic acid 3.0 mg; vitamin B_6 0.35 mg; vitamin B_{12} 2.2 µg; folic acid 100 µg; biotin 10 µg; retinol 20 µg; and vitamin C 1–10 µg (Campbell-Platt, 1987).

24.3.5 Butter

Butter is an important dairy product in India, although relatively little is utilized directly for consumption (Rangappa & Achaya, 1974). It is used in small quantities as table butter and the rest for melting into ghee or clarified butterfat.

(a) Preparation. Milk is boiled and cooled and inoculated with either a starter or a little curd from the previous day. After overnight fermentation it is churned with a wooden churn or metal churn, during which butter rises to the top and is removed. Unlike the trained dairy farmer in the West, the Indian farmer knows very little or nothing of the correct processing of milk, the right type of starter, the exact time and temperature of souring, washing of butter or its preservation. All his knowledge of dairying is traditional, handed down from ancient times, like the Vedas. After the cooperative movement in Gujarat, butter is now being manufactured by many dairies and is commercially available in different quantities.

(b) Microbiology. Krishnaswamy & Laxminarayana (1948) have studied the microflora of butter and found that lactobacilli are the most predominant bacteria, followed by streptococci, colifoms and aerobic spore-formers. Yeasts and moulds are also present and contribute considerably to the development of acidity, colours and off-flavours. *S. lactis* subsp. *diacetylactis* produces the flavour compounds diacetyl, acetaldehyde, acetoin and carbon dioxide.

(c) Composition and nutritive value. Butter has a moisture of 18–20%; fat 78–81%; non-fatty solids 0.7–1.5%; acidity 0.5–0.8% (as lactic acid); energy 680–760 Kcal or 2.8–3.2 MJ; Ca 400 mg; Fe 0–15 mg; Mg 30 mg; thiamine 0.1 mg; P 300 mg; Na 140 mg; riboflavin 0.5 mg; and niacin 2.5 mg.

24.3.6 Ghee

Ghee is the usual Indian name for clarified butter fat and is usually made from cow's, buffalo's, goat's or sheep's milk. It keeps well without refrigeration. It forms an essential part of ceremonial religious offering among Hindus; it figures prominently in wedding rites, and is poured over the funeral pyre by the son or closest male relative.

(a) Preparation. Butter, cream or churned curd is heated to 100–140°C to melt and to allow water to evaporate; then filtered through muslin to clarify and cooled to solidify. When buffalo's milk is used, ghee is white in colour, but is yellow if cows' milk is used (due to the presence of carotenoids).

(b) Microbiology and biochemistry. Lactic acid bacteria, namely *Streptococcus*, *Lactobacillus* and *Leuconostoc* are used in butter or yoghurt. There is hardly any further fermentation in ghee process. The severe heating kills the lactic flora. The product contains low levels of free fatty acids (<0.5% as oleic acid) and has a peroxide value <1 mEq/kg.

(c) Composition and nutritive value. Typical analysis of ghee is: moisture 0.1–1%. Per 100 g dry matter; protein 0.5–1 g; fat 97–98 g; carbohydrate 0; fibre 0; ash 1–2 g; energy 880 Kcal or 3.7 MJ; Na 10 mg; K 20 mg; Ca 20 mg; Mg 2 mg; P 30 mg; Fe 0.2 mg; Cu 0.03 mg; Zn 0.2 mg; S 10 mg; Cl 12 mg; niacin 0.1 mg; retinol 900 µg; carotene 550 µg; vitamin D 0.9 µg; vitamin E 1.8 µg; thiamine, riboflavin, folic acid, vitamins B_6, B_{12}, and C are all zero (Campbell-Platt, 1987).

24.3.7 Shrikhand

This is a popular lactic fermented product widely consumed in different parts of the country, more so in western and northern India, mainly Maharashtra, Gujarat and some parts of Karnataka. The name is derived from the Sanskrit word 'Shikharini'. It is a summer specialty which is served as a dessert and is popular for its refreshing taste, cooling effect and low cost. Its preparation involves production of curd by lactic fermentation of whole milk (cow's or buffalo's) followed by draining out the whey from the curd through a suspended muslin cloth bag for 6–8 hours. The resulting solid curd mass (chakka) is uniformly mixed with sugar (45–55%) and made into a semi-solid mass to which flavouring substances such as saffron and cardamom are added. Garnishing with shredded nuts like almonds and cashew nuts is a common practice. The product has a long shelf life of 35–40 days at 5°C, while at ambient temperature it survives only 2–3 days. The commonly used starter bacteria are strains of *L. lactis* ssp. *lactis*, *L. lactis* ssp. *cremoris*, *L. lactis* ssp. *lactis* biovar. *diacetylactis*, *L. delbreucki* ssp. *bulgaricus* and *S. thermophilus* used either singly or in combination of two cultures. Little work has been done on this product, except for a few investigations on the technology (Reddy *et al.*, 1984; Atreja & Deodhar, 1987).

Studies on the mutant strains of *L. lactis* ssp. *lactis* biovar. *diacetilactis* (parent strain and mutant strain 95 and PM) for the preparation of chakka showed that the mutant strain PM gave a better rated product (Khanna *et al.*, 1982). In another study different starter culture combinations have been employed such as: (i) *L. lactis* ssp. *lactis*, *L. lactis* ssp. biovar. *diacetylactis*, *L. mesenteriodes* ssp. *cremoris* (ratio 1:1); (ii) *L. mesenteriodes* and *L. delbruecki* ssp. *cremoris* (1:1); (iii) *S. thermophilus* and *L. delbruecki* ssp. *bulgaricus* (1:1); and (iv) *L. lactis* ssp. *lactis* and *L. lactis* ssp. *lactis* biovar. *diacetylactis* (1:1). Combination (iv), known as LF 40, was found to be the most suitable by many Shrikhand manufacturers (Patel & Abdel-Salam, 1986).

With reference to hygienic standards and practices, the situation is similar to other fermented products. Studies on the ability of an enterotoxigenic strains of *Staphylococcus aureus* to grow and produce enterotoxins in Shrikhand revealed that the occurrence of food poisoning bacteria as pre-processing contaminants in milk used for product making would have a serious impact on public health aspects. The inoculated culture of *S. aureus* was able to grow and produce enterotoxins in varying degrees in the presence of cultures of lactic acid bacteria. However, the activity of *S. aureus* was dependent upon the type of strains used (Varadaraj & Ranganathan, 1988).

Like other indigenous products, this product has remained neglected by the organized dairy section for a number of reasons. For example, tedious time- and labour-consuming conventional process, absence of suitable equipment for mechanizing its production, regional status of the product, etc. With a view to overcoming some of the limitations of the traditional method and to partially mechanize the shrikhand-making process, the Baroda dairy has adopted a process developed at NDDB which involves the use of an unperforated basket centrifuge for speedy drawing of whey and a planetary mixer for kneading and mixing purpose. In this system, curd is obtained from pasteurized skim milk within 8 hours, chakka is prepared in 90 minutes, and the kneading of chakka with cream, ground sugar, flavouring and garnishing is completed in 40 minutes. In addition to these labour-saving advantages, the product, in being untouched by hand, is of better quality and shelf life (Aneja *et al.*, 1977).

A close look into the microbiological aspects of fermented milks would indicate the importance of starter cultures in obtaining fermented milk products of desirable sensory attributes, good keeping quality and freedom from hazardous organisms.

24.4 Cereal legume or legume foods

Next to dairy products, this is the most dominant group of fermented foods consumed in India. In some areas religious and cultural requirements demand that only plant materials be eaten, and this requirement is not restricted to low-income groups. Even among the non-vegetarians, during certain times because of religious sentiments, only vegetarian food is consumed. Asians and Africans are said to be pioneers in utilizing fermentation as a means of preparing a wide variety of foods from cereals and pulses (Campbell-Platt, 1987). Most of these fermentation processes were developed long before the existence of microorganisms was recognized and therefore methods of preparation are arbitrary. They are valued for their palatability and wholesomeness. This type of acidified and leavened foods contribute a very interesting group of cereal/legume-based foods of considerable magnitude in the developing and underdeveloped world.

However, in contrast to the established fermented milk products, cereal/legume-based traditional foods have not made any significant impact on the population other than that which depends heavily on cereal products as staple diet. The addition of legume gives a higher protein content and provides a better balanced ratio of amino acids, overcoming the danger of lysine deficiency from cereals alone or sulphur-containing amino acid deficiency from legumes alone (Aykroyd & Doughty, 1964; Aliya & Geervani, 1981). *Idli*, *dosa*, *vada*, *dhokhla*, *bhallae*, *warri/vadian* and *pappadam* are some of the commonly consumed fermented foods belonging to this group. The microbiological and other parameters of some of these products are summarized in Table 24.2.

24.4.1 Idli

Idli is the most popular among the fermented rice–legume-based products of India. As to how it originated and who discovered and brought it to common usage remains a mystery. The credit for introducing it into general dietary should go to housewives of South India, particularly Tamil Nadu. The products have been in use since 1100 AD (Ramakrishnan, 1979).

(a) Appearance and general characteristics. *Idli* is a small, white, acid-leavened steamed cake made by microbial fermentation of a thick batter of carefully washed and soaked rice and dehulled blackgram dhal (*Vigna mungo* or *Phaseolus mungo*). They are soft and spongy with a desirable sour flavour and are eaten with coconut chutney, pickles, lentil curry or chutney powder which is a mixture of roasted and coarsely ground seasonings such as sesame seeds, corriander seeds, red chillies and split legumes.

(b) Preparation. The traditional method of preparation consists of carefully washing and soaking separately 2 parts of rice and 1 part of blackgram dhal for 2–4 hours at ambient temperatures and grinding them separately. The rice is coarsely ground while blackgram dhal is finely ground, and they are then mixed together with water to give a smooth, but rather coarse but uniform batter. Approximately 1% salt is added and the batter is allowed to ferment overnight, i.e. 12–14 hours. The quantity of water required to prepare the batter of desirable, uniform consistency varies from 1.5 to 2.2 times the dry weight of the ingredients (Wang & Hesseltine, 1982). The fermented batter is then poured with a ladle on to greased metal plates or into round-shaped moulds, and steam cooked for

Table 24.2 Microbiological, physicochemical, biochemical and nutritional parameters in some Indian fermented foods

Parameter		Bhallae	Dosa	Idli	Papadam	Vadai	Warri
Bacteria	(CFU/g)	$1.0–9.0 \times 10^9$	$1.8–5.6 \times 10^3$	$2.0–6.0 \times 10^9$	$2.5–5.5 \times 10^{10}$	$3.0–5.0 \times 10^{10}$	$3.0–8.0 \times 10^{10}$
Yeasts	(CFU/g)	$0.0–1.5 \times 10^5$	$0.0–2.0 \times 10^6$	$0.0–8.2 \times 10^5$	$0.0–4.5 \times 10^4$	$0.0–4.0 \times 10^6$	$0.0–11.0 \times 10^6$
pH		4.87 ± 0.10	4.0 ± 0.14	46.92 ± 0.21	5.20 ± 0.28	4.10 ± 0.10	3.20 ± 0.29
Total acids	(%)	0.56 ± 0.08	0.90 ± 0.11	0.60 ± 0.14	0.70 ± 0.20	0.76 ± 0.08	1.50 ± 0.10
Volume	(ml)	340 ± 50	400 ± 40	390 ± 20	250 ± 15	425 ± 20	420 ± 20
Soluble solids	(%)	3.40 ± 1.00	6.00 ± 1.20	6.40 ± 0.08	7.10 ± 0.20	6.80 ± 0.50	15.00 ± 2.50
Reducing sugars	(mg/g)	1.40 ± 0.60	4.00 ± 0.80	5.80 ± 1.20	14.00 ± 4.00	1.00 ± 0.20	5.00 ± 0.50
Non protein nitrogen	(%)	4.30 ± 0.10	2.90 ± 0.40	2.10 ± 0.05	0.35 ± 0.05	4.34 ± 1.00	0.68 ± 0.21
Total nitrogen	(%)	0.46 ± 0.08	0.31 ± 0.07	0.12 ± 0.05	4.00 ± 0.25	0.56 ± 0.06	4.90 ± 0.50
Total proteins	(%)	24.00 ± 0.60	16.00 ± 0.40	12.00 ± 0.05	23.00 ± 1.00	24.00 ± 1.00	27.00 ± 2.00
Soluble nitrogen	(%)	2.20 ± 0.05	1.00 ± 0.07	0.60 ± 0.12	–	2.70 ± 0.20	1.50 ± 0.13
Soluble proteins	(mg/g)	39.00 ± 3.00	38.00 ± 7.00	19.00 ± 6.00	38.00 ± 4.00	20.00 ± 2.00	20.00 ± 6.00
Amylases	(IU/g)	7.51 ± 1.52	33.20 ± 2.10	25.00 ± 5.00	5.00 ± 0.50	13.30 ± 2.50	3.05 ± 0.80
Proteinases	(IU/g)	5.02 ± 1.47	2.70 ± 0.70	2.10 ± 0.20	5.40 ± 0.50	6.00 ± 1.50	6.04 ± 1.12
Free amino acids	(mg/g)	16.48 ± 2.68	17.80 ± 2.00	12.97 ± 1.20	13.50 ± 2.50	20.60 ± 3.16	45.15 ± 9.30
Vitamin B_1	(mg/100g)	0.72 ± 0.06	0.30 ± 0.06	0.74 ± 0.06	0.37 ± 0.06	0.80 ± 0.04	0.95 ± 0.60
Vitamin B_2	(mg/100g)	0.40 ± 0.08	0.80 ± 0.02	0.78 ± 0.06	0.38 ± 0.09	0.68 ± 0.06	1.20 ± 0.20
Vitamin B_{12}	(µg/100g)	0.20 ± 0.01	0.30 ± 0.02	0.30 ± 0.04	0.20 ± 0.02	0.28 ± 0.10	0.80 ± 0.30

10–20 minutes to give a fine textural idli ready for consumption with chutney or curry or jam.

(c) *Microbiology of fermentation*

Effect of composition of ingredients. As already mentioned, the typical idli is prepared from a mixture of rice and blackgram. Radhakrishnamurthy *et al.* (1961) reported that the best acceptable texture and quality idli is obtained when the rice and blackgram are in the proportion of 2:1. The proportion of water used is between 2.1–2.2 parts per weight of the mixture. Steinkraus *et al.* (1967) have however, reported that acceptable idlis can be prepared using different proportions of the cereals and legume. Specifically, in designing diabetic foods, they found that the proportion of blackgram dhal could be increased. Attempts were made to substitute the blackgram dhal with other legumes such as common bean (Steinkraus & Van Veen, 1967), soybean (Soni & Sandhu, 1989a,c; Akolkar & Parekh, 1983; Ramakrishnan *et al.*, 1976), great northern bean (Sathe & Salunkhe, 1981) and mungbean (Soni & Sandhu, 1989a,c) but none was found to be as good as blackgram dhal. The special value of black gram for idlies is due to the mucilaginous substance that it contains which is not present in any other edible legume. It is believed that this mucilaginous substance helps in the retention of carbon dioxide during fermentation of the thick batter and is thus responsible for the soft, spongy, honey-comb texture of the idli. It is also believed that the traditional wet grinding of the dhal into a viscous batter would entrap air bubbles inside. The physical entrapment of air bubbles will not only exert a leavening action, but also assist the fermentation itself by aiding liberation of more gases for leavening. Subsequent studies by the same authors have revealed the role of mucilage to give sufficient adhesiveness to the batter, without which the idlis prepared would be brittle. Therefore, it may be possible to substitute blackgram dhal by contribution of other food materials containing mucilage, protein, or a surface-active principle. The rice variety is also important (Purushothaman *et al.*, 1977). Sarasa *et al.* (1985) substituted rice with bajra and rajmah, blackgram and horsegram for blackgram. The products obtained from Bajra and either bajra and either rajmah or blackgram were satisfactory, but bajra combined with horsegram did not yield an acceptable product.

24.4.2 Dosa

Dosa is a thin, crisp, fried, pancake-like staple food of southern India, eaten mainly as a breakfast food but becoming popular as a snack food throughout India. It is made of rice and dehulled split blackgram.

(a) Preparation. Dosa is prepared by soaking 1–3 parts rice and 1 part blackgram dhal (*Phaseolus mungo* L.) or yellow lentil in water for 4–6 hours at ambient temperature 27–30°C, grinding to a fine paste, and mixing the two together to make a free-running batter. About 1% salt is added and the mixture is allowed to ferment overnight for 10–16 hours. A longer time is required in colder months. The natural microflora of the ingredients act as the starter and brings about the fermentation. About 60–80 ml of batter is poured on to a hot, greased pan for a few minutes until crisp and brown on one side; the dosa is then turned, cooked on the other side, rolled and served with coconut chutney or lentil–vegetable curry.

(b) Microbiological changes. Studies on traditional dosa batter fermentation have indicated the role of several bacteria and yeasts (Padhye & Salunkhe, 1978; Soni *et al.*, 1985, 1986; Sandhu *et al.*, 1986; Sandhu & Soni, 1988; Soni & Sandhu, 1989c, 1990). Rice contributes bacteria belonging to five genera: *Leuconostoc*, *Streptococcus*, *Lactobacillus*, *Micrococcus* and *Bacillus*, and yeasts belonging to three genera: *Saccharomyces*, *Debaryomyces* and *Trichosporon.* Blackgram dhal contributes bacteria belonging to six genera: *Leuconostoc*, *Enterobacter*, *Flavobacter*, *Bacillus*, *Lactobacillus* and *Streptococcus*, and yeasts belonging to the genera *Debaryomyces*, *Torulopsis*, *Candida* and *Kluyveromyces* (Soni *et al.*, 1986). The commercial samples contained bacteria in the range of 10^4–10^8 per gram belonging to six genera and nine species, while the laboratory fermented batters had bacteria in the range of 10^7–10^8 per gram. *Leuconostoc mesenteroides* was the most commonly encountered bacterium, followed by *Streptococcus faecalis*, *Lactobacillus fermentum*, *Bacillus amyloliquefaciens*, *Lactobacillus delbrueckii*, *Bacillus subtilis*, *Pediococcus cerevisiae*, *Bacillus polymyxa* and *Enterobacter* sp. On the other hand, yeasts were found to vary from 0–10^7 per gram and were present in 70% commercial and 53% laboratory-fermented batters. *S. faecalis*, *B. amyloliquefaciens*, *L. delbrueckii*, *B. subtilis* and *B. polymyxa* were the principal bacteria observed during summer, whereas the remaining species were isolated during both the seasons. With regard to yeasts, *Oosporidium margaritiferum*, *Kluyveromyces marxianus*, *Candida kefyr* and *C. kruseii* were the main yeasts present in winter with the remaining species present during both seasons. Both bacteria and yeasts increased significantly with the progress of fermentation accompanied by increase in acidity and leavening. The bacteria appearing initially during dosa batter fermentation belonged to *Leuconostoc mesenteroides* along with bacilli and micrococci. Streptococci developed subsequently and persisted until the end of fermentation, while the other types slowly disappeared. *Debaryomyces hansenii*, *S. cerevisiae* and *Trichosporon beigelii* were found to be predominant among the yeasts appearing first. Later on, only *S. cerevisiae* persisted.

(c) Physical and biochemical changes. With the progress in fermentation, acidity is increased, and carbon dioxide is produced by heterofermentative *Leuconostoc mesenteroides*, with a rise in volume. As in case of idli batter, the pH declined from 5.0 ± 0.36 to 4.0 ± 0.14, while the total acids increased from 0.40 ± 0.09 to 0.9 ± 0.11%. Specific activity of amylase showed an increase during the early phase from 1.41 ± 0.18 to 1.63 ± 0.11 iu/mg soluble protein and finally declined to 0.7 ± 0.11 iu at the end. Proteinase activity did not exhibit any appreciable change of statistical significance, although a slight increase (0.02 ± 0.01 to 0.03 ± 0.01 iu/mg soluble protein) was observed. Fermentation brought about an increase in the total nitrogen, soluble protein and reducing sugar content. The rise in nitrogen level was significant. Although soluble protein declined initially during the early phase, it tended to increase after 6 hours and attained the level of 3.96 ± 0.76% at the end of fermentation. Reducing sugar levels fell from 0.25 ± 0.04 to 0.15 ± 0.06% during initial 18 hours and rose afterwards to the level of 0.4 ± 0.06%, exhibiting an overall statistical significance.

(d) Nutritive value. B vitamins, thiamine, riboflavin and niacin increased during fermentation. Proteins were partially converted to amino acids. With a moisture of 38–45% (dry matter basis) dosa has a protein content of 15–20 g/100 g; fat 15–25 g; carbohydrate 40–50 g, fibre 6–8 g; ash 5–7 g; and a calorific value of 400–420 Kcal (1.7 MJ).

(e) Improvements. Attempts have been made to prepare dosa-like products by replacing the conventional blackgram component with other legumes and also replacing rice with sorghum, maize and bajra (Raghavendra Rao *et al.*, 1979; Soni & Sandhu, 1990a). Batter types containing soybeans were most nutritious in terms of total nitrogen, protein, amylase, proteinase and vitamin B contents, followed by mungbean batters, while the conventional blackgram product was organoleptically most preferred. Maize dosa offered difficulty in preparation when compared with sorghum or bajra, as it sticks to the pan, lacked strength and broke while turning over. It had an overmoist and slimy texture, while dosa from sorghum did not have the above defects. At equal slurry concentration, the cold paste viscosity of batter from maize was much higher than that of others except for blackgram dhal which is known to have very high viscosity. The high viscosity of the batter may be responsible for sticking of the maize dosa to the pan. The peak viscosity as well as the viscosity on cooling to 50°C were lower for maize than for others, which indicated that, on boiling or baking, maize slurry did not set to a viscous gel and had relatively greater free-flowing characteristics. This explains the rather overmoist slurry texture of maize dosa. Reduction of water content in the batter reduced this tendency but affected spreadability and made dosa very thick. Longer baking at slightly

lower temperature and frequent turning over of the dosa slightly improved the texture due to greater evaporation of water. The best combination of physicochemical parameters was incubation at 28°C with an initial pH of 4.5 and supplementation of the batter with 1–2% sucrose.

The possibilities of improving the texture and organoleptic quality by yeast enrichment have been investigated by adding commonly occurring isolated strains of *S. cerevisiae*, *D. hansenii*, *Hansenula anomala*, *Trichosporon beigelii* and *Torulopsis candida* without eliminating the natural flora of the ingredients. *S. cerevisiae* enrichment gave best reults in terms of degree of leavening, final levels of amylases, reducing sugars, free amino acids, total proteins, vitamin B_1 and B_2 and organoleptic quality. *T. candida* and *H. anomala* also produced good leavening and other organoleptic characteristics, but *D. hansenii* enrichment did not give any significant rise in vitamin levels as in the other cases, although leavening, amylolytic activity and total proteins were quite good. *T. bagelii* did not produce any significant difference except for having good enzyme activity.

Idli and dosa have now attained an international reputation and therefore there is lot of scope for developing such widely accepted foods into a well-organized industrial sector on modern footings.

Of late, there have been attempts to make many of these products available in convenient mix form commercially, so that all the drudgery of soaking and grinding are avoided. The reconstitution will require adding a specified quantity of water, keeping for 10–15 minutes and making them in the griddle in the case of idli, deep-frying in the case of vadai and bhallae, or by spreading on a heated, fat-smeared griddle in the case of dosa. Idli, with a shelf life of 3 months is being produced in single serve packs. (See Table 24.1 for information on vadai and bhallae. Further details are in Campbell-Plalt, 1987).

24.4.3 Dhokla

This is a fermented food commonly used in Gujrat, prepared from a batter made of coarsely ground rice and bengalgram (*Cicer arietinum*) dhal powder in the proportion of 3:1, mixed with one part of curd and adequate water and fermented overnight at room temperature. Green chillies and table salt are added and the batter is poured on greased plates and steamed in a suitable pan for 20 minutes. After cooling, the dhokla is cut into squares, seasoned with corriander leaves and green chillies, mustard and asafoetida (Joshi *et al.*, 1989).

(a) Microbiology. Joshi *et al.* (1989) isolated *L. fermentum*, *L. lactis*, *L. delbrueckii*, *Leuconostoc mesenteroides* and the yeast *Hansenula silvicola* from the fermenting batter. Of these, *L. lactis* and *L. delbrueckii* could not be isolated after 10th and 12th hour of fermentation.

(b) Biochemical changes. As the fermentation progressed, the pH of the batter fell from 5.25 to 4.00 and that of the cooked product from 5.5 to 4.5. Change in volume of batter commenced at the 10th hour of fermentation and gradually increased from 0 to 95.6% (Joshi *et al.*, 1989). Total titratable acidity of batter increased from 0.33 to 1.2 g/l and that of the product from 0.07 to 0.5 g/l, indicating production of acid by the proliferating microorganisms. The diacetyl content of batter increased from 0.03 to 5.2 μg/g but it was absent from the product because of its total loss during steaming. The acetoin content of the batter rose from 0.7 to 9.6 μg/g during 0–12 hours, followed by a decline on prolongation of incubation. This may be due to conversion of acetoin to 2,2-butanediol under acidic conditions (Hammer & Babel, 1943). The acetoin content of the product increased from 0.07–0.95 μg/g. Total volatile fatty acids of the batter showed a gradual rise from 54.8 to 582.8 mg/L and that of product from 66.3 to 696.4 mg/g. The flavour of the product could be due to acetoin and the volatile fatty acids, namely acetic acid, propionic acid, isobutyric and isovaleric acid which showed a rise during fermentation (Wiseblatt & Kohn, 1960). The higher volatile fatty acid content of the product could be due to the release of particle (flour)-bound volatile fatty acid during steaming.

As regards sensory qualities of the products, the characteristic pleasant flavour was obtained with a 12-hour fermentation while a medium-sour taste was experienced with a 14-hour fermentation. A highly spongy texture was obtained with 16 hours when the percent increase in volume was 84.8. The optimum conditions are fermentation for 16 hours at ambient conditions when total titratable acidity was 0.95 g/l, acetoin 8.6 μg/g, and total volatile fatty acid content 479.6 mg/l.

(c) Nutritive value. The protein efficiency ratio (PER) of dhokla was found to be 2.97 as compared with that of skim milk powder 4.15 (Khader 1979).

24.4.4 Bhallae

Bhallae are Snack foods prepared from soaked blackgram paste and fermented for 5–6 hours. In many parts, this is eaten as snack food either as such or soaked in spiced tamarind water or curd.

(a) Microbiology. The microflora involved generally comprise yeasts (Sandhu & Waraich, 1984) and bacteria (Sandhu Soni *et al.*, 1986). The yeasts reported are *Candida famata*, *C. membranaefaciens*, *C. curvata*, *Trichosporon pullulans*, *C. parapsilosis*, *Rhodotorula polymarina*, *Wingea robertsii* and *Hansenula polymorpha.* Bacterial flora was in the range of 10^6–10^9 per gram with yeasts numbering up to 10^5 per gram. *Lactobacillus fermentum* and *Leuconostoc mesenteroides* were the

predominant bacteria. *B. subtilis, S. faecalis, Flavobacter* and *Achromobacter* were also present. As in other fermented foods, there was variation in their number with seasons. Summer favoured bacteria, while winter favoured yeasts.

24.4.5 Vada

This is another blackgram-based snack food of India prepared from unspiced legume paste and is deep fried in oil as in bhallae, but the main difference is that the dough is comparatively finer and the incubation period is 18–24 hours.

(a) Preparation. Legume beans, usually *Phaseolus mungo*, are soaked in water, then ground into paste, salt, chilies and onions (sometimes) added and left to ferment at ambient for few hours. The paste is then made into balls or flat cakes and deep-fried in vegetable oil.

(b) Microbiology and biochemistry. The microflora involved in fermentation include bacteria alone or in combination (Soni, 1987). Soni studied a total of 36 samples of which 20 were market samples and 16 laboratory fermented. The market samples had bacteria in the range of 10^7–10^{10} per gram belonging to *Leuconostoc mesenteroides, Lactobacillus fermentum, S. faecalis* and *B. subtilis*. Yeasts were in the range of 0–10^6 per gram, identified as *Debarymyces hansenii, Trichosporon beigelii, Hansenula anomala, S. cerevisiae, Kluyveromyces marxianus, Trichosporon beigelii, Pichia membranaefaciens* and *Candida curvata*. The laboratory-fermented samples had comparatively higher microbial load in the range of 10^8–10^{10} per gram.

Seasonal variations were also observed as in other cases, with higher numbers in summer. The successional prevalence and various changes were similar to those in *bhallae*. The pH declined from 6.35 to 4.1, while total acids increased from 0.26 to 0.76%. Fermentation brought about an increase in the volume and total soluble solids. Reducing sugars however, declined significantly following the 24-hour incubation. Total nitrogen and total protein did not vary much, although non-protein nitrogen rose from 0.30 to 0.56%. Soluble nitrogen and free amino acids exhibited a rise from 0.88 to 2.7% and 5.10 to 20.60 mg/g, respectively, whereas the levels of soluble proteins fell significantly. The amylolytic and proteolytic activities were also increased as fermentation progressed, along with water-soluble B vitamins, including B_1, B_2 and B_{12}.

(c) Composition and nutritive value. Typical analysis: moisture 30–45%. Per 100 g dry matter; protein 30–40 g; fat 20–30 g; carbohydrates 15–25%;

fibre 6–9 g; ash 5–8 g; energy 430–450 Kcal (1.8–1.9 MJ) (Campbell-Platt, 1987).

24.4.6 *Papadam*

These are thin, crisp, dried circular tortilla-like wafers made of legume, cereal or starch crop flour eaten as snack or with meals either roasted or deep fried.

(a) Preparation. The dough is prepared and spiced as in the case of *warries*, but without fenugreek and ginger, rolled into wafers of 10–20 cm in diameter and 0.5 mm thickness. These are then fermented and dried in the open air for 4–6 hours. After drying they are stored in the open.

(b) Microbiology. Both bacteria and yeasts have been isolated from papadam (Batra & Millner, 1974; Soni & Sandhu, 1990a). Market samples contained bacteria in the range of 10^8–10^{10} per gram belonging to three types. *Leuconostoc mesenteroides* was the most common, followed by *Lactobacillus fermentum* and *S. faecalis*. Yeasts varied from 0–10^3 per gram and belonged to six types. Of these, S. *cerevisiae* was the most predominant, followed by *Hansenula anomala*, *Trichosporon beigelii*, *Candida krusei*, *Pichia membranefaciens* and *Debaryomyces hansenii*. As in other cases, there were seasonal variations, bacteria being high in summer and yeasts in winter.

(c) Biochemical and nutritional data. The rise in bacterial cell counts and yeasts whenever present was followed by several changes and leavening. pH declined from 6.1 to 5.2 and total acids increased from 0.4 to 0.7%. Fermentation also brought about an increase in soluble solids and a decrease in reducing sugars. Total nitrogen and total proteins remained fairly constant, although non-protein nitrogen, soluble nitrogen and free amino acids increased slightly. Amylase activity generally exhibited a slight decrease while proteinase and vitamins B_1, B_2 and B_{12} increased.

24.4.7 *Adai*

This is made of rice and split legumes like dosai and eaten with sauce or chutney or sugar.

(a) Preparation. Ten parts rice and three parts each of blackgram dhal, bengalgram dhal and redgram dhal are cleaned well and soaked in water for 2–3 hours, coarsely ground, salt added and left to ferment for 2–3 hours in summer and overnight during cooler seasons. On completion of this, 1.5

parts grated coconut, 1–2 parts finely chopped onions and 0.6 parts chilli powder or green chopped chillies are added with enough water to make a batter. About 40 g batter is spread on to an oily hot plate (about 10 mm thick), browned on both sides, and served hot with coconut chutney or chutney powder with oil.

24.4.8 Punjabi warri/wadiyan

Warries are spiced, hollow, brittle friable balls 5–8 cm in dia meter and somewhat like Japanese miso. They are very popular in Northern India, particularly in Amritsar and are used as condiments in cooking with vegetables, legumes or rice.

(a) Preparation. Dehulled blackgram dhal is washed and soaked overnight in water and then ground to a fine paste. The paste is mixed with asafoetida (*Ferula foetida*, 0.5–1%), caraway (*Carum carvi*, 0.4–0.5%), cardamon (*Elletaria* cardamon, 1%), cloves (0.4–0.5%), fenugreek (*Trigonella foenumgraecum*, 1%), ginger (8%) and red chillies (*Capsicum annum*, 1%) and moulded into small balls (Batra & Millner, 1974) and allowed to undergo fermentation and drying in open air for 4–8 days.

As in other products, the microbial flora of the raw materials and the surroundings are associated with warri fermentation. Several authors (Batra & Millner, 1974, 1976; Reddy *et al.*, 1982; Sandhu & Waraich, 1984) suggested the involvement of yeasts, especially *Candida* sp., *Saccharomyces cerevisiae*, *Rhodotorula lactosa*, *Candida famata*, *Wignea robertsii*, *Debaryomyces hansenii*, *Hansenula anomala* and *C. parapsilosis* in warri fermentation. Batra & Millner (1974) carried out inoculation studies with *Candida* sp. on autoclaved material and found that the inoculated material was similar in all respects to the commercial samples. The involvement of both bacteria and yeasts have also been reported (Batra, 1981; Sandhu *et al.*, 1986; Sandhu & Soni, 1988). Batra (1981) reported that bacteria and yeasts generally occur in the unfermented paste in the range of 10^3–10^6 per gram, which increased to 10^6–10^7 per gram as the fermentation progressed. He found *Leuconostoc mesenteroides* to be the most abundant, along with *Lactobacillus fermentum*, *Candida krusei*, *Saccharomyces cerevisiae*, *C. rouxii* and *Candida* spp.

More recent studies indicate that bacteria, alone or in combination with lesser proportion of yeasts, bring about the fermentation (Sandhu & Soni, 1989; Soni & Sandhu, 1990b). Sandhu & Soni (1989) reported the presence of bacteria in all the commercial and laboratory samples while only half of the samples had yeasts. Commercial samples had 10^9–10^{12} bacteria per gram with *Leuconostoc mesenteroides* as the most predominant, followed by *Streptococcus faecalis*, *Lactobacillus fermentum* and *Bacillus subtilis*. Yeasts ranged from nil to 10^7 per gram, with *Saccharomyces cerevisiae* and *Pichia membranefaciens* as the most predominant, followed by *Candida vartiovaarai, Kluyveromyces marxianus, Trichosporon beigelii* and *Hansenula anomala*. Laboratory-fermented samples had higher bacterial load of 10^{10}–10^{12} per gram while yeasts were in the range of 0–10^{6} per gram. The additional microflora isolated were *Enterobacter* sp., *Flavobacter* sp., *Candida aquatica, Cryptococcus humicolus* and *Geotrichum candidum*.

There was seasonal variation in the development and prevalence of microorganisms, with a higher range in summer. As the fermentation progresses, both bacteria and yeasts, whenever present, increase in number bringing about a fall in pH and an increase in volume. Sandhu & Soni (1989) reported that the bacteria which appeared initially were *Leuconostoc mesenteroides, Lactobacillus delbruecki, L. fermentum* and *B. subtilis*, along with *Flavobacter* sp. which mainly produce acid and gas from various carbohydrates, thus making the environment unfit for the contaminants (Soni & Sandhu, 1990b). Most of the bacteria disappeared slowly with the progress of fermentation except *Leuconostoc mesenteroides* and *Lactobacillus plantarum*. Among the yeasts, although *Trichosporon beigelii, Saccharomyces cerevisiae, Candida krusei, Pichia membranaefaciens* and *Hansenula anomala* were present in abundance initially, eventually only *S. cerevisiae* and *Trichosporon beigelii* persisted.

(b) Biochemical changes. pH declined from 5.65 to 3.20, while the total acids increased from 0.50 to 1.50%. Fermentation also brought about an increase in volume and soluble solids. Reducing sugars decreased considerably following fermentation and drying. Total nitrogen and total proteins did not vary significantly, although non-protein nitrogen increased from 9.79 to 45.15 mg/g. Amylase activity increased initially and declined thereafter. On the other hand, proteinase activity increased from 4.82 to 6.04 iu/g.

(c) Nutritive value. Warries are rich in water-soluble B vitamins, including thiamine, riboflavin and cyanocobalamin.

The physicochemical, biochemical and microbiological changes of some of these fermented foods are summarized in Table 24.2.

24.5 Cereal foods

Several fermented cereal products are made on the Indian subcontinent. The practice of preparing and consuming in a particular way depends mostly upon the traditional method inherited from generation to generation, in many cases as a closely guarded secret.

24.5.1 Appam/kallappam/vellayappam

Raw rice is soaked for 3–6 hours and ground to a fine paste. Sugar and toddy (palm sugar) are added to taste and water is added to make a free-flowing batter; batter is allowed to ferment for 4–6 hours. At the end of this, the fermented batter is poured into moulds and steamed, resulting in idli-like pancakes. In the modern method, toddy is replaced by yeast and the pH is controlled, which gives the product a shelf life of 2–4 days. These are available along with other bakery products in a packaged form in many bakeries in Kerala. No scientific data are available on this product except its ingredient composition in recipe books.

24.5.2 Ambali

Ragi-millet flour is combined with water to make a thick batter and fermented for 14–16 hours. The batter is added with continuous stirring to partially (3/4) cooked rice, and cooking is completed. When cooled, yoghurt or curd is added and *ambali* is ready to eat (Ramakrishnan, 1979). The microorganisms isolated were *Leuconostoc mesenteroides* (1.6×10^9/g), *Lactobacillus fermentum* (1.6×10^9/g) and *Streptococcus faecalis*. The pH decreased from 6.4 to 4.0 and the volume increased by about 20% as a result of CO_2 production (Ramakrishnan 1979).

24.5.3 Rabadi/rabdi

Cooked maize flour is cooled and combined with buttermilk and the mixture allowed to ferment overnight. The resultant product is rabdi and is eaten as such. The microorganisms in the ready-to-serve *rabdi* are *Pediococcus acidilactis* (3.6×10^5/g), *Bacillus* spp. (1.1×10^6/g) and *Micrococcus* spp. (7.9×10^5/g). During fermentation, pH changed from 6.7 to 6.4 with slight increase of 5% in volume. There was no change in amino nitrogen or free sugar (Ramarkrishnan, 1977).

24.5.4 Nan

Nan is a leavened flat bread with central pouch. It may be eaten with centre filled with meat or other savoury filling. Nan originated in the Middle East and India, but is now produced world-wide.

(a) Preparation. To 1 kg of 70–80% extraction white wheatflour are added 500 ml water, 20 g yeast or 350 g sourdough from a previous batch, and mixture is kneaded vigorously for about 15 minutes; usually, some vegetable oil is added to soften. Divided dough balls are left to ferment at 25–32°C for 2–2.5 hours, mechanically 'knocked back' and flattened to discs

5–7 mm thick, dusted with flour, covered and allowed to ferment at 25–32°C for 30–60 minutes.

Nan are baked rapidly in hot oven (230–260°C) either on a metal tray, for 5–10 minutes, or directly in contact with the hot oven wall at 460–500°C for 2–3 minutes. The intense heat causes the centre of the dough to expand rapidly and to create a central pouch or pocket.

(b) Microbiology and biochemistry. *Saccharomyces cerevisiae* yeast and lactic acid bacteria are active, with lactic organisms, particularly *Lactobacillus* sp., dominating if sourdough from a previous batch or yoghurt are used. Yeast is mainly responsible for leavening by carbon dioxide production.

24.5.5 Bhatura and kulcha

These are made from white wheat flour and are similar to nan except that they are deep-fried. Few data on the microbiology and biochemistry are available.

24.5.6 Jalebi

This is a traditional Indian confection made from soft, smooth fermented wheatflour, deep-fried and crisp, about 5.7 cm in diameter, and soaked in sugar syrup; it is eaten as a snack, particularly on festive occasions.

(a) Preparation. Fine wheat flour (1 kg), dahi (200 g) and water (650 ml) are mixed together to form a thick batter and fermented overnight. Saffron may be added to give a good colour. The fermented batter is squeezed through cloth with a hole of about 4 mm diameter and deposited as continuous spirals into heated oil or ghee (160–180°C). In about 1 minute these spirals become light brown and are removed from the frying pan. The excess fat is drained away and the jalebi is immediately immersed into sugar syrup of about the consistency of honey for 2–3 minutes and then removed. Often, rose water or any other flavouring agent is added to the syrup (Batra, 1981).

(b) Microbiology and biochemistry. Mixed fermentation is by lactic acid bacteria, *L. fermentum, L. lactis and L. buchneri*, with *S. lactis* and *S. faecalis* producing lactic acid, lowering the pH from an initial 4.3–4.5 to a final 3.2–3.4. The yeast *Saccharomyces cerevisiae* is also active, producing CO_2 and causing about a 10% increase in volume during fermentation. Amino nitrogen and total sugars were found to decrease during fermentation (Ramakrishnan, 1977; Batra, 1981).

(c) Composition and nutritive value. Moisture 32–38% per 100 g, dry matter; protein 4–7 g; fat 15–20 g; carbohydrate 75–78 g; fibre 2–4 g; ash 2–3 g; calorific value 460–480 Kcal (1.9–2.0 MJ); Na 2 mg; K 90 mg; Ca 70 mg; Mg 10 mg; Fe 1 mg; Cu 0.1 mg; Zn 0.5 mg; thiamine, riboflavin 0.03 mg; niacin 2 mg; folic acid 14 μg; retinol, carotene, vitamin C and D are absent (Campbell-Platt, 1987).

24.6 Fermented fruits and vegetables

Fermentation is used throughout the world, often in conjunction with chemical preservation, using salt and acid to preserve fruits and vegetables. Vegetables are highly perishable commodities which begin to deteriorate shortly after harvest. Until the development of curing and freezing during the past 180 years, fermentation was one of the most important methods for preserving foods. In these products, salt-resistant lactic acid bacteria, initially heterofermentative *Leuconostoc* spp. and *Lactobacillus brevis*, are generally replaced with homofermentative *L. plantarum* and *Pediococcus* spp. As acids are produced and the pH drops, yeasts become more important. Fermentative pickling of vegetables enhances the shelf life, producing characteristic taste and aroma. Many vegetables are traditionally preserved by this method at household levels to preserve them for the period between crops (Nema, 1993). Preservation of vegetables depends upon the reduction of the activity of the enzymes present, inhibition of certain oxidative chemical changes, and the inhibition of growth of microorganisms that may produce deteriorative changes in the food (Arjisiri, 1980). During fermentation, sugar is converted to lactic acid and acetic acid, alcohol, carbon dioxide and several other metabolites.

24.6.1 Pickles

Pickles from vegetables and fruits like mango, gooseberry and lemons are prepared and used throughout India. More than their nutritive value, they act as food adjuncts and appetizers (Tandon, 1959; Giridhari Lal *et al.*, 1967) as they add palatability to Indian foods. Indian pickles is a big industry now and the products are popular throughout the world. The spices in the pickle aid digestion by stimulating the flow of gastric juices. Though long ago the housewife was the food processor, the standard recipes have now been modified to have universal acceptance and also to suit industrial processing conditions. The other vegetables being used for pickles are carrot, cucumber, cabbage, bitter gourd, beans, chilies, pepper, jackfruit, garlic, ginger, brinjal and onion. Although a majority of the pickles are unfermented, several of the vegetables pickles are fermented products. Salting or curing is done by introducing dry salt or brine to the raw materials which contain plenty of water as in the case of cucumber, cabbage, turnip and olives. The weak brine that is formed contains nutrients like proteins, carbohydrates and minerals and this helps an active fermentation of lactic cultures, which comes as surface contamination from raw materials. The organisms involved are salt-resistant lactic acid bacteria, initially of heterofermentative *Leuconostoc* spp. and *Lactobacillus brevis* which are generally replaced by homofermentative *Lactobacillus plantarum* and *Pediococcus* spp. Coliform bacteria, particularly *Enterobacter* and *Klebsiella* spp., are also involved. As acids are produced and the pH drops, yeasts become more important. Along with lactic acid, small quantities of acetic acid, alcohol, and sometimes hydrogen and carbon dioxide are also produced during fermentation. There is a considerable degree of change in the physical appearance of the fruit or the vegetable during the salting or curing process. The green or yellow colour of the fruit turns olive green. The opaque raw material becomes translucent, and a mild fermented flavour sets in.

24.7 Meat and fish foods

In the case of meat and fish, being highly perishable foods, lactic fermentations have been used since antiquity to preserve meat and for silage of offals from fish, poultry and animals. The low pH produced due to lactic acid inhibits spoilage and pathogenic microorganisms (Murali, 1984). Murali *et al.* (1982, 1984), Murali & Sankaran (1985) and Sankaran *et al.* (1984) were able to prolong the shelf life of fresh raw meat by the use of lactic cultures.

Among such meat products, fermented sausages are the most important group. Fermented sausage is a finely chopped meat mixed with salt, curing agents, spices and a few additives and stuffed into casings and dried for adequate time. The organisms responsible are *Pediococcus acidilactici*, *Pediococcus pentosaceus* and *Lactobacillus plantarum* as a mixture of two or three of these. Physical, microbiological and biochemical changes take place during fermentation, with lactic acid production resulting in a lowering of pH, a decrease in a_w, inhibition of spoilage and pathogenic microorganisms, proteolytic enzymes breaking down myofibrillar and sarcoplasmic proteins, salt-soluble myofibrillar proteins gellifying and providing firm consistency and texture to the product, and development of aroma compounds due to lipolytic activity of lactic acid becteria (Narasimha Rao, 1995). Late starters containing lactic acid bacteria have been developed for use in meat and meat products.

Fermentation methods have been extended to fish viscera and slaughter house offals and poultry offals, as high protein sources of animal feeds and were found to be beneficial in not only conservation, but as a control of environment pollution and health hazard (Ahmed *et al.*, 1993; Shaw *et al.*, 1994).

Varieties of pickles from fish and meat have been made in India, but except for their method of preparation being listed in recipe books, no scientific data are available on these products.

24.8 Miscellaneous

Besides the well-developed alcoholic beverages available commercially throughout the world, there are a few undistilled beverages which are specific to certain interior regions in India. Apong, toddy, kanji and chaang are some examples. A few other fermented foods which do not come under other groups are also included here.

24.8.1 Apong

This is a fermented sweet–sour alcoholic beverage common for all age groups in some tribal populations of Assam and Arunachal Pradesh.

(a) Preparation. Cooked rice is spread on a bamboo tray and, on cooling, is inoculated with powdered Epop (Pitha). This starter is a dry cake or ball prepared from rice dough with spices (consisting of ginger, black pepper and chillies) and powdered dry plant leaves. The inoculated substrate is filled into an earthen pot or packed in banana leaves and kept at room temperature for fermentation for 5 days. The ripening of the mash is judged by the juiciness of the substrate and its characteristic sweet–sour alcoholic aroma. A required quantity of water is added and the mash thoroughly stirred. Using a funnel-shaped bamboo basket with paddy straw at the bottom, it is filtered and the filtrate, which is a turbid, whitish liquid, is the apong. The left-over residue is used as pig feed. Peeled ripe bananas and millets can also be used as a substrate.

(b) Microbiology and biochemistry. This is a mixed fermentation involving lactic acid bacteria and yeasts. Apong is a good relaxing drink after the day's exhausting hard work. Curative values are not known, but there is a common belief among folk that it can provide instant relief for diarrhoea (Taye, 1983).

24.8.2 Pulimkanji

This is a traditional method of preserving left-over foods in festival seasons in Kerala. The biggest festival of Kerala – Onam – in olden times used to continue for 5–6 days, during which several vegetable preparations are made each day. The left-over food each day, comprising cooked rice and different types of cooked curried vegetables, is transferred into a large earthenware pot along with a little salt. The process is repeated every day and the next day after the festival is over seasoning the entire mixture with oil, dry chillies, mustard and curry leaves is done, the whole thing is boiled and served hot (Lakshmi, 1996; Saraswathy Vasudevan, 1996). It serves as a delicious breakfast food. No scientific work has been carried out on this product and it is becoming extinct now as the joint family system is slowly disappearing. As families are becoming small, festivals are becoming of shorter duration, and families economize in the quantity of food they cook, the result is that large amounts of food are not left over for pulimkanji production.

24.8.3 Kanji

This is a strong-flavoured, red-coloured alcoholic beverage made from carrot or beet juice. It is usually consumed in hot weather as a refreshing drink, principally in North India.

(a) Preparation. Roots of raw purple carrot or beet are washed, scraped and shredded. To 100 parts root may be added, 5–6 parts salt, 3–4 parts mustard seeds, *Brassica* species, and 400–500 parts water. The mixture is left to ferment at 26–34°C for 4–7 days, after which the pink liquid is drained off for drinking. Portion of previous Kanji may be added as starter, or in southern India, torani, a yeast-containing fermented rice gruel used as starter. May be bottled and consumed within 2–3 weeks.

(b) Microbiology and biochemistry. Yeasts *Hansenula anomala*, *Candida guilierrmondii*, *C. tropicalis* and *Geotrichum candidum* are active in the fermentation, converting sugars in roots to ethanol and other flavour components (Campbell-Platt, 1987).

24.9 Conclusion

Over the years, lactic acid bacteria have become an integral part of the human food chain through the continuous use of a variety of lactic-fermented foods. The food habits of mankind has been a challenging area of scientific and technological development. In recent years, there has been a growing trend in consumer awareness towards the non-acceptability of foods with chemical preservatives, leading to the preference for minimally processed foods which provide an array of health benefits and safety to mankind. These developments have led to the recognition of lactic acid bacteria as a potential source of biopreservatives in foods. Besides imparting desirable attributes to the femented products, the primary role of lactic acid bacteria has been preservation, attributed to the production of lactic acid, hydrogen peroxide, and secondary metabolites, including bacteriocins. Bacteriocins are known to have the potential to inhibit closely related organisms as well as a wide range of other undesirable microorganisms.

India has been the home of a wide variety of fermented foods which have developed over a period of many years. These foods have resulted from the need to utilize the easily available substrates of those regions of the country, thus making the food highly regionalized with unique sensory characteristics. Most of these traditional fermented foods involve natural fermentation such that on many occasions, the ultimate products may have undesirable attributes. Such a situation results from the inability of the inherent lactic flora to grow and establish over the other competing organisms. Besides, the standards of hygiene and sanitation followed in the preparation of many of these traditional fermented foods give ample scope for undesirable food contaminants to occur within them.

Thus, it is seen that fermented foods add pleasure and bring variety to monotonous diets. The microorganisms involved in these products modify the original material physically, chemically, nutritionally and organoleptically. The development of an acceptable flavour is an important contribution of fermentation. It makes unappealing foods more acceptable; and it stimulates the appetite and thereby increases the consumption, thus making a nutritional contribution in the case of people who do not consume enough food.

In general the advantages of fermentation are that it:

- produces desirable enzymes;
- destroys undesirable flavour/odour/toxins;
- adds flavours and odours;
- preserves;
- synthesizes desirable constituents such as vitamins;
- increases digestibility;
- changes the physical state of food; and
- produces desirable colour in food.

Whether eating for fun or longevity – or both – there will be a plentiful supply of health foods with the aid of fermentation. Technology will thus make healthy eating and living longer more attainable goals.

Consumer demand for natural, safe and environmentally friendly products is increasing, as is the demand for eco-friendly products, and ecologically absolute disclosures on the contents and processing of food is already strong. From ancient times biotechnology has been the traditional tool for processing and preserving food products, the biological equilibrium in products having been successfully and safely manipulated with the help of microorganisms and other natural ingredients. The heightened awareness of the relationship between health and diet will become more intense as consumers become even more demanding. Since many of these food products involve natural fermentation, very often, they fail to meet desirable quality attributes. Thus, it becomes important to develop processes with defined microbial cultures and which function under optimal conditions besides utilizing the beneficial properties of starter bacteria.

Thus, in these days of energy crisis, many of the traditional food fermentation technologies used in India could be adapted world-wide, as they are generally low-cost and energy-saving processing methods. The added attraction is on health grounds, in view of the recent emphasis upon the incorporation of plant materials into food.

References

Achaya, K.T. (1974) *Indian Dairy Products*, Asia Publishing House, Delhi, Lucknow.

Achaya, K.T. (1994) *Indian Food – A Historical Companion*, Oxford University Press, Bombay, Calcutta, Madras.

Ahmed, J., Mahendrakar, N.S., Dani, N.P. & Ramesh, B.S. (1993) Optimisation of the fermentation process for ensiling the tropical fresh water fish viscera. Poster session, Abstracts, 3rd International Food Convention, CFTRI, Mysore (India) 7–12 September 1993.

Akolkar, P.N. & Parekh, L.J. (1983) Nutritive value of soy *idli*. *Journal of Food Science and Technology*, **20**, 1–4.

Aliya, S. & Geervani, P. (1981) An assessment of the protein quality and vitamin B content of commonly used fermented products of legumes and millets. *Journal of the Science of Food and Agriculture*, **32**, 837–42.

Amar, M.A. & Lammerding, A.M. (1983) Health maintenance benefits of cultured dairy products. *Cult. Dairy Produce Journal*, **18**(2), 6–10, 19.

Ananthachar, T.K. & Desikachar, H.S.R. (1962) Effect of fermentation on the nutritive value of *Idli*. *Journal of Science and Industrial Research*, **21**, 191–2.

Aneja, R.P., Vyas, M.N., Nanda, K. & Thareja, V.K. (1977) Development of an industrial process for the manufacture of *Shrikhand*. *Journal of Food Science and Technology*, **14**, 159–63.

Arjisiri, A. (1980) *Preservation of vegetables by lactic fermentation.* Thesis in partial fulfilment for the award of M. Sc in Food Technology, Mysore University, Mysore.

Atreja, S.K. & Deodhar, A.D. (1987) Nutrient status during *Shri Khand* making. *Journal of Food Science and Technology*, **24**, 266–7.

Attaie, R., Whalsaw, P.J., Shahani, K.M. & Amer, M.A. (1987) Inhibition of growth of *Staphylococcus aureus* during production of acidophilus yoghurt. *Journal of Food Protection*, **50**, 224–8.

Aykroyd, W.R. & Doughty, J. (1964) Legumes in human nutrition. *FAO Publication No. 19*, United Nations, Rome.

Baisya, R.K. & Bose, A.N. (1975) Role of inoculating organisms on the physico-chemical change in milk and final curd quality. *Indian Journal of Dairy Science*, **28**, 179–83.

Balasubramanyam, B.V. (1995) *Studies on the application of antagonistic lactic acid bacteria in the biopreservation of selected indigenous milk and cereal pulse-based foods.* PhD Thesis, University of Mysore.

Banerjee, G.C. (1960) Fermented milks. *Indian Dairyman*, **12**, 357–61.

Banks, J.G., Board, R.G. & Sparks, N.H.C. (1986) Natural antimicrobial systems and their potential in food preservation of the future biotech. *Applied Biochemistry*, **8**, 103–47.

Batish, V.K., Grover, S. & Kamath, S.N. (1993) Improving shelf-life and safety of fermented milk products through genetically improved microorganisms. *Indian Journal Dairy Science*, **45**, 51–7.

Batra, L.R. (1981) Fermented cereals and gram legumes of India and vicinity, in *Advances in Biotechnology* (eds M. Moo Young & C.W. Robinson), Pergamon Press, Toronto, pp. 547–54.

Batra, L.R. & Millner, P.D. (1974) Some Asian fermented foods and beverages and associated fungi *Mycologia*, **66**, 942–50.

Batra, L.R. & Millner, P.D. (1976) Asian fermented foods and beverages. *Developments in Industrial Microbiology*, **17**, 117–28.

Bhatia, A.K., Singh, R.P. & Atal, C.P. (1977) Chaang – the fermented beverage of Himalayan folk. *Indian Food Packer*, **31**(4), 32–9.

Bottazzi, Y. & Vescovo, M. (1979) Carbonyl compounds produced by yoghurt bacteria. *Netherlands Milk Dairy Journal*, **23**, 71–8.

Breslaw, E.S. & Kleyn, D.H. (1973) *In-vitro* digestibility of protein in yoghurt at various stages of processing. *Journal of Food Science*, **38**, 1016–21.

Bureau of Indian Standards. Manak Bhavan, 9, Bahadur Shah Zafar Marg, New Delhi. Is:9617-1980. *Specification for Dahi.*

Campbell-Platt, G. (1987) *Fermented Foods of The World – A Dictionary and Guide,* Butterworths, London, pp. 294.

Cheeptongkum, N. (1975) *Fermented food products based on rice.* MSc Thesis, Mysore University.

Dahiya, D.S. & Prabhu, K.A. (1977) Indian Jack Fruit wine Symposium on Indigenous fermented foods (SIFF). Bangkok, Thailand, 21–27 November, 1977.

Daly, G., Sandine, W.E. & Elliker, P.R. (1972) Interactions of food starter cultures and food borne pathogens *Streptococcus diacetilactis* versus food-borne pathogens. *Journal of Milk and Food Technology*, **35**, 349–57.

Deeth, H.C. & Tamime, A.Y. (1981) Yoghurt: nutritive and therapeutic aspects. *Journal of Food Protection*, **44**, 78–86.

Desikachar, H.S.R., Radhakrishnamurthy, R., Ramarao, G., Kadkol, S.B., Srinivasan, M. & Subramanyan, V. (1960) Studies on Idli fermentation – Part 1. Some accompanying changes in the batter. *Journal of Science and Industrial Research* **19**C, 168–72.

Galloway, J.H. & Crawford, R.J.M. (1985) Cheese fermentations, in *Microbiology of Fermented Foods* (ed. B.J.B. Wood), Elsevier Applied Sci. Publishers, London, pp. 111–65.

Gandhi, D.N. & Nambudiripad, V.K.N. (1981) Isolation and characterisation of antibiotic producing culture from dahi (curd). *Indian Journal of Dairy Science*, **34**, 334–6.

Gilliland, S.E. (1979) Beneficial interrelationships between certain microorganisms and humans; *Candida* in microorganisms for use as dietary adjuncts. *Journal of Food Protection*, **42**, 164–7.

Gilliland, S.E. (1985) Role of starter culture bacteria in food preservation, in *Bacterial starter cultures for foods*, (ed. S.E. Gilliland) CRC Press, Boca Raton, FL, pp. 175.

Giridhari Lal *et al.* (1967).

Hanmer, B.W. & Babel F.G. (1943) Baeteriology of butter cultures: a review. *Journal of Dairy Science*, **26**, 83–168.

Hesseltine, C.W. (1965) A millenium of fungi, food and fermentation. *Mycologia*, LVII, 149–97.

Hesseltine, C.W. (1979) Some important fermented foods in mid-Asia, the Middle-East and Africa. *Journal of the American Oil Chemists Society*, **56**, 367–74.

Hesseltine, C.W. (1981) Future of fermented foods. *Process Biochemistry*, 1**6**(3).

Hesseltine, C.W. & Wang, H.L. (1979) Fermented foods. *Chemistry and Industry*, June, 393–9.

Hutchinson, C.M. & Ram Ayyar, C.S. (1925) Bakhar – the Indian Rice Beer fermentation Memoirs of the Indian Department of Agriculture and Bacteriology, **1**, 137–68.

Joseph, K., Narayana Rao, M.V., Indiramma, K., Swaminathan, M. & Subrahmanyan, V. (1961) Studies on nutritive value of *Idli* fortified with IMF. *Journal of Science Industrial Research*, **20**, 269.

Joshi, N., Gadhole, S.H. & Pradnya Kanekar (1989) Microbial and biochemical changes during dhokla fermentation with special reference to flavour compounds. *Journal of Food Science and Technology*, **26**, 113–15.

Kadkol, S.B., Desikachar, H.S.R. & Srinivasan, M. (1961) The mucilaginous principles in black gram (*Phaseolus mungo*) dhal. *Journal of Science Industrial Research*, **20**C, 252–3.

Khandewala, P.K., Ambegaokar, S.D., Patel, S.M. & Radhakrishna, M.V. (1962) *Journal of Science Industrial Research*, **21**, 275.

Khader, V. (1979) Studies on the nutritive value of dhokla based on rice, bergalgram & curd. *Ind. Nutr. Diet.*, **16**(8), 316–19.

Khanna et al (1982)

Kissinger, J.C. & Zaika, L.L. (1978) Effect of major species in Lebanon Bologna on acid production by starter cultures. *Journal of Food Protection*, **41**, 429–31.

Krishnamurthy, N. & Siddappa, G.S. (1965) Canning of Idli. *Journal of Food Science and Technology*, **2**, 132–53.

Krishnaswamy, M.A. (1948) Microflora of Indian butter. *Indian Journal of Dairy Science*, **1**, 127–9.

Krishnaswamy, M.A. & Laxminarayana, H. (1948) Microflora in butter. *Current Science*, **17**, 159.

Kroger, M., Kuruman, J.A. & Rasic, J.L. (1989) Fermented milks – past, present and future. *Fd Technol.*, **43**(1), 92–9.

Lakshmi, C.R. (1996) Personal communication.

Lakshminarayana, H. (1948) *Current Science*, **17**, 159.

Lekshmi, R., Garg, S.K. & Yadav, R.K. (1989) Incidence of *Staphylococci* in milk and milk products. *Indian Journal of Dairy Science*, **42**, 136–8.

Laxminarayana, H. & Shankar, P.A. (1980) Fermented milk in human nutrition. *Indian Dairyman*, **32**, 121–9.

Laxminarayana, H., Nambudiripad, V.K.N., Lakshmi, N., Anantharamaiah, S.N. & Sreenivasa Murthy, V. (1952) Studies on *dahi* II Survey of the quality of market *dahi. Indian Journal of Veterinary Science, Anim. Husb.*, **22**, 13–25.

Lewis, Y.S. & Johar, D.S. (1953) Microorganisms in fermented grain mashes used for food preparations. *CFTRI Bulletin*, **2**, 287.

Lewis, Y.S., Johar, D.S. & Subrahmanyan, V. (1955) Studies on process simplification in the preparation of a fermented type of food stuff-*idli. Bull. Central Food Technol. Res. Inst. Mysore, India*, **5**, 257.

Mann, G.V. (1977) Nutritional and therapeutic aspects of Industrious and related western fermented milk products. A review *Indian Journal of Dairy Science*, **84**, 393–404.

Matta, H., Kalra, M.S. & Ajit Singh (1991) Survival of pathogenic bacteria in yoghurt and *Dahi. Journal of Food Science and Technology*, **28**, 240–3.

Metchnikoff, E. (1908) *The prolongation of life*, First edition, G.P. Putmans and Sons, New York, NY.

Misra, R.K., Das, K.L. & Roy, P.K. (1993) Bacteriological study of curd around Bhubaneswar. *Indian Journal of Dairy Science*, **45**, 309–10.

Mohanan, K.R., Shankar, P.A. & Laxminarayana, H.L. (1984) Microflora of *dahi* prepared under house-hold conditions of Bangalore. *Journal of Food Science Technology,* **21**, 45–6.

Mukherjee, S.K., Albury, M.N., Pederson, C.S., Van Veen, A.G. & Steinkraus, K.H. (1965) Role of *Leuconostoc mesenteroides* in leavening the batter of idli – a fermented food of India. *Applied Microbiology*, **13**, 227–31.

Murali, H.S. (1984) *Studies on the improvement of raw mutton quality using microorganisms*, MSc Thesis, University of Mysore.

Murali, H.S. & Sankaran, R. (1985) *Retention of fresh meat colour and enhancement of flavour and tenderness by proteolytic and lipolytic microorganisms.* Technical Report, (69/85). Defence Food Research Laboratory, Mysore.

Murali, H.S., Leela, R.K., Sankaran, R. & Sharma, T.R. (1982) Studies on the effect of lactic culture on the improvement of raw meat colour. Presented at 23rd Annual Conference of AMI, CFTRI, Mysore.

Murali, H.S., Leela, R.K., Sankaran, R. & Sharma, T.R. (1984) Investigation on the retention of colour and improvement in texture of raw mutton by the use of lactic organisms. Presented at 4th Indian Convention of Food Scientists and Technologists, CFTRI.

Murali, H.S., Leela, R.K., Sankaran, R. & Sharma, T.R. (1985) Effect of some lactic cultures on the natural microflora of mutton. *Chem. Mikrobiol Technol Lebensm.*, **9**, 19–23.

Narasimha Rao, D. (1995) Lactic fermentation in the preservation of animal products, in *Proceedings of Micon International-94*, (eds R. Sankaran & K.S. Manja), Association of Microbiologists, India.

Narayana Rao, D. (1972) Predigested protein rich foods based on Indian oilseed meals and pulses. *Journal of Food Science and Technology*, **9**(3), 57–62.

Nema, S.K. (1993) *Concept of new product development with specific reference to Indian traditional foods.* Thesis in partial fulfilment of MSc in Food Technology, Mysore University.

Neerja, Dankher & Chauhan, B.M. (1987) Technical note: preparation, acceptability and B vitamin content of rabadi – a fermented ddpeaslmillet food. *International Journal of Food Science and Technology*, **22**, 173–6.

Nyunt, Win (1983) MSc. (Food Technology) Thesis (in part fulfilment) Mysore, D.K. (1978) *Journal of Food Biochemistry*, **2**, 327.

Oberman, H. (1985) Fermented milks, in *Microbiology of Fermented Foods*, (ed. B.J.B. Wood), Volume 1, Elsevier Applied Science Publishers, London, pp. 167–95.

Padhye, V.W. & Salunkhe, D.K. (1978) Biochemical studies on black gram (*Phaseolus mungo* L) III fermentation of the black gram and rice blend and its influence on the *in vitro* digestibility of the proteins. *Journal of Food Biochemistry*, **2**, 327–47.

Patel, R.S. & Abdel-Salam, M.H. (1986) Shrikhand – an Indian analogue of western quarg. *Cult. Dairy Produce Journal*, **21**(1), 6–7.

Pillai, R.A.V., Khan, M.M.H. & Reddy, V.P. (1993) Incidence of aerobic spore formers in *Lassi. Journal of Food Science and Technology*, **30**, 141–2.

Prasad, G., Khan, B.L. & Kulshreshla, D.C. (1984) Survival of *E. coli* and *Ent. aerogenes* in *dahi* II. contamination after curdling of milk. *Indian Journal of Dairy Science*, **37**, 261–3.

Prevention of food Adulteration Act (1954–1995) EBC Publishing Pvt. Ltd., Lucknow.

Purushothaman, D., Dhanapal, N. & Rangaswami, G. (1977) *Symposium on indigenously fermented foods.* Bangkok, Thailand.

Radhakrishna Rao, M.V. (1961) National Academy of Sciences. National Research Council, Washington, DC. Publication No. 843, 291.

Radhakrishnamurthy, R., Desikachar, H.S.R., Srinivasan, M. & Subramanyan, V. (1961) Studies on Idli fermentation II Relative participation of blackgram flour and rice senodina in the fermentation. *Journal of Science and Industrial Research*, **20**, 342–4.

Raghavendra Rao, S.N., Mallesh, N.G., Sreedharamurthy, S., Viraktamath, C.S. & Desikachar, H.S.R. (1979) Characterstics of roti, dosa & Vermicclli from maize, sorghum and bajra. *Journal of Food Science and Technology*, **16**(1), 21–4.

Rajalakshmi, R. & Vanaja, K. (1967) Chemical and biological evaluation of the effects of fermentation on the nutritive value of foods prepared from rice gram. *British Journal of Nutrition*, **21**, 467–73.

Ram Ayyar, C.S. (1928) Microorganisms and production of *dahi. Agricultural Journal of India*, **23**, 101.

Ramakrishnan, C.V. (1977) The use of fermented foods in India, in *Symposium on Indigenous fermented foods*, Bangkok, Thailand.

Ramakrishnan, C.V. (1979) Studies on Indian fermented foods. *Baroda Journal of Nutrition*, **6**, 1–7.

Ramakrishnan, C.V., Parekh, L.J., Akolkar, P.N., Rao, G.S. & Bhandari, S.D. (1976) Studies on soy idli fermentation. *Plant Foods for Man*, **2**, 1–2, 15–33.

Rangamannar, C. (1994) *Flavours from lactic acid bacteria.* Thesis, MSc in Food Technology (in part fulfillment), Mysore University.

Rangappa, K.S. & Achaya, K.T. (1974) *Indian Dairy Products*, Asia Publishing House, Bombay, Calcutta, Lucknow.

Reddy, G.V., Freind, B.A., Sahani, K.M. & Farmer, R.E. (1993) Antitumor activity in yoghurt components. *Journal of Food Protein*, **46**, 8–11.

Reddy, K.K., Ali, M.P., Rao, B.V. & Rao, T.J. (1984) Studies on production and quality of *Shrikhand* from buffalo milk. *Indian Journal of Dairy Science*, **37**, 293–7.

Reddy, K.P., Sahani, K.M. & Kulkarni, S.M. (1976) B. complex vitamins in acidified and cultured milk. *Journal of Dairy Science*, **59**, 191–5.

Reddy, N.R. & Salunkhe, D.K. (1980) Effect of fermentation on phytate phosphorus and mineral content in black gram, rice and blackgram and rice blends. *Journal of Food Science*, **45**, 1708–12.

Reddy, N.R., Sathe, S.K., Pierson, M.D. & Salunkhe, D.K. (1981) *Idli* – an Indian fermented food. A review. *Journal of Food Quality*, **5**, 89–101.

Reddy, N.R., Pierson, M.D., Sathe, S.K. & Salunkhe, D.K. (1982) Legume based fermented foods: Their preparation and nutritive quality. *CRC. Critical Reviews in Food Science and Nutrition*, **17**, 335–70.

Rehana, F., Vijaya Sarathy Reddy, O., Venkataraman, O. & Basappa, S.C. (1991) Preliminary microbiological & biochemical studies on Chaang fermentation. 9th Indian Convention of Food Scientists and Technologists, CFTRI, Mysore, India.

Rose, A.H. (1982) *Fermented Foods*. Economic microbiology series, Academic Press, New York.

Sahani, K.M. & Chandan, R.C. (1979) Nutritional and healthful aspects of cultured and culture-containing Dairy Foods. *Journal of Dairy Science*, **62**, 1685–94.

Sandhu, D.K. & Soni, S.K. (1988) Optimization of physico-chemical parameters for Indian *dosa* batter fermentation. *Biotechnology Letters*, **10**, 227–82.

Sandhu, D.K. & Soni, S.K. (1989) Microflora associated with Indian Punjabi *warri* fermentation. *Journal of Food Science and Technology*, **26**, 21–5.

Sandhu, D.K. & Soni, S.K. (1991) Microbial prevalence and succession associated with *idli* fermentation. *Indian Journal of Microbiology*, **31**, 285–9.

Sandhu, D.K. & Waraich, M.K. (1984) Distribution of yeasts in indigenous fermented foods with a brief review of literature. *Kawaka*, **12**, 73–85.

Sandhu, D.K., Soni, S.K. & Vikhu, K.S. (1986), in *Yeast Biotechnology* (eds P. Tauro & R.K. Vashishta), Haryana Agricultural University Press, Hisar.

Sandine, W.E., Muralidhara, K.S., Elliker, P.R. & England, D.C. (1972) Lactic acid bacteria in food and health: a review. *Journal of Milk and Food Technology*, **35**, 691–702.

Sankaran, R., Murali, H.S., Leela, R.K. & Sharma, T.R. (1984) Enhancement of colour and texture of mutton by the use of lactic acid bacteria. *Journal of Food Science and Technology*, **23**, 172–5.

Sarasa, M.S., Ingle, A.O. & Nath, N. (1985) Microbiological examination of certain new *idli* batters during fermentation. *Indian Journal of Microbiology*, **25**, 212–20.

Saraswathy Vasudevan (1996) (Personal communication).

Sathe, S.K. & Salunkhe, D.K. (1981) Fermentation of the great northern bean (*Phaseolus vulgaris* L) and rice blends. *Journal of Food Science*, **46**, 1374.

Shaw, D.M., Narasimha Rao, D. & Mahendrakar, N.S. (1994) Rapid fermentation process for ensiling of poultry wastes. Poster Session, Abstracts, 16th International Congress of Biochemistry & Molecular biology, New Delhi, India, 19–22 September 1994.

Soni, S.K. (1987) *Studies on some Indian fermented foods – microbiological and chemical aspects.* PhD Thesis, Guru Nanak Dev University.

Soni, S.K. & Sandhu, D.K. (1990a) Indian fermented foods. Microbiological and biochemical aspects. *Indian Journal of Microbiology*, **30**, 135–57.

Soni, S.K. & Sandhu, D.K. (1989a) Fermentation of *idli*: Effects of changes in raw materials and physico-chemical conditions. *Journal of Cereal Science*, **10**, 227–38.

Soni, S.K. & Sandhu, D.K. (1989b) Microflora associated with Indian Punjab Wari fermentation. *Journal of Food Science and Technology*, **26**, 21–5.

Soni, S.K. & Sandhu, D.K. (1989c) Nutritive improvement of Indian *dosa* batters by yeast enrichment and black gram replacement. *Journal of Fermentation and Bioengineering*, **68**, 52–5.

Soni, S.K. & Sandhu, D.K. (1990b) Biochemical and nutritive changes associated with Indian punjabi *warri* fermentation. *Journal of Food Science and Technology*, **27**, 82–5.

Soni, S.K. & Sandhu, D.K. (1991) Role of yeast domination in Indian *Idli* batter fermentation. *World Journal of Microbiology and Biotechnology*, **7**, 505–7.

Soni, S.K., Sandhu, D.K. & Vilkhu, K.S. (1985) Studies on *dosa* – an indigenous Indian fermented food: some biochemical changes accompanying fermentation. *Food Microbiology*, **2**, 175–81.

Soni, S.K., Sandhu, D.K., Vilkhu, K.S. & Kamra, N. (1986) Microbiological studies on *Dosa* fermentation. *Food Microbiology*, **3**, 45–53.

Srinivasan, M.R. & Ananthakrishnan, C.P. *Milk products of India.*

Srinivasan, M.R. & Banerjee, B.N. (1946) Studies on the methods of preparation of ghee II Souring process. *Indian Journal of Veterinary Science and Animal Husbandry*, **16**, 72–83.

Steinkraus, K.H. (1967) Studies on Idli – an Indian fermented food of blackgram Rice Food. *Food Technology*, **21**, 110–13.

Steinkraus, K.H. (1983) *Handbook of Indigenous fermented foods.* Microbiology Series, vol. 9, Marcel Dekker Inc., New York.

Steinkraus, K.H. & Van Veen, A.G. (1967) *Food Technology*, **91**, 21.

Steinkraus, K.H., Van Veen, A.G. & Thiebeau, D.B. (1967) Studies on Idli. An Indian fermented blackgram rice food. *Food Technology*, **27**(6), 110–13.

Susheelamma, B.S. & Rao, M.V.L. (1961) Surface active principle in blackgram and its role in the texture of leavened foods. *Journal of the Science of Food and Agriculture*, **74**, **25**(6), 665.

Tamang, J.P., Sarkar, P.K. & Hesseltine, C.W. (1988) Traditional fermented foods and beverages of Darjeeling and Sikkim – A review. *Journal of the Science of Food and Agriculture*, **44**, 375–84.

Tamime, A.Y. & Deeth, H.C. (1980) Yoghurt, Technology and Biochemistry. *Journal of Food Protein*, **43**, 939–77.

Tamime, A.Y. & Robinson, R.K. (1985) *Yoghurt: Science and Technology*, Pergamon Press, Oxford.

Tamime, A.Y. & Robinson, R.K. (1988) Fermented milks and their future trends. Part II. Technological aspects. *Journal of Dairy Research*, **55**, 281–307.

Taye, L. (1983) *Biochemical and microbiological changes during Apong fermentation.* MSc Thesis in Food Technology (in part fulfilment) Mysore University.

Thyagaraja, N., Otani, H. & Hosono, A. (1991) Microflora in *idli* – a traditional fermented cereal pulse product from India. *Journal of Food Science and Technology*, **24**, 504–7.

Tiwari, N.P. & Singh, P. (1964) Survival of pathogens in *Dahi.* I Contamination of milk before the preparation of *Dahi. Indian Journal of Dairy Science*, **17**: 97–101.

Tiwari, N.P. & Singh, I.P. (1966) Survival of pathogens in *dahi.* II Contamination after the curdling of milk. *Indian Journal of Dairy Science*, **19**: 162–4.

Tiwari, N.P., Madanlal, Nambudripad, V.K.N. & Laxminarayana, H.L. (1969) Studies on diacetyl producing *Streptococcus* isolated from *dahi. Indian Journal of Dairy Science*, **22**, 100–2.

Van Veen, A.G. & Steinkraus, K.H. (1970) Nutritive value and Wholesomeness of fermented foods. *Journal of Agricultural and Food Chemistry*, **18**, 576.

Van Veen, A.G., Hackler, L.R., Steinkraus, K.H. & Mukherjee, S.K. (1967) Effect of fermentation on the nutritive value of *idli. Journal of Food Science*, **32**, 339–41.

Vedamuthu, E.R. (1982) Fermented milk, in *Fermented Foods. Economic Microbiology*, Vol. 7, (ed. A.H. Rose), Academic Press Inc.

Vedamuthu, E.R. (1985) What is wrong with cultured buttermilk today? *Dairy and Food Sanitation*, **5**, 8–13.

Venkatasubbaiah, P., Dwarakanath, C.T. & Sreenivasamurthy (1984) Microbiological and physicochemical changes in *idli* batter during fermentation. *Journal of Food Science and Technology*, **22**, 59–63.

Venkatasubbaiah, P., Dwarakanath, C.T. & Sreenivasamurthy, V. (1985) Involvement of yeast flora in *idli* fermentation. *Journal of Food Science and Technology*, **22**, 88–90.

Wang, H.L. & Hesseltine, C.W. (1969) *Cereal Chemistry*, **63**, 563.

Wang, H.L. & Hesseltine, C.W. (1970) *Journal of Agriculture and Food Chemistry*, **18**, 572.

Wang, H. & Hesseltine, C.W. (1981) Use of microbial cultures: Legume and cereal products. *Fd. Technol*, **35**: 79–83.

Wang, H.L. & Hesseltine, C.W. (1982) Prescott & Dunn's Industrial Microbiology (ed. J. Read), AVI Publishing Co. Inc., Westport.

Weber, G.H. & Broich, W.A. (1986) Shelf life extension of cultured dairy foods. *Cul. Dairy Produce Journal*, **21**, 19–23.

Wiseblatt, L. & Kohn, F.W. (1960) Some solatile aromatic compounds in fresh bread. *Cereal Chemistry*, **37**, 55–6.
Wood, B.J.B. (1985) *Microbiology of Fermented Foods*, Vols I & II, Elsevier Applied Science Publishers, London, New York.
Yadav, P.L., Sanyal, M.K. & Bhat, P.N. (1989) Indigenous dairy products. *Indian Dairyman*, **41**, 211–14.
Zaika, L.L. & Kissinger, J.C. (1984) *Journal of Food Science*, **49**, 5.
Zaika, L.L., Zell, T.E., Palumbo, S.A. & Smith, J.L. (1978) *Journal of Food Science*, **43**, 186.

25 Fermented weaning foods

M.R. ADAMS

25.1 Breast-feeding and weaning

Maternal breast milk is universally recognized as the best food for an infant. It is nutritionally balanced and generally free from pathogens and other substances that may be hazardous to health.

Breast-feeding has also been shown to protect infants from infectious diseases, particularly diarrhoea (Feachem & Koblinski, 1984; Kovar *et al.*, 1984; Victoria *et al.*, 1987). It does this both indirectly, through shielding the child from contaminated sources of food and water, and more directly, through the anti-infective propeties of breast milk itself. Milk has been shown to contain a number of antimicrobial factors; Immunoglobulin A (IgA) is found in the greatest concentration, but others include lactoferrin, lysozyme and lactoperoxidase and oligosaccharides that are believed to act as receptor analogues for various enteropathogens and their toxins (Mata, 1986; Facon *et al.*, 1993). In primates, placental transfer of antibodies occurs and transmisson of passive immunity to infection is largely prenatal, but the value of milk antimicrobials has been demonstrated against specific enteropathogens such as *Campylobacter*, *Escherichia*, *coli*, *Salmonella* and *Vibrio cholerae* (Holmgren *et al.*, 1976; Glass *et al.*, 1983; Cruz *et al.*, 1988; Megraud *et al.*, 1990; Ruiz-Palacios *et al.*, 1990; Haddock *et al.*, 1991).

For these reasons, bodies, such as the World Health Organization (WHO), the United Nations Children's Fund (UNICEF) and the United Nations Educational, Scientific and Cultural Organization (UNESCO) recommend that breast milk alone is the best possible food and drink for a baby during its first 4–6 months of life, and that breast-feeding should continue well into the second year of a child's life, and for longer if possible.

However, when a child reaches an age of about 4–6 months, most mothers are unable to produce sufficient milk to sustain adequate growth. At this stage, during the so-called weaning period, breast milk is replaced or supplemented with other foods. In many situations, the premature termination of breast-feeding and the early introduction of other foods can be precipitated by a variety of social and economic factors such as the need for the mother to work away from the child.

25.2 Weaning, diarrhoeal disease and malnutrition

Weaning is a particularly hazardous time for the infant, involving as it does the loss or dilution of the anti-infective properties of maternal breast milk, its replacement with foods often of low energy and nutrient density, and the exposure of an immature immune system to new sources of infection in the environment. It is at this stage that the incidence of diarrhoea in children is highest, increasing even in those children still receiving breast milk (Mata, 1983). This conjunction of weaning and illness has long been known. One early example comes from studies in Russia at the end of the 19th century, which reported increases in infant mortality during the summer, when mothers often worked long hours in the fields, generally leaving their children at home where they could not be nursed (Kunitz, 1993). More recent reports have highlighted how this problem of weanling diarrhoea persists and is particularly serious in developing countries. In South-eastern China the annual incidence of diarrhoeal disease was reported to be highest in children under 5 (Kangchuan *et al.*, 1991). In the Guatemalan village of Cauqué the frequency of diarrhoeal disease in children increased after the first 6 months of life, reaching a peak at weaning, in this case at around 18 months (Figure 25.1) (Mata, 1983). Rowland & McCollum (1977) found

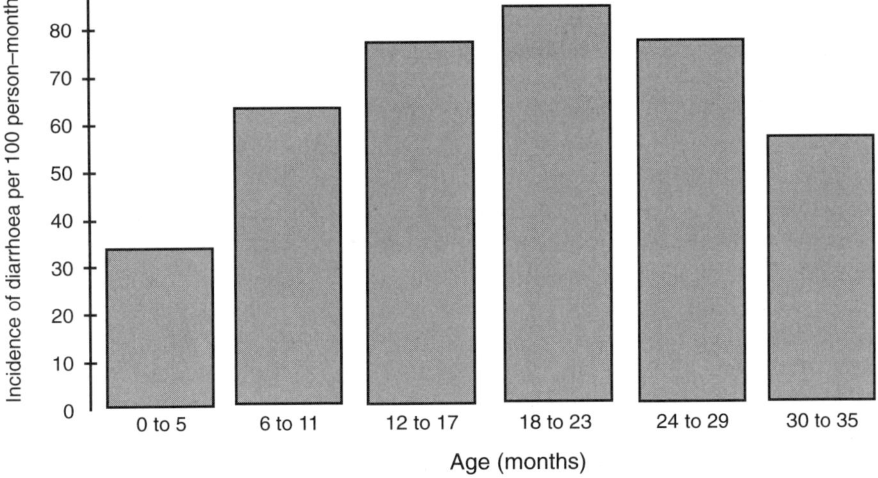

Figure 25.1 Incidence of childhood diarrhoeal disease in Cauque, Guatemala. (From Mata, 1983.)

that the incidence of diarrhoea in children in the Gambia was particularly high between the ages of 7 and 18 months with a peak at 9 months, while work in Bangladesh showed that diarrhoea occurred most frequently during the second 6 months of life (Black *et al.*, 1982a,b).

For otherwise healthy adults in the developed world, diarrhoea is an unpleasant but generally transient and infrequent condition with no long-term adverse effects. However, its consequences for children in developing countries are far more severe. It has been estimated that throughout the world, excluding China, there are 1400 million episodes of diarrhoea in children under 5 each year and that in 1990 more than 3 million children died as a result (Motarjemi *et al.*, 1993). About 80% of the deaths from diarrhoea occur during the first 2 years of life (WHO, 1992).

Infectious diarrhoea is not a single clinical syndrome; three different types are defined reflecting differences in their pathogenesis and requiring different approaches to their treatment.

- Acute watery diarrhoea is characterized by a sudden onset and relatively short duration; generally less than 7 days. It can be caused by a number of agents including enterotoxigenic *E. coli*, *Shigella*, *Campylobacter jejuni*, rotavirus and others.
- Dysentery is a diarrhoea caused by invasive organisms, particularly *Shigella*, which damage the intestinal mucosa, leading to the presence of blood, mucus and pus in the stools.
- Persistent diarrhoea is the term given to diarrhoea that begins acutely as watery diarrhoea or dysentery but is of unusually long duration – more than 14 days. It too can be caused by a number of different organisms.

The main cause of death from watery diarrhoea is the dehydration caused by the loss of fluid and electrolytes in the diarrhoeal stools. Simple, cheap and effective measures are available for correcting dehydration. These are based around oral rehydration therapy (ORT) which manages the dehydration by correcting initial deficits of fluid and electrolytes and then replacing further losses as they occur. Such approaches are by no means new and were first tried in the 1830s to combat cholera. It was not until the mid-20th century though, when glucose was included in rehydration fluids to improve salt and water absorption, that their success really became apparent (Coetzer & Kroukamp, 1989). Defined mixtures of glucose and electrolytes have been recommended for oral rehydration therapy (Table 25.1), but other solutions such as soup, rice water, yoghurt drinks and plain water can be more appropriate and nearly as effective.

ORT is able to correct dehydration in at least 95% of episodes of watery diarrhoea (WHO, 1992) and, as a result of its widespread promotion, mortality rates from diarrhoea appear to have declined. In the early 1980s for example it was estimated that 4.6 million children were dying each year as

Table 25.1 Composition of oral rehydration salts solution recommended by WHO and UNICEF

Ingredient	Amount (g/l)	Ions	Concentration (mmol/l)
Sodium chloride	3.5	Sodium	90
Trisodium citrate, dihydrate	2.9[a]	Potassium	20
Potassium chloride	1.5	Citrate	10[b]
		Chloride	80
Glucose (anhydrous)	20.0	Glucose	111

[a] Or 2.5 g sodium bicarbonate.
[b] Or 30 mmol bicarbonate.

a result of diarrhoea (Snyder & Merson, 1982) compared with the more recent estimate of 3.2 million cited above. Such rates are still offensively high and there appears to be an increased association of mortality with persistent diarrhoea where a poor response to ORT may be seen (Tomkins & Watson, 1989).

The availability of successful treatments for some types of diarrhoea does not however eliminate the cause of infection and prevent its reoccurrence. The same child can have repeated attacks throughout the year. The overall average in developing countries is estimated at 3.3 episodes a year, the average African child will have diarrhoea 4.9 times in a year, and in some areas of the world this figure can be more than 9 (Coetzer & Kroukamp, 1989; WHO, 1992). In the exclusively breast-fed infant there is little evidence of diarrhoea having any impact on the child's growth and development since regular breast-feeding will compensate for the fluid and nutrient losses that occur (Tomkins & Watson, 1989). In the weaned child, however, diarrhoea can lead to malnutrition as a result of anorexia, malabsorption of nutrients, losses of nutrients with stools and vomitus, and the widespread practice of witholding food from the sick child. Nutritional deficiencies can, in their turn, affect the immune status of a child and their resistance to infection. This can predispose them to further more severe diarrhoeal episodes and other serious infections, establishing what has been described as the malnutrition – infection cycle (Figure 25.2); a topic that was well reviewed by Tomkins & Watson (1989).

Even where the child survives this vicious circle, their growth and development can be severely impaired. In a study of 126 infants in the Gambia, diarrhoea was identified as the main cause of growth faltering observed between the ages of 6 months and 2 years (Rowland *et al.*, 1988), but other infections can also contribute to this (Figure 25.3).

A number of organisms have been identified as important causes of diarrhoea in young children in developing countries, including bacteria,

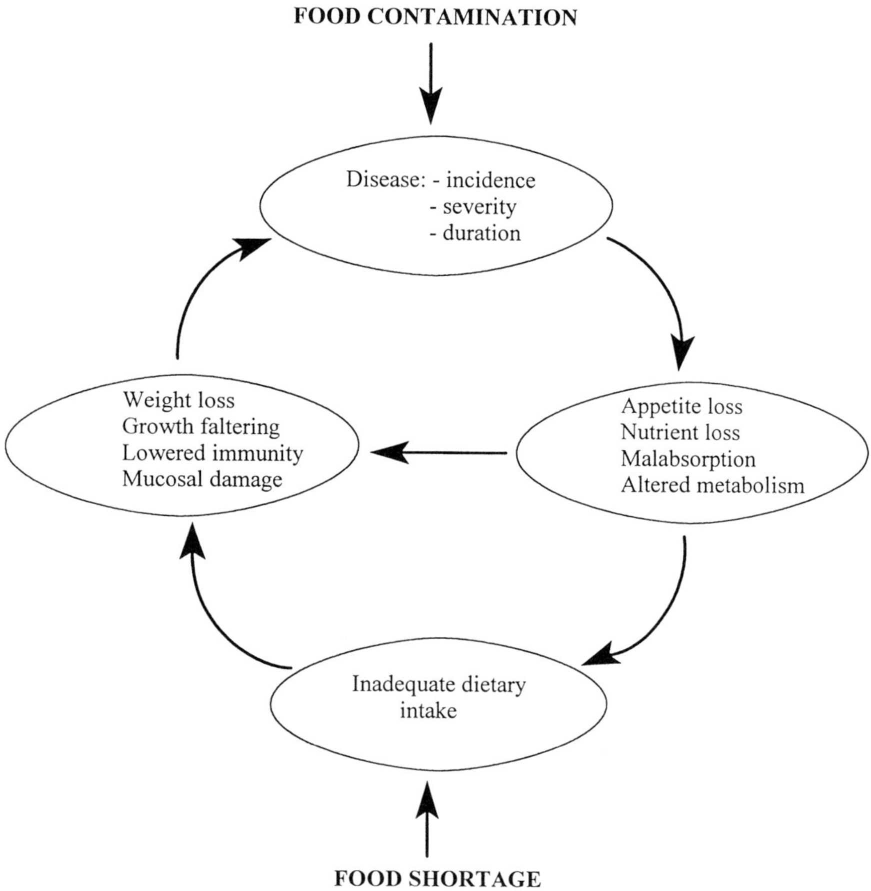

Figure 25.2 The malnutrition–infection cycle.

viruses and protozoa (Table 25.2). Many others, such as *Bacillus cereus*, *Yersinia enterocolitica*, and *Clostridium perfringens* are known causes of diarrhoea, but their association with this particular problem is either of minor significance or remains to be established (WHO, 1992). Most are spread by the faecal–oral route and can be acquired by a variety of routes, through flies, fomites, poor personal hygiene, and especially contaminated food and water. Precise data on the extent to which food is responsible for childhood diarrhoea are not available and published figures which estimate that between 15% and 70% of all diarrhoea episodes in young children are foodborne reflect this uncertainty (Esrey & Feachem, 1989).

The spread of illness by water and food are clearly closely interrelated since water used in food preparation, as an ingredient or for washing can itself often be the source of enteric pathogens in the food. Contamination of food poses particularly serious problems though, since foods have the

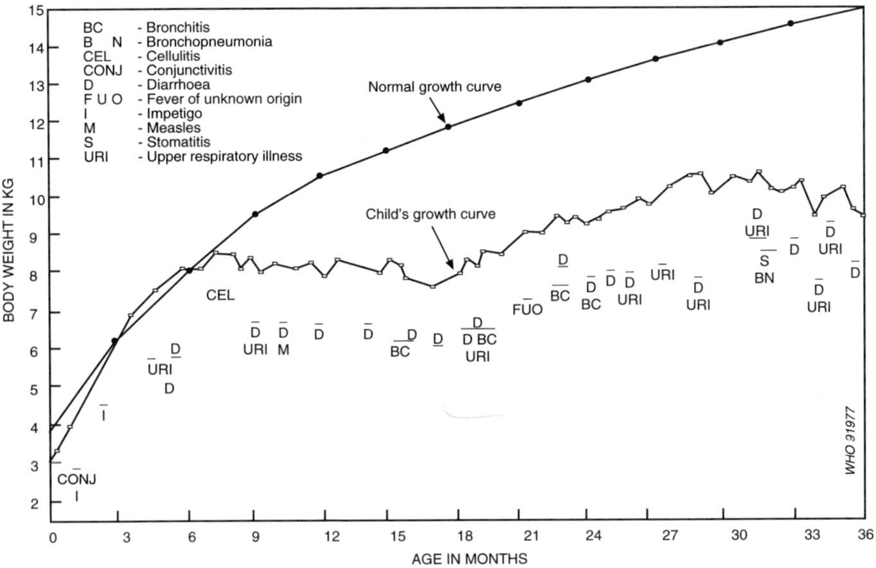

Figure 25.3 Growth pattern of a child with frequent episodes of diarrhoea and other infections. (From Mata, 1971.)

Table 25.2 Pathogens frequently identified in children with acute diarrhoea seen at treatment centres in developing countries

	Pathogen	Percentage of cases
Viruses	Rotavirus	15–25
Bacteria	Enterotoxigenic *Escherichia coli*	10–20
	Shigella	5–15
	Campylobacter jejuni	10–15
	Vibrio cholerae O1	5–10[a]
	Salmonella (non-typhoid)	1–5
	Enteropathogenic *Escherichia coli*	1–5
Protozoa	*Cryptosporidium*	5–15
No pathogen found		20–30

[a] In endemic areas; may be higher during epidemics.
Adapted from WHO (1992)

potential to support bacterial growth, thereby increasing the risk of illness. Under the right conditions, bacterial numbers can increase in a very short time from a low, possibly harmless level to numbers sufficient to cause infection when the food is consumed.

There is considerable evidence that the conditions under which weaning foods are prepared and stored in developing countries result in them being heavily contaminated with enteric indicator bacteria and pathogens (Motarjemi *et al.*, 1993). One study in Bangladesh examined samples of foods fed to children of weaning age and found 41% contained *E. coli*. These foods were more frequently and more heavily contaminated than foods for adults and this was related to the fact that weaning foods were stored longer at high ambient temperatures (Black *et al.*, 1982c). This was later confirmed independently by other work in Bangladesh which found the highest levels of faecal coliforms occurred in foods that made up a large proportion of the weanling's diet. Panta bhat, a boiled rice dish to which water is added, was the most frequently contaminated, with almost 90% of samples containing faecal coliforms and a median count of 3.9 log cfu/ml. All samples were more contaminated in the rainy season and numbers of faecal coliforms increased 10-fold when there was a delay of more than 4 hours between preparation and consumption. Some 90% of samples of panta bhat were eaten more than 12 hours after preparation (Henry *et al.*, 1990).

More than 80% of the feeds given to a sample of 90 children attending well-baby clinics in Kingston, Jamaica had coliform and enterococci counts in excess of 4 log cfu/ml, more than two-thirds had counts higher than 5 log cfu/ml (Hibbert & Golden, 1981). In Guatemala, tortillas were found to contain *E. coli* at levels between 3 and 7 log cfu/g and *Staphylococcus aureus* and *Bacillus cereus* at even higher levels (Caparelli Mata, 1975).

Work in Liberia found a higher prevalence of diarrhoea in an urban slum area studied (33.8%) when compared with poor rural areas (22.1%). The quality of the local water supply was better in the urban area, but water quality deteriorated during storage in large open vessels in the home. Two-thirds of the food samples examined were contaminated with enteric bacteria and levels of contamination were significantly higher in the urban area (Mølbak *et al.*, 1989). In the Gambia, baby milks were found to be heavily contaminated as a result of poor hygiene during their preparation, though the traditional cooked local millet gruel seemed less vulnerable in this respect, presumably as a result of the heat treatment it received. Extensive microbial growth was seen in all products when they were stored over 8 hours at ambient temperature (Barrell & Rowland, 1980).

A similar picture has been found in South East Asia. In Myanmar, in a survey of 775 samples of foods for children aged 6–29 months, 505 contained *E. coli*, 28 contained non-O1 *V. cholerae* and two contained *Salmonella* (Khin Nwe *et al.*, 1991). However, somewhat surprisingly, when children were classified according to the level of contamination found in the foods they were eating – high, medium and low – the risk of diarrhoea was not significantly different between the different groups (Aung *et al.*, 1991). In rural northern Thailand the mean total bacterial count of more than a hundred samples of weaning food was 4.6 log cfu/g (Imong *et al.*, 1989).

A feature that recurs in many of these studies conducted in different parts of the world is that comparable foods for adults are often less contaminated than weaning foods and that prolonged storage at ambient temperatures after cooking is frequently a contributory factor in the high counts observed. A case-control study of diarrhoea in children in rural Malaysia, for example, identified storage of cooked food before consumption and bottle feeding as risk factors, along with drinking unboiled water (Knight *et al.*, 1992).

Clearly, this is a multifactorial problem and a number of measures are required for its solution. Among these, improvements in sanitation and water supplies which will help limit opportunities for food contamination are of fundamental importance. So too is the promotion of safe food handling practices to control both the sources of contamination and the growth and/or survival of pathogens in the food. To this end, a number of national bodies and international agencies such as the WHO have taken a very active role in public education in food safety, promoting the application of modern, systematic approaches to food safety such as the Hazard Analysis Critical Control Point (HACCP) concept which identifies where hazards arise and how best they can be controllled (see, for example, Bryan, 1992). They have also described those basic food safety principles that are particularly important for the preparation of food for infants and young children (Table 25.3).

Unfortunately, progress is often hampered by socioeconomic and cultural factors which prevent sound food hygiene practices from being followed. Where people are poor and food is in short supply then the requirements of food hygiene may often be regarded as rather a superfluous luxury and, in many cases, it may be simply impossible to follow many of its basic rules. For example, the only water supplies that are available to a community may be contaminated. There may be insufficient time to prepare meals immediately before consumption so that food has to be prepared in quantity some time in advance and stored. There may be a shortage of fuel for boiling water, cooking or reheating stored food. Refrigeration is very unlikely to be available in the poorest areas.

25.3 Lactic fermentation

Given these limitations, attention has also focused on the potential of traditional technologies such as lactic fermentation to enhance the safety of foods, for adults and for infants and young children in particular. Its value as a household technology to improve food safety was the topic of a workshop convened in 1995 under the auspices of the FAO and WHO in collaboration with the South African Department of Health (WHO, 1996).

Table 25.3 Basic principles for the preparation of safe food for infants and young children

- ***Cook food thoroughly***
  Many raw foods, notably poultry, raw milk and vegetables, are very often contaminated with disease-causing organisms. Thorough cooking will kill these organisms. For this purpose all parts of the food must become steaming hot, which means they must reach a minimum temperature of 70°C.
- ***Avoid storing cooked food***
  Prepare food for infants and young children freshly, and give it to them immediately after preparation when it is cool enough to eat. Foods prepared for infants and young children should preferably not be stored at all. If this is impossible, food could be stored only for the next meal, but kept cool (at temperatures below 10°C) or hot (at temperatures near or above 60°C). Stored food should be reheated thoroughly. Again, this means that all parts of the food must reach at least 70°C.
- ***Avoid contact between raw foodstuffs and cooked foods***
  Cooked food can become contaminated through even the slightest contact with raw food. This cross-contamination can be direct, as, for example, when raw food comes into contact with cooked food. It can also be indirect and subtle: for example, through hands, flies, utensils or unclean surfaces. Thus, hands should be washed after handling high-risk foods, e.g. poultry. Similarly, utensils used for raw foods should be carefully washed before they are used again for cooked food. The addition of any new ingredients to cooked food may again introduce pathogenic organisms. In this case, food needs to be thoroughly cooked again.
- ***Wash fruits and vegetables***
  Fruits and vegetables, particularly if they are given to infants in raw form, must be washed carefully with safe water. If possible, vegetables and fruits should be peeled. In situations when these foods are likely to be heavily contaminated – for example, when untreated waste water is used for irrigation or untreated nightsoil is used for soil fertilization – fruits and vegetables which cannot be peeled should be thoroughly cooked before they are given to infants.
- ***Use safe water***
  Safe water is just as important in preparing food for infants and young children as it is for drinking. Water used in preparing food should be boiled, unless the food to which the water is added has subsequently to be cooked (e.g. rice, potatoes). Remember that ice made of unsafe water will also be unsafe.
- ***Wash hands repeatedly***
  Wash hands thoroughly before you start preparing or serving food and after every interruption – especially if you have changed the baby, used the toilet, or been in contact with animals. It should be remembered that household animals often harbour germs that can pass from hands to food.
- ***Avoid feeding infants with a bottle***
  Use a spoon and cup to give drinks and liquid foods to infants and young children. It is usually difficult to get bottles and teats completely clean. Spoons, cups, dishes and utensils used for preparing and feeding infants should be washed straight after use. This will facilitate their thorough cleaning. If bottles and teats must be used, they should be thoroughly washed and boiled after every use.
- ***Protect foods from insects, rodents and other animals***
  Animals frequently carry pathogenic organisms and are potential sources of contamination of food.
- ***Store non-perishable foodstuffs in a safe place***
  Keep pesticides, disinfecting agents and other toxic chemicals in labelled containers and separate from foodstuffs. To protect against rodents and insects, non-perishable foodstuffs should be stored in closed containers. Containers which have previously held toxic chemicals should not be used for storing foodstuffs.
- ***Keep all food preparation premises meticulously clean***
  Surfaces used for food preparation must be kept absolutely clean in order to avoid food contamination. Scraps of food and crumbs are potential reservoirs of germs and can attract insects and animals. Garbage should be kept in safe, covered places and be disposed of quickly.

From Motarjemi *et al.* 1994.

Lactic acid fermentation has attracted attention in this respect because of the well-established ability of lactic acid bacteria to inhibit the growth and survival of other organisms. It is a property that has led to their widespread use to increase both food safety and keeping quality in a huge range of traditional fermented foods.

A number of factors have been identified as contributing to this antimicrobial activity (Table 25.4) and these have been reviewed on a number of occasions (Lindgren & Dobrogosz, 1990; Ray & Daeschel, 1992; Holzapfel *et al.*, 1995). Their relative contributions will differ considerably, but it is clear that for such a range of different lactic acid bacteria to be used in this way the principal antimicrobial must be something that is central and common to them all. This feature is the fermentative pathways they use to derive cellular energy, which mean that wherever they grow they produce organic acids, principally lactic acid. In a lactic fermentation, levels of lactic acid can exceed 100 mM (1%, w/v) and, depending on the buffering capacity of the surrounding medium, the ultimate pH is generally in the range 3.5–4.5. As a weak organic acid, the antimicrobial activity of lactic acid is due mainly to the ability of the undissociated acid to penetrate the cell and as such will depend on both the amount of acid produced and the prevailing pH. In a typical fermented plant product, very roughly about half of the lactic acid present will be undissociated (see also Chapter 1, Vinegar). Though they can tolerate acid conditions, sensitivity to acidity does vary within the lactic acid bacteria and this, in part, accounts for the sequence of different organisms often seen as fermentation develops in products such as sauerkraut and yoghurt.

Bacteriocins are polypeptide antimicrobials which are inhibitory to other bacteria, normally those very closely related to the producing organism. Bacteriocins produced by lactic acid bacteria have attracted considerable attention in recent years though the generally limited range of activity probably means that most have little direct effect on fermented food safety, though they can influence it in other ways by determining which strains of lactic acid bacteria dominate (Olsen *et al.*, 1995).

Table 25.4 Antimicrobial factors associated with lactic acid bacteria

- Low pH
- Organic acids
- Bacteriocins
- Carbon dioxide
- Hydrogen peroxide
- Ethanol
- Diacetyl
- Low redox potential
- Nutrient depletion
- Crowding

One exception to this general picture is the bacteriocin nisin, which is produced by strains of *Lactococcus lactis*. It shows activity against a broad range of Gram-positive bacteria and is an approved and widely used food preservative (Delves-Broughton, 1990). Hitherto its use has been confined mainly to inhibiting spore-forming spoilage organisms which are particularly sensitive, but numerous other possible applications are under development. Nisin is active against the vegetative Gram-positive pathogens *Staphylococcus aureus* and *Listeria monocytogenes*, though the former is one of the most nisin-resistant Gram-positives and *L. monocytogenes* has been shown to acquire resistance to high levels of nisin quite readily (Harris *et al.*, 1991; Ming & Daeschel, 1993; Davies & Adams, 1994). Gram-negative bacteria, which include most of the diarrhoeagenic bacteria of concern, are unaffected by nisin unless their outer membrane is damaged in some way, impairing its barrier function (Stevens *et al.*, 1992).

In recent years numerous new bacteriocins have been reported, though many are, as yet, only poorly characterized. Studies on new antimicrobials from lactic acid bacteria have to take great care to eliminate interference from the strong background inhibition due to lactic acid and failure to do so can lead to misleading results. Occasional reports appear in the literature of unidentified antibiotic factors from lactic acid bacteria showing activity against Gram-negatives, but the precise nature of these have yet to be confirmed.

25.4 Fermented weaning foods

Many of the foods given to children at weaning are based around cooked gruels or porridges of the local starchy staple such as maize, millet, rice, plantain or cassava. Carbohydrate-rich plant products show a particular propensity to undergo fermentation by lactic acid bacteria and there is a huge range of traditional products prepared in this way. Many of these differ in the raw material used or some aspect of their processing but others are essentially the local names of similar products. A survey in just one province of Mozambique recorded six names for fermented porridge (Lechtig & Srivastava, 1988, cited by Ashworth & Draper, 1992). In many African countries it is customary for children to be given these products at weaning and some examples are presented in Table 25.5. These traditional practices do appear under some threat. A survey in Kenya found that while the majority of mothers in rural locations (66%) gave their young children fermented foods, only 40% did so in urban areas (Watson *et al.*, 1996). Paradoxically, one of the reasons given for this observation was that since fermented foods by their very nature are not freshly prepared, health workers described them as bad.

Table 25.5 Some fermented starchy crops of Africa

Name	Substrate	Country/region
Abreh	Sorghum	Sudan
Agidi	Maize	Nigeria
Ogi	Maize, sorghum, millet	Nigeria, West Africa
Akasa	Maize	Ghana
Uji	Maize, sorghum, millet, cassava	Kenya, East Africa
Banku	Maize	Ghana
Ting	Sorghum	Botswana
Kenkey/Koko	Maize	Ghana
Magai	Sorghum	Tanzania
Mahewu	Maize	South Africa
Kocho	Ensete	Ethiopia
Injera	Sorghum, teff	Ethiopia
Humulur	Sorghum	Sudan
Nasha	Sorghum	Sudan
Mawe	Maize	Benin
Obusera	Millet	Uganda
Gari	Cassava	Nigeria, West Africa
Fufu	Cassava	Nigeria, West Africa
Lafun/Kokonte	Cassava	Nigeria, West Africa
Chikawangue	Cassava	Zaire
Cingwada	Cassava	East and Central Africa

Data from Ashworth & Draper, 1992; Nche, 1995; Oyewole, 1995.

25.5 Monitoring and challenge studies

In contrast, numerous laboratory and field studies have demonstrated the inhibitory effect of lactic fermentation on the growth/survival of various pathogenic bacteria in this type of food. In some cases the inhibitory effect may have been over-estimated, since the microbiological procedures used would not recover cells that were sub-lethally injured but still potentially infective. Nevertheless the balance of evidence overall is quite compelling.

Odugbemi and coworkers showed that enteropathogenic *Escherichia coli, Salmonella typhi* and *S. paratyphi* could not grow in the fermented maize product, ogi (Odugbemi *et al.*, 1991). A comparison of unfermented and fermented maize dough in Ghana (pH 5.9 and 3.9 respectively) found that all the samples of the unfermented product contained Gram-negative bacteria, including some *E. coli* bearing enterotoxin encoding plasmids. Gram-negatives were found in only 16 out of 51 samples of fermented dough and the counts were significantly lower (4.0 as opposed to 5.9 log cfu/g). Porridges made from fermented and unfermented doughs and stored for 6 and 12 hours, in accordance with local custom, also showed significantly higher levels of Gram-negative bacteria in the porridge made from unfermented dough (4.2 versus 3.8 log cfu/g). Enterotoxigenic *E. coli* were found in both, though fewer were isolated from the fermented porridge (Mensah *et al.*, 1990). In this work, *Salmonella* and *Shigella* were not found in any sample but in laboratory studies, Mensah and co-workers demonstrated the ability of fermented maize dough to reduce numbers of *Shigella* and enterotoxigenic *E. coli* significantly after 4 hours storage (Mensah *et al.*, 1991).

Campylobacter jejuni is particularly susceptible to adverse conditions such as reduced pH (Adams & Moss, 1995) and an initial level of 6–7 log cfu/ml was undetectable after 30 minutes in the fermented cereal gruel, mahewu with a pH of 3.6 (Simango & Rukure, 1991). Strains of enteropathogenic and enterotoxigenic *E. coli* inoculated into mahewu were still detectable after 24 hours, though none had increased in numbers and some strains had decreased by one or two log cycles.

During production of mawè, a fermented maize dough from Benin, counts of lactic acid bacteria increased to more than 9 log cfu/g within 6 hours and the pH had decreased from 6.13 to 4.12. Enterobacteriaceae counts increased slightly over the first 24 hours of fermentation to 3.4 log cfu/g, but during the following 24 hours decreased to undetectable levels (<1.7 log cfu/g) (Hounhouigan *et al.*, 1994). With the fermented maize product, kenkey, moulds and catalase-positive, Gram-negative and Gram-positive bacteria present at levels of 5–6 log cfu/g at the start of steeping were not detectable after 24 hours fermentation when the pH had decreased from 6.5 to 3.7 (Halm *et al.*, 1993). The pH of sorghum fermented with a starter derived from the Tanzanian product, togwa, dropped below 4.0 within 12 hours (Kingamkono *et al.*, 1995). Added *Campylobacter* strains were not detectable after 6 hours and *Salmonella*, *Shigella* and *Staphylococcus aureus* were not detectable after 12 hours. In contrast, the natural fermentation of tef (*Eragrostis tef*) was far slower, taking 40 hours to reach a pH of 5.0. During this time the pathogens *Salmonella typhimurium* and *Staphylococcus aureus* were both able to multiply to high levels, though by 96 hours when the pH was 4.0 both were undetectable (<10 cfu/g) (Girma *et al.*, 1989).

In all cases, effective inhibition of competing bacteria appears to depend on achieving numbers of lactic acid bacteria sufficient to decrease the pH rapidly to levels where the growth of the pathogen is prevented. In a laboratory study using a rice-based weaning food, it was shown that when the lactic acid bacteria and pathogens were present initially in comparable numbers then inhibition of the pathogens was slight . Reduction of the stationary phase population of *E. coli* by more than 1 log cycle only occurred when the inoculum level of the starter exceeded that of the *E. coli* by more than 3 log cycles. Even when the starter outnumbered the *E. coli* by a factor of 10^5, the *E. coli* still grew for 5 hours, increasing in numbers by almost 2 log cycles (Yusof *et al.*, 1993). This work also supported the contention that acidity is the most important antimicrobial factor. Using single strain pure cultures, the greatest inhibition was achieved with the best acid producers. When a nisin-producing strain of *Lactococcus lactis* was used no additional inhibition was apparent, even against Gram-positive pathogens, and the antimicrobial effect observed could be accounted for entirely by the acid produced.

Another way of achieving high initial levels of the desirable starter organism is by recycling material from previous successful batches. A comparison of sorghum fermentation with and without the addition of a starter derived from previous batches showed a significantly lower pH was achieved with the starter (<4.0 within 24 hours compared with >5.2 in the controls) and this was reflected in a significantly greater inhibition of pathogens (Kingamkono *et al.*, 1994). Recycling portions of a naturally fermented, formulated weaning food containing a mixture of cereals, legumes, roots and tubers gave a rapid and reliable fermentation that could establish a pH below 4.0 within 24 hours incubation. Cooking to produce a porridge eliminated Enterobacteriaceae from both fermented and unfermented concentrates, but post-process contamination with *Salmonella* led to substantial growth of the pathogen in the unfermented porridge but a rapid decline in numbers in the fermented product (Nout *et al.*, 1989a). Studies with fermented porridges made from rather simpler mixtures of ingredients gave similar results where strains of *Salmonella typhimurium, Shigella* and *E. coli* died at rates of 1.2, 0.9 and 0.6 log cycles/hour respectively (Nout *et al.*, 1989b).

A starter culture of two lactobacilli isolated from local fermented foods and displaying strong antibacterial activity was used to produce an improved version of the Nigerian product, ogi which the authors named DogiK (Olukoya *et al.*, 1994).

Little comparable work has been done on the survival of enteric viruses in fermented foods, but the available evidence suggests that they are relatively unaffected by lactic fermentation unless it is combined with a heating step. Using the bacteriophage MS-2 as a model for human viruses, Nout and coworkers (1989b) noted that the titre declined only slowly, by approximately 0.1 log cycle/hour, in a fermented porridge at pH 3.8. Even greater survival was found in work which used the simian rotavirus SA-11 as a model for the unculturable human rotavirus (Wood & Adams, 1992). In 24 hours at 30°C, the virus titre decreased by 0.25 log cycles at pH 4.0 and by 1 log cycle at pH 3.3, though the heat sensitivity was considerably increased at the lower pH.

25.6 Epidemiological evidence

Compared with the wealth of microbiological data available, the epidemiological evidence for a reduced incidence of diarrhoea when fermented weaning foods are used is very sparse. A prospective study in Tanzania looked at samples of 105 children aged under 5 in each of two villages 4 km apart. In one, it was the custom to consume a fermented maize gruel, magai, while in the other fermented products were not consumed. In other respects the villages were considered to be well-matched in terms of location, available water quality, hygienic conditions and nutritional status. Over a period of 9 months, the mean number of episodes of diarrhoea in children eating fermented gruel was significantly less than that in children from the neighbouring village, 2.1 compared with 3.5. Within both diet groups the frequency of diarrhoea was age-dependent, occurring more frequently in children under 3 years of age than in older children (Lorri & Svanberg, 1994). These are interesting findings but there is an urgent need for more work to confirm them elsewhere if scarce resources are to be targeted to the promotion of this technique.

25.7 D- and L-lactate

A potential problem identified with lactic fermentation concerns the isomer of lactic acid produced. The principal lactic acid bacteria associated with natural cereal fermentations produce a mixture of D- and L-lactate isomers. For example, studies on traditional African products have identified *Lactobacillus fermentum*, *Lb. brevis* and *Lb. plantarum* as major components of the microflora (Adegoke & Babalola, 1988; Achi, 1990) and in one fermented weaning food, D-lactate comprised 42% of the total acidity (Nout *et al.*, 1989a). D-Lactate is a non-physiological isomer in mammals, is poorly metabolized, and will accumulate in the blood, especially if there is thiamine deficiency (Goldman *et al.*, 1960; Anon, 1990). Excessive consumption of D-lactate or over-production *in vivo* as a result of surgical resection or bypass of the small bowel can result in D-lactate acidosis (Gautier, 1969; Stolberg *et al.*, 1982; Mason, 1986; Anon, 1990). The FAO/WHO Joint Committee on Food Additives reviewed the toxicological evidence available and concluded that there was evidence that babies in their first 3 months of life have difficulties in utilizing small amounts of DL or D-lactate and that neither should be used for infant foods (WHO, 1974). Whether D-lactate would pose a problem for a young child receiving a fermented product as a major component of their diet, particularly if that child is malnourished to some degree, is not known and requires further investigation.

Some lactic acid bacteria produce predominantly L-lactate, but in fermentations conducted without strict asepsis they are likely to be outgrown by those producing a mixture of the D and L-isomers. One strategy that has been proposed to ensure a robust but predominantly L-lactate-producing fermentation is the use of a nisin-producing strain of *Lactococcus lactis*, where the nisin would inhibit competitive lactic acid bacteria that might otherwise eventually dominate the fermentation (Yusof *et al.*, 1995).

25.8 Nutritional improvements

Fermentation has also been associated with improvement of the nutritional value of plant foods by removing or reducing levels of toxic or anti-nutritional factors (see, for example Chavan & Kadam, 1989). Often this is a result of other aspects of the process such as the chopping, slicing or maceration. For example, the removal of the cyanogenic glucoside, linamarin, during the production of gari from cassava has been shown to be due to the cellular damage caused by grating which releases the enzyme linamarase (Vasconcelos *et al.*, 1990). In the case of a related product, fufu where whole cassava roots are soaked and undergo fermentation, endogenous linamarase activity still seems to be the main mechanism of cyanogen removal, but in this case microbial activity is essential for softening the tissues and allowing the enzyme access to its substrate (Westby & Choo, 1994).

Phytic acid is an abundant plant constituent that can interfere with mineral metabolism in the body. Its levels have been shown to be reduced during fermentation, mainly as a result of endogenous phytase activity. Levels of phytase are increased in germinated cereals (see below) but the fermentation process can provide optimum pH conditions for enzyme activity (Svanberg *et al.*, 1993).

25.9 Germination/malting

Malting is another traditional food processing technology which is often considered of value in the preparation of weaning foods. The use of malted cereals or legumes in weaning food formulations has been described on a number of occasions and efforts have been made to produce them commercially (Malleshi & Desikachar, 1982; Malleshi *et al.*, 1989; Suhasini Livingstone *et al.*, 1993). Its use in this context is primarily to improve the nutritional value of weaning foods, though it also impacts on their safety.

Malting is a controlled germination in which a range of hydrolytic enzymes are produced to mobilize the seed's nutrient and energy reserves for the growing plant. This process is arrested by drying the sprouting seed in order to preserve both the enzyme activity and much of the nutrient content. Traditionally it has been applied mostly to cereal grains but has also been used with legumes where it is known to improve their flavour, nutritional qualities and cooking properties (Vanderstoep, 1981; Ashworth & Draper, 1992; Akpapunam *et al.*, 1996).

Malted cereals are often used in conjunction with fermentation. Since most lactic acid bacteria are unable to utilize starch directly, extensive starch degradation by malt has the advantage of releasing soluble sugars, promoting lactic fermentation and a rapid acidification of the product. Use of malt also gives an added importance to the lactic fermentation, since the malt would generally be added after the cooking/gelatinization step and could itself be a source of pathogens. Microbial counts were found to increase dramatically during the preparation of malted wheat and chickpea for use in a weaning food, up to levels of 8–9 log cfu/g during the germination stage (Susahini Livingstone *et al.*, 1992). Subsequent kilning reduced microbial numbers back to levels comparable with those in the starting material, but the counts in final malted weaning food were higher than in similar products made by popping or roller drying of wheat and chickpea. Where sun-drying is used rather than kilning, the reduction in microbial load is likely to be much less and the need for high hygienic standards during malt preparation even greater.

A particularly valued property of malted cereals in weaning food preparation is their ability to reduce the dietary bulk of foods while maintaining or even enhancing their nutrient density. A starchy porridge with 30% flour can have an energy density of about 5 kJ/g (the energy density of breast milk is about 3.2 kJ/g), but its thickness can prevent children from consuming sufficient to meet their energy needs. In some cultures it is customary to overcome this problem of acceptability by thinning the normal gruel consumed by adults with water. This of course decreases its energy density and increases the amount a child must consume (Lorri & Svanberg, 1995). When a malted cereal, generally in the form of a flour, is used in a cereal gruel, the enzyme activity degrades the gelatinized starch, reducing the product's viscosity and improving its acceptability (Mosho & Svanberg, 1990). Though the acid developed during fermentation may also contribute to the thinning of cereal gruels, the reduction in viscosity is mainly a result of amylase activity in the added malt (Lorri & Svanberg, 1993).

Using both fermented and non-fermented maize–soybean porridges to which 1–5% of a cereal malt flour had been added, it was possible to feed infants recovering from diarrhoea in Nigeria and Ghana a porridge with 20–25% total solids, approximately double the level present in the traditional fermented maize porridge (Mensah *et al.*, 1995). In Tanzania, the use of malted millet or sorghum, known as kimea or 'power flour' has been promoted. Improved energy intakes were seen in children admitted to a hospital clinic with acute diarrhoea when they were fed a 'power flour'-digested weaning food (Darling *et al.*, 1995), but studies with the non-hospitalized have shown no significant differences in food and energy intake and pointed to the need to examine local attitudes and practices carefully before attempting to promote new child feeding techniques (Kibona *et al.*, 1995).

25.10 Conclusions

It is clear that fermentation and associated technologies such as malting can improve both the nutritional value and bacteriological safety of weaning foods. They are not however, a complete solution to problems of malnutrition and diarrhoeal disease, though in fairness this has never been claimed. Malted grain prepared in unhygienic conditions and used in a fermented weaning food could itself be a potent source of pathogens. Fermentation seems relatively ineffective against viral pathogens and its antibacterial effect, while often impressive, does not give the same assurance of safety as thorough heating. Heat processing remains the single most effective safety measure available and the need for good basic food hygiene remains of paramount importance. Fermentation should be seen as a useful adjunct to these rather than as a substitute.

Other questions do remain. If we recognize its value, how do we preserve and develop a traditional technology that appears to be disappearing with increasing urbanization and how do we introduce or re-introduce it to societies where it has fallen into disuse? More particularly, the cost and likely impact of programmes to extend the use of fermentation must be weighed against alternative strategies to combat diarrhoeal disease which may be more cost-effective.

References

Achi, O.K. (1990) Microbiology of 'obiolor': a Nigerian fermented non-alcoholic beverage. *Journal of Applied Bacteriology*, **69**, 321–5.

Adams, M.R. & Moss, M.O. (1995) *Food Microbiology*, Royal Society of Chemistry, Cambridge.

Adegoke, G.O. & Babalola, A.K. (1988) Characteristics of microorganisms of importance in the fermentation of fufu and ogi – two Nigerian fermented foods. *Journal of Applied Bacteriology*, **65**, 449–53.

Akpapunam, M.A., Igbedioh, S.O. & Aremo, I. (1996) Effect of malting time on chemical composition and functional properties of soybean and bambara groundnut flours. *International Journal of Food Sciences and Nutrition*, **47**, 27–33.

Anon (1990) The colon, the rumen and D-lactate acidosis. *Lancet*, **336**, 599–600.

Ashworth, A. & Draper, A. (1992) The potential of traditional technologies for increasing the energy density of weaning foods. *WHO/CDD/EDP/92.4*, WHO, Geneva.

Aung Myo Han, Khin Nwe Oo, Tin Aye & Thein Hlaing (1991) Bacteriologic studies of food and water consumed by children in Myanmar, 2: Lack of association between diarrhoea and contamination of food and water. *Journal of Diarrhoeal Diseases Research*, **9**, 91–3.

Barrell, R.A. & Rowland, M.G.M. (1980) Commercial milk products and indigenous weaning foods in a rural West African environment: a bacteriological perspective. *Journal of Hygiene (Cambridge)*, **84**, 191–202.

Black, R.E., Brown, K.H., Becker, S. & Yanus, M. (1982a) Longitudinal studies of infectious diseases and physical growth of children in rural Bangladesh. I Patterns of morbidity. *American Journal of Epidemiology*, **115**, 305–14.

Black, R.E., Brown, K.H., Becker, S., Abdul Alim, A.R.M. & Huq, I. (1982b) Longitudinal studies of infectious diseases and physical growth of children in rural Bangladesh. II

Incidence of diarrhoea and association with known pathogens. *American Journal of Epidemiology*, **115**, 315–24.

Black, R.E., Brown, K.H., Becker, S., Abdul Alim, A.R.M. & Merson, M.H. (1982c) Contamination of weaning foods and transmission of enterotoxigenic *E. coli* diarrhoea in children in rural Bangladesh. *Transactions of the Royal Society of Tropical Medicine and Hygiene*, **76**, 259–64.

Bryan, F.L. (1992) *Hazard Analysis Critical Control Point Evaluations: a guide to identifying hazards and assessing risks associated with food preparation and storage.* WHO, Geneva.

Caparelli, E. & Mata, L. (1975) Microflora of maize prepared as tortillas. *Applied Microbiology*, **29**, 802–6.

Chavan, J.K. & Kadam, S.S. (1989) Nutritional improvement of cereals by fermentation. *Critical Reviews in Food Science and Nutrition*, **28**, 349–400.

Coetzer, P.W.W. & Kroukamp, L.M. (1989) Diarrhoeal disease – epidemiology and intervention. *South African Medical Journal*, **76**, 465–72.

Cruz, J.R., Gill, L., Caceres, P. & Pareja, G. (1988) Breast milk anti-*Escherichia coli* heat-labile toxin IgA antibodies protect against toxin-induced infantile diarrhea. *Acta Paediatrica Scandinavica*, **77**, 658–62.

Darling, J.C., Kitundu, J.A., Kingamkono, R.P., Msengi, A.E., Mduma, B., Sullivan, K.R. & Tomkins, A.M. (1995) Improved energy intakes using amylase-digested weaning foods in Tanzanian children with acute diarrhoea. *Journal of Pediatric Gastroenterology and Nutrition*, **21**, 73–81.

Davies, E.A. & Adams, M.R. (1994) Resistance of *Listeria monocytogenes* to the bacteriocin nisin. *International Journal of Food Microbiology*, **21**, 341–7.

Delves-Broughton, J. (1990) Nisin and its use as a food preservative. *Food Technology*, **44**, 100–17.

Esrey, S.A. & Feachem, R.G. (1989) Intervention for the control of diarrhoeal disease among young children: promotion of food hygiene. *Report Document WHO/CDD 89/30 Diarrhoeal Disease Control Programme.* WHO, Geneva.

Facon, M., Skura, B.J. & Nakai, S. (1993) Potential for immunological supplementation of foods. *Food and Agricultural Immunology*, **5**, 85–91.

Feachem, R.G. & Koblinski, M.A. (1984) Interventions for the control of diarrhoeal diaseases among young children: promotion of breast-feeding. *Bulletin of the World Health Organization*, **62**, 271–91.

Gautier, E. (1969) Lactic acidosis in infancy. *Helvetica Medica Acta*, **35**, 423–32.

Girma, M., Gashe, B.A. & Lakew, B. (1989) The effect of fermentation on the growth and survival of *Salmonella typhimurium*, *Staphylococcus aureus*, *Bacillus cereus* and *Pseudomonas aeruginosa* in fermenting tef (*Eragrostis tef*). *MIRCEN Journal*, **5**, 61–6.

Glass, R.I., Svennerholm, A.-M., Stoll, B.J., Khan, M.R., Belayet Hossain, K.M., Imdadul Huq, M. & Holmgren, J. (1983) Protection against cholera in breast-fed children by antibodies in breast milk. *New England Journal of Medicine*, **308**, 1389–92.

Goldman, H.I., Karelitz, S., Seifter, E., Acs, H.E. & Schell, N.B. (1960) Lactic acid acidosis. *American Journal of Diseases in Children*, **100**, 705–7.

Haddock, R.L., Cousens, S.N. & Guzman, C.C. (1991) Infant diet and salmonellosis. *American Journal of Public Health*, **81**, 997–1000.

Halm, M., Lillie, A., Sørensen, A.K. & Jakobsen, M. (1993) Microbiological and aromatic characteristics of fermented maize doughs for kenkey production in Ghana. *International Journal of Food Microbiology*, **19**, 135–43.

Harris, L.J., Fleming, H.P. & Klaenhammer, T.R. (1991) Sensitivity and resistance of *Listeria monocytogenes* ATCC 19115, Scott A and UAL 500 to nisin. *Journal of Food Protection*, **54**, 836–40.

Henry, F.J., Patwary, Y., Huttley, S.R.A. & Aziz, K.M.A. (1990) Bacterial contamination of weaning foods and drinking water in rural Bangladesh. *Epidemiology and Infection*, **104**, 79–85.

Hibbert, J.M. & Golden, M.H.N. (1981) What is the weanling's dilemma? Dietary faecal bacteria ingestion of normal children in Jamaica. *Journal of Tropical Pediatrics*, **27**, 255–58.

Holmgren, J., Hanson, L.A., Carlson, B., Lindblad, B.S. & Rahimtoola, J. (1976) Neutralizing antibodies against *Escherichia coli* and *Vibrio cholerae* enterotoxins in human milk from a developing country. *Scandinavian Journal of Immunology*, **5**, 867–71.

Holzapfel, W.H., Geisen, R. & Schillinger, U. (1995) Biological preservation of foods with reference to protective cultures, bacteriocins and food-grade enzymes. *International Journal of Food Microbiology*, **24**, 343–62.
Hounhouigan, D.J., Nout, M.J.R., Nago, C.M., Houben, J.H. & Rombouts, F.M. (1994) Microbiological changes in *mawè* during natural fermentation. *World Journal of Microbiology and Biotechnology*, **10**, 410–12.
Imong, S.M., Rungruengthanakit, K., Ruangyuttikarn, C., Wongsawasdii, L., Jackson, D.A. & Drewett, R.F. (1989) The bacterial content of infant weaning foods and water in rural Northern Thailand. *Journal of Tropical Pediatrics*, **35**, 14–18.
Kangchuan, C., Chenshui, L., Qingxin, Q., Ningmei, Z., Guokui, Z., Gongli, C., Yijun, X. Yiejie, L. & Shifu, Z. (1991) The epidemiology of diarrhoeal diseases in Southeastern China. *Journal of Diarrhoeal Disease Research*, **9**, 94–9.
Khin Nwe Oo, Aung Myo Han, Thein Hlaing & Tin Aye (1991) Bacteriologic studies of food and water consumed by children in Myanmar, 1: the nature of contamination. *Journal of Diarrhoeal Disease Research*, **9**, 87–90.
Kibona, N., Doland, C., Watson, F.E., Alnwick, D. & Tomkins, A. (1995) An evaluation of a project to improve child nutrition in Tanzania. *International Journal of Food Sciences and Nutrition*, **46**, 233–9.
Kingamkono, R., Sjögren, E., Svanberg, U. & Kaijser, B. (1994) pH and acidity in lactic-fermenting cereal gruels: effects on viability of enteropathogenic microorganisms. *World Journal of Microbiology and Biotechnology*, **10**, 664–9.
Kingamkono, R., Sjögren, E., Svanberg, U. & Kaijser, B. (1995) Inhibition of different strains of enteropathogens in a lactic-fermenting cereal gruel. *World Journal of Microbiology and Biotechnology*, **11**, 299–303.
Knight, S.M., Toodayan, W., Caique, W.J.C., Kyi, W., Barnes, A. & Desmarchelier, P. (1992) Risk factors for the transmission of diarrhoea in children: a case-control study in rural Malaysia. *International Journal of Epidemiology*, **21**, 812–18.
Kovar, M.G., Serdula, M.K., Marks, J.S. & Fraser, D.W. (1984) Review of the epidemiologic evidence for an association between infant feeding and infant health. *Pediatrics*, **74**, 615–38.
Kunitz, S.J. (1993) Diseases and the European mortality decline 1700–1900, in *The Cambridge World History of Human Disease* (ed. K.F. Kiple), Cambridge University Press, Cambridge, pp. 287–92.
Lindgren, S.E. & Dobrogosz, W.J. (1990) Antagonistic activities of lactic acid bacteria in food and feed fermentations. *FEMS Microbiology Reviews*, **12**, 149–64.
Lorri, W. & Svanberg, U. (1993) Lactic acid-fermented cereal gruels: viscosity and flour concentration. *International Journal of Food Sciences and Nutrition*, **44**, 207–13.
Lorri, W. & Svanberg, U. (1994) Lower prevalence of diarrhoea in young children fed lactic acid-fermented cereal gruels. *Food and Nutrition Bulletin*, **15**, 57–63.
Lorri, W. & Svanberg, U. (1995) An overview of the use of fermented foods for child feeding in Tanzania. *Ecology of Foods and Nutrition*, **34**, 65–81.
Malleshi, N.G. & Desikachar, H.S.R. (1982) Formulation of a weaning food with low hot paste viscosity based on malted ragi and green gram. *Journal of Food Science and Technology*, **19**, 193–7.
Malleshi, N.G., Daodu, M.A. & Chandrasekhar, A. (1989) Development of weaning food formulations based on malting and roller drying of sorghum and cowpea. *International Journal of Food Science and Technology*, **24**, 511–19.
Mason, P.D. (1986) Metabolic acidosis due to D-lactate. *British Medical Journal*, **292**, 1105–6.
Mata, L.J. (1971) Nutrition and infection. Protein Advisory Group Bulletin, **11**, 18–21.
Mata, L. (1983) Epidemiology of acute diarrhea in childhood: an overview, in *Acute Diarrhea, its Nutritional Consequences in Children* (ed. J.A. Bellani), Nestlé Nutrition Workshop Series, Raven Press, New York, pp. 3–22.
Mata, L. (1986) Breast-feeding and host defense. *Frontiers in Gastrointestinal Research*, **13**, 119–33.
Megraud, F., Boudraa, G., Bessaoud, K., Bensid, S., Dabis, F., Soltani, R. & Touhami, M. (1990) Incidence of *Campylobacter* infection in infants in Western Algeria and the possible protective role of breast feeding. *Epidemiology and Infection*, **105**, 73–8.
Mensah, P.P.A., Tomkins, A.M., Drasar, B.S. & Harrison, T.J. (1990) Fermentation of cereals for reduction of bacterial contamination of weaning foods in Ghana. *Lancet*, **336**, 140–3.

Mensah, P., Tomkins, A.M., Drasar, B.S. & Harrison, T.J. (1991) Antimicrobial effect of fermented Ghanaian maize dough. *Journal of Applied Bacteriology*, **70**, 203–10.

Mensah, P., Ndiokwelu, C.I., Uwaegbute, A., Ablordey, A., van Boxtel, A.M.G.A., Brinkman, C., Nout, M.J.R. & Ngoddy, P.O. (1995) Feeding of lactic acid-fermented high nutrient density weaning formula in paediatric settings in Ghana and Nigeria: acceptance by mother and infant and performance during recovery from acute diarrhoea. *International Journal of Food Sciences and Nutrition*, **46**, 353–62.

Ming, X. & Daeschel, M.A. (1993) Nisin resistance of foodborne disease and the specific resistance responses of *Listeria monocytogenes* Scott A. *Journal of Food Protection*, **56**, 944–8.

Mølbak, K., Højlyng, N., Jepsen, S. & Gaarsley, K. (1989) Bacterial contamination of stored water and stored food: a potential source of diarrhoeal disease in West Africa. *Epidemiology and Infection*, **102**, 309–26.

Mosha, A-C & Svanberg, U. (1990) The acceptance and intake of bulk-reduced weaning foods: the Lungaga village study. *Food and Nutrition Bulletin*, **12**, 69–74.

Motarjemi, Y., Käferstein, F., Moy, G. & Quevedo, F. (1993) Contaminated weaning food: a major risk factor for diarrhoea and associated malnutrition. *Bulletin of the World Health Organization*, **71**, 79–92.

Motarjemi, Y., Käferstein, F., Moy, G. & Quevedo, F. (1994) Contaminated food, a hazard for the very young. *World Health Forum*, **15**, 69–71.

Nche, P.F. (1995) *Innovations in the production of kenkey, a traditional fermented maize product of Ghana: nutritional and safety aspects.* PhD Thesis, Wageningen, The Netherlands.

Nout, M.J.R., Rombouts, F.M. & Hautvast, G.J. (1989a) Accelerated natural lactic fermentation of infant food formulations. *Food and Nutrition Bulletin*, **11**, 65–73.

Nout, M.J.R., Rombouts, F.M. & Havelaar, A. (1989b) Effect of accelerated natural lactic fermentation of infant food ingredients on some pathogenic microorganisms. *International Journal of Food Microbiology*, **8**, 351–61.

Odugbemi, T., Odujirin, H.T., Akitoye, C.O., Oyerinde, J.P.O. & Esumeh, F.I. (1991) Study on the pH of ogi, Nigerian fermented weaning food and its effect on enteropathogenic *Escherichia coli, Salmonella typhi* and *Salmonella paratyphi. Journal of Tropical Medicine and Hygiene*, **94**, 219–23.

Olsen, A., Halm, M. & Jakobsen, M. (1995) The antimicrobial activity of lactic acid bacteria from fermented maize (kenkey) and their interactions during fermentation. *Journal of Applied Bacteriology*, **79**, 506–12.

Olukoya, D.K., Ebigwei, S.I., Olasupo, N.A. & Ogunjimi, A.A. (1994) Production of DogiK: an improved ogi (Nigerian fermented weaning food) with potentials for use in diarrhoea control. *Journal of Tropical Pediatrics*, **40**, 108–13.

Oyewole, O.B. (1995) Lactic fermented foods in Africa and their benefits. Paper presented at the WHO/FAO Workshop on the Assessment of Fermentation as a Household Technology for Improving Food Safety. 11–15 December 1995, South Africa.

Ray, B. & Daeschel, M. (eds) (1992) *Food Biopreservatives of Microbial Origin*, CRC Press Inc., Boca Raton, Florida.

Rowland, M.G.M. & McCollum, J.P.K. (1977) Malnutrition and gastroenteritis in the Gambia. *Transactions of the Royal Society of Tropical Medicine and Hygiene*, **71**, 199–203.

Rowland, M.G.M., Rowland, S.G.J.G. & Cole, T.J. (1988) Impact of infection on the growth of children from 0 to 2 years in an urban West African community. *American Journal of Clinical Nutrition*, **47**, 134–8.

Ruiz-Palacios, G.M., Calva, J.J., Pickering, L.K., Lopez-Vidal, Y., Volkow, P., Pezzarossi, H. & Stewart-West, M. (1990) Protection of breast-fed infants against *Campylobacter* diarrhea by antibodies in human milk. *Journal of Pediatrics*, **116**, 707–13.

Simango, C. & Rukure, G. (1991) Survival of *Campylobacter jejuni* and pathogenic *Escherichia coli* in *mahewu*, a fermented cereal gruel. *Transactions of the Royal Society of Tropical Medicine and Hygiene*, **85**, 399–400.

Snyder J.D. & Merson, M.M. (1982) The magnitude of the global problem of acute diarrhoeal disease: a review of actrive surveillance data. *Bulletin of the World Health Organization*, **60**, 605–13.

Stevens, K.A., Sheldon, B.W., Klapes, N.A. & Klaenhammer, T.R. (1992) Effect of treatment conditions on nisin inactivation of Gram negative bacteria. *Journal of Food Protection*, **55**, 763–7.

Stolberg, L., Rolfe, R., Gitlin, N., Merit, J. Mann, L., Linder, J. & Finegold, S. (1982) D-Lactic acidosis due to abnormal gut flora. *New England Journal of Medicine*, **306**, 1344–8.

Suhasini Livingstone, A., Sandhu, J.S. & Malleshi, N.G. (1992) Microbiological evaluation of malted wheat, chickpea, and weaning food based on them. *Journal of Tropical Pediatrics*, **38**, 74–7.

Suhasini Livingstone, A., Feng, J.J. & Malleshi, N.G. (1993) Development and nutritional quality evaluation of weaning foods based on malted, popped and roller dried wheat and chickpea. *International Journal of Food Science and Technology*, **28**, 35–43.

Svanberg, U., Lorri, W. & Sandberg, A.-S. (1993) Lactic fermentation of non-tannin and high-tannin cereals: effects on *in vitro* estimation of iron availability and phytate hydrolysis. *Journal of Food Science*, **58**, 408–12.

Tomkins, A. & Watson, F. (1989) Malnutrition and infection: a review. *United Nations ACC/SCN State of the Art Series Nutrition Policy Discussion Paper* No. 5, 136pp.

Vanderstoep, J. (1981) Effect of germination on the nutritive value of legumes. *Food Technology*, **35**, 83–5.

Vasconcelos, A.T., Twiddy, D.R., Westby, A. & Reilly, P.J.A. (1990) Detoxification of cassava during gari preparation. *International Journal of Food Science and Technology*, **25**, 198–203.

Victoria, C.G., Vaughan, J.P., Lombardi, C., Fuchs, S.M.C., Gigante, L.P., Smith, P.G., Nobre, L.C., Teixeira, A.M.B., Moreira, L.B. & Barros, F.C. (1987) Evidence for protection by breast-feeding against infant deaths from infectious diseases in Brazil. *Lancet*, **2**, 319–22.

Watson, F.E., Ngesa, A., Onyang'o, J., Alnwick, D. & Tomkins, A.M. (1996) Fermentation – a traditional anti-diarrhoeal practice lost? The use of fermented foods in urban and rural Kenya. *International Journal of Food Sciences and Nutrition*, **47**, 171–9.

Westby, A. & Choo, B.K. (1994) Cyanogen reduction during the lactic fermentation of cassava. *Acta Horticulturae*, **375**, 209–15.

WHO (1974) Toxicological evaluation of some food additives including anticaking agents, antimicrobails, antioxidants, emulsifiers and thickening agents. *WHO Food Additives Series*, No. **5**, pp. 461–5.

WHO (1992) *Readings on diarrhoea: student manual,* World Health Organization, Geneva.

WHO (1996) Fermentation: assessment and research. *Report of a FAO/WHO Workshop on fermentation as a technology to improve food safety* FAO/FNU/FOS/96.1.

Wood, G.W. & Adams, M.R. (1992) Effect of acidification, bacterial fermentation, and temperature on the survival of rotavirus in a model weaning food. *Journal of Food Protection*, **55**, 52–5.

Yusof, R.M., Morgan, J.B. & Adams, M.R. (1993) Bacteriological safety of a fermented weaning food containing L-lactate and nisin. *Journal of Food Protection*, **56**, 414–17.

Yusof, R.M., Baker, T.A., Morgan, J.B. & Adams, M.R. (1995) Effect of *ragi* and L-lactate producing cultures on enteric pathogens in a rice-based weaning food. *World Journal of Microbiology and Biotechnology*, **11**, 654–7.

26 Potential infective and toxic microbiological hazards associated with the consumption of fermented foods

N.J. ROWAN, J.G. ANDERSON AND J.E. SMITH

26.1 Introduction

While the introduction and consumption of many solid-state fermented foods in developed countries are generally considered to be in their infancy (i.e. where these foods are often perceived as novel or exotic), fermented ethnic products, such as tempe, miso, tef injerra, togwa, ergo, rice tape, siljo etc., have been traditionally the principal source of nutrients for people residing in many developing countries world-wide. The traditional fermentation of cereal products widely practised in Africa and other developing countries usually involves a spontaneous development of different lactic acid-producing bacteria and the final bacteriological status of the product is influenced in part by the raw materials and process method (Steinkraus, 1983). This natural lactic fermentation process is considered to be an effective method of preserving these foods, thus providing the population with a safe nutritious food supply (Smith & Palumbo, 1983). Its use for safer weaning foods is discussed in Chapter 25.

From a historical perspective it is not difficult to understand how such preservation processes could have arisen in primitive societies driven by the need to optimize the use of scarce food resources. An attestament to the intrinsic value of the approach is the extent to which the operating principles underlying solid-state food fermentations have been discovered and developed in many regions of the world. While the microbiological changes involved are generally highly reproducible and robust, nevertheless the use of natural substrates with their associated microflora – together with the potential for operator 'mistakes' or alterations in processing procedures – can introduce some degree of variation in the microbiological quality of the final product.

Not all products carry the same degree of risk. For many, there appears to be little if any risk, but for some the results can be disastrous, as in the classical case of tempe bongkrek poisoning which has caused many deaths over the years (van Veen, 1967; Steinkraus, 1983). Fortunately, such dramatic consequences of process failure are rare with indigenous fermented foods; however, the extent to which other less acute microbiological problems might occur is difficult to assess because of the difficulties of establishing cause and effect with some types of food-related illness and the general lack of good epidemiological surveillance data.

This chapter describes how the multifaceted lactic acid fermentation process normally renders indigenous fermented foods safe for consumption; describes several studies which attempt to demonstrate the inhibitory nature of many natural and starter culture fermented foods against established and emerging bacterial pathogens; and highlights the real threat of emerging enteropathogens in indigenous lactic-fermented foods prepared under poor sanitary and hygiene conditions.

In addition to potential problems caused by pathogenic and toxigenic bacteria, it has long been realized that certain moulds and their toxic metabolic products (i.e. mycotoxins) can pose a threat with fermented food products. Although mould strains used in these processes have generally been found to be non-toxigenic, if unknown mould contaminants develop during the fermentation there is always the possibility that mycotoxins will be produced. Moreover, since the raw materials used are good substrates for mould growth and mycotoxin formation, these relatively stable toxic compounds may be present in commodities before the fermentation, and persist into the final product. Examples of mycotoxin contamination of fermented food products will be given in this chapter, together with a discussion of the health implications associated with the ingestion of various problematic mould mycotoxins.

26.2 Fermented foods and opportunist microbial pathogens

A number of processes exist by which solid-state fermented foods may be rendered safe from opportunist microbial pathogens. The lactic acid bacteria (represented mainly by the genera *Streptococcus*, *Lactobacillus*, *Pediococcus*, *Leuconostoc* and *Bifidiobacterium*), play a dominant role in solid-state fermented foods. Lactic acid bacteria (LAB) are collectively assigned to the family Lactobacteriaceae, where its members are Gram-positive, non-spore-forming rods (with the exception of *Sporolactobacillus inulinus*) or cocci that can grow in the presence of oxygen and form lactate by the fermentation of carbohydrates (glucose, lactose). In contrast to the Enterobacteriaceae, which also produce lactic acid, they are obligate fermenters. They do not contain haemins (cytochromes, catalase). The LAB produce a number of antimicrobial agents including metabolic products such as short-chain organic acids, carbon dioxide, hydrogen peroxide, lactoperoxidase, diacetyl and other more specifically inhibitory substances such as antibiotics and bacteriocins (Spelhaug & Harlander, 1989). It is the cumulative or holistic effect of their action which dictates whether competing microorganisms are inhibited or survive and proliferate in foods of concern. Apart from their antagonistic behaviour towards unwanted spoilage and opportunist pathogenic organisms, LAB produce lactic acid which imparts the unique tangy flavour to fermented products, lowers the pH, denatures the protein and releases the moisture uniformly.

26.2.1 Low pH and organic acids

Acidity (pH) may be the primary factor in the preservation and sanitation of fermented foods. Although there are many bacteria which can grow in acidic conditions, the great majority which are of medical significance are neutrophilic, i.e. they grow over the mid pH range from about pH 5.0 to 9.0. Although the external pH can vary over this range, homeostasis of internal pH is essential for neutrophils. By a mechanism which is still only partially understood, neutrophilic bacteria maintain relatively constant internal pH (pH_i) values in the range 7.5 to 8.0.

During food fermentations, when the acidity is sufficiently great, this overcomes the homeostatic pH control mechanism of neutrophilic bacteria and growth ceases completely when the pH_i drops to 6.6–6.8 (Ingram, 1987). This external acidity (pH) effect is magnified in the presence of organic acids which can cross the cell membrane. Thus, the antimicrobial effects of organic acid production which occur during food fermentations is due both to the depression of pH below the growth range of certain spoilage and pathogenic bacteria and to metabolic inhibition by the undissociated acid molecules.

For acetic acid the growth of most food-poisoning organisms and spore-forming bacteria is inhibited by 0.1% and mycotoxigenic moulds by 0.3% of the undissociated acid. Citric and lactic acids have moderate antimicrobial activities and inhibit microbial growth only at low pH values. A concentration of 0.001% undissociated citric acid is reported to inhibit the growth of *Staphylococcus aureus* under anaerobic conditions (Hobbs & Roberts, 1993a,b).

The effects of lactic acid bacteria and organic acids on *Bacillus cereus* have been reported by Wong & Chen (1988). Acetate was found to be more inhibitory to the growth of vegetative cells while formate was more inhibitory to spore germination. Acetate, formate and lactate (all at 0.1 M) completely inhibited *B. cereus* at pH 6.1, 6.0 and 5.6 respectively. Endospores were more resistant to these organic acids than vegetative cells.

Apart from their inhibitory effects on food-poisoning bacteria, the LAB also have antifungal properties. In particular, it has been reported that LAB can inhibit aflatoxin biosynthesis, though the extent to which this inhibition is due to lactic acid and/or other LAB metabolites is unclear. Nevertheless, it has been suggested that the antimycotic and antimycotoxigenic potential of the LAB might eventually find commercial application as a biopreservative (Gourama & Bullerman, 1995).

26.2.2 *Bacteriocins*

While the antagonistic nature of many of these antimicrobial compounds is well known, previously, the role of bacteriocins (produced during many lactic fermentations) in the inhibition of foodborne bacterial pathogens has often been overlooked. In recognition of the discovery that antibiotic substances of the colicin type (prototype bacteriocins produced by various members of the family Enterobacteriaciae) may also be produced by non-coliform bacteria, the term 'bacteriocin' was coined by Jacob *et al.* in 1953. Bacteriocins are specifically defined as protein antibiotics of the colicin type, i.e. molecules characterized by lethal biosynthesis, predominant intra-species killing activity, and adsorption to specific receptors on the surface of bacteriocin-sensitive cells. Many of these proteinaceous antimicrobial compounds (i.e. ribosomally synthesized, catonic, low molecular weight and heat-stable) are inhibitory to food poisoning, spoilage and in some cases, pathogenic bacteria (Jack *et al.*, 1995). Although it is not a common phenomenon, LAB are known to produce two classes of bacteriocins; (i) those that are cidal to a narrow range of target organisms (usually closely related to the producer strain and thus closely related to the bacteria that are indigenous to the foods of concern); and (ii) those that inhibit a broad spectrum of Gram-positive bacteria (Kalchayanand *et al.*, 1992). In general however, bacteriocins produced by LAB tend to be active against a wide range of Gram-positive bacteria, and some have also been reported to inhibit Gram-negative species including *Aeromonas hydrophilia* (Vignolo *et al.*, 1993).

It appears that two major classes of killing activities are displayed by bacteriocins; some form ion channels in the cytoplasmic membrane, and others exhibit nuclease activity upon gaining entry to a sensitive cell (Pugsley, 1984a,b). However, the antibacterial action of low-molecular weight bacteriocins (synthesized by LAB) is produced principally by destabilization of the cytoplasmic membrane functions (Jack *et al.*, 1995). The lantibiotic (e.g. nisin Z produced by *Lactococcus lactis*) and non-lanthionine-containing bacteriocins (e.g. pediocin PA-1 produced by *Pediococcus acidilactici*) produced by some LAB appear to affect the cell membrane permeability barrier by forming water-filled membrane channels or pores, probably by a barrel-stave mechanism (Jack *et al.*, 1995). These barrel-stave poration complexes have been proposed to be formed between one, two, or probably even more species of amphipathic peptides, resulting in ion leakage, loss of proton motive force and, ultimately, cell death (Klaenhammer, 1993).

26.2.3 *Other factors*

With some fermented foods/beverages, ethanol is a major product of the fermentation. Generally food spoilage and disease-producing microorganisms cannot develop in either an acidic or an alcoholic environment. When both products are produced, as is the case with some fermentations, then the combination of acid and alcohol gives double protection. Whereas excessive acidity interferes with cell pH homeostasis, alcohol is considered to act by denaturing proteins and possibly interaction with membrane lipid components. It is known that ethanol produces a rapid release of intercellular constituents from exposed bacteria, and disorganization of the membrane probably results from penetration of the solvents into the hydrocarbon interior of the cytoplasmic membrane (Russell & Chopra, 1990).

With other processes, acid fermentation combined with salting remains one of the most practical methods of preserving and often enhancing the organoleptic and nutritional qualities of fresh vegetables, cereal gruels and milk–cereal mixtures. Salting increases the osmotic pressure and this, together with acidity and some chemical components, is considered to be responsible for the bactericidal properties of soy sauce (Ujii & Yokoyama, 1956).

With sauerkraut, in addition to salt and acidity, it is considered that certain cabbage substances are of value in the fermentation because they inhibit the development of certain undesirable Gram-negative bacteria (Steinkraus, 1983).

26.3 Acid resistance of some foodborne pathogens

An emerging pathogen which has caused concern with some fermented food products is *E. coli* 0157:H7. Although meat products and milk have been the main foodstuffs implicated in transmission of this organism, nevertheless some outbreaks have involved acidic foods, *viz.* apple cider (Besser *et al.*, 1993), dry fermented sausage (Centers for Disease Control and Prevention, 1994), yoghurt (Morgan *et al.*, 1993) and mayonnaise (Weagant *et al.*, 1994). These occurrences have drawn attention to the acid-tolerance properties of *E. coli* 0157:H7 (Arnold & Kaspar, 1995). In addition to promoting survival in foods, acid tolerance may play a role in *E. coli* 0157:H7 pathogenesis by promoting survival during passage through the stomach (Gorden & Small, 1993).

In studies to understand the special properties of *E. coli* serotype 0157:H7 it has been found that growth can occur at pH levels ranging from 4.0 to 9.0 (Glass *et al.*, 1992), which is not significantly different from the pH range of 4.4 to 9.0 reported for other *E. coli* (Jay, 1992). The essential difference is the ability of the 0157:H7 serotype to ***survive*** (not grow) in highly acidic foods, e.g. at pH 2.0 there was a minimal drop of CFU after 24 hours (Miller & Kaspar, 1994). It is also known that this sustained acid tolerance in *E. coli* 0157:H7 is not dependent upon adaptation but is induced by stationary phase or starvation (Cheville *et al.*, 1996). Clearly, *E. coli* 0157:H7 poses a new threat to fermented food products should these become contaminated at any stage of preparation.

With certain other pathogenic bacteria, it has now been established that exposure to certain conditions can result in a marked increase in acid tolerance. This so-called acid tolerance response (ATR) was shown to occur with *Listeria monocytogenes* cells which, following a 1 hour exposure to mild acid (pH 5.5) were then capable of withstanding severe acid stress (pH 3.5). Susceptibility to pH 3.5 acid was growth phase-dependent since stationary-phase *Listeria* cultures were naturally resistant to the pH 3.5 challenge while exponential-phase cultures required adaptation to pH 5.5 to induce acid tolerance (O'Driscoll *et al.*, 1996).

Full expression of the ATR in *L. monocytogenes* required *de novo* synthesis, since chloramphenicol, a protein synthesis inhibitor, prevented full induction of acid tolerance. Analysis of protein expression indicated that the ATR in *L. monocytogenes* is a complex response involving a modulation in the expression of a large number of genes (Davis *et al.*, 1996).

The survival of adapted and non-adapted *L. monocytogenes* has been examined in a variety of acid foods. Acid adaptation enhanced the survival of *L. monocytogenes* in acidified dairy products, including cottage cheese, yoghurt and whole-fat cheddar. Acid-adapted *L. monocytogenes* cultures also demonstrated increased survival during active milk fermentation by a lactic acid culture (Gahan *et al.*, 1996).

Salmonella typhimurium has also been shown to exhibit an ATR (Hall & Foster, 1996). In *S. typhimurium*, the ATR has been characterized as a complex defence system that can minimize the lethal effects of extremely low pH (pH 3). The study by Baik *et al.* (1996) illustrates that the ATR can also defend against weak acids such as butyric, acetic and propionic acids.

Undoubtedly, ATRs will be found and characterized in other medically significant bacteria and this phenomenon may help to explain the occasional persistence of some pathogens in acidic foods.

26.4 Spoilage of popular fermented foods

A number of spoilage bacteria and/or opportunist foodborne microbial pathogens are encountered during commonly used solid-state fermentations of food. Lactic fermented food provide protection against foodborne illnesses, where children regularly consuming lactic fermented products (e.g. togwa, a commonly used weaning food in Tanzania with a final pH < 4) show a significantly lower number of diarrhoeal episodes than non-users (Lorri & Svanberg, 1994). While fermented foods, the mainstay of many developing countries, have long been perceived as safe for consumption, a number of researchers have investigated the fate of many prevalent foodborne bacterial pathogens during the production of such foods.

The growth and survival of different enteropathogenic microorganisms (i.e. *Bacillus cereus*, *Campylobacter jejuni*, enterotoxigenic *Escherichia coli* (ETEC) *Salmonella typhimurium* and *Shigella flexneri*) during lactic acid fermentation of cereal gruels (e.g. togwa) prepared from low-tannin (white) and high-tannin (red) sorghum varieties were studied by Kingamkono *et al.* (1994). The authors used fermented gruel (starter culture of *Lactococcus lactis* and *Candida krusei*, SC) recycled daily or stored for 7, 14 or 28 days, germinated cereal flour (power flour, PF), or a combination of PF and SC (PF + SC) as starters. At 24 hours, the pH of all gruels with added starter was ≤4; the pH in the control indigenously fermented gruels without starter was ≥5.2.

In control gruels, the enteropathogens remained at the inoculation level or increased in number, while these organisms were inhibited within 24 to 48 hours in the fermented gruels (PF + SC, SC), in the order: *Bacillus*, *Campylobacter*, *Escherichia*, *Salmonella* and finally *Shigella*. While the use of starter cultures rendered togwa free of these enteropathogens after 48 hours, an indigenous fermentation was not sufficient to inhibit their development and proliferation where these organisms were present as contaminants. The slower inhibition of ETEC, *Salmonella typhimurium* and *Shigella flexneri* (Gram-negative bacteria) than *Bacillus cereus* (Gram-positive bacterium) was attributed to fact that Gram-negative bacteria had a cell wall which required breaching before cell death (Andersson, 1986) and partly to the acid tolerance response, an inducible pH-homeostatic function protecting the cells from acid stress (Foster & Hall, 1991).

While the contrasting behaviour of *Campylobacter jejuni* compared with other Gram-negative bacteria may be attributed in part to the fact that *Campylobacter* is microaerophilic, although other factors may be having an effect (Kingamkono *et al.*, 1994). The faster pH decrease in gruels inoculated with PF + SC that had been recycled daily, was explained by the fact that the starter cultures produced through daily recycling, undergo selection of microrganisms that grow best in acid and near-anaerobic conditions, making them dominant in the media (Nout *et al.*, 1987b).

The fates of many enteropathogens and spoilage organisms have been investigated in other fermented foods such as tef injerra, a pancake-like fermented bread which is the mainstay of many African countries (e.g. Ethiopia). Gashe *et al.* (1982) identified *Enterobacter*, *Klebsiella*, *Proteus* and *Citrobacter* as the most prominent microorganisms during the first 18 to 24 hours of fermentation. *Leuconostoc mesenteroides* and *Streptococcus faecalis*, followed by *Pediococcus cerevisease* and *Lactobacillus* spp., were also isolated at a later stage of fermentation. According to Gifawesen & Besrat (1982) the yeasts isolated from fermenting tef were *Saccharomyces*, *Torulopsis*, *Candida* and *Pichia* species. Gashe (1985) also showed that members of the Enterobacteriaceae are responsible for the pH falling to 5.8 during the first 18 hours of fermentation. The lactic acid bacteria then reduced the pH to 3.8 at a later stage in the fermentation.

Girma *et al.* (1989) studied the effect of fermentation and baking on the survival of *Salmonella typhimurium*, *Staphylococcus aureus*, *Bacillus cereus* and *Pseudomonas aeruginosa* in fermenting tef. All enteropathogens were inoculated into the tef dough at a level of 10^3 cells g^{-1} and the dough was allowed to ferment naturally (i.e. without the addition of starter cultures) for up to 96 hours. *Staphylococcus aureus* grew well until the pH dropped to 5. Thereafter the viability decreased as the pH lowered; however, even after 72 hours when the pH was 4.2, viable cells were still present. *Bacillus cereus* on the other hand grew poorly below pH 5.8. Unlike studies carried out by Kingamkono *et al.* (1994) on the survival of *B. cereus* in fermenting togwa (described earlier), Girma *et al.* (1989) showed that most vegetative cells reverted to spores which remained resistant to the acid produced. As a result, the spores were recovered from fermented tef dough after 96 hours when the pH was about 4.0. *Salmonella typhimurium* also grew until the pH dropped to 5.5, thereafter its viability significantly declined and at pH 4.0 only a few viable cells were detected. *Pseudomonas aeruginosa* also grew well until the pH of the dough was reduced to 5 and thereafter the population decreased and at pH 4 only a few viable organisms were isolated from the dough.

Since *B. cereus* endospores and some of the other asporogenic enteropathogens were consistently isolated from the fermenting dough at pH 4, the authors investigated the survival of these bacteria during the baking of the dough. The baking process took about 1 to 2 minutes where the temperature of the dough in the hot oven reached 90–93°C. The process killed all the asporogenous cells, but *B. cereus* spores survived. The authors did not study *B. cereus* or *Staphylococcus aureus* enterotoxin production and/or inactivation; however, they felt that the survival of *B. cereus* spores may be of great significance to food hygiene.

The growth and survival of these and other opportunistic microbial pathogens, e.g. *Clostridium botulinum* (Yokotsuka, 1983a), *Clostridium perfringens* (Aryanta *et al.*, 1991), *Listeria monocytogenes* (Ashenafi, 1991), *Aeromonas hyrophilia* (Kim, 1996), *Vibrio cholera* O1 (Spelhaug & Harlander, 1989), *Vibrio parahaemolyticus* (Aryanta *et al.*, 1991), *Salmonella enteriditis* (Ashenafi, 1993), *Salmonella sofia* (Aryanta *et al.* 1991), *Yersinia enterocolitica* (Girma, 1986) and *Klebsiella* spp. (Nigatu & Gashe, 1994), in many other solid-state fermented foods has been an area of intense research over the last decade. For many of these enteropathogens, ingestion of a few hundred cells can induce illness (Bryan, 1979; Black *et al.*, 1988). Many highly susceptible individuals, e.g. those with a compromised (HIV/AIDS patients) or an underdeveloped (infants) immune system, may suffer bouts of illness after consuming a few cells, e.g. only three *Salmonella ealing* cells kg^{-1} of infant milk formulae were implicated in the 1985 UK outbreak (Rowe, 1987). While there is a general agreement that the lactic acid fermentation process improved the keeping quality and nutritional value of these foods, it did not eliminate the presence of most of these enteropathogens (especially when these problematic test organisms were subjected to a natural or indigenous fermentation without the use of starter cultures).

Svanberg *et al.* (1992) showed that a natural lactic fermentation of a weaning gruel (final pH of 3.8) was unable to completely inhibit the growth of *Staph. aureus*. Ashenafi (1993) demonstrated the survival of both *Salmonella typhimurium* and *S. enteritidis* after the lactic fermentation of ergo, a traditional Ethiopian sour milk, while *B. cereus* spores survived tempe (Nout *et al.*, 1987a; Tunçel *et al.*, 1990) and other solid-state fermentation processes (Cook *et al.*, 1991; Nigatu & Gashe, 1994). The use of starter cultures during the fermentation of fish sausage resulted in a significant reduction in *E. coli*, *S. typhimurium*, *S. sofia* and *Staph. aureus* cell numbers where the pH was reduced quickly to <4 within 48 hours. The high population of *Pediococcus acidilactici* and the rapid increase in acidity during inoculated fermentations were strong deterrents to the survival and growth of the various microbial pathogens. While this was sufficient to inhibit the growth of these enteric pathogens, it did not eliminate their presence.

By contrast, during the indigenous fermentation (i.e. without the use of starter cultures) of these fish sausage mixes, LAB rapidly grew to populations of 10^8 cfu g^{-1}, but this was not sufficient to prevent the growth of *Salmonella*, *Staph. aureus* and *E. coli* to unacceptable levels (e.g. 10^6–10^7 cfu g^{-1}). Consequently, these and other authors (Smith & Palumbo, 1983; Bacus, 1984) advocated the use of starter cultures in the production of these fermented sausages, as rapid growth of LAB to 10^8 cfu g^{-1} with concomitant reduction of pH to below 5 were the critical factors identified in controlling the growth of pathogenic bacteria. While the use of LAB starter cultures is the recommended approach to preserving fermented foods, the real problem of the survival of pathogenic enteric bacteria and the emergence of new acid-tolerant enteropathogens may have been previously underestimated and not properly addressed. Underestimation of the number of survivng bacteria may have occurred with the majority of these fermentation studies, where many researchers used selective laboratory-based media to recover these heat- and/or acid-stressed pathogenic microorganisms (Aryanta *et al.*, 1991; Kingamkono *et al.*, 1994; Nigatu & Gashe, 1994). It is well known that the use of antibiotics and dyes used in these selective media are inhibitory to the resuscitation of stressed organisms and a true reflection of the number of survivors is not obtained via recovery on these media (Rowan, 1996). In many of these studies the lower detection limit for recovery of acid-stressed organisms was remarkably high, e.g. 5×10^2 cfu g^{-1} (Tunçel *et al.*, 1990), whereby enteropathogenic bacteria may have survived in low numbers and were not detected due to the poor sensitivity of the experimental procedure.

Potentially pathogenic bacteria which form dormant endospores (e.g. *Bacillus* spp., *Clostridium* spp.) appear to tolerate and survive lactic acid fermentations (Tunçel *et al.*, 1990). In the past, these organism have been dismissed as possible threats to food safety, as a popular opinion was that these spore-forming organisms (particularly *B. cereus*), present in these foods as contaminants, required to reach high population levels (e.g. 10^7–10^8 cfu g^{-1}) in order to produce enterotoxins which may elicit a bout of food poisoning (Granum *et al.*, 1993a). Due to their ubiquitous nature, the raw materials of many solid-state fermented foods are known to be frequently contaminated with *Bacillus* spores (Tunçel *et al.*, 1990). While the infectious dose of *B. cereus* is thought to be in the range 10^5–10^8 cells, diarrhoeal enterotoxin is produced before the cells reach a level of 10^7 cells g^{-1} (Granum *et al.*, 1993a). Granum *et al.* (1993b) suggested that the food industry should be concerned about levels as low as 10^3–10^4 cells g^{-1} food, as infants may be particularly vulnerable to low numbers of pathogenic organisms (due to their immature immune system) and as food intoxication may be caused by ingestion of *B. cereus* cells or spores which may subsequently form enterotoxins in the ileum. In Norway, *B. cereus* was the most common cause of foodborne outbreaks in 1990 (Aas *et al.*, 1992).

In the past, it has been the practice of many clinical laboratories to simply disregard isolates of *Bacillus* spp. – often described as 'inconsequential aerobic spore-forming bacteria' – but these were later found to be of clinical importance (Berkowitz, 1994). Several reports have implicated other members of the genus *Bacillus* (i.e. *B. subtilis*, *B. licheniformis*, *B. pumilus*, *B. brevis*, *B. thuringiensis* and *B. sphaericus*) as the aetiological agents in proven foodborne illness outbreaks (Rowan, 1996). Furthermore, these and other *Bacillus* spp. (i.e. *B. piliformis*, *B. coagulans*, *B. circulans*, *B. mycoides*, *B. macerans*, *B. panotheniticus*, *B. alvei* and *B. laterosporus*) have been known to cause opportunist infections in immunocompromised hosts (Rowan, 1996).

The fate of *Clostridium botulinum* has been examined during the production of Japanese soy sauce. *C. botulinum* was tested in shoyu for 3 months at 30°C. Neither *C. botulinum* type A nor type B grew. Spores of type A remained viable, but those of type B decreased slightly in number after 3 months (Yokotsuka, 1983a). The possible danger of *C. botulinum* developing in improperly handled fish sauce–paste fermentations has been discussed (Steinkraus, 1983). A relatively recent incident in the UK serves as a timely reminder of this deadly food poisoning organism (Hobbs & Roberts, 1993b). Hazelnut yoghurt was shown to be the aetiological agent responsible for 27 cases of botulism (produced by *C. botulinum* type B) in June 1988. Fortunately, the antidote was available and only one person (aged 74) died. The producer had used contaminated hazelnut conserve, a low-acid product insufficiently heated to kill spores.

In view of the serverity of its infection, studies on the growth and survival of the enterohaemorrhagic *E. coli* 0157:H7 in popular lactic acid-fermented foods is recommended. Another emerging pathogen that warrants investigation is the emetic form of *Bacillus cereus*, as its enterotoxins are more heat and acid stable than the diarrhoeagenic form of this organism (Granum, 1994).

26.5 Factors increasing susceptibility of the consumers of fermented foods to microbial hazards

The study and control of established and new emerging pathogens in solid-state fermented foods must always be examined under the conditions which exist in the environment where these products are prepared, fermented and consumed. Atopic people (particulary children) may suffer bouts of food-related infections or poisonings depending on a number of inter-related factors co-existing. These factors include, preparation of foods under unhygienic conditions, poor personal hygiene, contaminated water supply, the presence of enteropathogens in raw materials or fermented foods as contaminants, the type and virulence of the enteropathogens present, the immune status of the individual being compromised (either from an earlier food or waterborne infection/poisoning and/or due to the atopic individual suffering from AIDS) or underdeveloped (i.e. children), consumption of inadequately fermented foods (i.e. possibly an indigenous fermentation process without the involvement of a LAB starter culture), etc.

World-wide (excluding China), it is estimated that 1400 million episodes of diarrhoea occur annually in children under 5 years. In 1990, over 3 million of such children died as a result. Up to 70% of diarrhoeal episodes could be due to pathogens transmitted through food. Diarrhoea is among the major causes of child morbidity and mortality in Tanzania and also contributes to a high rate of malnutrition (Kingamkono *et al.*, 1994). On average, children under 5 years of age experience five episodes of diarrhoea per year in Tanzania (Kingamkono *et al.*, 1994) and other developing countries (Chen, 1983). While breast-feeding protects young children from morbidity and mortality caused by diarrhoea, the protection is reduced during weaning (i.e. infants reaching 4–6 months of age) where breast milk needs to be supplemented, and later (>2 years of age) substituted by appropriate foods (JHU/WHO, 1989). Foodborne disease can cause severe and/or long-lasting damage to health, including acute, watery and bloody diarrhoea (leading to severe dehydration or ulceration), meningitis, as well as chronic diseases affecting the renal, articular, cardiovascular, respiratory and immune system (Archer, 1984; Archer & Young, 1988). However, the most serious implications of foodborne infections are their effects on nutritional status. Despite the complex nature of the interaction between infectious diseases and malnutrition, it is generally accepted that infectious diseases can affect children's growth once weaning is initiated (Mata, 1971). Indeed, a poor food intake, aggravated by loss of nutrients from vomiting, diarrhoea, malabsorption and fever over an extended period leads to nutritional deficiencies with serious consequences for the growth and immune systems of infants and children. Thus, a child whose resistance is suppressed becomes vulnerable to other diseases and is subsequently caught in a vicious circle of infection and malnutrition (Figure 26.1) (Motarjemi *et al.*, 1993). Annually, about 13 million children under the age of 5 years die in developing countries; in the majority of cases, these deaths are due to infections and associated malnutrition (Motarjemi *et al.*, 1993). Since studies have confirmed that cereal-based weaning foods and water are important sources of pathogens, it is recommended that this particular food product be subjected to a lactic fermentation using a starter culture which is renewed daily.

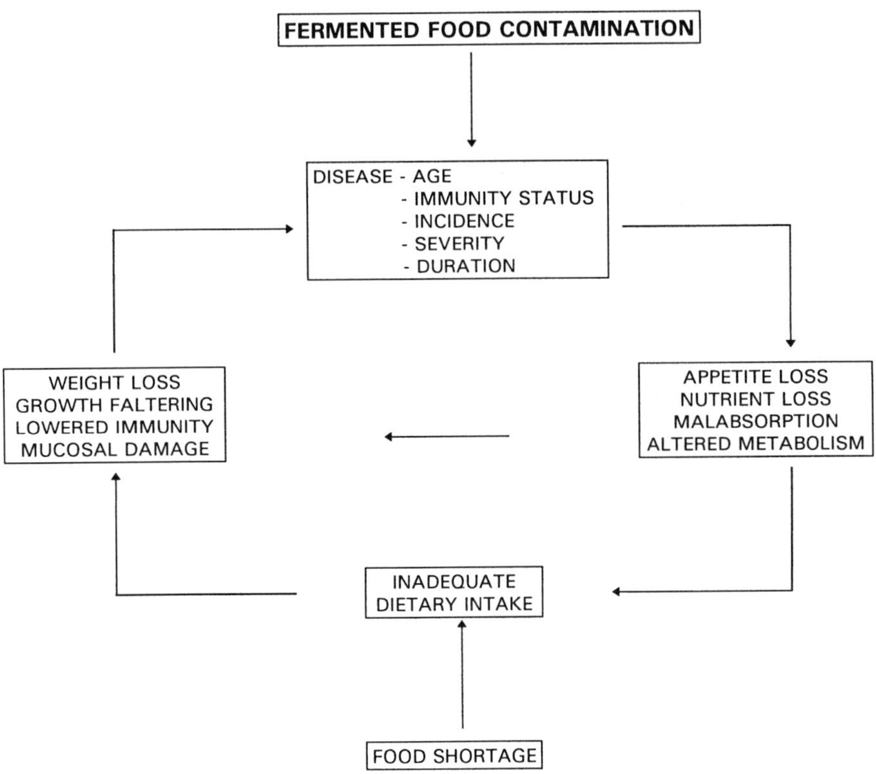

Figure 26.1 Malnutrition and foodborne infection cycle. (Adapted from Motarjemi *et al.*, 1993.)

Overall, it is of paramount importance that these fermented products be prepared under good sanitary and hygiene conditions. In developing countries, most foodborne illnesses are of a bacterial nature. This is due to poverty, low level of education, poor sanitation and hygiene practices, poor methods of food preservation and keeping, unavailability of potable drinking water and absence or scarcity of health facilities (Nigatu & Gashe, 1994). Sources of food contamination are numerous: nightsoil, polluted water, flies, pests, domestic animals, unclean utensils and pots, dirty hands, and a polluted environment caused by lack of sanitation, domestic animal droppings, dust and dirt, etc. (Motarjemi *et al.*, 1993). In order to ensure that solid-state fermented foods such as tempe, tef, etc. (which are the mainstay of many developing communities) are microbiologically safe, foodhandlers, particularly mothers, must be educated about the above dangers.

The susceptibility of the consumer population to low numbers of enteric bacterial pathogens in naturally fermented foods will depend in part on the immunological status of the consumer. The majority of people infected with the human immunodeficiency virus (HIV) or AIDS have a compromised immune system (Mueller & Williams, 1995). Considering the involvement of opportunist bacterial pathogens in food-related illnesses, individuals suffering from HIV/AIDS are particularly at risk. As of the end of 1995, and following an extensive country-by-country review of HIV/AIDS data, a cumulative total of 6 million AIDS cases were estimated to have occurred in adults and children world-wide and currently 2.1 million adults are estimated to be active and infected with HIV or AIDS (Mertens & Lowbeer, 1996; Quinn, 1996). Of the total prevalent HIV infections, the majority remain concentrated in eastern, central and southern Africa (Mertens & Lowbeer, 1996).

At present, HIV and AIDS account for more than 50% of adult medical admissions into some of the national and provincial hospitals as well as 10–15% of paediatric admissions (Mhalu & Lyamuya, 1996). Diarrhoeal episodes occur in 90% of African AIDS patients and 'slim disease' (prolonged diarrhoea and wasting usually due to enteropathogenic infection) is pathognomic of AIDS in central Africa (O'Keffe & Wood, 1996). By the year 2000, it is estimated (WHO project) that close to 40 million persons will be infected with HIV, more than 90% of whom will be in developing countries (Mertens & Lowbeer, 1996). Thus, the bacteriological quality of fermented foods is of paramount importance to these atopic individuals.

The extent of enteropathogen survival and subsequent proliferation in indigenously fermented foods (which may be accompanied with frequent episodes of foodborne illness) in developing countries may be currently underestimated (Quinn, 1996). This is due in part, to the inadequate epidemiological surveillance and reporting of diarrhoeal/vomiting cases in these problematic countries. Considering the foregoing information, it is recommended that solid-state fermented foods be prepared under good hygiene and sanitary conditions using starter cultures prepared and recycled daily.

26.6 The role of fungal toxins (mycotoxins) in fermented foods

Provision for a safe food supply, including fermented foods and beverages, is a fundamental requirement of any modern society. In the previous section, attention was drawn to the potential contamination of fermented products by bacterial pathogens. In this section, the presence of fungal contaminants or their toxic metabolic products will be examined. In each case the contaminant fungi exist freely in the environment and can enter the fermented food chain by natural methods.

Mycotoxins are secondary metabolites of certain filamentous fungi that can cause toxicity to humans and animals when low concentrations are ingested or inhaled (Smith *et al.*, 1995). In nature, such toxins are primarily derived from agricultural crops such as cereals and oil seeds and products derived from them and from animal-derived foods such as milk. Mycotoxins can enter the human dietary system by indirect or direct contamination (Table 26.1). Indirect contamination can occur when an ingredient of a food or beverage fermentation (e.g. cereals, legumes, etc.) has previously been contaminated with toxin-producing moulds and although the mould may be killed or removed during processing, mycotoxins will often remain in the final product (Miller & Trenholm, 1994; Smith *et al.*, 1994). Direct contamination can occur in two ways: (i) the fermentation may involve a fungus essential for the process but also capable of producing mycotoxins; and (ii) the process or the final product can become infected with a toxigenic mould with subsequent toxin formation. Thus, almost all fermented foods and beverages have the potential to be affected by toxigenic moulds at some stage during their production, processing, transport or storage. The reality of this assumption will now be considered.

Mycotoxins are a structurally diverse group of mainly small molecular weight compounds produced mostly by five genera of fungi, *viz*: *Aspergillus*, *Penicillium*, *Fusarium*, *Alternaria*, and *Claviceps* (Smith &

Table 26.1 Possible routes for mycotoxin contamination of fermented foods and beverages

a) Agricultural products, e.g. cereals, oilseeds, legumes, fruits and vegetables.
b) Secondary infection by toxigenic moulds of processing stages and of the final product.
Residues in animal tissues and animal products, e.g.
milk, dairy produce, meat.
Mould-ripened foods, e.g. involving toxigenic starter cultures
cheeses, meat products, oriental products.

(Adapted from Jarvis, 1976.)

Moss, 1985). Mycotoxins can elicit a wide spectrum of toxicological effects (a mycotoxicosis) which have been extensively studied in various animal species and increasingly confirmed in humans (Smith *et al.*, 1995). While over 300 mycotoxins have been characterized under laboratory conditions, only about 20 mycotoxins have been shown to occur naturally in agricultural raw materials at significant levels and frequency to be considered to represent a food safety concern.

The principal and most regularly documented mycotoxins produced by the five genera are *Aspergillus* toxins: aflatoxins B_1, G_1, and M_1, ochratoxin A, sterigmatocystin and cyclopiazonic acid; *Penicillium* toxins: cyclopiazonic acid, citrinin and patulin; *Fusarium* toxins: deoxynivalenol, nivalenol, zearalenone, T-2 toxin, diacetoxyscirpenol, moniliformin and fumonisins; *Alternaria* toxins: tenuazonic acid, alternariol and alternariol methyl ether; *Claviceps* toxins: ergot alkaloids (Steyn, 1995).

Toxicologically induced diseases of humans associated with mould-contaminated foodstuffs have long been suspected, but in most cases it has been difficult to determine the true extent of the problem and to obtain accurate corroboration. To achieve a positive acceptance of human toxicity, the following aspects must be considered, *viz*:

1. Mycotoxins usually occur in very low concentrations and can be difficult to detect. There may be more than one mycotoxin present with possible synergism in toxic potential.
2. By the time a mycotoxicosis is suspected, the foodstuff will no longer be available.
3. Clinical signs and/or symptoms are usually present as vague, subacute or chronic conditions (e.g. cancer).
4. Physicians are usually not familiar with the symptomatology of a mycotoxicosis.
5. The concept of 'dose–response' in toxicology is difficult and often impossible to apply to the diagnosis of a mycotoxin-related disease. Most human mycotoxicoses are chronic in nature, sporadic, seasonal and geographic and can also interact with other toxins and nutrients in foodstuffs.
6. In complex diets multiple mycotoxin insult can possibly occur (Prelusky *et al.*, 1994).

Consumption of fermented foods and beverages heavily contaminated with mycotoxins is not likely to occur to any real extent in most advanced societies because of the existence of strict food regulations. However, the widespread presence of fumonisins in maize especially in humid, developing countries, could be a notable exception (Sydenham *et al.*, 1994). High-quality agricultural practices with improved storage and transportation facilities, have much reduced toxigenic mould growth in raw agricultural materials destined for the human food chain. Where poor agricultural practices and warm, hot climates prevail, such as in many developing countries, higher endemic levels of mycotoxins must be anticipated in the raw food materials such as maize, rice and peanuts.

Concern still exists over the long-term exposure to low levels of mycotoxins in the food chain. There is still insufficient understanding of the effects of varying levels of ingestion of a single mycotoxin or a mixture of mycotoxins, the susceptibility of individuals, and the effects of other types of toxins in the environment, or synergy or even antagonism (Smith *et al.*, 1995). There is also little known about the effects of mycotoxins on the most vulnerable individuals in society such as children, pregnant women, the aged, the immunocompromised and especially individuals receiving nutritionally inadequate diets.

Fungal growth and mycotoxin production are dependent on environmental factors with the limits for mycotoxin production usually being more narrow than those for growth (Smith & Moss, 1985). The substrate composition of the foodstuff will exert a major influence on the type of microflora and subsequent mycotoxin formation (Table 26.2). Similarly, temperature, water activity, pH and redox potential will have marked effects on mycotoxin formation (for an extensive review of these factors, see Frisvad & Samson, 1991; Moss, 1991).

In practice, all ingredients destined for food and beverage fermentations will have been exposed at some time to mould contamination. The nature and extent of toxigenic mould contamination, together with relevant environmental factors, will determine the presence or absence of

Table 26.2 Possible *Penicillium* mycotoxin occurrence under storage conditions of various food materials

Cereals	Penicillic acid, verrucosidin, verrucofortine, xanthomegnin, viomellein, ochratoxin A, citrinin, roquefortine C, penitrem A, nephrotoxic glycopeptides in colder climates; citreoviridin, cyclochlorotine, islanditoxin, luteoskyrin, rugulosin, griseofulvin, viridicatumtoxin, nephrotoxic glycopeptides in warmer regions.
Nuts and high-lipid grains and fruits	Penitrem A, terrestric acid, roquefortine C, cyclopiazonic acid, cyclopaldic acid, compactin, chaetoglobosin C, oxaline, secalonic acid D, citrinin.
Fruits and vegetables	Patulin, terrestric acid, penitrem A, compactins, oxaline, roquefortine C, citrinin, citromycetin.
Meat and eggs	Cyclopiazonic acid, penitrem A, terrestric acid, roquefortine C, rugulosin, ochratoxin A, cyclopaldic acid, citromycetin, rugulosin, botryodiploidin, patulin, citrinin.
Cheese	Cyclopiazonic acid, cyclopaldic acid, compactin, botryodiploidin, ochratoxin A, patulin.

(Adapted from Frisvad & Samson, 1991.)

mycotoxins in the final products. Mycotoxins associated with the agricultural raw materials can be formed during growth in the field, during and after harvest, especially if the material is inadequately dried or becomes damp in storage. While the identification of contaminating toxigenic fungi can be of diagnostic value when a mycotoxicosis is suspected, positive conclusions can only be deduced by extraction and identification of the suspected toxin since:

- the presence of the toxigenic fungus is no assurance that it was producing the toxin;
- a given toxin can persist in a product when the fungus is no longer there;
- a given fungus may be capable of producing more than one toxin; and
- a given toxin may be produced by different genera of fungi (Smith & Moss, 1985).

There is an extensive literature documenting the toxic effects of mycotoxins on a wide range of animals (see Smith & Henderson, 1991, for references). Experimental studies with animals have compiled useful knowledge on modes of action, pathological conditions and degrees of toxicity. Genotoxicity resulting in carcinogenicity, mutagenicity and/or teratogenicity has long been associated with the aflatoxins in animals and humans, while the carcinogenicity of other mycotoxins such as sterigmatocystin, ochratoxin A, T-2 toxin, patulin and zearalenone is well proven in animals but still in most cases not yet definitively demonstrated in humans (IARC, 1993). Is there a realistic health risk to humans from the consumption of mycotoxin-contaminated fermented foods and beverages, particularly when consumed regularly over a long period of time with perhaps occasional high levels of mycotoxin present?

Establishing a causal relationship between mycotoxin exposure and human disease is a complicated procedure because of many uncertainties associated with human epidemiological studies. A significant uncertainty associated with such studies is establishing an accurate assessment of human mycotoxin exposure or intake. Recent studies have shown that when aflatoxin B_1 is consumed by humans, reactive electrophilic epoxides are formed which can covalently react with certain nucleophilic centres in macromolecules like DNA, RNA and proteins, forming aflatoxin adducts with a longer half-life in the human system. It is now easily possible to identify such adducts in urine and blood, giving a greater confidence in assessing actual dietary intake (Groopman, 1994). Analysis of such adducts in human populations can act as a surrogate for human genotoxicity testing. While most of the human chronic related disease studies have been focused on the aflatoxins, there is now increasing evidence linking ochratoxin A with human disease. Ochratoxin A has been regularly detected in human blood in several European countries and the frequency seems to indicate continuous exposure in humans (Dirheimer, 1996). The presence of ochratoxin A in human blood could be considered as an indicator for indirect assessment of exposure to this mycotoxin which is increasingly being considered as a possible human carcinogen (IARC, 1993; Schlatter *et al.*, 1996).

The immunotoxic effects of several mycotoxins is now well documented (Pestka & Bondy, 1994). A mycotoxin can suppress one or more functions of the immune system and the net outcome would be increased susceptibility to infection or neoplasia. A mycotoxin could also stimulate immune function resulting in hypersensitivity or an autoimmune-like disorder. Of particular interest is the capacity of some mycotoxins to alter normal immune function when present in foods at levels below observable overt toxicity. Aflatoxins, patulin, citrinin and zearalenone experimentally alter immunity and recent immunological studies with ochratoxin A and certain trichothecenes could have significance to human health (Pestka & Bondy, 1990, 1994).

It is recognized that the significance of mycotoxins as the cause of human disease (especially chronic disease) is not easily determined (Smith *et al.*, 1995). Possible adverse human health effects related to dietary exposure to mycotoxins will reflect the level of mycotoxins in the food consumed, on the amount of food consumed, on the body weight and physiological state of the individual, and the bioavailability and toxicity of the compound(s) to humans (Kuiper-Goodman, 1994). By the use of risk assessment it is possible to determine whether mycotoxins present a health concern when present in the human diet, while risk management deals with the prevention of human mycotoxicoses. Thus, risk assessment, together with statutory governmental regulatory efforts (which can vary from country to country) are applied towards ensuring that mycotoxins in foods and beverages are restricted to levels well below those that may represent potential acute or chronic health exposure (van Egmond & Dekker, 1995). In practice, risk asssessment reflects exposure to mycotoxins obtained through assessment relative to the dose that is considered to be safe, which has been calculated through hazard assessment (Kuiper-Goodman, 1996).

26.6.1 Mycotoxins and fermented foods

In Asia, many mould fungi including *Aspergillus*, *Rhizopus*, *Mucor*, *Actinomucor*, *Amylomyces*, *Neurospora*, and *Monascus* are widely used in the manufacture of fermented foods and beverages. Such products (see elsewhere in this volume) are normally derived from vegetable raw materials such as soya beans, rice, wheat, groundnut, cassava, etc. Furthermore, many of these fermentations will have mixed microbial stages involving yeasts and bacteria. It is widely accepted that there have been no examples of overt mycotoxicity related to the fermentations generated by *Rhizopus*, *Mucor*, *Actinomyces*, *Amylomyces*, *Neuropora* and *Monascus*.

Such fungi appear to be truly non-toxigenic. Also, there is considerable evidence that species of *Rhizopus* can actively break down and detoxify aflatoxin (Bol & Smith, 1990).

The yellow-green aspergilli (koji moulds) used in Asia soybean fermentations have been extensively examined for their potential to form mycotoxins in the koji process and for their possible carry over into subsequent stages of the fermentations. The relationship between the two main koji moulds, *A. oryzae* and *A. sojae* and the toxigenic *A. flavus* and *A. parasiticus* has been extensively examined. While early studies were based mainly on morphological characteristics, increasingly this has been replaced by a chemometric approach involving DNA relatedness, electrophoretic comparisons of proteins and patterns of secondary metabolites (Samson & Frisvad, 1991; Samson, 1993). Strong linkages exist between *A. oryzae* and *A. flavus* and *A. sojae* and *A. parasiticus*.

With the discovery of the aflatoxins in the 1960s the koji moulds were studied extensively for any mycotoxin production (Yokotsuka *et al.*, 1966). While no aflatoxins have ever been demonstrated in commercial production strains, it is now recognized that *A. oryzae*, *A. sojae* and *A. tamari* can all produce other mycotoxins when exposed to specific stress environmental conditions (Frisvad, 1986; Trucksess *et al.*, 1987) (Table 26.3). While these mycotoxins can be formed under unique conditions, there is little evidence that they are produced in commercial production processes.

Several reasons could explain the obvious lack of evidence of mycotoxin presence in *Aspergillus* fermentations. The most relevant factor is the duration of the koji fermentation. Under normal growing conditions with excess nutrients, toxigenic fungi usually take several days to commence toxin formation while most koji fermentations will seldom exceed 48–72 hours (Yokotsuka, 1983a,b). Most experimental evidence of mycotoxin formation by koji moulds only occurred after prolonged incubation. Another important reason for lack of mycotoxin formation is the well-documented observation that the soybean is only weakly supportive of toxin formation by fungi (Gupta & Venkitasubramanian, 1975).

A further consideration could be possible detoxification of any formed toxin or actual inhibition of toxin formation by the presence of other microorganisms. The koji fermentation is not a pure monoculture of *Aspergillus*, other contaminating microorganisms will always be present. Similarly, the moromi microorganisms may also contribute to the metabolism of formed toxins. Microbial detoxification of aflatoxin and other

Table 26.3 Mycotoxins reported to be produced by koji moulds

Aspergillus oryzae	Cyclopiazonic acid, kojic acid, maltoryzine, beta-nitropropionic acid
A. sojae	Aspergillic acid, kojic acid
A. tamarii	Kojic acid

mycotoxins is now well documented and must be highly relevant to all fermented food and beverage production processes.

There are a number of reports that indicate that LAB can affect mould growth and aflatoxin production. While *Lactobacillus* spp. have been shown to inhibit aflatoxin biosynthesis, others such as *Lactococcus lactis* can actively stimulate aflatoxin production. It has been considered that the inhibition of aflatoxin is due to lactic acid or lactic acid bacterial metabolites and that these metabolites are heat-stable, low-molecular weight compounds (for fuller details, see Gourama & Bullerman, 1995).

Finally, selection of commercial starter strains of *Aspergillus* for non-toxin production is of paramount importance. However, at local village level this may not always be the case and introduction of toxigenic strains cannot be totally dismissed. Industrial strain selection programmes must regularly check for any possible genetic changes that could lead to potential mycotoxin formation (Kalayanamitr *et al.*, 1987).

In Europe and north America, mould-ripened foods are primarily cheeses and meats ripened by starter cultures from strains of *Penicillium roquefortii* (blue cheese), *P. camembertii* (white cheese) and *P. nalgiovense* (raw, cured and dry meats). Since most *Penicillium* spp. are potentially toxigenic, it has been imperative to pursue selection processes to produce non-toxigenic starter cultures (Leistner, 1984). In establishing suitable non-toxigenic, but technically suitable starter cultures, the following criteria must be applied:

1. Toxicological screening, with chemical and biological methods, of isolates obtained from high-quality products.
2. Sensory evaluation of foods inoculated with potentially suitable strains and ripened under the usual conditions.
3. Extraction of the finished product and investigation of the extracts for mycotoxins (Leistner, 1990).

At present, totally non-toxigenic strains of *P. roqueforti* and *P. camembertii* are not available. Most strains of *P. roquefortii* produce variable traces of patulin, PR toxin or roquefortine C, while *P. camembertii* mostly produce small levels of cyclopiazonic acid. A limited number of *P. nalgiovense* strains which appear to be non-toxigenic have been isolated. Recombinant gene technology is now being used at an experimental level to improve the overall technical properties of these strains, including non-toxicity, but none is yet available at the commercial level. The ability of some *Penicillium* starter cultures to inhibit pathogenic microorganisms such as *Listeria monocytogenes* is less actively pursued and development here could have immense implications for the future safety of fermented foods.

Again, it should be stressed that mycotoxin production by these *Penicillium* starter cultures is usually only obtained under laboratory stress conditions and if mycotoxins are found in the final cheese products, the levels are extremely low.

26.6.2 Mycotoxins and fermented alcoholic beverages

Fermented alcoholic beverages are produced in most parts of the world in many different forms and tastes. The types of beers produced in particular regions and countries will generally reflect the types of cereal crops in cultivation (Hesseltine, 1988; Campbell-Platt, 1989). Thus, cooler regions will normally produce and consume beers and lagers from barley, whereas warmer climates will have a higher production and consumption of beers from maize and sorghum, and wines from grapes.

In practice, there are five major stages in the manufacture of beers from grains, *viz*: malting, mashing, fermentation, maturation, and finishing. The potential for mycotoxins or their residues to occur in beers will depend upon the initial cereal, the conditions of storage of the cereal and the malt and the fate of the mycotoxins during malting, brewing and fermentation (Flannigan, 1996).

Initial soaking stages of barley before germination can lead to considerable loss of certain mycotoxins, especially deoxynivalenol, but less so with zearalenone (Smith *et al.*, 1994). Various mycotoxins, particularly trichothecenes and fumonisins produced by *Fusarium* spp., regular contaminants of barley, are inhibitory to barley germination during malting and can affect malt characteristics (Flannigan, 1996). Deoxynivalenol and zearalenone can be produced by several fusarial species during barley malting, while *A. clavatus* can also produce toxins during the malting stage. Toxins unique to maize can enter the barley malt mashing stage when maize and other starch adducts are added to the mashing process, and may be carried through the entire process into the final product. Several mycotoxins, especially T-2 toxin, can adversely affect the *de novo* synthesis of α-amylase during malting (Flannigan, 1996).

Beer produced from malt heavily contaminated with *Fusarium* can show reduced gas stability and gushing out explosively when the bottle or can is opened.

In most European and North American beers mycotoxins in the finished products occur only rarely and then at trace levels, and so do not constitute a serious health hazard. The occurrence of mycotoxins in beers is most likely in local or regional beers in parts of Africa and South America. Toxins of relevance include zearalenone, fumonisin B_1 and aflatoxin B_1, and these have all been detected, albeit at low levels. For further details on the occurrence of mycotoxins in beers worldwide, see Flannigan (1996) and Baxter (1996).

It is salutary to note that the main concern with mycotoxins and the brewing process is with brewing by-products; farm animals fed on contaminated malting and brewing by-products risk exposure to deoxynivalenol, aflatoxins, fumonisins, zearalenone and ochratoxin A.

References

Aas, N., Gondrosen, B. & Langeland, G. (1992) *Norwegian Food Control Authority's Report on Food Associated Diseases in 1990*, SNT Report 3, Oslo.

Andersson, R. (1986) Inhibition of *Staphylococcus aureus* and spheroplasts of Gram-negative bacteria by an antagonistic compound produced by a strain of *Lactobacillus plantarum. International Journal of Food Microbiology*, **3**, 149–60.

Archer, D. (1984) Diarrhoeal episodes and diarrhoeal disease: acute disease with chronic implications. *Journal of Food Protection*, **47**, 322–8.

Archer, D. & Young, F. (1988) Contemporary issues: diseases with a food vector. *Clinical Microbiology Reviews*, **1**, 377–98.

Arnold, K.W. & Kaspar, C.W. (1995) Starvation and stationary phase induced acid tolerance in *Escherichia coli* 0157:H7. *Applied and Enviornmental Microbiology*, **61**, 2037–9.

Aryanta, R.W., Fleet, G.H. & Buckle, K.A. (1991) The occurrence and growth of microorganisms during the fermentation of fish sausage. *International Journal of Food Microbiology*, **13**, 143–56.

Ashenafi, M. (1991) Growth of *Listeria monocytogenes* on fermenting tempeh made of various beans and its inhibition by *Lactobacillus plantarum. Food Microbiology*, **8**, 303–10.

Ashenafi, M. (1993) Fate of *Salmonella enteritidis* and *Salmonella typhimurium* during the fermentation of ergo, a traditional Ethiopian sour milk. *Ethiopian Medical Journal*, **31**, 91–8.

Bacus, J.N. (1984) Utilisation of microorganisms in meat processing, in *A Handbook for Meat Plant Operators*, John Wiley & Sons, New York.

Baik, H.S., Bearson, S., Dunbar, S. & Foster, J.W. (1996) The acid tolerance response of *Salmonella typhimurium* provides protection against organic acids. *Microbiology-UK*, **142**, 3195–200.

Baxter, E.D. (1996) The fate of ochratoxin A during malting and brewing. *Food Additives and Contaminants*, **13**(supplement), 23–4.

Berkowitz, F.E. (1994) The Gram-Positive bacilli: a review of the microbiology, clinical aspects, and antimicrobial susceptibilities of a heterogeneous group of bacteria. *Paediatric and Infectious Diseases Journal*, **13**, 1126–38.

Besser, R.E., Lett, S.M., Weber, J.T., Doyle, M.P., Barrett, T.J., Wells, J.G. & Griffin, P.H. (1993) An outbreak of diarrhea and hemolytic uremic syndrome from *Escherichia coli* 0157:H7 in fresh-pressed apple cider. *Journal of the American Medical Association*, **269**, 2217–20.

Black, R.E., Levine, M.M., Clements, M.L., Hughes, T.P. & Blaser, M.J. (1988) Experimental *Campylobacter* infections in humans. *Journal of Infectious Diseases*, **157**, 472–9.

Bol, J. & Smith, J.E. (1990) Biotransformation of aflatoxin. *Food Biotechnology*, **3**, 127–44.

Bryan. F.L. (1979) Infections and intoxications caused by other bacteria, in *Foodborne Infections and Intoxications*, 2nd edn, (eds H. Riemann & F.L. Bryan), Academic Press, New York, pp. 211–97.

Campbell-Platt, G. (1989) *Fermented Foods of the World: A Dictionary and Guide*, Butterworth, London.

Centers for Disease Control and Prevention (1994) Preliminary report: *Escherichia coli* 0157:H7 outbreak linked to commercially distributed dry-cured salami – Washington and California, 1994. *Morbid Mortality Weekly Report*, **44**, 157–60.

Chen, C.W. (1983) Interactions of diarrhoea and malnutrition – mechanisms and interventions, in *Diarrhoea and Malnutrition* (eds L.C. Chen & N.S. Scrimshaw), Plenum Press, New York, London, pp. 3–19.

Cheville, A.M., Arhold, K.W., Buchrieser, C., Cheng, C.-M. & Kaspar, C.W. (1996) rpoS regulatin of acid, heat and salt tolerance in *Escherichia coli* 0157:H7. *Applied and Environmental Microbiology*, **62**, 1822–4.

Cooks, P.E., Thomba, M. & Campbell-Platt, G. (1991) Growth of *Bacillus cereus* during rice tape fermentations. *Letters in Applied Microbiology*, **13**, 78–81.

Davis, M.J., Coote, P.J. & Obyrne, C.P. (1996) Acid tolerance in *Listeria monocytogenes* – the adaptive tolerance response (ATR) and growth-phase-dependent acid resistance. *Microbiology-UK*, **142**, 2975–82.

Dirheimer, G. (1996) Mechanistic approaches to ochratoxin toxicity. *Food Additives and Contaminants*, **13**, 45–8.

Flannigan, B. (1996) Mycotoxins in malting and brewing, in *Mycotoxins in Cereals: An Emerging Problem?* (ed. J.P.F. D'Mello), Scottish Agricultural College, Edinburgh, pp. 45–55.

Foster, J.W. & Hall, H.K. (1991) Induced pH homeostasis and the acid tolerance response of *Salmonella typhimurium. Journal of Bacteriology*, **173**, 5129–35.

Frisvad, J.C. (1986) Taxonomic approaches to mycotoxin identification (taxonomic indication of mycotoxin content in foods), in *Modern Methods in the Analysis and Structural Fluidation of Mycotoxins* (ed. R.J. Cole), Academic Press, New York, pp. 415–36.

Frisvad, J.C. & Samson, R.A. (1991) Filamentous fungi in foods and feeds: ecology, spoilage and mycotoxin production, in *Handbook of Applied Mycology, Vol. 3. Foods and Feeds* (eds D.K. Arora, K.G. Mukerji & E.H. Marth), Marcel Dekker Inc., New York, pp. 31–68.

Gahan, C.G.M., Odriscoll, B. & Hill, C. (1996) Acid adaptation of *Listeria monocytogenes* can enhance survival in acidic foods and during milk fermentation. *Applied and Environmental Microbiology*, **62**, 3128–32.

Gashe, B.A. (1985) Involvement of lactic acid bacteria in the fermentation of tef (*Eragrostis tef*), and Ethiopian fermented food. *Journal of Food Science*, **50**, 800–1.

Gashe, B.A., Girma, M. & Besrat, A. (1982) Tef Fermentation. I. The role of microorganisms in fermentation and their effect on the nitrogen content of tef. *SINET: Ethiopian Journal of Science*, **5**, 69–75.

Gifawesen, C. & Besrat, R. (1982) Yeast flora of fermenting tef (*Eragrostis tef*) dough. *SINET: Ethiopian Journal of Science*, **5**, 21–5.

Girma, M. (1986) Microbiology of mayonnaise-based salads and cottage cheese varieties. MSc Thesis, University of East Anglia, UK.

Girma, M., Gashe, B.A. & Lakew, B. (1989) The effect of fermentation on the growth and survival of *Salmonella typhimurium*, *Staphylococcus aureus*, *Bacillus cereus* and *Pseudomonas aeruginosa* in fermenting tef (*Eragrostis tef*). *MIRCEN Journal*, **5**, 61–6.

Glass, K.A., Loeffelholz, J.M., Ford, J.P. & Doyle, M.P. (1992) Fate of *Escherichia coli* 0157:H7 as affected by pH or sodium chloride, and in fermented dry sausage. *Applied and Environ mental Microbiology*, **58**, 2513–16.

Gorden, J. & Small, P.L.C. (1993) Acid resistance in enteric bacteria. *Infectious Immunology*, **61**, 364–7.

Gourama, H. & Bullerman, L.B. (1995) Antimycotic and antiaflatoxigenic effects of lactic acid bacteria: a review. *Journal of Food Protection*, **57**, 1275–80.

Granum, P.E. (1994) *Bacillus cereus* and its toxins. *Journal of Applied Bacteriology*, **76**, 615–65.

Granum, P.E., Brynestad, S., & Kramer, J.M. (1993a) Analysis of enterotoxin production by *Bacillus cereus* from dairy products, food poisoning incidents and non-gastrointestinal infections. *International Journal of Food Microbiology*, **17**, 269–79.

Granum, P.E., Brynestad, S., O'Sullivan, K. & Nissen, H. (1993b) The enterotoxin from *Bacillus cereus*: production and biochemical characterisation. *Netherlands Milk Dairy Journal*, **47**, 63–70.

Groopman, J.D. (1994) Molecular dosimetry methods for assessing human aflatoxin exposures, in *The Toxicology of Aflatoxin, Human Health, Veterinary and Agricultural Significance* (eds D.L. Eatson & J.D. Groopman), Academic Press, New York, pp. 314–41.

Gupta, S.K. & Venkitasubramanian, T.W. (1975) Production of aflatoxin on soybeans. *Applied Microbiology*, **29**, 834–6.

Hall, H.K. & Foster, J.W. (1996) The role of fur in the acid tolerance response of *Salmonella typhimurium* is physiologically and genetically separable from its role in iron acquisition. *Journal of Bacteriology*, **178**, 5683–91.

Hesseltine, C.W. (1988) Mycotoxins and alcohol production: a review, in *Perspectives of Mycopathology* (ed. A.W. Malhatra), Malhotra Publishing House, New Delhi, pp. 1–23.

Hobbs, B.C. & Roberts, D. (eds) (1993a) *Food Poisoning and Food Hygiene*, 6th Edn, Edward Arnold, London.
Hobbs, B.C. & Roberts, D. (1993b) Outbreaks of food poisoning and other food-borne discasc, in *Food Poisoning and Food Hygiene* (cds B.C. Hobbs & D. Roberts), Edward Arnold Publishers, London, pp. 90–124.
IARC (1993) Some naturally occurring substances, food items, and constituents, heterocyclic aromatic amines and mycotoxins, in *IARC Monographs on the Evaluation of Carcinogenic Risks to Humans*, Vol. 56, Lyons, International Agency for Research on Cancer, World Health Organization.
Ingram, J. (1987) Effect of temperature, pH, water activity, and pressure on growth, in *Escherichia coli* and *Salmonella typhimurium*: *Cellular and Molecular Biology* (ed. F.C. Neidhart), American Society for Microbiology, Washington, DC, pp. 1543–54.
Jack, R.W., Tagg, J.R. & Ray, B. (1995) Bacteriocins of Gram-positive bacteria. *Microbiological Reviews*, **59**, 171–200.
Jacob, F., Lwoff, A., Siminoitch, A. & Wollman, E. (1953) Définition de quelque terms relatifs à la lysogénie. *Ann ales d'Institut Pasteur (Paris)*, **84**, 222–4.
Jarvis, B. (1976) Mycotoxins in food, in *Microbiology in Agriculture, Fisheries and Food* (eds F.A. Skinner & J.G. Carr), Academic Press, London, pp. 251–67.
Jay, J.M. (1992) *Modern Food Microbiology*, 4th edn, Van Nostrand Reinhold, New York.
JHU/WHO (1989) Research on improving infant feed practices to prevent diarrhoea or reduce its severity: memorandum from a JHU/WHO meeting. *Bulletin of the World Health Organization*, **67**, 27–33.
Kalchayanand, N., Hanilin, M.B. & Ray, B. (1992) Sublethal injury makes Gram negative and resistant Gram positive bacteria sensitive to the bacteriocins pediocin AcH and nisin. *Letters in Applied Microbiology*, **15**, 239–43.
Kim, W.J. (1996) Screening of bacteriogenic lactic acid bacteria and their antagonistic effects in sausage fermentation. *Journal of Microbiology and Biotechnology*, **6**, 461–7.
Kingamkono, R., Sjögren, E., Sranberg, U. & Kaijser, B. (1994) pH and acidity in lactic-fermenting cereal gruels: effects on viability of enteropathogenic microorganisms. *World Journal of Microbiology and Biotechnology*, **10**, 664–9.
Klaenhammer, T.R. (1993) Genetics of bacteriocins produced by lactic acid bacteria. *FEMS Microbiological Reviews*, **12**, 39–86.
Kalayanamitr, A., Bhumiratana, A., Flegel, T.W., Glensukm, T. & Skinmyo, A. (1987) Occurrence of toxicity among protease, amylase and color mutants of a non-toxic soy sauce koji mold. *Applied and Environmental Microbiology*, **53**, 1980–2.
Kuiper-Goodman, T. (1994) Prevention of human mycotoxicoses through risk assessment and risk management, in *Mycotoxins in Grain: Compounds other than Aflatoxin* (eds J.D. Miller & H.L. Trenholm), Eagon Press, St Paul, Minnesota, pp. 439–69.
Kuiper-Goodman, T. (1996) Risk assessment of ochratoxin A: an update. *Food Additives and Contaminants*, **13**(Supplement), 53–7.
Leistner, L. (1984) Toxigenic penicillia occurring in feeds and foods: a review. *Food Technology, Australia*, **36**, 404–6.
Leistner, L. (1990) Mould-fermented foods: recent developments. *Food Biotechnology*, **4**, 433–41.
Lorri, W. & Svanberg, U. (1994) Lower prevalence of diarrhoea in young children using lactic-acid fermented cereal gruel. *Food and Nutrition Bulletin*, **15**, 57–63.
Mata, L.J. (1971) Nutrition and infection. *Protein Advisory Group Bulletin*, **11**, 18–21.
Mertens, T.E. & Lowbeer, D. (1996) HIV and AIDS – where is the epidemic going? *Bulletin of the World Health Organization*, **74**, 121–9.
Mhalu, F.S. & Lyamuya, E. (1996) Human-immunodeficiency-virus infection and AIDS in East Africa – challenges and possibilities for prevention and control. *East African Medical Journal*, **73**, 13–19.
Miller, J.D. & Trenholm, H.L. (eds) (1994) *Mycotoxins in Grains: Compounds other than Aflatoxin*, Eagan Press, St Paul, Minnesota.
Miller, L.G. & Kaspar, C.W. (1994) *Escherichia coli* 0157:H7 acid tolerance and survival in apple cider. *Journal of Food Protection*, **57**, 460–4.
Morgan, D., Newman, C.P., Hutchinson, D.N., Walker, A.M., Rowe, B. & Majid, F. (1993) Verotoxin producing *Escherichia coli* 0157 infections associated with the consumption of yoghurt. *Epidemiological Infections*, **11**, 181–7.

Moss, M.O. (1991) The environmental factors controlling mycotoxin formation, in *Mycotoxins and Animal Foods* (eds J.E. Smith & R.S. Henderson), CRC Press Inc., Boca Raton, F1., pp. 37–56.

Motarjemi, Y., Käferstein, F., Moy, G. & Quevedo, F. (1993) Contaminated weaning food: a major risk factor in diarrhoea and associated malnutrition. *Bulletin of the World Health Organization*, **71**, 79–92.

Mueller, G.P. & Williams, R.A. (1995) Surgical infections in AIDs patients. *American Journal of Surgery*, **169**, 34–8.

Nigatu, A. & Gashe, B.A. (1994) Inhibition of spoilage and food-borne pathogens by lactic acid bacteria isolated from fermenting tef (*Eragrostis tef*) dough. *Ethiopian Medical Journal*, **32**, 223–9.

Nout, M.J.R., Beirnick, G. & Bonants-Van Laarhoven, T.M.G. (1987a) Growth of *Bacillus cereus* in soybean tempeh. *International Journal of Food Microbiology*, **4**, 293–302.

Nout, M.J.R., de Dreu, M.A., Zuurbier, A.M. & Bonants-Van Laarhoven, T.M.G. (1987b) Ecology of controlled soybean acidification for tempe manufacture. *Food Microbiology*, **4**, 165–72.

O'Driscoll, B., Gahan, C.G.M. & Hill, C. (1996) Adaptive acid tolerance response in *Listeria monocytogenes*. Isolation of an acid-tolerance mutant which demonstrates increased virulence. *Applied and Environmental Microbiology*, **62**, 1693–8.

O'Keefe, E.A. & Wood, R. (1996) AIDS in Africa. *Scandinavian Journal of Gastroenterology*, **31**, 147–52.

Pestka, J.J. & Bondy, G.S. (1990) Alteration of immune function following dietary mycotoxin exposure. *Canadian Journal of Physiology and Pharmacology*, **68**, 1009–16.

Pestka, J.J. & Bondy, G.S. (1994) Immunotoxic effects of mycotoxins, in *Mycotoxins in Grain: Compounds other than Aflatoxin* (eds J.D. Miller & H.L. Trendholm), Eagon Press, St Paul, Minnesota, pp. 339–58.

Preluska, D.B., Rotter, B.A. & Rotter, R.G. (1994) Toxicology of mycotoxins, in *Mycotoxins in Grain: Compounds other than Aflatoxin* (eds J.D. Miller & H.L. Trenholm), Eagon Press, St Paul, Minnesota, pp. 359–403.

Pugsley, A.P. (1984a) The ins and outs of colicins. I. Production and translocation across membranes. *Microbiological Science*, **1**, 168–75.

Pugsley, A.P. (1984b) The ins and outs of colicins. II. Lethal action, immunity and ecological implications. *Microbiological Science*, **1**, 203–5.

Quinn, T.C. (1996) Global burden of the HIV pandemic. *Lancet*, **348**, 99–106.

Rowan, N.J. (1996) *Studies on the growth, survival, interaction and detection of potentially pathogenic* Listeria *and* Bacillus *species in infant milk formulae*, PhD Thesis, University of Strathclyde, Glasgow.

Rowe, B. (1987) *Salmonella ealing* associated with consumption of infant dried milk. *Lancet*, **10**, 900–3.

Russell, A.D. & Chopra, I. (1990) *Understanding Antibacterial Action and Resistance*, Ellis Horwood, New York.

Samson, R.A. (1993) Taxonomy – current concepts of *Aspergillus* systematics, in *Biotechnology Handbooks, Vol. 7. Aspergillus* (ed. J.E. Smith), Plenum Press, New York, London, pp. 1–22.

Samson, R.A. & Frisvad, J.C. (1991) Current taxonomic concepts in *Penicillium* and *Aspergillus*, in *Cereal Grain, Mycotoxins, Fungi and Quality in Drying and Storage* (ed. J. Chelkowski), Elsevier, Amsterdam, pp. 405–39.

Schlatter, C.H., Studer-Rohr, J. & Rasonyi, T.H. (1996) Carcinogenicity and kinetic aspects of ochratoxin A. *Food Additives and Contaminants*, **13**, 43–4.

Smith, J.E. & Henderson, R.S. (eds) (1991) *Mycotoxins and Animal Foods*, CRC Press Inc., Boca Raton, F1.

Smith, J.E. & Moss, M.O. (1985) *Mycotoxins: Formation, Analysis and Significance*, John Wiley & Sons, Chichester, New York.

Smith, J.E., Solomons, G.L., Lewis, C.W. & Anderson, J.G. (1994) *Mycotoxins in Human Nutrition and Health*. Directorate-General XII, Science, Research & Development. EUR 16048EN.

Smith, J.E., Solomons, G., Lewis, C. & Anderson, J.G. (1995) Role of mycotoxins in human and animal nutrition and health. *Natural Toxins*, **3**, 187–92.

Smith, J.L. & Palumbo, S.A. (1983) Use of starter cultures in meats. *Journal of Food Protection*, **46**, 997–1006.
Spelhaug, S. & Harlander, S.K. (1989) Inhibition of food-borne bacterial pathogens by bacteriocins from *Lactococcus lactis* and *Pediococcus pentosaceus. Journal of Food Protection*, **52**, 856–62.
Steinkraus, K.H. (1983) *Handbook of Indigenous Fermented Foods*, Marcel Dekker, New York.
Steyn, P.S. (1995) Mycotoxins, general view, chemistry and structure. *Toxicology Letters*, **82/83**, 843–51.
Svanberg, U., Sjögren, E., Lorri, W., Svennerholm, A-M. & Kaijser, B. (1992) Inhibited growth of common enteropathogenic bacteria in lactic-fermented cereal gruels. *World Journal of Microbiology and Biotechnology*, **8**, 601–6.
Sydenham, E.W., Shepherd, G.S., Gelderblom, W.C.A., Thiel, P.G. & Marasas, W.F.O. (1994) Fumonisins: their implications for human and animal health, in *Occurrence and Significance of Mycotoxins* (ed. K.A. Scudamore), Central Science Laboratory, London, pp. 42–8.
Trucksess, M.W., Mislevec, P.B., Young, K., Bruce, V.E. & Page, S.W. (1987) Cyclopiazonic acid production by cultures of *Aspergillus* and *Penicillium* species isolated from dried beans, corn meal, macaroni and pecans. *Journal of the Association of Analytical Chemists*, **70**, 123–6.
Tunçel, G., Nout, M.J.R., Brimer, L. & Göktan, D. (1990) Toxicological, nutritional and microbiological evaluation of tempe fermentation with *Rhizopus oligosporus* of bitter and sweet apricot seeds. *International Journal of Food Microbiology*, **11**, 337–44.
Ujii, F. & Yokoyama, K. (1956) Studies in the bactericidal action of soy sauce for pathogenic bacteria. *Reports of Food Sanitation*, **6**, 1.
van Egmond, H.P. & Dekker, W.H. (1995) Worldwide regulations for mycotoxins in 1994. *Natural Toxins*, **3**, 332–6.
van Veen, A.G. (1967) The bongkrek toxins, in *Biochemistry of Some Foodborne Microbial Toxins* (eds R.I. Mateles & G.N. Wogen), M.I.P. Press, Cambridge, Massachusetts. (cited in Steinkraus, K.H., *Handbook of Indigenous Fermented Foods*, p. 93).
Vignolo, G.M., Suriani, F., de Ruiz Holgado, A.P. & Oliver, G. (1993) Antibacterial activity of *Lactobacillus* strains isolated from dry fermented sausages. *Journal of Applied Bacteriology*, **75**, 344–9.
Weagant, S.D., Bryant, J.L. & Bark, D.H. (1994) Survival of *Escherichia coli* 0157:H7 in mayonnaise-based sauces at room and refrigerated temperatures. *Journal of Food Protection*, **57**, 629–31.
Wong, H.-C. & Chen, Y.-L. (1988) Effects of lactic acid bacteria and organic acids on growth and germination of *Bacillus cereus. Applied and Environmental Microbiology*, **54**, 2179–84.
Yokotsuka, T. (1983a) Patterns of production and consumption of Japanese Shoyu, in *Handbook of Indigenous Fermented Foods* (ed. K.H. Steinkraus), Marcel Dekker, Inc., New York, pp. 438–51.
Yokotsuka, T. (1983b) Scale-up of traditional fermentation technology. *Korean Journal of Applied Microbiology and Bioengineering*, **11**, 353–71.
Yokotsuka, T., Sasaki, M., Kikuchi, T., Assao, Y. & Nabukora, A. (1966) Production of fluorescent compounds other than aflatoxin by Japanese industrial molds, in *Biochemistry of Some Foodborne Microbial Toxins* (eds R.I. Mateles & G.N. Wogan), MIT Press, Cambridge, Mass., pp. 111–52.

27 The impact of genetic engineering on food and beverage fermentations

J.E. SMITH

27.1 Introduction

Food production is the largest world-wide industry. The modern food industry now serves the function of supplying high-quality, wholesome foods, all the year round and at a distance in time and location from the place of primary production. The origin of the food chain is in production agriculture with the planting of the seed or the rearing of animals and concludes with utilization of the food products by the consumer. The products of the farm are linked to the consumer mainly through the food processing industry where the agricultural products are manufactured into shelf-stable, convenient and palatable foods and beverages (Angold *et al.*, 1989; OECD, 1992).

Fermented foods and beverages have a long historical commitment in most societies and arise from the action of microorganisms or enzymes on a wide range of agricultural raw materials with associated desirable biochemical changes producing significant organoleptic improvements in the final product (Campbell-Platt, 1989). Consequently, these final products are usually more nutritious, more digestible, have improved flavour, and are microbiologically and toxicologically safer. Such foods and beverages are an accepted and essential part of the diet in most parts of the world, incorporating a wide diversity of raw materials as substrates, involving levels of technology from the most primitive to the most advanced, and achieving an unbelievable range of sensory and textural qualities in the final products. Fermented foods include bread, sauerkraut, cheeses, yoghurts, soy sauce, tempeh, mushrooms, etc., while fermented beverages include alcoholic beers, wines, brandy, whisky and non-alcoholic coffee, tea and cocoa. While many food and beverage fermentations have remained at the level of a village art, a significant number have achieved major commercial application and produce significant financial returns to most national economies.

In essence, all forms of food and beverage fermentations are examples of biotechnological processes. Biotechnology can be viewed as a series of enabling technologies involving the practical application of organisms or their cellular components to manufacturing and service industries and environmental management (EFB, 1995; Smith, 1996). Historically, biotechnology was an art rather than a science, exemplified in the production of wines, beers, cheeses, etc. where the practicalities of manufacture were well understood and reproducible while the molecular mechanisms went unknown for centuries. Biotechnology is not something new but represents a developing and expanding series of techniques dating back in some cases thousands of years when humans first began unwittingly to use microbes to produce foods and beverages. It is only in relatively recent times that these traditional biotechnological processes have been subjected to rigorous scientific scrutiny and analysis, though many still operate with large aspects of empiricism (OECD, 1992).

The biotechnological production of foods and beverages has advanced with the developments in modern science and biochemical engineering. Greater understanding of the microorganisms involved, the production of stable, pure cultures, the development of new strains of microorganisms, improved understanding of enzyme technology, aseptic techniques and advances in bioprocess technology have all led to major advances in the many forms of food and beverage fermentations which are more fully described elsewhere in this volume (Smith, 1994).

However, a new aspect of biotechnology arose in the 1970s and early 1980s when it became possible to alter precisely the genetic constitution of living organisms by processes outwith traditional breeding practices (Old & Primrose, 1990). These new processes now generally termed 'genetic engineering' or 'genetic modification' have had, and will continue to have, major effects on all aspects of traditional biotechnologies. Their initial impact has been mainly in medical-related areas producing new healthcare products and, more recently, gene therapies to treat certain diseases. However, increasingly genetic engineering is being directed to the food and agricultural sectors.

While these new aspects of biotechnology have been in practice for several years producing, among other things, insulin from genetically manipulated microorganisms, it is only with the arrival of foods derived from or including genetically manipulated organisms (GMOs) that there has been increased public awareness, recognition, and some concern of this aspect of biotechnology (EFB, 1994).

27.2 Applied genetics and recombinant DNA technology

Genetic manipulation has actually been performed for centuries by selective breeding of plants, animals and microorganisms superimposed on natural variations. However, the potential for genetic variation has been limited to close taxonomic relatives. In contrast, recombinant DNA techniques or genetic engineering offer potentially unlimited opportunities for creating new combinations of genes that presently do not exist under natural conditions. Genetic engineering can be viewed as the formation of new combinations of heritable material by the insertion of nucleic acid molecules, produced by whatever means outside the cell, with a virus, bacterial plasmid or other vector system so as to allow their incorporation into a host organism by the use of recombinant DNA technology (Old & Primrose, 1990).

Natural variety in organisms arises from an array of organismal operating mechanisms including mutation and sexual recombination. Man has created new variations by deliberate processes such as plant and animal breeding and microbial strain selection, and has steadily enhanced the quality and quantity of the world's food supply. While traditional methods of genetic manipulation will continue to dominate world-wide food production technologies (including food and beverage fermentations), the newer cellular and molecular genetical techniques will increasingly become more utilized and relevant.

Genetically engineered plants and microorganisms are moving from the laboratory to the field and food factory and are steadily appearing in increasing numbers at consumer outlets. This article will not be concerned with the scientific details of how GMOs are created, but will be concerned with matters related to their safety for human consumption when used in food and beverage fermentations. Techniques that are considered to constitute genetic engineering are shown in Table 27.1, while Table 27.2 outlines those techniques which do not constitute genetic engineering.

While the future benefits of genetic engineering in the food and beverage industries are potentially immense, the inherent dangers of tampering with genetic processes should always be recognized. In no small way, the implementation of these new molecular techniques will need acceptance from the consumers. At present, the public seem to have a somewhat negative attitude to manipulation of foods and in some instances show some hostility to genetic engineering of foodstuffs. As stated in the ACOST (1990) report

Table 27.1 Definition of genetic modification

Examples of the techniques which constitute genetic modification are:

1. Recombinant DNA techniques consisting of the formation of new combinations of genetic material by the insertion of nucleic acid molecules, produced by whatever means outside the cell, into a virus, bacterial plasmid or other vector system so as to allow their incorporation into a host organism in which they do not occur naturally but in which they are capable of continued propagation.
2. Techniques involving the direct introduction into an organism of heritable material prepared outside the organism, including micro-injection, macro-injection and micro-encapsulation.
3. Cell fusion (including protoplast fusion) or hybridization techniques where live cells with new combinations of heritable genetic material are formed through the fusion of two or more cells by means of methods that do not occur naturally.

Table 27.2 Techniques which do not constitute genetic modification

The following techniques do not constitute genetic modification if they do not involve the use of recombinant DNA molecules or genetically modified organisms

1. *In vitro* fertilization.
2. Conjugation, transduction, transformation or any other natural process, and polyploidy induction.
   The Regulations shall not apply to the following techniques of genetic modification if they do not involve the use of genetically modified organisms as recipient or parental organisms:
3. Mutagenesis.
4. The construction and use of somatic hybridoma cells (i.e. for the production of monoclonal antibodies).
5. Cell fusion (including protoplast fusion) of plant cells where the resulting organisms can also be produced by traditional breeding methods.
6. Self-cloning of non-pathogenic, naturally occurring microorganisms which fulfil the criteria of general safety for recipient microorganisms.
7. Self-cloning of non-pathogenic, naturally occurring organisms other than microorganisms. Such organisms (i.e. plants or animals) must be as safe in the containment facility as any recipient or parental organism.

Developments in Biotechnology, 'Public perception of biotechnology will have a major influence on the role and direction of development and there is growing concern about genetically modified products. Associated with genetic manipulation are diverse questions of safety, ethics and welfare'.

27.3 Genetically engineered plants and microorganisms

Genetically modified foods could appear in many final forms, *viz*. fresh products (e.g. fruit and vegetables, live bacteria in bio-yoghurts) in which the modified DNA remains intact; processed products (e.g. tinned meat, fruit and vegetables, and tomato paste) in which the DNA has been denatured; products prepared using processing aids modified by biotechnology (e.g. the genetically modified enzyme chymosin for cheese production); products obtained from a genetically engineered source but the product is unmodified or chemically identical (e.g. sugar from genetically modified sugar beet plants); and plants modified for disease resistance or herbicide tolerance (e.g. glyphosate tolerant soybeans) (OECD, 1992; EFB, 1994). Crops in which single purified genes have been introduced thus creating plant GMOs are listed in Table 27.3 while some of the products of genetic engineering approved for food use in the UK are listed in Table 27.4.

Increasingly, most agriculturally important crop plants are being subjected to some degree of genetic engineering (Table 27.3). Of particular significance is the ability to use genetic engineering methods to enhance resistance to the diseases and pests associated with specific crop plants and also to develop plants that can tolerate specific herbicides which allows

Table 27.3 Food-based crops into which single purified genes have been introduced[a]

Cereals	Beverage plants	Legumes/oil seeds	Vegetables, fruits
Rice	Coffee	Linseed	Carrot
Maize		Soybean	Tomato
		Rape	Strawberry
			Raspberry
Wheat		Sunflower	Celery
Rye		Palm	Cauliflower
			Tobacco
			Sugar beet
			Cucumber
			Potato
			Melon
			Apple

[a] This list is constantly being added to.
Adapted from OECD, 1992.

Table 27.4 Products of gene technology approved by the ACNFP[a] for food use in the UK

Product/feature	Company	Approval date
Baker's yeast (*Saccharomyces cerevisiae*) faster liberation of CO_2	Gist Brocades	March 1990
Bovine chymosin from *Kluyveromyces lactis* (yeast)	Gist Brocades	January 1991
Bovine chymosin from *Aspergillus niger* var. *awamori* (filamentous fungus)	Genencor	May 1991
Bovine chymosin from *Escherichia coli* K-12 (bacterium)	Pfizer	March 1992
Brewer's yeast (*Saccharomyces cerevisiae*) ability to break down starch	BRF International	February 1994
Tomato paste products from GM tomato/ thicker tomato pastes	Zeneca	February 1995
Protein and oil from GM soybean/herbicide (glyphosate) tolerant	Monsanto	February 1995
Oil from GM oilseed rape/male sterile/ fertility restorer	Plant Genetic Systems	February 1995
Oil from GM oilseed rape/herbicide (glufosate)-tolerant	AgrEvo	May 1995
Flavr Savr® tomato	Calgene	October 1995

[a] ACNFP, Advisory Committee on Novel Foods and Processes – a UK Government-appointed committee.

their survival while the targeted weeds are killed. Several programmes now seek to improve the nutritional value of food plants, e.g. oil crops modified to alter their saturated fat content, high protein crops, etc. Many soft fruits suffer from extensive damage which is often associated with the softening of their wall structures linked with the ripening process. The new Flavr Savr® tomatoes have been modified by engineering the switching-off of the gene encoding the polygalacturanase enzyme. Increasingly, such traditional crop plants as soybean, wheat, maize and rice will be genetically manipulated for some beneficial reason, and subsequently be used as a substrate in traditional food or beverage fermentations.

Microorganisms which have been genetically engineered include *Saccharomyces cerevisiae* for baking and brewing purposes. The genetically engineered yeast produces CO_2 faster than conventional yeasts, ensuring that during breadmaking the dough rises more quickly. This was achieved by inserting *Saccharomyces*-derived genes for maltose permease and maltase with a promotor that is not subject to catabolite repression by glucose. While the GMO has been available now for several years to the producer company it has not yet been applied to commercial production!

A GMO brewer's yeast has operated only at pilot scale. Its novel engineered feature is a yeast gene for amylase that allows digestion of starch and residual glucans in the beer, yielding a low-carbohydrate beer.

As cheese production has increased world-wide, the supply of rennet or chymosin has become problematic. While microbially derived chymosin has partly reduced the shortage it has become possible to clone the calf chymosin gene into various microorganisms which are then able to produce the chymosin enzyme in consistent, reproducible quantities.

Kluyveromyces lactis, *Aspergillus niger* var. *awamori* and *Escherichia coli* K12 have been successfully engineered for bovine chymosin production; the bovine chymosin produced from the various microbial GMOs is now being successfully used by some cheese producers and the products successfully sold as vegetarian cheeses. In this example of genetic engineering there is no live GMO present, rather a chymosin enzyme, chemically identical to the calf rennet enzyme, is the utilized product. Full use and acceptance of this method of chymosin production could eventually lead to a reduction in calf production and slaughter. (For further details of these GMOs, see ACNFP Annual Reports 1990–1995.)

Throughout the EC there have been hugely funded programmes on the genetic engineering of the lactic acid bacteria and related species, but as yet there are no examples of such GMOs achieving commercial application.

27.4 The safety of genetically engineered foods

Much scientific effort and funding has been committed to the development of genetically engineered plants and microorganisms, in each case to produce a better end-product for the consumer. Much debate is now taking place on the safety and ethical aspects of such products destined for public consumption. Can such products with 'unnatural' gene alteration lead to unforeseen problems for the present and future generations? (Straughan, 1989)

The safety of the human food supply is of paramount importance to most nations and all foods should be fit for consumption, i.e. not injurious to health or contaminated (OECD, 1993). When foods or food ingredients are derived from GMOs they must be seen to be as safe as, or safer than, their traditional counterparts. The concept of **substantial equivalence** is widely used in the determination of safety by comparison with analogous conventional food products together with intended use and exposure (WHO, 1991). When substantial equivalence can be shown, then normally no further safety considerations are necessary. When substantial equivalence is not clearly established then the points of difference must be subjected to further safety scrutiny.

When such novel products are entering into the marketplace the consumer must be assured of their quality and safety. Thus, there needs to be toxicological and nutritional guidance in the evaluation of novel foods and ingredients to highlight any potential risks which can then be dealt with appropriately. Any concept to achieve the safety assessment of novel foods and food ingredients must satisfy the needs of the producer, manufacturer, legislator and consumer. The concept should be in line with accepted scientific considerations, the results of the safety assessment must be reproducible and acceptable to the responsible health authorities, and the outcome must satisfy – and above all, convince – the consumer (ILSI, 1996).

In the UK, a Government-appointed Advisory Committee on Novel Foods and Processes (ACNFP) with a wide complement of independent experts who are only influenced by the scientific factors and ultimate safety of the product, has now been operating for 9 years. The ACNFP, while giving advice to the Government concerning novel foods and processes, also gives guidance to the food industry on the information concerning safety that it expects to see in any submission (HMSO, 1991). While the decisions arrived at by the Committee rely strongly on scientific expertise, due recognition to consumer needs and matters related to ethics are strongly adhered to.

The International Life Sciences Institute Europe has recently produced a set of guidelines for the Safety Assessment of Novel Foods (ILSI, 1996). Furthermore, a Special EU Committee has been set up to consider the nature and safety of Novel Foods within the EU.

In the context of food and beverage fermentations, the following aspects related to the use of genetic engineering will be discussed. The novel modified food may be a genetically engineered microorganism growing on unmodified plant material; unmodified microorganism growing on genetically engineered plant or combinations of both; or a product derived from such combinations. The ACNFP developed and published a decision-tree scheme which, by answering a series of questions, indicates the type of information likely to be required for individual submissions (Figure 27.1) (HMSO, 1991; ACNFP Annual Report, 1994). The information requirements for the scheme are listed in Table 27.5. By working through the decision-tree scheme a novel food or process will fall into one of 19

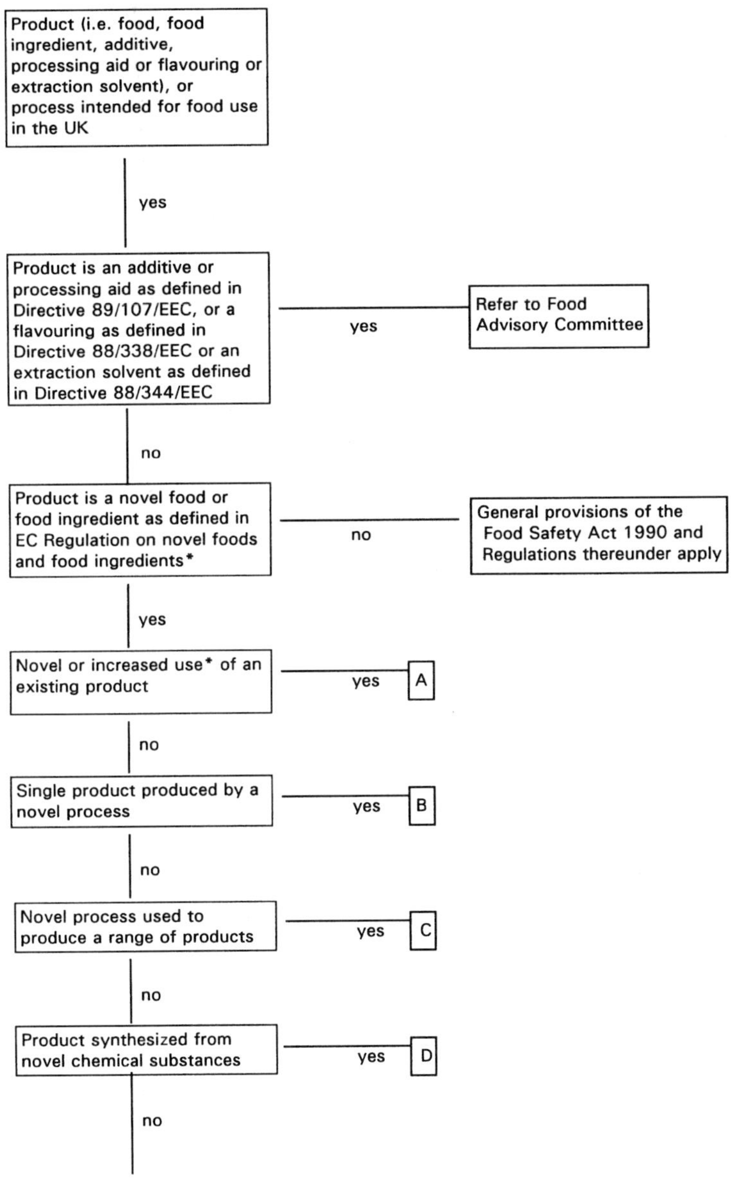

Figure 27.1 Revised decision tree for assessing a novel food or process and the information required for its safety evaluation. (Adapted from ACNFP Report, 1994.)

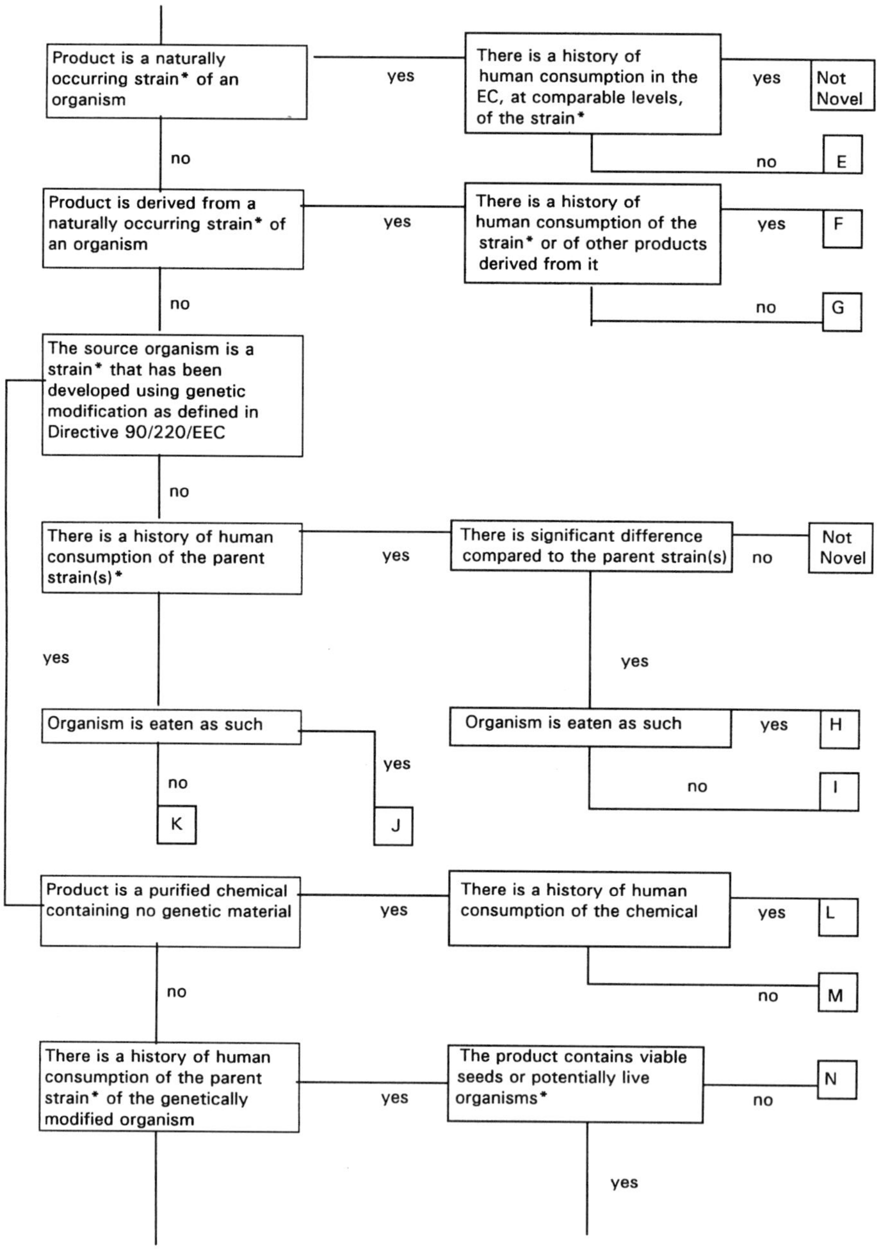

Figure 27.1 *Continued*

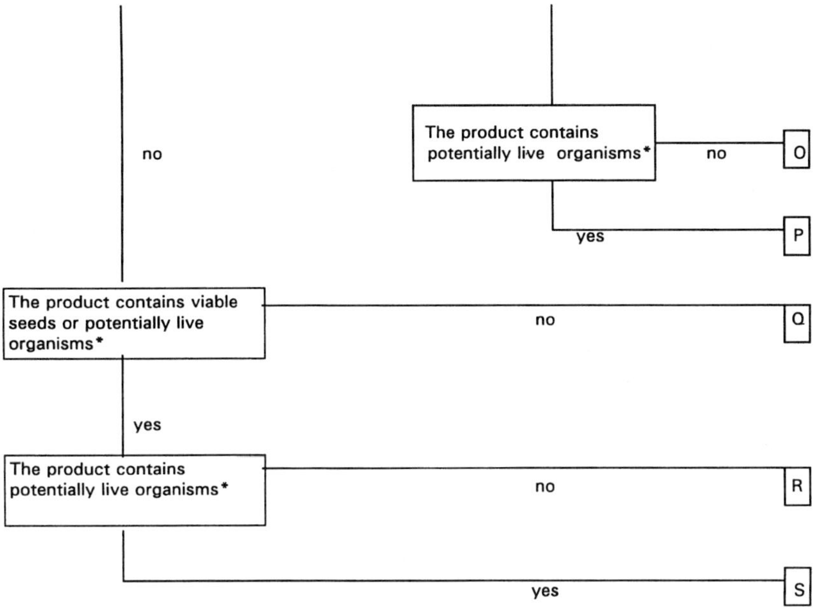

Revised decision tree: Explanatory notes

EC Regulation: until the EC Regulation on novel foods and food ingredients is adopted and/or comes into force, this box should be read as 'Product is a novel food or is produced by a novel process' as defined by the ACNFP in its 1991 Guidelines on the Assessment of Novel Foods and Processes.

Increased use: increased introduction and use of an existing product (e.g. an ethnic food ingredient) which has not previously formed part of the traditional EC diet, and/or food ingredients proposed for use in specialised diets, for example for infants, where such foods/ingredients have no history of consumption in this target group.

Strain: this term is used to describe 'strain' in the case of microorganisms, 'variety' in the case of plants and 'lines' in the case of animals.

Parent strain: the origin(s) of the genetic material in the source organism derived by non-GM (i.e. traditional breeding) techniques.

Potentially live organisms: this term is used to include all organisms which are potentially live, other than seeds.

Figure 27.1 *Continued*

categories (A to S). Of special interest are foods that are, or contain, the products of genetic engineering which exist in categories K to S. For each of the categories there is an information requirement of up to 15 items of information (Table 27.5). As a result of these deliberations it is possible to arrive at a valid judgement on the safety of the novel food. In the context of microbial-based food fermentations and the use of the decision-tree, two examples are worked through in Table 27.6, viz:

1. Chymosin from genetically modified food organisms.
2. Genetically modified baker's yeast.

Table 27.5 Information requirements for use in association with the decision-tree

Exit point	Information requirements
A	I, II, IV, V, VIII
B	II, III, IV, V, VIII, XV
C	II, III, V, VIII, XV
D	II, IV, V, VIII, IX, XV
E	II, V, VI, VIII, IX, XV
F	I, II, III, IV, V, VI, XV
G	II, III, IV, V, VI, VIII, IX, XV
H	I, II, V, VI, VII, XV
I	I, II, III, IV, V, VI, VII, XV
J	II, V, VI, VII, VIII, IX, XV
K	II, III, IV, V, VI, VII, VIII, X, XI, XV
L	I, II, III, IV, V, VI, VII, VIII, X, XI, XV
M	II, III, IV, V, VI, VII, VIII, IX, X, XI, XV
N	I, II, III, IV, V, VI, VII, VIII, X, XI, XII, XV
O	I, II, V, VI, VII, VIII, X, XI, XII, XIII, XV
P	I, II, V, VI, VII, VIII, X, XI, XII, XIII, XIV, XV
Q	II, III, IV, V, VI, VII, VIII, IX, X, XI, XII, XV
R	II, V, VI, VII, VIII, IX, X, XI, XII, XIII, XV
S	II, VI, VII, VIII, IX, X, XI, XII, XIII, XIV, XV
I	Evidence of previous human exposure
II	Intake and extent of use
III	Technical details of the process
IV	Product specification
V	Nutritional assessment
VI	History of the organism
VII	Characterization of derived strain in comparison with the parent strain
VIII	Toxicological assessment
IX	Human data
X	Effect of the genetic modification on the known properties of the parent organism
XI	Genetic stability of the modified organism
XII	Site of expression of any novel genetic material
XIII	Transfer of the novel genetic material
XIV	Assessment of a modified organism for survivability, replication and colonization/amplification in the human gut
XV	Safety information

Adapted from ACNFP Annual Report (1994).

Table 27.6 Structural approach for the safety assessment of novel foods[a]

1. Chymosin from a genetically modified food microorganism	
Questions:	
Product is an additive, flavouring or extraction solvent?	No
Novel food as defined in EC regulations?	Yes
Novel or increased use of an existing product?	No
Single product produced by a novel process?	No
Novel process used to produce a range of products?	No
Product synthesized from novel chemical substances?	No
Product is a naturally occurring strain of a organism?	No
Product derived from naturally occurring strains of organism?	No
Strain has been developed using genetic modification?	Yes
Product is a purified chemical containing no genetic material?	Yes
History of human consumption of chemical?	Yes
Therefore, L[b] – information requirements: I, II, III, IV, V, VI, VII, VIII, XI, XV.	
2. Genetically modified baker's yeast	
Questions:	
Product is an additive, flavouring or extraction solvent?	No
Novel food as defined in EC regulations?	Yes
Novel or increased use of an existing product?	No
Single product produced by a novel process?	No
Novel process used to produce a range of products?	No
Product synthesized from novel chemical substances?	No
Product is a naturally occurring strain of a organism?	No
Product derived from naturally occurring strains of organism?	No
Strain has been developed using genetic modification?	No
Product is a purified chemical containing no genetic material?	Yes
History of human consumption of chemical?	Yes
Product contains viable seeds or potential live organisms?	Yes
Product contains potential live organisms?	Yes
Therefore, P[b] – information requirements: I, II, VI, VII, VIII, X, XI, XII, XIII, XIV, XV	

[a] Adapted from ACNFP Annual Report (1994).
[b] Exit points from Figure 27.1 and Table 27.5.

The soybean (*Glycine max*) is an important food crop and is processed mainly for oil and protein, and occurs in a wide array of food products. It is also extensively used as a source of protein in animal feedstuffs. However, in the context of the present volume it is substantially used world-wide as the substrate for several microbially derived fermented foods, e.g. tempeh, soy sauce, miso, etc. Recently, a genetically manipulated herbicide tolerant variety of soybean has been approved by the ACNFP and will be commercially available. The pathway of the eventual clearance and safety assessment of this variety will be briefly outlined and can be considered as a model for the eventual arrival of other substrates for food fermentations.

Recent advances in plant biotechnology have made it possible to insert into plants genes which can protect the recipient plant from the harmful effects of a particular herbicide. The present variety of soybean has been modified to be tolerant to the herbicide glyphosate (GTS) (ACNFP Annual Report, 1994). The GTS plant was derived from a commercial soybean cultivar genetically modified to contain a gene from the bacterium *Agrobacterium* sp. strain CP4. The gene encodes the enzyme 5-enolpyruvyl glutamate-3-phosphate synthase (EPSPS) involved in the synthesis of aromatic amino acids. Plant EPSPS is inhibited by glyphosate, preventing the synthesis of essential amino acids and consequently leading to the death of the plant.

The modified beans also contain part of an EPSPS gene from *Petunia hybrida* which encodes a chloroplast transit peptide (CTP) responsible for delivering the bacterial EPSPS (CP4 EPSPS) to the chloroplasts, the actual site of aromatic amino acid synthesis in plants. When the enzyme reaches the chloroplast the attached transit peptide is cleaved and degraded. The CP4 EPSPS which is insensitive to glyphosate thus allows the plant to synthesize aromatic amino acids in the presence of glyphosate. The genetic modificiation was achieved by using the **particle gun bombardment** method.

The GM soybeans will be used in exactly the same way as the unmodified beans. Soybeans are not used in an unprocessed form because they contain toxic factors, including trypsin inhibitors, which must be denatured and detoxified by heat treatment, e.g. cooking before use for koji production. Major uses include extracted oil and defatted flakes.

When assessing the safety of the genetically engineered soybean, the ACNFP compared it with existing strains and gave special attention to the following:

- the intentional changes;
- any unintentional changes arising from the modification;
- the stability of the GMO under the intended conditions of use; and
- the likelihood of genetic transfer.

(a) Intentional changes. The only differences between the modified and unmodified varieties is the presence of the EPSPS gene and the CTP sequence. There were no food safety concerns over the DNA in the EPSPS gene and the CTP sequence as they are composed of DNA which is non-toxic. The only protein present in the modified beans is the very low levels of the CP4 EPSPS enzyme which is unlikely to be a health hazard. The protein was found to be non-toxic in an acute toxicity study with mice.

(b) Unintentional changes. The modification procedure enabled the marker gene to be jettisoned and polymerase chain reaction and Southern blot analysis were used to characterize and confirm the inserted DNA as only the EPSPS gene and nothing else. No significant changes were noted after compositional data and nutritional studies were concerned with unmodified soybean strains. Furthermore, there was no evidence that the allergenic potential of the GM soybeans and their products were different from other soybean varieties.

(c) Stability of the GMO. The glyphosate phenotype was studied over several fermentations and in crosses between the GMO and other soybean lines. The inserted DNA behaved as a single dominant gene and was inherited in a Mendelian fashion. Together with other molecular analyses, it was confirmed that the EPSPS gene in the GMO was stable. However, long-term observations must continue to be made by the Company to determine if there is any genetic drift.

(d) Genetic transfer. The risk of genetic transfer from the modified bean and products to the human or their gut microflora was considered negligible, since the beans would not be consumed in a viable form and the processes used to produce the soybean products would destroy DNA.

Thus, the GM soybean line and other lines derived from subsequent crosses with commercial soybean cultivars and products derived from these beans are equivalent to and as safe for human consumption as beans from other conventional soybean strains, and products derived from them. There are no consumer safety concerns associated with the modified beans and soya product derived from them.

The application of genetic engineering to food production is intended to enhance the useful and desirable characteristics of the organisms and to eliminate the undesirable. The consumer has always shown a willingness to pay more for better and more convenient products and to reject products that do not achieve their expectations.

Is it possible that plants transgenic for herbicide resistance could transfer this property to close field relatives such as weeds? Under normal field conditions, gene transfer between close relatives is a very rare phenomenon and there is little evidence that this will change with transgenic organisms. However, released transgenic plants will continue to be monitored to validate these viewpoints.

While the scientific case for further use of genetic engineering in the manipulation of food organisms is strong and persuasive, public perception is somewhat ambivalent. Genetic engineering is seen by some as 'unnatural' and unnecessary in food production. Public confidence must be achieved for the success of this new technology. Consumer information and education of new biotechnology must progress together, with food-related proposals. Consumers must come genuinely to believe they will benefit more directly by an increase in real food quality, better flavour, better texture and better nutrient content.

References

ACOST (Advisory Committee on science and Technology) (1990) *Developments in Biotechnology*, HMSO, London.

Advisory Committee on Novel Foods and Processes (ACNFP). Annual Reports 1990–1995. Ministry of Agriculture and Fisheries, London (free on request).
Angold, R., Beech, G. & Taggart, J. (1989) *Food Biotechnology*, Cambridge University Press, Cambridge.
Campbell-Platt, G. (1989) *Fermented Foods of the World: A Dictionary and Guide*, Butterworth, London.
EFB (European Federation of Biotechnology) (1994) Biotechnology in Foods and Drinks. Task Group on Public Perceptions of Biotechnology.
EFB (European Federation of Biotechnology) Newsletter (1995) October, DECHEMA, Dusseldorf.
HMSO (Her Majesty's Stationery Office) (1991) *Guidelines on the Assessment of Novel Foods and Processes*. Report on Health and Social Subjects. No. 38. HMSO, London.
ILSI (International Life Sciences Institute) (1996) *The Safety Assessment of Novel Foods*, ILSI Europe, Brussels.
OECD (1993) *Safety Evaluation of Food Derived by Modern Biotechnology*, OECD, Paris.
OECD (Organization for Economic Co-Operation and Development) (1992) *Biotechnology, Agriculture and Food*, OECD, Paris.
Old, R.W. & Primrose, S.B. (1990) *Principles of Gene Manipulations – An Introduction to Genetic Engineering*, Blackwell Scientific Publications, Oxford.
Smith, J.E. (1994) New opportunities in food biotechnology. *Australian Biotechnology*, **4**, 288–91.
Smith, J.E. (1996) *Biotechnology*, 3rd edn, Cambridge University Press, Cambridge.
Straughan, R. (1989) *The Genetic Manipulation of Plants, Animals and Microbes. The Social and Ethical Issues for Consumers*. A Discussion Paper. National Consumer Council, London.
WHO (World Health Organization) (1991) *Strategies for Assessing Safety of Foods*, WHO, Geneva.
Wood, B.J.B. (1985) *Microbiology of Fermented Foods*, Vols 1&2, Elsevier Applied Science Publishers, London.

Index

Abbreviations The following abbreviations are used in this index:

EFA, essential fatty acids (also indexed as Fatty acids, essential)
FD, flow diagram
LAB, lactic acid bacteria
MO, microorganism
q.v., "which see"
SSF, solid substrate fermentation

Tables are indicated by page numbers printed in italic.
Figures are indicated by page numbers printed in bold.

Major contents listings are not listed here, and the Index should therefore be used in conjunction with the Table of Contents.